Library of
Davidson College

Titles in This Series

12 **David S. Johnson and Catherine C. McGeoch, Editors,** Network Flows and Matching: First DIMACS Implementation Challenge
11 **Larry Finkelstein and William M. Kantor, Editors,** Groups and Computation
10 **Joel Friedman, Editor,** Expanding Graphs
9 **William T. Trotter, Editor,** Planar Graphs
8 **Simon Gindikin, Editor,** Mathematical Methods of Analysis of Biopolymer Sequences
7 **Lyle A. McGeoch and Daniel D. Sleator, Editors,** On-Line Algorithms
6 **Jacob E. Goodman, Richard Pollack, and William Steiger, Editors,** Discrete and Computational Geometry: Papers from the DIMACS Special Year
5 **Frank Hwang, Clyde Monma, and Fred Roberts, Editors,** Reliability of Computer and Communication Networks
4 **Peter Gritzmann and Bernd Sturmfels, Editors,** Applied Geometry and Discrete Mathematics, The Victor Klee Festschrift
3 **E. M. Clarke and R. P. Kurshan, Editors,** Computer-Aided Verification '90
2 **Joan Feigenbaum and Michael Merritt, Editors,** Distributed Computing and Cryptography
1 **William Cook and Paul D. Seymour, Editors,** Polyhedral Combinatorics

DIMACS
Series in Discrete Mathematics
and Theoretical Computer Science

Volume 12

Network Flows and Matching

First DIMACS
Implementation Challenge

David S. Johnson
Catherine C. McGeoch
Editors

NSF Science and Technology Center
in Discrete Mathematics and Theoretical Computer Science
A consortium of Rutgers University, Princeton University,
AT&T Bell Labs, Bellcore

American Mathematical Society

This DIMACS volume resulting from the DIMACS Implementation Challenge contains revised and refereed versions of papers presented at the Challenge Workshop held at DIMACS from October 14, 1991, through October 16, 1991, along with supplemental material about the Challenge and the Workshop.

1991 *Mathematics Subject Classification.* Primary 68-04, 68-06, 90B10; Secondary 68Q20, 68Q25, 90C05, 90C06.

Library of Congress Cataloging-in-Publication Data

Network flows and matching: first DIMACS implementation challenge/David S. Johnson, Catherine C. McGeoch, editors.
  p. cm.—(DIMACS series in discrete mathematics and theoretical computer science; v. 12)
  "NSF Science and Technology Center in Discrete Mathematics and Theoretical Computer Science, a consortium of Rutgers University, Princeton University, AT&T Bell Labs, Bellcore."
  Based on papers presented at the Challenge Workshop, held at DIMACS, Oct. 14–16, 1991.
  Includes bibliographical references.
  ISBN 0-8218-6598-6 (acid-free)
  1. Systems theory. 2. Computer science—Mathematics. I. Johnson, David S., 1945– . II. McGeoch, Catherine C., 1959– . III. DIMACS (Group) IV. Challenge Workshop (1991:DIMACS) V. Series.
QA76.9.M35N48   1993   93-28698
511'.8–dc20   CIP

Copying and reprinting. Individual readers of this publication, and nonprofit libraries acting for them, are permitted to make fair use of the material, such as to copy an article for use in teaching or research. Permission is granted to quote brief passages from this publication in reviews, provided the customary acknowledgment of the source is given.

Republication, systematic copying, or multiple reproduction of any material in this publication (including abstracts) is permitted only under license from the American Mathematical Society. Requests for such permission should be addressed to the Manager of Editorial Services, American Mathematical Society, P.O. Box 6248, Providence, Rhode Island 02940-6248.

The appearance of the code on the first page of an article in this book indicates the copyright owner's consent for copying beyond that permitted by Sections 107 or 108 of the U.S. Copyright Law, provided that the fee of $1.00 plus $.25 per page for each copy be paid directly to the Copyright Clearance Center, Inc., 27 Congress Street, Salem, Massachusetts 01970. This consent does not extend to other kinds of copying, such as copying for general distribution, for advertising or promotional purposes, for creating new collective works, or for resale.

Copyright ©1993 by the American Mathematical Society. All rights reserved.
The American Mathematical Society retains all rights except those granted
to the United States Government.
Printed in the United States of America.
The paper used in this book is acid-free and falls within the guidelines
established to ensure permanence and durability. ∞
Printed on recycled paper. ♻

This volume was prepared by the authors using $\mathcal{A}_{\mathcal{M}}\mathcal{S}$-TEX and $\mathcal{A}_{\mathcal{M}}\mathcal{S}$-LATEX, the American Mathematical Society's TEX macro system.

10 9 8 7 6 5 4 3 2 1   98 97 96 95 94 93

Contents

Foreword	ix
Preface	
D. S. JOHNSON AND C. C. MCGEOCH	xi
Goldberg's Algorithm for Maximum Flow in Perspective: A Computational Study	
R. J. ANDERSON AND J. C. SETUBAL	1
Implementations of the Goldberg-Tarjan Maximum Flow Algorithm	
Q. C. NGUYEN AND V. VENKATESWARAN	19
Implementing a Maximum Flow Algorithm: Experiments with Dynamic Trees	
T. BADICS AND E. BOROS	43
Implementing the Push-Relabel Method for the Maximum Flow Problem on a Connection Machine	
F. ALIZADEH AND A. V. GOLDBERG	65
A Case Study in Algorithm Animation: Maximum Flow Algorithms	
G. E. SHANNON, J. MACCUISH, AND E. JOHNSON	97
An Empirical Study of Min Cost Flow Algorithms	
R. G. BLAND, J. CHERIYAN, D. L. JENSEN, AND L. LADÁNYI	119
On Implementing Scaling Push-Relabel Algorithms for the Minimum-Cost Flow Problem	
A. V. GOLDBERG AND M. KHARITANOV	157
Performance Evaluation of the MINET Minimum Cost Netflow Solver	
I. MAROS	199

A Speculative Contraction Method for Minimum Cost Flows:
Toward a Practical Algorithm
- S. Fujishige, K. Iwano, J. Nakano, and S. Tezuka — 219

An Experimental Implementation of the Dual Cancel and Tighten
Algorithm for Minimum-Cost Network Flow
- S. T. McCormick and L. Liu — 247

A Fast Implementation of a Path-Following Algorithm for
Maximizing a Linear Function over a Network Polytope
- A. Joshi, A. S. Goldstein, and P. M. Vaidya — 267

An Efficient Implementation of a Network Interior Point Method
- M. G. C. Resende and G. Veiga — 299

On the Massively Parallel Solution of Linear Network Flow Problems
- S. Neilsen and S. Zenios — 349

Approximating Concurrent Flow with Unit Demands and Capacities:
An Implementation
- J. M. Borger, T. S. Kang, and P. N. Klein — 371

Implementation of a Combinatorial Multicommodity Flow Algorithm
- T. Leong, P. W. Shor, and C. Stein — 387

Reverse Auction Algorithms for Assignment Problems
- D. A. Castañon — 407

An Approximate Dual Projective Algorithm for Solving Assignment
Problems
- K. G. Ramakrishnan, N. K. Karmarkar, and A. P. Kamath — 431

An Implementation of a Shortest Augmenting Path Algorithm for
the Assignment Problem
- J. Hao and G. Kocur — 453

The Assignment Problem on Parallel Architectures
    M. Brady, K. K. Jung, H. T. Nguyen, R. Raghavan, and
    R. Subramonian   469

An Experimental Comparison of Two Maximum Cardinality Matching Programs
    S. T. Crocker   519

Implementing an $O(\sqrt{N}M)$ Cardinality Matching Algorithm
    R. B. Mattingly and N. P. Ritchey   539

Solving Large-Scale Matching Problems
    D. Applegate and W. Cook   557

Appendix A: Electronically Available Materials
    C. C. McGeoch   577

Appendix B: Panel Discussion Highlights
    D. S. Johnson   583

Foreword

This DIMACS volume resulting from the Implementation Challenge of November 1990 through August 1991 contains refereed versions of papers presented at the culminating workshop held at DIMACS October 14–16, 1991. We would like to thank the organizers, David S. Johnson and Catherine C. McGeoch, for all the work they did on the workshop and this first DIMACS Implementation Challenge.

Fred Roberts, Acting Director
Robert Tarjan, Co-Director
Diane Souvaine, Associate Director

Preface

In recent years there has been growing interest in the application of computational and statistical tools to problems in the analysis of algorithms. In many algorithmic domains worst-case bounds are too pessimistic and tractable probabilistic models too unrealistic to provide meaningful predictions of practical algorithmic performance. Experimental approaches can provide useful knowledge where purely analytical methods fail. Moreover, experiments can provide new insights into algorithmic mechanisms and therefore motivate and guide deeper analytical results. Despite the potential value of the experimental approach, however, there has not been a strong experimental tradition in the field of algorithm analysis.

The DIMACS Implementation Challenge was organized for the purpose of encouraging more work in this area. Sponsored by DIMACS, the Challenge specifically encouraged experimental work in the area of network flows and matchings. Participants at sites in the United States, Europe, and Japan undertook projects during November 1990–August 1991 to test and evaluate algorithms for these problems. The Challenge culminated in a three-day workshop, held October 14–16, 1991, at the DIMACS Center at Rutgers University. This Proceedings contains revised and refereed versions of 22 of the papers presented at the Workshop, along with supplemental material about the Challenge and the Workshop.

One goal of the Challenge was to promote the active interchange of information and sharing of test data among participants. A repository of test instances, sample codes, and even draft papers was maintained at DIMACS, and made electronically available to all participants. Extended abstracts of the workshop papers were circulated before the workshop, and authors were given several months after the workshop to continue their experiments before the Proceedings submissions were due. These submissions were then refereed. In most cases the papers were then substantially revised in response to requests for more experimental data and more detailed explanations of the algorithmic issues involved. The papers are now in final form (although in several cases authors have continued their research and will be writing follow-up papers).

As a result of the cooperative approach taken by the authors and the long gestation period for their papers, this Proceedings should prove more informative than would a random collection of experimental papers on the same topic. Many papers report results on the same classes of instances, and calibrate the speed of their computers using the same set of benchmark tests. Many authors cite each other and make use of lessons learned by the other participants (and, in a few cases, point out the existence of contradictory results that had to be left for future studies to resolve). Thus readers will have an easier task of comparing results and drawing their own conclusions.

The papers are divided into five groups, according to the algorithmic problems they address. The first group of five papers concerns the maximum network flow problem. All touch on the recent push/relabel approach of Goldberg and Tarjan, which workshop participants agreed is the method of choice for practically all instance classes. Anderson and Setubal implement Goldberg-Tarjan and compare it to many of the well-known algorithms that preceded it. Nguyen and Venkateswaran study the effects of various algorithmic choices one can make in an implementation of Goldberg-Tarjan, and Badics and Boros examine the value of using sophisticated data structures in implementing the algorithm (and an alternative algorithm due to Cheriyan and Hagerup). Alizadeh and Goldberg report on experiments in which the Goldberg-Tarjan algorithm is implemented on a massively parallel SIMD computer. Finally Shannon, MacCuish, and Johnson report on lessons learned by animating key aspects of the algorithm's inner workings.

The second group of papers concerns the minimum cost network flow problem. Here there was no clear "champion," as is amply demonstrated in a paper by Bland et al., which compares four distinct approaches (a cost scaling algorithm of Bland and Jensen, a relaxation algorithm of Bertsekas, a network simplex code of Grigoriadis, and a scaling push/relabel algorithm of Goldberg and Tarjan). The paper concludes that each algorithm can either win or lose by large amounts, depending on the instance class. The remaining papers in this group focus more on individual algorithms. Goldberg and Kharitonov investigate a variety of heuristic ideas for speeding up the Goldberg-Tarjan approach. Maros reports on experiments applying his own network simplex code to the problem. Two papers, by Joshi, Goldstein, and Vaidya and by Resende and Viega, study interior point approaches to the linear programming formulation of the problem. Two other papers, by Fujishige et al. and by McCormick and Liu, examine some new algorithmic ideas in the context of the graph-theoretic formulation of the problem. The final paper in this group, by Neilsen and Zenios, reports on experiments comparing implementations of a relaxation algorithm and a new algorithm of the authors on a massively parallel machine.

The third group of papers considers the multicommodity flow problem. There has recently been much interesting theoretical work on combinatorial approximation schemes, which by settling for merely getting *close* to optimal, offer the possibility of running much faster than traditional optimization approaches (such as the simplex algorithm). Papers by Borger, Kang, and Klein and by Leong, Shor, and Stein report on experiments with these theoretically interesting algorithms. Their results suggest that one needs to sacrifice very little in terms of accuracy to obtain impressive speedups, assuming the number of commodities is large.

The fourth group of papers deals with the assignment problem, where again the algorithm of choice may depend heavily on the type of instance under consideration. Castañon considers algorithmic methods for fine-tuning the auction approach to the problem. Hao and Kocur study a new implementation of an algorithm based on shortest augmenting paths. Ramikrishnan, Karmarkar, and Kamath report on extensive tests with a mild specialization of their "approximate dual projective" interior point method for linear programming. Brady et al. look at parallel implementations of several assignment algorithms, and in particular, implementations of the auction algorithm on a variety of parallel architectures.

The fifth and final group of papers considers general graph matching. For the unweighted case, an algorithm of Micali and Vazirani, long viewed as the theoretical champion, had never been seriously studied from an experimental point of view. Papers by Crocker and by Mattingly and Ritchey present alternative implementations of this algorithm, comparing it for instance to Gabow's efficient version of Edmonds' blossom algorithm. Crocker also studies the rate of blossom formation as a way to gain insight into algorithmic performance. A tentative conclusion is that the Micali-Vazirani algorithm is not yet generally competitive with modern variants of Edmonds' 1965 algorithm, although there are certain types of graphs on which it seems to win, and in general its running time appears to be more stable. Applegate and Cook consider weighted matchings, reporting on an implementation of Edmonds' algorithm that can solve geometric instances with over 100,000 vertices.

This Proceedings contains two appendices. Many of the instance generators used in the studies, along with several solution algorithms, bibliography files, and related material, are currently stored on a remotely-accessible computer at DIMACS. Appendix A summarizes the items available and explains how they may be obtained. (It is hoped that the files will remain accessible for several years after the publication of this Proceedings.) Appendix B reports on three panel discussions that were held at the Challenge workshop, summarizing the issues raised and conclusions drawn.

One of those conclusions was that there should be further DIMACS Implementation Challenges, and we are happy to report that this has come to pass. As we write this Preface, a second DIMACS Implementation Challenge, concerning problems of clique finding, graph coloring, and satisfiability testing, is well underway. A third Challenge, devoted to parallel computation, is in the planning stages for 1993-94. Proceedings of these and succeeding Challenges should appear in the AMS DIMACS Series in future years.

We gratefully acknowledge the National Science Foundation, which provided grant support for Cathy McGeoch to visit DIMACS and coordinate the Challenge (RUI Grant No. 9013079). We would also like to thank DIMACS and the DIMACS staff for their considerable assistance. In particular, we are grateful to the late Daniel Gorenstein, founding director of DIMACS, for his early and continuing support of the project. In addition, Pat Toci's help was indispensable in organizing and running the workshop, Virginia Moore provided much-needed advice on preparing the preliminary proceedings, and David Zimmerman was a key player from the start, helping to set up the DIMACS directories, arranging for their availability via ftp, and providing general-purpose computing advice and support. Thanks to Tamas Badics for providing programming support and documentation, and for helping transcribe the videotapes of the panel discussions on which Appendix B is based. Thanks also to Kyshon Michaux of AT&T Bell Laboratories, who provided invaluable assistance in collecting and organizing the papers, and to the other members of the Steering Committee (Mike Grigoriadis of Rutgers, Bob Tarjan of Princeton, and Clyde Monma of Bellcore), who not only steered, but also acted as the program committee for the workshop. Finally, thanks to all who participated in the Challenge and supported the idea.

May 1993

DAVID S. JOHNSON
CATHERINE C. MCGEOCH

Goldberg's algorithm for maximum flow in perspective: a computational study

RICHARD J. ANDERSON AND JOÃO C. SETUBAL

ABSTRACT. We present a computational study of Goldberg's algorithm and five other algorithms for the maximum flow problem. Goldberg's algorithm with periodic global relabeling was found to be by far the fastest. Four variations of Goldberg's algorithm were tested, differing in the selection rule used to discharge active vertices. We obtained the best results with FIFO and highest-label-first selection rules. We also show that the performance of Goldberg's algorithm on the input distributions used is very much better than on specially designed bad graphs.

1. Introduction

We describe our current research on the implementation of sequential algorithms for the maximum flow problem. We assume the reader is familiar with this problem and the basics of maximum flow algorithms; a good introduction can be found in [**Tar83**, chapter 8]. We use n for number of vertices and m for number of edges.

The aim of this research was to study the performance in practice of Goldberg's algorithm [**Gol87, GT88**] (also known in the literature as the *preflow-push algorithm* or the *push-relabel algorithm*). At the same time we wanted to get a general picture of the successive improvements that have been achieved on maximum flow algorithms over the years. Therefore we implemented five other algorithms, starting with Ford and Fulkerson's [**FF62**].

1991 *Mathematics Subject Classification*. Primary 90B10, 05C85, 68-04.

Key words and phrases. network flow, maximum flow, Goldberg's algorithm, Dinic's algorithm, algorithm implementation.

Research of the first and second authors was supported in part by NSF Presidential Young Investigator Award CCR-8657562, Digital Equipment Corporation, NSF CER grant CCR-861966, and NSF/Darpa grant CCR-8907960.

Research of the second author was supported in part by Brazilian Agency FAPESP grant 87/1385-7.

The second author was on leave from State University of Campinas (UNICAMP, Brazil).

Some other computational studies of maximum flow algorithms have been done before. Those of Cheung [**Che80**] and Imai [**Ima83**] were done before the appearance of Goldberg's algorithm, and basically concluded that Dinic's algorithm [**Din70**] was the fastest for the inputs used. Derigs and Meier [**DM89**] then compared Dinic's to Goldberg's, and found Goldberg's algorithm to be faster on the two input classes tested. Our study differs from that in [**DM89**] on three aspects: (1) our implementation is different from theirs on a key part, and we show that Goldberg's algorithm can be made to run even more efficiently; (2) we ran the implementation on a more varied group of input classes; (3) the studies cited dealt with graph instances of no more than 2^{12} vertices and we present results of instances up to 2^{16} vertices. In addition we present a simple family of graphs that makes Goldberg's algorithm achieve its worst-case running time.

The outline of the paper is as follows: in section 2 we give brief descriptions of all algorithms that we implemented, and particulars of our implementations. In section 3 we describe the test data and methodology used, and present and analyze results. Also in this section we compare Derigs and Meier's implementation to ours. In section 4 we describe the family of graphs that causes worst-case running time for Goldberg's algorithm. We give conclusions and suggest further work in section 5.

2. Implementation descriptions

Ford and Fulkerson: This algorithm [**FF62**] simply keeps looking for augmenting paths using depth-first search. Worst-case running time is not polynomial.

Edmonds and Karp(breadth first search): This algorithm [**EK72**] finds augmenting paths by breadth-first search and runs in $O(nm^2)$ time.

Edmonds and Karp(maximum gain): This algorithm [**EK72**] augments the flow through paths of maximum capacity. These paths are found using a variant of Dijkstra's algorithm for shortest paths. Running time is $O(m^2 \log n \log C)$, where C is the maximum edge capacity.

Dinic: This algorithm [**Din70**] repeatedly builds a layered graph based on the unsaturated edges and finds a *blocking flow* on this layered graph by augmenting the flow through as many paths as possible. Our implementation follows that presented in [**Tar83**]. There are basically two routines: one that builds the layered graph using breadth-first search, and another that finds a blocking flow on the layered graph. There are at most n blocking flow phases and worst-case running time is $O(n^2 m)$. A layered graph is not actually built; distance labels are computed and flags are set on the edges that belong to the (implicit) layered graph.

Karzanov: This is similar to Dinic's algorithm but it has an improved method for finding a blocking flow [**Kar74**]. This method uses the concept of *preflow*, which allows vertices to receive more flow than they can send. Our implementation follows that proposed by Tarjan [**Tar84**], with the addition of a vertex-pruning routine that cuts off vertices that are unreachable at the start of each blocking flow phase. The running time is $O(n^3)$.

Goldberg: This algorithm [**Gol87, GT88**] also uses the preflow concept, but does not find blocking flows. Instead, vertices that have excess flow (the *active vertices*) use the basic *push operation* to send flow to neighbor vertices. *Labels* associated with every vertex are used to determine the direction of the flow. The value of a vertex label represents a lower bound on the distance in number of edges from the vertex to the sink, or to the source (in case the vertex cannot reach the sink). Labels are updated throughout the algorithm execution by the basic *relabel operation*. The operation of selecting an active vertex, pushing flow from it and possibly relabeling it is called a *discharge*. The algorithm presents flexibility in the order in which active vertices are selected; the running time depends on this order. We implemented four rules for active-vertex selection, obtaining four Goldberg variations as described below (in parenthesis is the short name used to describe each variant in this paper):

(i) least recently updated active vertex, using a FIFO queue (queue).
(ii) active vertex with highest label, using buckets (highest-label).
(iii) active vertex with largest excess, using a priority queue (largest-excess).
(iv) most recently updated active vertex, using a stack (stack).

A worst-case running time bound valid for all variants is $O(n^2 m)$. Goldberg and Tarjan [**GT88**] proved the bound $O(n^3)$ for variant queue and Cheriyan and Maheshwari [**CM89**] proved the bound $O(n^2 m^{1/2})$ for highest-label.

In all these variants, we chose to discharge an active vertex until its excess flow is zero. In addition, the implementations use the *global relabeling heuristic*. This heuristic, also suggested in [**Gol87, GT88**] to improve the performance of the algorithm in practice, turns out to be an essential part of any implementation. The improvements in performance may reach two orders of magnitude for graphs with just a thousand vertices, and are even larger with larger graphs. This global relabeling is done by performing a periodic backwards breadth-first search on the unsaturated edges first from the sink, and then from the source. The labels are updated to be the exact distance in number of edges from each vertex to the sink or, when a vertex cannot reach the sink, the exact distance to the source plus n. Our implementations start with exact labels, and global relabeling is then called every n discharge operations, if $m \leq 10n$. If $m > 10n$, global relabeling is called every $2n$ discharge operations. These rules are empirical but they seem to be close to the "optimum frequency" of global relabeling, which seems to vary depending on the implementation and on graph structure.

The relabeling heuristic used by Derigs and Meier [**DM89**] is different than the one described above. It requires finding a *gap* in the values of the labels. Let us denote the value of vertex v's label by $d(v)$. Derigs and Meier prove that if there exists a value $l < n$ such that for all vertices v, $d(v) \neq l$, then all v such that $l < d(v) < n$ can be relabeled to n. This is easy to implement, since it simply requires maintaining an array of existing labels, which is checked and possibly updated at every relabel operation. Note, however, that for vertices with labels above n this heuristic does nothing. As we shall see, this turns out to be a disadvantage whenever the minimum cut is close to the sink and a lot of excess flow has to be returned to the source. We have compared these two heuristics and present the results in section 3.7.

Finally, we would like to point out that the coding effort required in implementing these four variants of Goldberg's algorithm was fairly small and similar to that required by Dinic's algorithm. In particular, program strucuture follows closely that described in [**GT88**, section 4].

3. Experiments

3.1. Environment. We ran the experiments on a DECstation 5000/200 based on the RISC chip R3000, with 32 megabytes of main memory, running ULTRIX 4.1. The programs were written in C and compiled with ULTRIX's cc compiler, with the optimizer flag -O set. Running times were measured with the UNIX system call `getrusage` by selecting the field `ru_utime`.

3.2. Test data and methodology. We ran the programs on a variety of input graphs. We present results from the input classes described in the DIMACS document "The Benchmark Experiments" as well as two other classes from the DIMACS document "The Core Experiments", as shown below, together with the acronyms we use to designate them. The reader is directed to the DIMACS documents cited above for a complete description of these classes. We give below between square brackets a rough indication of edge density.

- Washington random level, sub-class wide (*wrlgw*) $[m \approx 3n]$.
- Washington square sparse (*wss*) $[m \approx 4n]$.
- Washington line moderate (*wlm*) $[m = O(n^{3/2})]$.
- Genrmf-long (*rmfl*) $[m \approx 6n]$.
- Genrmf-wide (*rmfw*) $[m \approx 6n]$.
- Acyclic dense (*ad*) $[m = n(n-1)/2]$.

Our experiments were divided in two phases. Initially we tested all programs on small instances of some input classes. From these studies it became clear that Goldberg's algorithm was far superior to all others in all input classes except *ad*, where Dinic's algorithm was fastest. The results for this phase are presented in section 3.3.1. In the second phase, we conducted more complete experiments on much larger graphs from the "benchmarks" document, but using only the four Goldberg variations plus Dinic's algorithm on input class *ad*. The methodology used was as follows.

- All programs tested used exactly the same set of inputs; this was assured through a file of seeds for the pseudo-random number generator, which was UNIX random().
- At the end of each run the code checks the solution for consistency and maximality.
- Running times reported exclude input, checking, and output time, but we do report a sample of input and output times in a section 3.4.
- For phase 1, means reported were computed based on 10 runs for each algorithm, size, and input class, except for Ford and Fulkerson's algorithm, where only one instance of each input class and size was solved.
- For phase 2, means reported were computed based on 20 runs for each size and input class, except *wlm*, where 30 runs were conducted. The standard deviation was also computed for each mean, and found to be under 20% of the mean in all cases except for class *ad*, where standard deviations were as high as 47% of the mean.
- In phase 2, for classes *wrlgw*, *rmfl* and *rmfw* values for n were approximately $2^{10}, 2^{12}, 2^{14}, 2^{16}$. For class *wlm* values for n were approximately $2^{10}, 2^{11}, 2^{12}, 2^{13}$. For *ad* values for n were approximately $2^{8.5}, 2^{9}, 2^{9.5}$. The highest values in the sparse classes represent the largest input graphs that our machine could handle. [1] Class *wss* was used only on phase 1.
- In phase 2, asymptotic performance was estimated by doing a power regression analysis of the data using a standard least-squares method for the fit.

3.3. Results.

3.3.1. *All algorithms.* Table 1 presents mean running times for all implementations on 10 instances and two sizes in each of five input classes. It is clear from the table that the best algorithm is Goldberg's. It was by far the fastest in all input classes except *ad*, where Dinic's was faster. The table makes also clear the importance of the global relabeling heuristic for the performance of Goldberg's algorithm. Note for instance that in class *wlm* with 1026 vertices it ran 100 times faster with the heuristic than without. Finally, it is striking to see the gains in performance achieved by Goldberg's algorithm in comparison with Ford and Fulkerson's even on these small graphs. Again in class *wlm* computation time went down from a day and a half to a fraction of a second. Edmonds and Karp [**EK72**], however, note that breadth-first search would likely "be incorporated innocently into a computer implementation" of Ford and Fulkerson's algorithm before the improved running time bound became known.

---

[1] Our implementation requires 28 bytes per edge. It is possible to lower the storage requirements to about 20 bytes per edge, but even that would not be enough to handle inputs with 2^{18} vertices in the sparse classes. We note that every edge (v, w) appears in the adjacency lists of both v and w even if (w, v) does not exist.

TABLE 1. Mean running times (in seconds, unless otherwise indicated) for all algorithms. Notes: (1) under each input class name is the number of vertices of the instances solved. (2) FF is Ford and Fulkerson's algorithm and its timings were on one instance only. (3) EK1 is Edmonds and Karp's algorithm using breadth-first search; EK2 uses maximum gain. (4) D is Dinic's algorithm and K is Karzanov's algorithm. (5) G1 and G2 are Goldberg's algorithm. G1 is queue without global relabeling; G2 is queue with global relabeling.

alg.	wss		rmfl		rmfw		wlm		ad	
	531	1026	575	1152	432	1024	514	1026	128	256
FF	69.82	11 m	8.71	9 m	3.22	8 m	18 m	39 h	3.10	25.19
EK1	8.91	35.51	1.90	9.86	0.69	8.50	4.56	27.08	3.94	35.30
EK2	4.29	12.49	0.86	3.03	3.01	15.98	0.97	3.27	6.94	56.85
D	0.37	1.09	0.48	2.00	0.27	1.93	0.26	0.84	0.37	1.63
K	1.60	5.86	3.04	12.86	2.64	14.93	1.20	3.56	2.98	27.10
G1	3.27	13.25	3.35	18.39	1.42	15.62	10.28	21.33	2.55	26.62
G2	0.12	0.27	0.10	0.28	0.11	0.53	0.08	0.20	0.56	3.47

3.3.2. *Goldberg.* Table 2 presents the mean running times of the Goldberg variations on the largest instances tested for each input class. Class *wlm* is subdivided in two since the times obtained have a bimodal distribution for three of the implementations. The reasons for this are explained in section 3.5.

Tables 3, 4, 5, and 6 present the mean number of nonsaturating pushes, saturating pushes, local relabels, and global relabelings, respectively.

Table 7 shows the slopes that estimate the asymptotic performance in terms of running time and number of nonsaturating pushes. As an illustration, figure 1 shows the actual running time growth curves for the four Goldberg variations on class *rmfw*.

3.3.3. *Goldberg and Dinic on acyclic dense graphs.* Table 8 compares Goldberg's and Dinic's algorithms on the one input class where Dinic's was the faster implementation. The Goldberg implementation used in the comparison was the fastest one using periodic global relabeling on that input class.

3.4. Input and output time samples. Table 9 shows the total running time for two instances of two different input classes, one sparse and one moderately dense. This table is intended to give an idea of how much time is spent in reading the input and writing a complete solution (the final flow on every edge) to disk.

3.5. Analysis of Goldberg's algorithm performance. The results in table 2 show that there is no clear winner among the four implementations tested. The fastest in three of the input classes was highest-label, but it was by

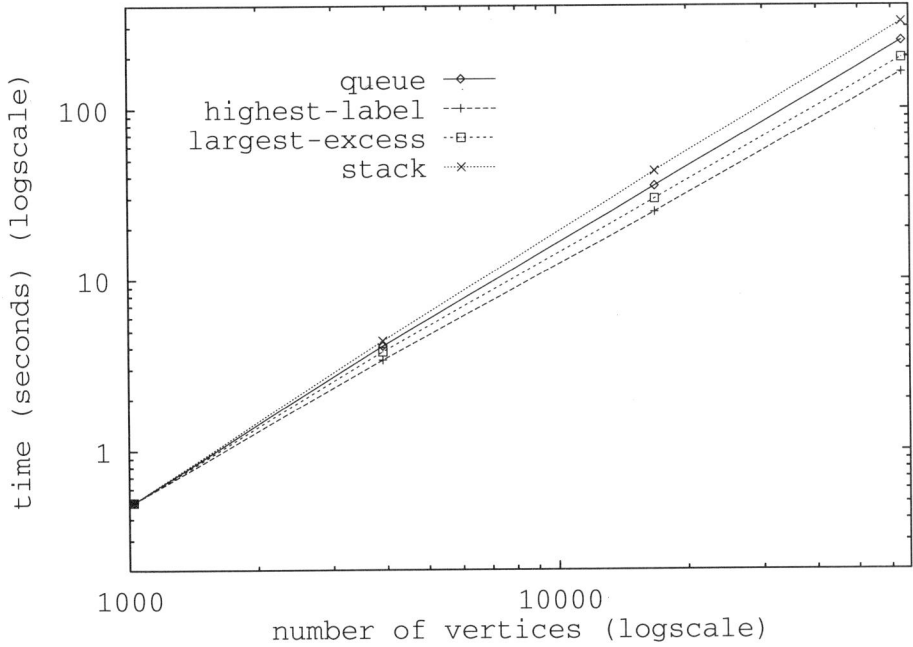

FIGURE 1. Growth of running time with size for input class *rmfw*.

TABLE 2. Mean running times (in seconds) on largest instance of each input class for Goldberg implementations. In parenthesis is the value of n for each class. In boldface is the best time in each class.

Goldberg variation	$wrlgw$ (65538)	$rmfl$ (65536)	$rmfw$ (63504)	wlm_1 (8194)	wlm_2 (8194)	ad (724)
queue	**41.6**	56.1	248.0	2.7	5.8	50.5
highest-label	1081.3	**55.0**	**162.1**	**1.7**	**4.3**	71.1
largest-excess	396.3	114.2	196.7	7.0	10.7	**40.1**
stack	435.1	309.7	319.3	20.1	21.2	50.1

TABLE 3. Mean number of nonsaturating pushes (in thousands) on largest instance of each input class for Goldberg implementations. In boldface is the smallest value in each class.

variation	$wrlgw$	$rmfl$	$rmfw$	wlm_1	wlm_2	ad
queue	**937**	1597	5872	17	34	30
highest-label	24516	**1190**	3478	**7**	**20**	29
largest-excess	4441	1988	**3407**	59	78	**11**
stack	11749	11477	9681	300	292	21

far the slowest in class $wrlgw$. Implementation queue seems to be more robust, since it was never outperformed by large margins. Another observation is that stack did not perform well compared to the others.

One has to be a little cautious in making these comparisons because of the way that the global relabeling heuristic is applied. As described in section 2, we use the frequencies n and $2n$ for global relabeling invocation, depending on graph density. However, experiments on a few instances showed that the optimum frequency (the one that yields the fastest time) is different depending on implementation, on input class, and on size. For example, for one instance of the $rmfw$ class, the best frequency for queue was $2n$, whereas for highest-label it was n. On the other hand variations in running time seem to be relatively small when the frequency is changed from n to $2n$ or to $n/2$. For the same $rmfw$ instance mentioned above, with 2^{12} vertices, the running time for the queue implementation was 4.07 seconds with frequency $n/2$, 3.54 with n, 3.50 with $2n$ and 3.70 with $2.5n$. A similar behavior was observed in other instances. In spite of these variations, we believe that the relative standing of the implementations shown in table 2 would not change even if the optimum frequency of global relabeling had been used on each implementation and instance.

An examination of tables 2, 3, 4, and 5 shows that the number of nonsaturating pushes (nsp) is a good predictor of running time. This also agrees with the theoretical analysis, where nsp is the dominant term. On the other hand, growth

TABLE 4. Mean number of saturating pushes (in thousands) on largest instance of each input class for Goldberg implementations.

variation	$wrlgw$	$rmfl$	$rmfw$	wlm	ad
queue	145	54	91	9	40
highest-label	275	75	72	10	20
largest-excess	137	61	70	7	12
stack	112	51	73	6	8

TABLE 5. Mean number of relabel operations (in thousands) on largest instance of each input class for Goldberg implementations.

variation	$wrlgw$	$rmfl$	$rmfw$	wlm	ad
queue	229	326	1951	6	7
highest-label	18794	969	2007	4	14
largest-excess	2973	492	1949	11	9
stack	5674	493	1538	15	9

TABLE 6. Mean number of global relabelings on largest instance of each input class for Goldberg implementations.

variation	$wrlgw$	$rmfl$	$rmfw$	wlm	ad
queue	14.2	23.9	92.2	1.1	20.1
highest-label	374.4	17.8	54.3	0.3	19.9
largest-excess	67.5	29.8	53.4	3.4	7.5
stack	179.3	175.1	152.2	17.9	14.5

TABLE 7. Slopes of $\log(n)$ versus $\log$(running time) and in parenthesis $\log(n)$ versus $\log$(number of nonsaturating pushes).

variation	$wrlgw$	$rmfl$	$rmfw$	ad
queue	1.3(1.2)	1.3(1.3)	1.5(1.4)	2.6(1.7)
highest-label	1.9(1.9)	1.3(1.3)	1.4(1.4)	2.7(1.6)
largest-excess	1.6(1.5)	1.4(1.3)	1.4(1.4)	2.5(1.3)
stack	1.5(1.4)	1.6(1.6)	1.6(1.5)	2.5(1.5)

TABLE 8. Comparison between Goldberg's (largest-excess) and Dinic's algorithms on acyclic dense graphs. Time is in seconds.

algorithm	time ($n = 724$)	slope
largest-excess	40.1	2.5
Dinic	12.3	2.0

TABLE 9. Running times (in seconds) of the queue implementation broken by program section on two instances.

times	$rmfw$ $n = 8214$ $m = 22829$	wlm $n = 8194$ $m = 187381$
input time	2.06	10.76
computation time	13.55	11.66
output time	1.16	2.51

rates for both running time and nsp in these input classes, shown in table 7, are much less than those predicted by the worst-case analysis. Table 7 also shows that in all input classes except ad the running time slope is very close to the nsp slope. For class ad this is not the case. The main reason seems to be the high edge density of instances in this class, which causes both the relabel operation and the global relabeling to take much longer than on sparse graphs with similar number of vertices.

Class ad presented a number of peculiarities. It was the one class where Goldberg's algorithm was outperformed by Dinic's. One might be tempted to blame this result on an excessive number of global relabelings. Indeed, in our preliminary work we thought that m would be a better general value for the global relabeling frequency, since we feared that in dense graphs a frequency of n would cause too many global relabelings. However, this was not the case for class ad, where $2n$ seems to be the best value. On the other hand, as shown in section 3.7 below, by using the gap relabeling strategy instead of periodic global relabeling the performance of Goldberg's algorithm on ad can be substantially improved, although not to the point that it can beat Dinic's algorithm. Finally Badics et al. [**BBC92**], using an implementation related to Goldberg's algorithm but very different from ours, obtained running times on class ad which seem to be significantly better than those presented here. A more detailed study of Goldberg's algorithm on this class could fully explain these peculiarities as well as indicate whether similar behavior should be expected in other classes of dense graphs.

Results for input class wlm were divided into two sub-classes because the running times for three of the implementations showed bimodal distribution. This is due to the fact that instances from this class have the minimum cut

TABLE 10. Mean running times (in seconds) for variations of Goldberg's algorithm using periodic global relabeling and gap relabeling. Notes: (1) under each input class name is the number of vertices of the instances solved. (2) running times are means over 10 runs. (3) Q is the queue implementation using global relabeling; QG is the queue implementation using gap relabeling; HLF is the highest-label implementation using global relabeling; HLFG is the highest-label implementation using gap relabeling.

alg.	wss		rmfl		rmfw		wlm		ad	
	4098	16386	4096	15488	3920	16807	1026	4098	256	512
Q	1.6	8.6	1.6	9.0	4.0	35.2	0.2	1.5	4.0	27.1
QG	3.3	29.8	3.7	35.6	7.1	83.1	0.7	7.8	2.3	16.4
HLF	7.2	45.7	1.6	8.0	3.4	25.7	0.2	1.0	4.6	34.6
HLFG	2.7	16.7	2.3	20.1	6.7	63.6	0.5	4.0	1.9	8.5

either close to the source (and the maximum flow can be found quickly) or close to the sink (maximum flow found more slowly). This behavior was more marked in highest-label, followed by queue and then largest-excess. Variation stack was not sensitive to the cut location, but it was also the slowest for this class.

3.6. Procedure profiles. A limited study of the run-time profiles of the Goldberg implementations provided the following rough breakdown, when the inputs were sparse graphs:

(i) Global relabeling: 25-40%
(ii) Push, Relabel, PushRelabel, and Discharge procedures: 45-65%
(iii) Data Structure operations: 5-15%

These results suggest that in order to improve the implementations more effort should be directed to optimize the manner in which global relabeling is applied. Note also that the cost of data structure operations is relatively small. This is true even for variation largest-excess, which uses a priority queue.

For dense graphs, the share of the relabel procedure was larger than on sparse graphs (up to 25% versus no more than 10%). Some technique for improving edge-scanning may give substantial savings in these cases.

3.7. A comparison between periodic global relabeling and gap relabeling. In a separate set of experiments we compared the performance of periodic global relabeling and gap relabeling. The gap relabeling implementations also start with exact labels. The results appear in table 10. These experiments show that the global relabeling heuristic is superior to the gap relabeling heuristic

TABLE 11. Running times (in seconds) and cut location for 10 instances of class wlm, with 4098 vertices. HLF is the highest-label implementation using global relabeling; HLFG is the highest-label implementation using gap relabeling.

instance	running time		cut location
	HLF	HLFG	
1	0.91	0.69	at source
2	0.61	0.62	at source
3	0.65	0.66	at source
4	1.65	6.33	at sink
5	1.03	8.52	at sink
6	0.62	0.66	at source
7	0.62	0.64	at source
8	1.08	11.08	at sink
9	0.62	0.63	at source
10	1.84	10.59	at sink

in every class except ad. [2] They also show that gap relabeling is more effective with the highest-label implementation than with the queue implementation, which is consistent with Derigs and Meier's results [**DM89**].

One of the main advantages of periodic global relabeling is that it speeds up not only the flow travel to the sink, but also the return of excess flow to the source. This is not the case of gap relabeling, which is of no help in returning flow to the source. The consequence is that implementations based on gap relabeling are even more sensitive to the cut location than those based on periodic global relabeling. This behavior is most clear in class wlm for the reasons alluded to above, as can be seen in table 11. There we present the actual running times on 10 instances for the highest-label implementation with periodic global relabeling and with gap relabeling, along with each instance's minimum cut location.

On the other hand, gap relabeling does not depend on the number of edges in the graph, and this was one of the factors that contributed to its superior performance in class ad. Even in sparse graphs the measured cost of gap relabeling is much lower than periodic global relabeling, standing at about 3% of the total running time in instances where many gap relabelings were performed. These observations suggest that some combination of periodic global relabeling and gap relabeling may prove to be the most robust approach to an implementation of Goldberg's algorithm, in particular with respect to dense graphs.

---

[2]Dinic's implementation was still faster on class ad even with this improved performance. On the set of instances with 512 vertices Dinic's mean running time was 6.1 seconds, compared with 8.5 for highest-label with gap relabeling.

4. A hard instance for Goldberg's algorithm

The major result that one can draw from our measurements is that on the families of input graphs tested Goldberg's algorithm performs far better than $\Theta(n^3)$, the bound given in [**GT88**].

Cheriyan and Maheshwari [**CM89**] have described a family of sparse graphs which require $\Theta(n^3)$ time for a queue-based implementation of Goldberg's algorithm. However, these bad case examples are very sensitive to the details of the algorithm, and in particular they assume that the algorithm performs a propagation of relabeling operations which is not done in our implementation. Therefore, our implementation performs substantially better than $\Theta(n^3)$ on these graphs. A similar but weaker result is that of Martel [**Mar89**], who presents a family of graphs that requires $\Omega(nm)$ time of Goldberg's algorithm.

After a substantial amount of effort, we discovered families of graphs that lead to $\Theta(n^3)$ performance for our queue-based implementation. One of these families has worst-case behavior even if labels are kept as exact distances at all times. Below we present this family. We assume that the graph is represented by adjacency lists where the vertices are ordered by increasing vertex number; the worst-case behavior is sensitive to this order.

Figure 2 illustrates our construction. It is based upon the constructions by Cheriyan and Maheshwari [**CM89**], but is substantially simpler. The graph is parameterized by k. Our figure indicates how vertices are numbered, and shows the capacities of the edges. The unmarked edges have infinite capacity. The algorithm begins by introducing a very large excess to vertex 2 from the source s. A large excess is moved back and forth between vertex 2 and vertex 1 passing through the vertices $5, 6, \ldots, k+5$.

The running time for the queue version of Goldberg's algorithm is $\Theta(k^3)$ on this graph. The running time can be explained by showing that k^2 single units of flow appear at separate times at vertex 4, and each one of these is routed down the chain of length k to the sink t. A very large excess is initially introduced to vertex 2. The large excess is routed through the following sequence of vertices: $2, 5, 2, 6, 2, 7, \cdots, k+4, 2, k+5, 1$. Each time the excess reaches a vertex in $5, 6, \ldots, k+4$, a single unit of flow is routed to 1 and then to 4. When the excess reaches 1, the edges $(5,1), \ldots, (k+4, 1)$ are saturated, as is the edge $(1, 4)$. The shortest path from 1 to the sink is now through the vertices 2 and 3. The flow is routed back to 2, and in the process the edges $(5, 1), \ldots, (k+4, 1)$ are unsaturated. When the flow reaches 2, and $(2, 3)$ is saturated, a situation similar to the original is reestablished. Flow is sloshed back and forth between 2 and 1 for k phases, leading to the $\Theta(k^3)$ running time.

To verify that the algorithm performs as outlined above, it is necessary to step through the operations, and show that the relabeling operations cause the flow to be routed as described. In this type of case, the most convincing approach to demonstrating the behavior is to look at actual performance results. Table 12

TABLE 12. Performance on bad case graphs. Time is in seconds.

n	time (Goldberg)	Nonsaturating Pushes	time (Dinic)
117	0.19	8515	0.03
227	0.97	57415	0.09
447	6.67	420415	0.36
887	49.84	3215215	1.55
1767	378.58	25144015	7.67

gives the results for five graphs in the family. Note that for the larger examples, the running time and number of nonsaturating pushes grows by roughly a factor of eight when the graph doubles in size, indicating cubic performance. The table also gives the performance of Dinic's algorithm on these graphs.

As mentioned above this family of graphs also gives $\Theta(n^3)$ performance for a version of Goldberg's algorithm that keeps labels as exact distances *at all times*. Unfortunately such is not the case if only *periodic* global relabeling is performed. In this case flow can be routed in a different manner than the one described. Tests on these graphs indicate that the performance of the algorithm is indeed very sensitive to exactly when global relabeling is performed. Nevertheless, we believe that the construction could be modified to ensure that the global relabelings do not change the performance of the algorithm, but we have not worked out all of the details. One particular difficulty in regulating the occurrence of global relabeling steps is that, as the algorithm progresses, the paths from the source to the sink get longer, so that there are an increasing number of vertices that are active.

5. Conclusions and Further work

Our experiments, together with previous studies, show that an implementation of Goldberg's algorithm using periodic global relabeling is the fastest method in practice to solve a large class of maximum flow problems. It clearly and soundly beats Dinic's algorithm on most input classes that we tested, except on one very dense class. Even though the computational study in this paper was not comprehensive, both in terms of input classes and of algorithms, we believe that Goldberg's algorithm with periodic global relabeling should be the method of choice for solving maximum flow problems in practice. Its performance is excellent, and the code is relatively simple. The only reservation that we have concerns the lack in our study of "real life" instances of the maximum flow problem.

We would also like to point out that Goldberg's algorithm lends itself to a natural and efficient parallel implementation. Motivated by the DIMACS Challenge, we designed and tested such an implementation for a shared-memory multiprocessor with 20 processors, and obtained results that we consider very good [**AS92**]. We achieved speed-ups of up to 5.8 with 16 processors with re-

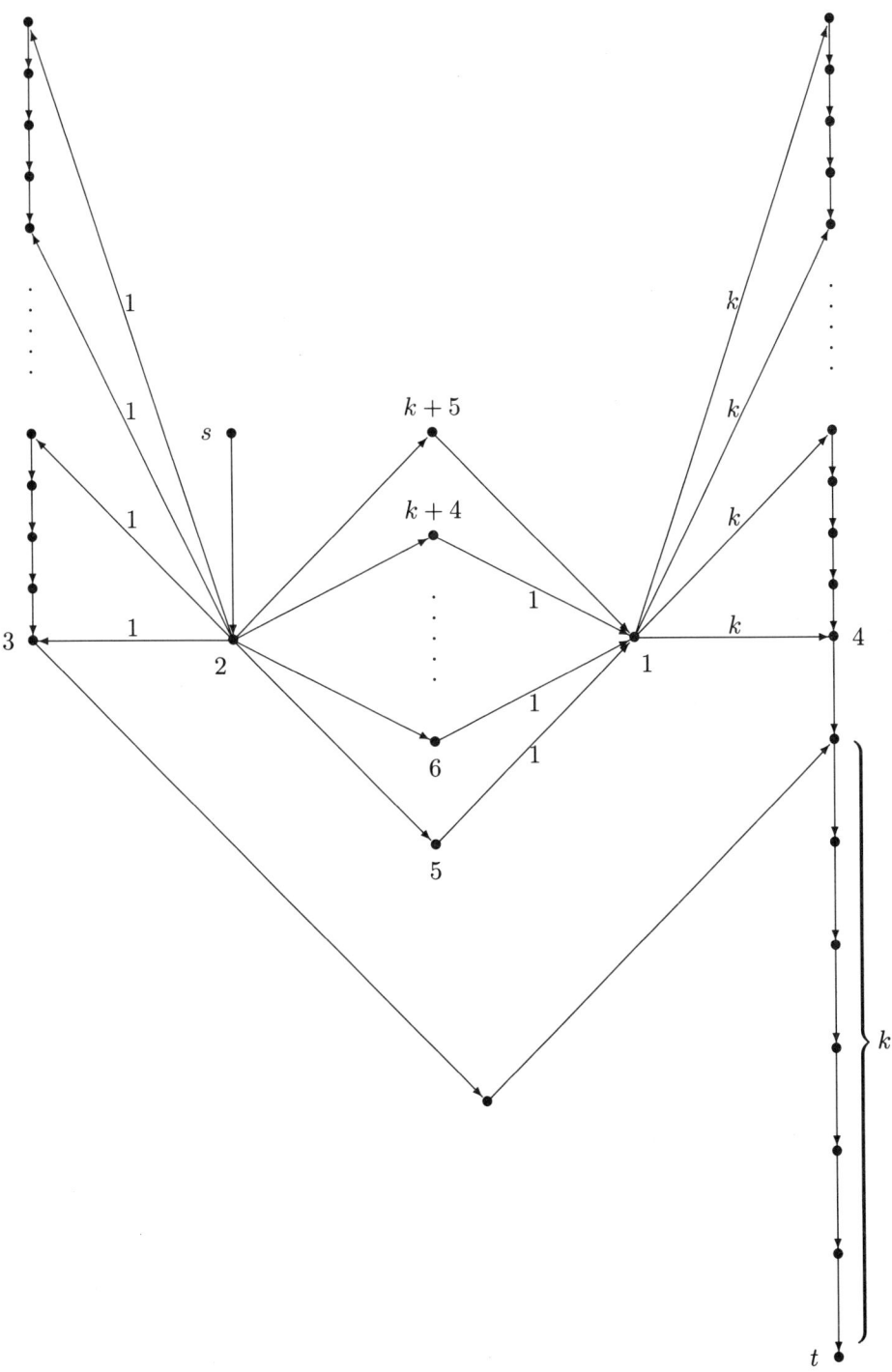

FIGURE 2. Bad case for Goldberg's algorithm

spect to the queue implementation described in this paper. We found evidence that hardware effects such as bus contention were the main obstacles in achieving better speed-ups. There were input classes, however, where not enough parallelism was available, resulting in speed-ups as low as 4 with 16 processors. In developing the parallel implementation the most difficult task was in discovering a method that allows global relabeling to be executed concurrently with the other operations.

We conclude by outlining some possible lines of research. One of the basic problems that should be understood is why the observed performance of Goldberg's algorithm is so much better than the worst-case bounds. A very nice theoretical result would be to determine what is the expected performance of Goldberg's algorithm on some of the input classes used.

Another interesting problem is to generate a class of graphs that causes worst-case behavior of Goldberg's algorithm irrespective of active-vertex selection rule and/or global relabeling schemes.

In implementing Goldberg's algorithm we concentrated on looking at the data structure used to keep track of active vertices. There are many other variants of the algorithm to look at in search of performance improvements. In particular, we would like to find a different way of applying the global relabeling heuristic that did not depend on the number of discharge operations. Ideally there should be a way to monitor the progress of the algorithm that would tell when to trigger global relabeling to obtain maximum benefit at the lowest cost. Another idea, as mentioned above, is to combine both periodic global relabeling and gap relabeling into one implementation. Other possible schemes are propagation of relabel operations [**GT88**] and periodic global relabeling based on time spent in each relabeling [**AG92**]. The rule used to determine how much flow is pushed from an active vertex also presents flexibility. We have implemented a method proposed in [**GT88**], but there are other possibilities such as ordering the edges by decreasing capacity.

6. Acknowledgments

We would like to thank two anonymous referees for valuable comments.

References

[AG92] F. Alizadeh and A. V. Goldberg, *Implementing the push-relabel method for the maximum flow problem on a connection machine*, Tech. Report STAN-CS-92-1410, Department of Computer Science, Stanford University, February 1992.

[AS92] R. J. Anderson and J. C. Setubal, *On the parallel implementation of Goldberg's maximum flow algorithm*, Proc. 4th Symp. on Parallel Algorithms and Architectures, 1992, pp. 168–177.

[BBC92] T. Badics, E. Boros, and O. Cepek, *Implementing a new maximum flow algorithm*, DIMACS Implementation Challenge Workshop – Algorithms for network flows and matching (David S. Johnson and Catherine C. McGeoch, eds.), DIMACS, January 1992, Technical Report 92-4, pp. 84–94.

[Che80] T. Cheung, *Computational comparison of eight methods for the maximum network flow problem*, ACM Trans. Math. Software **6** (1980), no. 1, 1–16.

[CM89] J. Cheriyan and S. N. Maheshwari, *Analysis of preflow push algorithms for maximum network flow*, SIAM J. Comput. **18** (1989), no. 6, 1057–1086.

[Din70] E. A. Dinic, *Algorithm for solution of a problem of maximum flow in a network with power estimation*, Soviet Math. Dokl. **11** (1970), 1277–1280.

[DM89] U. Derigs and W. Meier, *Implementing Goldberg's max-flow-algorithm — a computational investigation*, ZOR — Methods and Models of Operations Research **33** (1989), 383–403.

[EK72] J. Edmonds and R. M. Karp, *Theoretical improvements in algorithmic efficiency for network flow problems*, J. Assoc. Comput. Mach. **19** (1972), no. 2, 248–264.

[FF62] L. R. Ford, Jr. and D. R. Fulkerson, *Flows in networks*, Princeton Univ. Press, Princeton, N. J., 1962.

[Gol87] A. V. Goldberg, *Efficient graph algorithms for sequential and parallel computers*, Ph.D. thesis, Massachussetts Institute of Technology, Cambridge, Mass., January 1987.

[GT88] A. V. Goldberg and R. E. Tarjan, *A new approach to the maximum-flow problem*, J. Assoc. Comput. Mach. **35** (1988), no. 4, 921–940.

[Ima83] H. Imai, *On the practical efficiency of various maximum flow algorithms*, J. Oper. Res. Soc. Japan **26** (1983), no. 1, 61–82.

[Kar74] A. V. Karzanov, *Determining the maximal flow in a network by the method of preflows*, Soviet Math. Dokl. **15** (1974), 434–437.

[Mar89] C. Martel, *A comparison of phase and nonphase network flow algorithms*, Networks **19** (1989), 691–705.

[Tar83] R. E. Tarjan, *Data structures and network algorithms*, SIAM, Philadelphia, Pennsylvania, 1983.

[Tar84] R. E. Tarjan, *A simple version of Karzanov's blocking flow algorithm*, Oper. Res. Lett. **2** (1984), 265–268.

Appendix A. Machine Calibration Table

TABLE 13. "Machine calibration" results on a DECstation 5000/200. This table gives various running times (in seconds) of two programs distributed to DIMACS Challenge participants. It is intended to allow comparisons between the running times we present in this paper and those obtained in environments different from ours. "Loaded" means that another program of equal priority was running at the same time; 'nice" means the test was done in the background, at low priority; the other parameters refer to compiler switches.

wmatch test 1 (C)			
parameters	real	user	sys
–	1.9	1.9	0.0
-O	2.8	1.5	0.1
-O3	2.4	1.5	0.0
-O, loaded	3.6	1.6	0.1
-O, nice	2.6	1.5	0.0
wmatch test 2 (C)			
–	18.9	18.0	0.3
-O	26.9	15.1	0.5
-O3	27.9	15.0	0.3
-O, loaded	43.4	15.3	0.2
-O, nice	31.4	15.1	0.2
netflow test 1 (fortran)			
–	4.0	2.0	0.2
-O -Olimit 510	2.8	1.3	0.1
-O3 -Olimit 510	2.1	1.3	0.2
netflow test 2 (fortran)			
–	6.6	3.7	0.3
-O -Olimit 510	4.2	2.4	0.3
-O3 -Olimit 510	2.8	2.4	0.2

(R. Anderson) DEPARTMENT OF COMPUTER SCIENCE AND ENGINEERING, FR-35, UNIVERSITY OF WASHINGTON, SEATTLE, WASHINGTON 98195

E-mail address: anderson@cs.washington.edu

(J. Setubal) DEPARTMENT OF COMPUTER SCIENCE AND ENGINEERING, FR-35, UNIVERSITY OF WASHINGTON, SEATTLE, WASHINGTON 98195

Current address: DCC–IMECC, Universidade Estadual de Campinas (UNICAMP), Caixa Postal 6065, Campinas, SP, Brazil

E-mail address: setubal@dcc.unicamp.br

Implementations of the Goldberg-Tarjan Maximum Flow Algorithm

Q. C. NGUYEN   V. VENKATESWARAN

ABSTRACT. In this work, we investigate the use of various heuristics and data structures to carry out the operations of the Goldberg-Tarjan algorithm. A simple queue implementation for selecting nodes to discharge or a maximum distance strategy together with periodic global label updates is shown to have the best overall performance. Global updates are crucial; our strategy of making a global update once per $|V|$ local relabelings appears to be the right strategy for most problem types. It is frequent enough to detect optimality quickly, yet does not appear to be computationally excessive. We also have investigated ways to reduce the computation needed to perform global updates. These, however, do not appear promising, on the problems tested.

1. Introduction

In this work, we investigate the use of various heuristics and data structures to carry out the operations of the Goldberg-Tarjan algorithm, [1]. While this algorithm is perhaps the most promising currently available algorithm for this problem, its considerable generality makes possible different implementations, with varying performance characteristics. There has been a long felt need to investigate systematically the various implementation issues raised by the algorithm. Our research is aimed in this direction. For similar work in this area, see [2], [3].

1991 Mathematics Subject Classification. Primary 90B10, 68R10.

Our initial investigations confirmed that the basic Golberg-Tarjan algorithm with two stages outperforms previous algorithms, such as the implementation of Dinic's algorithm in [4], on a wide range of DIMACS specified core problems, [5], [6]. We also experimented with various termination heuristics and were led to the conclusion that global relabeling or updating is crucial to the algorithm performance. Thus, our motivation here has been to investigate different strategies for performing such global updates and for reducing the work involved in doing these operations. Also, the node selection strategy is an important determinant of performance. We report here on our tests of different strategies for node selection.

In the remainder of this section, we explain our notation. In Section 2, we describe our work with various heuristic techniques and in Section 3 some special data structures that we have experimented with. Section 4 describes the computational experiments, and Section 5 our conclusions.

Notations

Let $G = (V, E)$ be a directed graph and $c: E \rightarrow N^+$ be a capacity function on arcs. Let $s, t \in V$ ($s \neq t$) denote the source and sink respectively. We call $N = (V, E, s, t, c)$ a network. We denote the residual graph corresponding to a preflow $g: E \rightarrow R_0^+$ by R_g and its edge set by E_g. The function $e: V \rightarrow R_0^+ \cup \{\infty\}$ gives the excess at each node u ($u \neq t$). (For a definition of *residual graph, preflow, excess*, see, for example, [2]). We let $d_g: V \rightarrow N^+ \cup \{0\}$, or simply d, denote the distance function in the Goldberg-Tarjan algorithm. We will refer to $d(u)$, ($u \in V$), as the distance label of node u.

We use the terms *relabel, push, discharge* in the standard way, [1]. By a global update or global relabeling, we mean setting every distance label to the value of the distance of the node to t (s), accomplished typically by performing a breadth first search (BFS) from t (s) on incoming arcs in R_g.

2. Heuristics

In this section, we concentrate on several "heuristic" issues. These include questions such as: (i) How can the discharge operation be implemented efficiently? (ii) Is there an efficient termination criterion that would avoid spending significant amounts of time in the relabeling steps after the maximum flow has been reached? (iii) How often should we update the local distance labels? (iv) What are the relative performance of different node selection strategies in the discharge step?

The first question deals with how the push and relabel steps can be combined effectively to push flow from an active node v. The third question considers the trade-off between having "better" estimates of the distance labels and the cost of obtaining them. Essentially, performing periodic global updates is an answer to the second and third questions since it also determines if the maximum flow has been reached. The fourth question is interesting because the time bound of the Goldberg-Tarjan algorithm is directly influenced by the active node selection strategy.

The second question is especially significant for the following reason. Recall that the first phase of the algorithm terminates when there are no active vertices in the list. The algorithm can reach maximum flow long before the list of active vertices is exhausted. In that case, the push/relabel steps that are performed after maximum flow has been reached are entirely wasteful. The key to designing an efficient termination heuristic is to be able to detect maximum flow as soon as it has been obtained.

An active vertex becomes inactive if it can get rid of its excess flow to its neighbors or if its distance label has been driven to $|V|$. Now, if the capacity of the source for a given network (i.e., the sum of the capacities of all outgoing arcs from the source) is greater than the sink capacity, then the first phase will terminate with excess flow at some vertices. In this case, the first phase can only terminate if it has done a sufficient amount of push/relabel steps to drive the distance labels of these vertices (and other vertices) to $|V|$. This is basically wasted work. In general, if the minimum cutset is close to the source, then the amount of wasted push/relabel steps is small since the number of vertices that have to be relabeled to $|V|$ will be small. On the other hand, if the minimum cutset is closer to the sink, the wasted work may be large. Hence, an implementation of the Goldberg-Tarjan algorithm that does not employ an efficient termination strategy will exhibit significant bimodal behavior in run times if the input networks has the minimum cutsets either too close to the source or to the sink. (We, in fact, observe such behavior with some variants on the WL_Moderate network problems; Section 4.)

In this work, we have investigated heuristics for performing the following operations:

— The Discharge procedure,

— The Global Update procedure, and

— The Node Selection strategy.

2.1 Discharge Procedure The Discharge procedure involves repeated applications of the push/relabel steps to an active node v in an effort to push the flow closer toward the sink. We are interested in how the two steps, the push and the relabel step, can be combined effectively. We have considered three basic variants:

— Variant D0 - Apply the push/relabel step to an active node v until either $e(v)$ becomes 0 or v is relabeled. If a push from v to another node w causes $e(w)$ to become positive, add w to the list of active nodes. After the push/relabel step on v is completed, add v to the list if v remains active.

This variant involves at most *one* application of the relabel step.

— Variant D1 - Apply the push/relabel step to an active node v until either $e(v)$ becomes 0 or v is relabeled and at least one push has been made from v. The rest of the procedure is similar to that of Variant D0.

This variant is different from the first one in one aspect. If the push/relabel step on v causes v to be relabeled, then we want to make sure that we have already pushed some flow out of v. Otherwise, we apply the push/relabel step on v a second time (if v is still active.) This ensures that the work of the relabel step is not wasted. At most *two* relabel steps will be required before some flow is pushed out of v.

— Variant D2 - Apply the push/relabel step to an active node v until $e(v)$ becomes 0. The rest of the procedure is similar to that of Variant D0.

This variant may involve several relabelings of v before $e(v)$ becomes 0 or v becomes inactive (i.e., $d(v)$ becomes ∞.) This variant may be useful in early identification of nodes that have become dead.

2.2 Global Update Procedure

The basic question here is the trade-off between having "better" estimates of the distance labels and the cost of obtaining them. Obviously, the two extremes are (i) never to update distance labels globally or (ii) to maintain exact distance labels at all time. We implemented both extremes and quickly found them to be ineffective. The first fails because of the excessive number of push/relabel steps. The second extreme is ineffective because the cost of maintaining the exact distance labels is excessive. Obviously, a better strategy is somewhere in between, i.e., to periodically update the distance labels. The key is to schedule updates judiciously.

We have investigated two basic strategies:

— Variant U0 - Perform a global update (by a backward breadth-first search from the sink) after every $|V|$ relabels.

— Variant U1 - Perform a global update every time the sum of the distance increases (due to the relabel steps) is greater than or equal to $|V|$.

This variant is different from the first one in the following way. Basically, a global update is beneficial when the distance labels are far from the "exact" values. Hence we want a measure of how "inexact" the current distance labels are. In some sense, the size of the increase in the distance label of a node (due to the relabel step) is an indication of how inexact it is and thus, the desirability of a global update. Obviously, this variant will perform more updates than the other variant.

These variants are different from the update procedures used in [2] or [3]. Anderson and Setubal perform a global update after every $|E|/2$ discharge steps whereas Derigs and Meier do not update the distance labels exactly. Rather, they have developed a method wherein they check for the existence of a distance gap at the end of every relabel step. When such a gap is detected, they set the distance labels of the nodes disconnected from the sink to $|V|$.

In a way, the Derigs-Meier gap-search strategy is basically a termination procedure since it does not update the distances of all the nodes in the network. Thus, it does not really address the trade-off issue. On the other hand, Anderson and Setubal's update strategy is dependent on the number of arcs in the network, not nodes. For very dense networks where $|E|$ is much larger than $|V|$, this strategy is potentially equivalent to the one that does no global update at all.

2.3 Node Selection Strategy The node selection strategy is significant because it affects the worst-case time bound directly. None of the previously discussed heuristics has any effect on the theoretical time bound of the algorithm. Here, we have evaluated three node selection procedures:

— Variant S0 - First-In-First-Out selection (FIFO) using a queue.

— Variant S1 - Last-In-First-Out selection (LIFO) using a stack.

— Variant S2 - Maximum-Distance selection using buckets.

We use a "bucket" data structure instead of a priority queue for the following reason. Since we discharge excess flow from a node that has the highest distance label, any new active nodes that need to be included in the priority queue will have very high distance labels as well. In a priority queue, such nodes would be introduced at the left end of the lowest level and "percolated" up. Because of their high distance labels they would come to reside somewhere near the top of the tree. Hence, these nodes will traverse the entire height of the tree. On the other hand, since we know precisely the values that the distance labels may take, we can list nodes by distance labels more efficiently with a bucket data structure where each bucket represents a distance value from 1 to $|V|$.

3. Data Structures

In this section, we discuss some specialized operations that we have considered to speed up the basic algorithm, and the requisite data structures.

3.1 Restoring the BFS Tree In our implementation, we start the algorithm with the distance label of each node set to equal its distance to t. This is accomplished by doing a breadth first search (BFS) from t on incoming arcs. Periodically, as mentioned in Section 1, we reset distance labels to actual distances to t, on the residual network, in like manner. In fact, roughly 30% of the run time in Stage 1 is taken up by these "global relabeling" steps. However, if we maintain the BFS tree, it would simply suffice to reconnect the disconnected nodes back to the tree. Completely rebuilding the tree from scratch would not be required.

We keep T in terms of linked lists of entering edges for each node, overlaid on the edge list representing the network (see Fig. 1). Then, whenever a tree arc (u,v) is saturated, we delete all of the nodes in the subtree of u. When it comes time to globally reset node labels (see Section 2 for heuristic rules regarding this), we have the situation depicted in Fig. 2. The set $S \subset V$ of nodes is disconnected from T. The object is to reconnect S to T. The algorithm that we use to effect this is as follows:

1. $\forall u \in S$, set $d(u) := \min \{d(v) | v \in T, (u,v) \in R_g\} + 1$.

2. $R_{\min} = \min_{u \in S} d(u)$

    $R_{\max} = \max_{u \in S} d(u)$.

Set up $(R_{max} - R_{min} + 1)$ buckets and assign nodes in S to buckets according to their $d()$ value.

3. Set level ← R_{min}
    do
    {
        load (Q, level): that is, mark and enqueue all nodes in bucket "level" into queue Q.
        while $(Q \neq \phi)$
        {
          dequeue (v)
          if (d(v) > level)
            { level ← d (v)
              load (Q, level)
            }
          for all $(u,v) \in R_g$, $u \in S$, u not marked, mark u,
          set d(u) ← level + 1
          enqueue (u)
        }
        level = level + 1
    } while (level ≤ R_{max})

COMMENTS:

1. In the case of a directed graph, there is nothing to be gained by preserving the old tree edges among the nodes in S.

2. Correctness follows from the fact that we enqueue all unmarked nodes in a bucket simultaneously at the appropriate level; otherwise, the procedure is essentially the breadth first search method.

3. Let $m = \sum_{u \in S} |\delta(u)|$, $\Delta = R_{max} - R_{min} + 1$. Then Step 1 takes $O(m)$ time, Step 2 takes $O(|S|+\Delta)$ and Step 3 $O(m+\Delta)$.

In Section 4, we discuss the efficacy of this procedure for resetting labels.

3.2 Organizing Edge Lists to Facilitate Pushing Our basic data structure to represent the network requires the neighbors list $N(u)$ of each node, u, to contain nodes due both to forward as well as reverse arcs as in [1]. Clearly, Push-Relabel is facilitated if only nodes corresponding to forward arcs are present in each $N(u)$. However, the data structure to perform the breadth-first-research requires $N(u)$ in terms of incoming arcs (see Fig. 3). One strategy that we have used maintains the primary neighbors list in terms of outgoing arcs in R_g. Another set of neighbors lists, in terms of incoming arcs is set up, at an expense of $O(|E_g|)$ each time the breadth-first search procedure is called. The results from using such an approach are discussed below.

4. Computational Experiments

4.1 Computing Environment All of the computations were done on a SUN SPARC 2 Workstation with 64 megabytes of main memory, running UNIX®. All of our programs are written in C and were compiled using the cc command with the -O option (i.e., with the optimizer feature on). A few of the commonly used scalar variables, eg. $|V|$, $|E|$ were declared as register variables. The computation times shown are CPU seconds, and are exclusive of input/output times, and the time required to obtain the space for all of the requisite data structures (malloc() calls) and to initialize them. Processing times were determined using the C-function, clock(), initiated appropriately. The storage requirement for our variants of the Goldberg-Tarjan algorithm is roughly $5|V| + 10|E|$.

Benchmark timings on DIMACS specified problems are shown in Table 1.

4.2 Algorithm Versions Tested We have investigated the performance of 10 variants, in all. Six of these, denoted GT-000 through GT-212, are aimed at studying the effectiveness of various node selection and push-relabel strategies. We have used the following convention to denote these variants:

The first digit, following GT, describes the discharge (push/relabel) operation. The three variants are either 0: applying push/relabel steps to an active node v until either the excess of v becomes 0 or v is relabeled; or 1: applying push/relabel steps to an active node v until either excess of v becomes 0 or v is relabeled and at least one push has been made from v; or 2: applying push/relabel steps to an active node v until excess of v becomes 0.

The second digit describes the global update procedure. The variants are 0: performing a global update after every $|V|$ relabels; or 1: performing a global update every time the sum of the distance increases (due to the relabel steps) is greater than or equal to $|V|$.

The third digit describes the node selection strategy. The three variants are First-In-First-Out (0), Last-In-First-Out (1) and Maximum Distance (2).

The algorithm dubbed RST uses the tree restoration technique of Section 3.1. In other respects, it is similar to GT-210.

The algorithm AS is the the "gold-q" algorithm described in [3], and made available through DIMACS. It uses a queue to process excess nodes and performs a global update once per $|E|/2$ discharge operations.

The algorithm DRGS was designed to test the gap technique for identifying dead nodes, proposed in [2]. In that paper, the authors observed that if during the course of the algorithm, a gap should occur in the d-values, i.e., if there should occur an $S \subset V$ such that

$$\min \{d(u)|u \in S\} > \max \{d(u)|u \in V \setminus S\} + 1$$

then the nodes in S can be declared *inactive* and their d-values set to $|V|$ (or infinity). Our implementation, DRGS, uses the algorithm GT-210, and an array to list the frequency of nodes at each d-value. If a gap is detected, we scan through the nodes list, picking nodes in S and setting their labels to infinity. DRGS does not make use of global updates. Therefore, its performance should indicate if local relabeling coupled with the gap technique is competitive with periodically performing global updates.

The algorithm RES uses the data structures described in Section 3.2; otherwise, it is similar to GT-210. Only the current residual network, R_g, is kept and this is effected by maintaining lists of outgoing arcs for each node, to facilitate pushing. These lists are doubly linked lists. When an arc (u,v) drops out of R_g, it is deleted from $N(u)$; however, its reverse arc, (v,u) will be present in R_g, permitting (u,v) to be retrieved later, if it is needed. Also, when an arc (u,v) enters R_g, it is always inserted in $N(u)$ immediately in front of the current edge [1] of u.

4.3 Test Problems The problems tested are the ones prescribed in [7]. Test generators were obtained from DIMACS and from J. C. Setubal, at the University of Washington (a C-version of Waissi's acyclic graph generator). For more information on the generators, see [5].

We use the following abbreviations to denote the different problem types:

AC_Dense: Random dense acyclic network (generated using **ac_max.c**)

WL_Moder: Washington Line Moderate network (using **washington**)

RMFGEN_L: Long grid network (using **genrmf**)

RMFGEN_W: Wide grid network (using **genrmf**).

Results are based on a sample size of 25. All of the algorithms were run on the same set of test problems.

4.4 Test Results Mean run times are shown in Tables 2 and 3, along with standard deviations in parentheses. In most cases, these are within 10-20% of the mean. Maximum and minimum run times are shown in Tables 4 and 5. Table 6 shows mean run times for some additional test problems. Table 7 gives the mean operation count for the main variants. Table 8 gives the estimates of run time constants and exponents for GT-210 and GT-212. Table 9 shows the bimodal behavior of run times of some variants on WL_Moder problems.

We make the following observations:

1. The results shown in Tables 2 and 3 would seem to indicate that the variant GT-212 has the best total run time. Recall that GT-212 employs the maximum-distance selection approach (using bucket data structures). GT-212 is particularly effective for "hard" problems (such as RMFGEN_W and RMFGEN_L). However, when viewed against a larger set of problems, such as the ones originally proposed in [5], Table 6, GT-212 performs rather poorly against other variants, such as GT-210, for the following problem types: Washington Random Level Graphs (both wide and long), Washington Square Graphs (sparse), Washington Random 2 Level Graphs (both wide and long), and Waissi Transit Graphs. In fact, when viewed against the original

problems, GT-210 seems to perform the best. Other variants that are based on different discharge procedures or global updates, such as GT-000, GT-100, and GT-200, seem to differ very little. In fact, the biggest difference comes from the selection procedures for active nodes. The stack implementation, GT-211, performs poorly on all problem types (particularly, on Washington Line Graphs, which are not presented in Table 2, but proposed originally in [5]).

For GT-210 and GT-212, we hypothesize the functional relationship of run time, t, to $|V|$ to be:

$$t = \alpha_1 |V|^{\alpha_2}$$

and fit a least-squares line to the observed run times. As shown in Table 2, for AC_Dense problems, $|V|$ ranged from 2^8 to 2^{10}, for WL_Moder problems from 2^{10} to 2^{14} and for the RMFGEN problems from 2^{10} to 2^{16}. There were 25 observations at each $|V|$ setting. Estimates of α_1, α_2 and the R^2 values are shown in Table 8.

2. It is clearly important to perform global updates. DRGS, which in other respects is similar to GT-210, appears to be outperformed by the latter algorithm, particularly on "hard" problems.

3. The strategy of performing a global update per $|V|$ number of relabelings appears to be a sound strategy. The AS algorithm does a global update once per $|E|/2$ number of discharges. In the case of dense graphs, this implies considerably fewer updates (Table 7), and longer running times. This is exacerbated if the optimal cut is near the sink. Many more nodes are on the source side of the cut then and, many more relabelings are required before the active nodes among these are made inactive. Thus, run times for these problems are likely to exhibit a bimodal distribution — longer run times if the optimal cut is near the sink, shorter times otherwise. With our algorithm, we did not observe such a marked bimodal behavior. Table 9 show individual run times of different variants for WL_Moder problems ($|V| = 2^{13}$). It is clear that AS exhibits the most marked bimodal behavior among all the variants presented. Much less so are the GT-210 and GT-212 variants. The DRGS variant shows even less bimodal behavior. This is to be expected since DRGS's gap search strategy is very effective in determining termination. What is very interesting is that GT-211 shows absolutely no bimodal behavior at all. It is speculated that GT-211 takes much longer to get the maximum flow because of the problem network structure and as such is not affected by the cut location.

4. RST has no advantage over GT-210. Clearly, restoring the breadth first search tree, T, is only attractive if $|S|$ (Fig. 2) is small, relative to $|V|$. However, in all of the problem types tested, the ratio $|S|/|V|$ was about 0.7-0.8, at the time global update is invoked. Thus, typically, only about a tenth of the calls to global update warrant using the restore procedure. This is not enough to

recoup the overhead of keeping T and updating S. Also, we observed that in the problem types tested, only a few relabelings are required to disconnect a large fraction of the nodes from T. This precludes a strategy such as performing a restore operation each time $|S|/|V|$ exceeds a small preset limit, say 0.25.

5. RES shows improved performance only in the case of dense graphs. The improvement is significant (compared to GT-210). From Table 7, we see that the number of relabels for this algorithm is substantially lower than that for GT-210. The speed-up is perhaps due not only to the new data structure but also to the fact that the neighbor lists are reordered each time R_g changes.

5. Conclusions

It would appear that a simple queue implementation for selecting nodes to discharge or a maximum distance strategy with periodic global updates (once per $|V|$ relabelings) has the best overall performance. The importance of global updates cannot be overemphasized. Our investigations suggest the above rule for scheduling these updates. Further research is needed to discover if there are other, better strategies. Also, these global updates account for a significant portion of the total run time. Further research is needed to identify ways of trimming down this time, perhaps through performing approximate rather than exact global updates.

REFERENCES

[1] A. V. Goldberg and R. E. Tarjan, "A New Approach to the Maximum Flow Problem," Journal of the ACM 35(4), (1988), 921-940.

[2] U. Derigs and W. Meier, "Implementing Goldberg's Max-Flow-Algorithm - A Computational Investigation," ZOR-Methods and Models of Operations Research, V.33, (1989), 383-403.

[3] R. J. Anderson and J. C. Setubal, "Parallel and Sequential Implementations of Maximum-Flow Algorithms," Draft presented at the First DIMACS Implementation Challenge Workshop, Rutgers University, NJ, Oct. (1991).

[4] D. Goldfarb and M. Grigoriadis, "A Computational Comparison of the Dinic and Network Simplex Methods for Maximum Flow," Annals of Operations Research, Vol. 13, (1988), 83-123.

[5] C. McGeoch, "The First DIMACS International Algorithm Implementation Challenge: The Core Experiments," July 3, (1991).

[6] Q. C. Nguyen and V. Venkateswaran, "Implementations of the Goldberg-Tarjan Maximum Flow Algorithm," talk presented at the First DIMACS Implementation Challenge Workshop, Rutgers University, NJ, Oct. 14-16, (1991).

[7] C. McGeoch, "The First DIMACS International Algorithm Implementation Challenge: The Benchmark Experiments," January 7, 1992.

OPERATIONS RESEARCH DEPARTMENT, AT&T BELL LABORATORIES, HOLMDEL, NJ 07733

E-mail address: qcn@hocus.att.com, ven@hocus.att.com

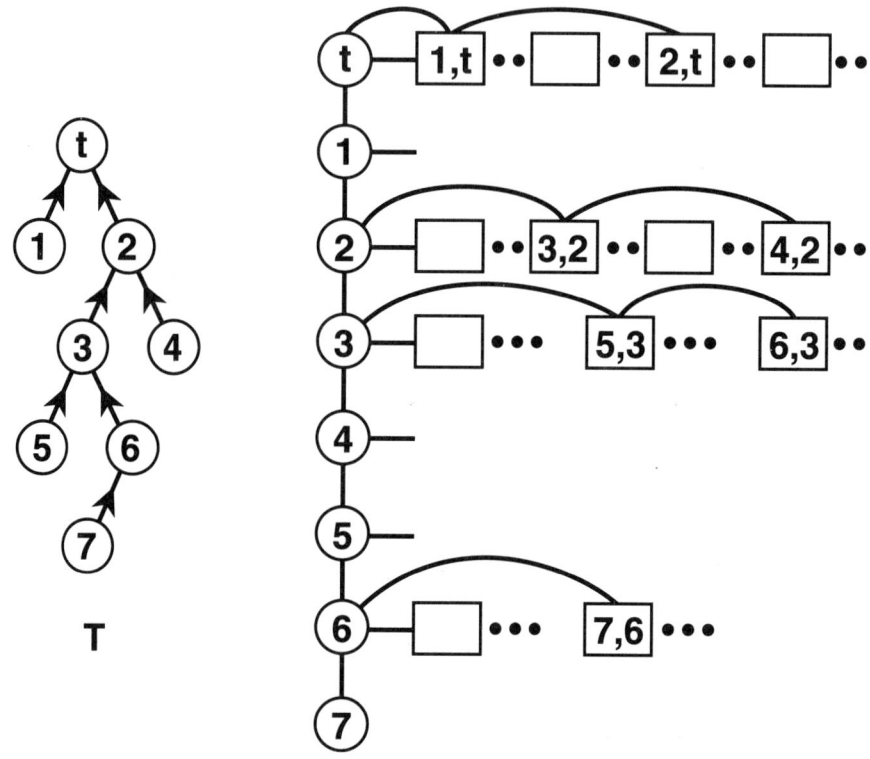

Figure 1. Storing the Breadth First Search Tree, T

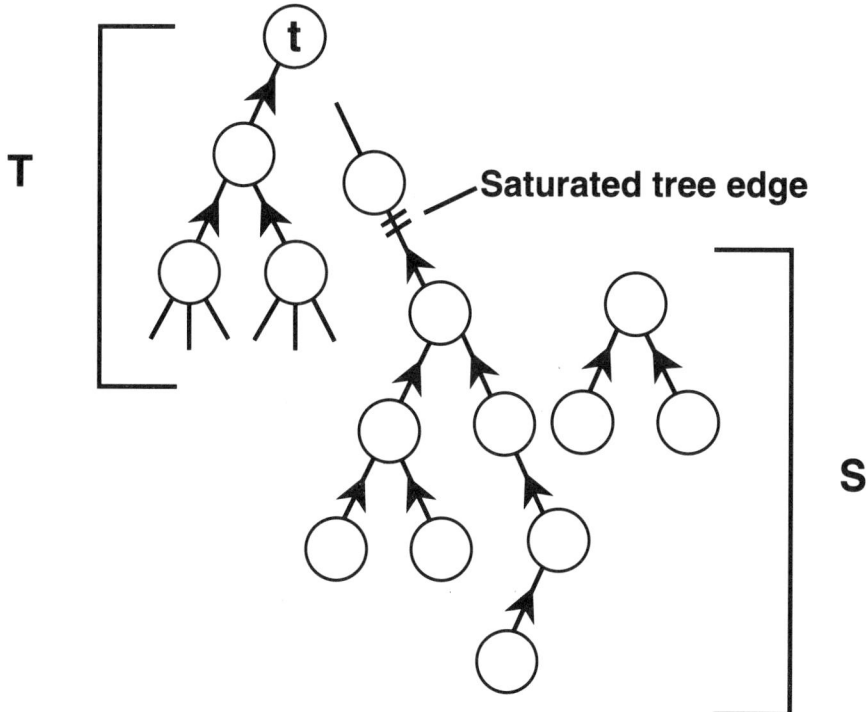

Figure 2. T and disconnected nodes, S

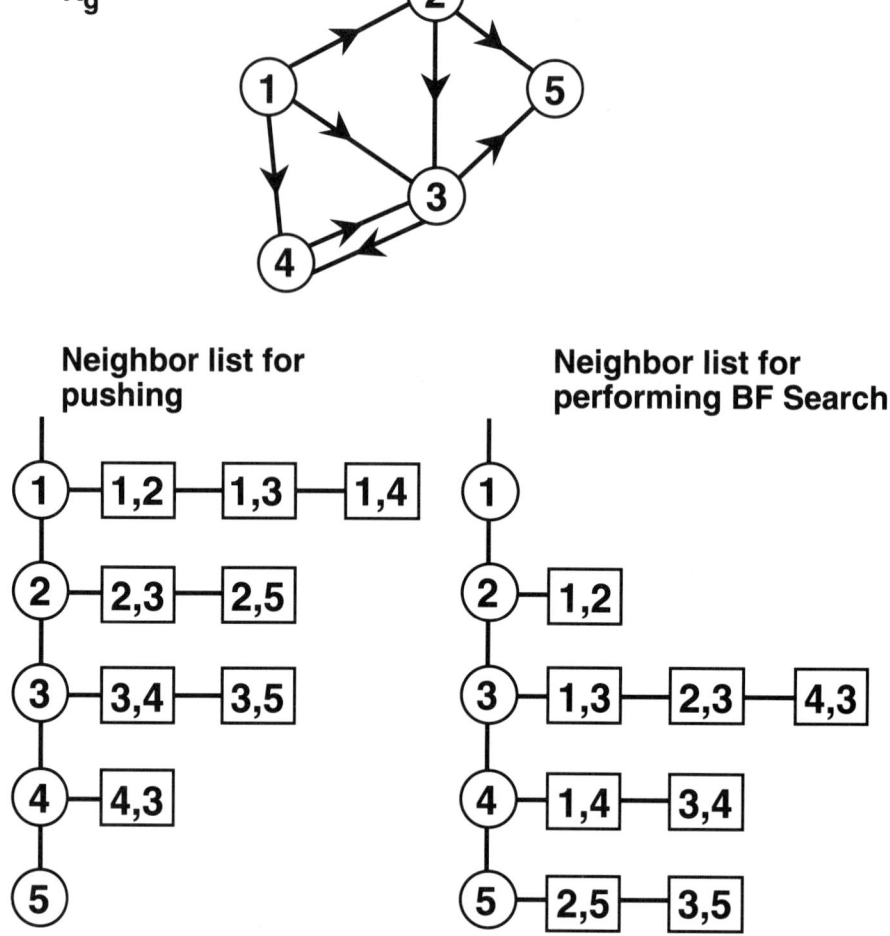

Figure 3. Ideal data structures for pushing, performing Breadth First Search

wmatch test 1(C)		
real	user	sys
3.08s	2.76s	0.10s
wmatch test 2(C)		
25.00s	24.43s	0.31s

Table 1. Benchmark run times

Problem Type	$|V|$	$|E|$	GT-000	GT-100	GT-200	GT-210	GT-211
AC_Dense	256	32,640	1.29 (0.34)	1.30 (0.38)	1.43 (0.42)	1.48 (0.39)	1.55 (0.52)
	512	130,816	6.84 (1.52)	6.91 (1.49)	7.50 (1.64)	8.03 (1.64)	8.35 (2.55)
	1024	523,776	41.24 (9.29)	42.02 (10.42)	47.34 (10.77)	50.08 (11.78)	50.28 (17.91)
WL_Moder	1026	8074	0.11 (0.02)	0.11 (0.02)	0.10 (0.02)	0.10 (0.02)	0.29 (0.04)
	2050	22,299	0.28 (0.06)	0.27 (0.06)	0.26 (0.06)	0.26 (0.06)	0.81 (0.09)
	4098	65,029	0.76 (0.16)	0.74 (0.17)	0.72 (0.16)	0.73 (0.16)	2.28 (0.22)
	16,386	522,261	5.13 (1.15)	5.04 (1.13)	4.99 (1.10)	5.03 (1.12)	19.10 (0.70)
RMFGEN_L	1152	4956	0.32 (0.04)	0.28 (0.05)	0.26 (0.04)	0.27 (0.03)	0.49 (0.05)
	4096	18,368	1.50 (0.18)	1.36 (0.17)	1.26 (0.16)	1.26 (0.16)	3.29 (0.39)
	15,488	71,687	9.11 (0.69)	7.99 (0.59)	7.41 (0.55)	7.42 (0.57)	28.27 (2.80)
	65,536	311,040	62.92 (5.21)	53.87 (4.66)	50.47 (4.20)	50.22 (4.14)	299.43 (22.71)
RMFGEN_W	1024	4608	0.59 (0.06)	0.58 (0.06)	0.57 (0.06)	0.55 (0.05)	0.52 (0.06)
	3920	18,256	4.23 (0.37)	4.06 (0.37)	3.96 (0.28)	3.74 (0.25)	4.52 (0.38)
	16,807	80,262	34.72 (1.65)	32.99 (1.82)	31.63 (1.84)	29.51 (1.74)	48.19 (3.72)
	65,025	314,840	220.40 (12.44)	211.39 (12.07)	199.91 (13.27)	187.66 (11.25)	365.05 (32.65)
Total			389.44	368.91	357.81	346.34	832.42

Table 2. Mean, standard deviation of run times

Problem Type	$\|V\|$	$\|E\|$	GT-212	AS	DRGS	RST	RES
AC_Dense	256	32,640	1.56 (0.46)	4.63 (1.71)	1.08 (0.38)	2.14 (0.52)	1.07 (0.25)
	512	130,816	8.43 (2.96)	36.28 (14.11)	5.61 (1.53)	11.92 (2.32)	5.04 (1.04)
	1024	523,776	51.90 (18.46)	328.67 (130.54)	40.06 (12.75)	71.66 (15.35)	24.40 (5.35)
WL_Moder	1026	8,074	0.09 (0.02)	0.15 (0.06)	0.10 (0.01)	0.14 (0.02)	0.13 (0.03)
	2050	22,299	0.24 (0.07)	0.47 (0.22)	0.24 (0.03)	0.37 (0.07)	0.34 (0.07)
	4098	65,029	0.65 (0.18)	1.70 (0.89)	0.64 (0.09)	1.03 (0.19)	0.93 (0.18)
	16,386	522,261	4.42 (1.19)	19.24 (13.23)	4.21 (0.36)	7.43 (1.27)	6.18 (1.15)
RMFGEN_L	1152	4956	0.24 (0.06)	0.27 (0.03)	0.32 (0.06)	0.34 (0.04)	0.34 (0.04)
	4096	18,368	0.99 (0.20)	1.39 (0.17)	1.71 (0.40)	1.69 (0.20)	1.58 (0.20)
	15,488	71,687	4.72 (1.13)	7.75 (0.53)	12.91 (3.22)	9.67 (0.73)	8.77 (0.81)
	65,536	311,040	24.94 (6.16)	50.68 (3.99)	151.18 (41.99)	64.66 (5.28)	56.25 (5.56)
RMFGEN_W	1024	4608	0.46 (0.05)	0.48 (0.05)	0.78 (0.08)	0.77 (0.08)	0.64 (0.05)
	3920	18,256	3.12 (0.27)	3.58 (0.26)	7.15 (0.57)	4.96 (0.37)	4.52 (0.25)
	16,807	80,262	22.10 (1.66)	31.03 (1.61)	80.95 (5.63)	38.16 (2.11)	38.27 (1.58)
	65,025	314,840	133.07 (10.77)	225.83 (9.98)	727.56 (55.94)	237.85 (14.45)	265.28 (8.61)
Total			256.93	712.15	1034.50	452.79	413.74

Table 3. Mean, standard deviation of run times (cont'd)

Problem Type	$|V|$	$|E|$	GT-000	GT-100	GT-200	GT-210	GT-211
AC_Dense	256	32,640	1.90, 0.63	1.98, 0.63	2.07, 0.63	2.22, 0.68	2.57, 0.60
	512	130,816	8.85, 2.80	8.68, 2.77	9.87, 3.25	10.67, 3.57	14.90, 3.63
	1024	523,776	54.30, 19.53	54.78, 18.75	62.55, 22.43	69.85, 23.08	84.45, 17.03
WL_Moder	1026	8074	0.15, 0.08	0.15, 0.08	0.15, 0.07	0.13, 0.08	0.38, 0.23
	2050	22,299	0.37, 0.22	0.37, 0.20	0.35, 0.20	0.37, 0.20	0.97, 0.62
	4098	65,029	0.98, 0.55	1.00, 0.52	0.97, 0.53	0.98, 0.52	2.62, 1.80
	16,386	522,261	6.77, 3.80	6.60, 3.77	6.52, 3.73	6.63, 3.72	20.52, 17.12
RMFGEN_L	1152	4956	0.42, 0.23	0.40, 0.20	0.35, 0.20	0.35, 0.18	0.60, 0.37
	4096	18,368	1.87, 1.22	1.63, 1.05	1.58, 1.00	1.57, 1.00	4.02, 2.60
	15,488	71,687	10.35, 7.70	8.87, 6.50	8.20, 6.13	8.18, 6.03	34.13, 22.42
	65,536	311,040	73.51, 52.96	63.15, 44.83	60.23, 42.10	59.30, 41.33	336.72, 247.14
RMFGEN_W	1024	4608	0.67, 0.47	0.68, 0.45	0.67, 0.45	0.65, 0.42	0.65, 0.37
	3920	18,256	5.02, 3.40	4.65, 3.28	4.60, 3.38	4.22, 3.30	5.18, 3.73
	16,807	80,262	38.23, 30.22	35.97, 28.43	34.42, 27.93	33.87, 26.52	54.23, 42.11
	65,025	314,840	235.59, 194.46	229.07, 172.68	225.02, 169.14	200.84, 157.86	426.32, 296.24

Table 4. Maximum and minimum run times

Problem Type	$|V|$	$|E|$	GT-212	AS	DRGS	RST	RES
AC_Dense	256	32,640	2.50, 0.73	6.28, 0.72	1.70, 0.47	3.02, 1.08	1.47, 0.60
	512	130,816	14.72, 4.23	46.71, 7.67	8.15, 2.07	16.07, 5.58	6.60, 2.48
	1024	523,776	84.40, 21.07	412.46, 22.68	60.66, 15.40	89.50, 35.88	34.12, 12.55
WL_Moder	1026	8,074	0.13, 0.05	0.30, 0.09	0.13, 0.08	0.18, 0.10	0.18, 0.10
	2050	22,299	0.33, 0.15	0.94, 0.24	0.32, 0.20	0.47, 0.28	0.43, 0.25
	4098	65,029	0.90, 0.43	3.36, 0.64	0.82, 0.52	1.32, 0.78	1.15, 0.68
	16,386	522,261	6.02, 3.03	33.08, 4.52	4.97, 3.80	9.57, 5.98	7.70, 4.80
RMFGEN_L	1152	4956	0.38, 0.13	0.34, 0.21	0.42, 0.20	0.43, 0.23	0.42, 0.23
	4096	18,368	1.47, 0.72	1.73, 1.14	2.88, 1.18	2.15, 1.33	2.00, 1.20
	15,488	71,687	7.23, 2.53	8.75, 6.73	22.37, 8.45	10.70, 7.90	10.45, 7.07
	65,536	311,040	42.35, 14.13	58.72, 42.97	243.52, 82.70	75.08, 53.33	66.90, 44.46
RMFGEN_W	1024	4608	0.57, 0.38	0.56, 0.40	0.88, 0.65	0.93, 0.63	0.72, 0.53
	3920	18,256	3.70, 2.70	4.01, 3.06	8.05, 5.83	5.62, 4.10	4.93, 4.08
	16,807	80,262	26.62, 19.30	33.83, 27.63	95.15, 70.31	42.61, 34.13	41.53, 35.17
	65,025	314,840	156.34, 112.81	240.28, 199.70	841.18, 626.31	270.11, 202.21	279.07, 243.64

Table 5. Maximum and minimum run times (cont'd)

Problem Type	$|V|$	GT-210	GT-212
Washington Random Level, Wide	2^{14}	4.74	39.48
Washington Random Level, Wide	2^{16}	28.59	651.58
Washington Random Level, Long	2^{14}	3.94	17.84
Washington Random Level, Long	2^{16}	21.23	76.99
Washington Sparse	2^{14}	5.30	28.07
Washington Sparse	2^{16}	29.38	291.31
Washington Random 2 Level, Low	2^{14}	3.48	12.48
Washington Random 2 Level, Low	2^{16}	18.35	86.92
Washington Random 2 Level, High	2^{14}	4.16	16.49
Washington Random 2 Level, High	2^{16}	22.30	110.86
Waissi Transit Low	2^{14}	4.97	34.80
Waissi Transit Low	2^{16}	29.51	268.85
Waissi Transit High	2^{14}	5.53	35.43
Waissi Transit High	2^{16}	32.92	294.18

Table 6. Mean run times for GT-210, GT-212, on additional DIMACS problems (sample size = 10).

Problem Type	Operation	AS	GT-210	GT-211	GT-212	DRGS	RES
AC_Dense $|V|=1024$ $|E|=523,776$	global updates	0.8	16.6	17.0	20.8	1.0*	9.4
	relabels	89,963	12,275	13,133	15,100	18,251	6985
	unsat. pushes	135,192	57,103	33,093	38,283	66,864	50,320
	sat. pushes	112,852	80,024	13,232	35,185	87,619	74,226
WL_Moder $|V|=16,386$ $|E|=522,261$	global updates	0.6	1.6	3.6	1.6	1.0	1.6
	relabels	68,395	11,373	45,175	9838	5253	11,215
	unsat. pushes	108,547	41,062	853,874	22,926	34,894	41,068
	sat. pushes	28,234	17,828	12,340	20,383	15,998	17,261
RMFGEN_L $|V|=65,536$ $|E|=311,040$	global updates	23.4	11.4	40.7	8.7	1.0	10.6
	relabels	366,587	621,890	2,350,156	433,829	3,458,814	560,931
	unsat. pushes	1,899,253	1,974,582	21,381,934	606,723	4,841,846	1,772,375
	sat. pushes	54,276	117,579	156,523	119,196	330,808	63,386
RMFGEN_W $|V|=65,025$ $|E|=314,840$	global updates	88.4	46.2	63.8	46.2	1.0	59.0
	relabels	1,961,074	2,106,355	2,960,722	1,940,443	15,336,997	2,426,227
	unsat. pushes	6,357,176	5,452,010	26,279,060	3,906,834	22,881,260	6,341,299
	sat. pushes	89,047	591,824	578,315	518,215	1,627,968	209,801

Table 7. Mean operation count

* We initiate DRGS with a single breadth-first search

Problem Type	GT-210			GT-212		
	α_1	α_2	R^2	α_1	α_2	R^2
AC_Dense	1.06×10^{-6}	2.54	0.96	1.32×10^{-6}	2.51	0.94
WL_Moder	5.73×10^{-6}	1.41	0.98	4.98×10^{-6}	1.41	0.96
RMFGEN_L	2.59×10^{-5}	1.30	0.99	1.00×10^{-4}	1.15	0.98
RMFGEN_W	3.17×10^{-5}	1.41	0.99	3.79×10^{-5}	1.36	0.99

Table 8. Run time constants and exponents for GT-210, GT-212

	AS	GT-210	GT-211	GT-212	DRGS
	8.53	2.52	6.58	2.20	1.95
	1.77	1.40	6.82	1.22	1.53
	1.91	1.50	6.90	1.23	1.57
	1.83	1.48	6.55	1.23	1.53
	1.82	1.40	6.27	1.17	1.45
	7.58	2.07	6.77	1.92	1.62
	8.00	2.32	6.47	2.28	1.90
	1.78	1.45	6.20	1.22	1.40
	7.65	2.22	7.40	2.08	1.75
	8.40	2.52	6.82	2.20	1.85
	8.19	2.13	7.10	1.95	1.78
	7.97	2.18	6.72	2.07	1.75
	1.79	1.43	6.57	1.18	1.48
	7.24	2.03	7.10	1.87	1.57
	1.76	1.42	6.25	1.15	1.43
	7.64	2.23	7.00	2.05	1.78
	1.68	1.38	5.87	1.15	1.42
	10.00	2.57	6.22	2.33	2.08
	7.76	2.17	6.55	1.98	1.72
	1.80	1.47	7.27	1.18	1.52
Total	105.10	37.89	133.43	33.66	33.08

Table 9. Run times on 20 random instances of WL_Moder ($|V|=2^{13}$) exhibiting bimodal behavior on some variants.

Implementing a Maximum Flow Algorithm: Experiments with Dynamic Trees

T. BADICS AND E. BOROS

August 15, 1992

ABSTRACT. This paper contains experimental results with a maximum flow algorithm developed by Cheriyan and Hagerup. Comparison was made to a preflow-push type maxflow implementation, and we examined also the effect of the dynamic trees data structure and several heuristic parameters.

1. Introduction

In this paper we report on an implementation of a maximum flow algorithm by Cheriyan and Hagerup [1]. Our aim was to test the behavior of this algorithm in practice, concerning it's good theoretical worst case bound. We were particularly interested in the effect of using theoretically well behaving data structures such as dynamic trees [15], and Fibonacci heaps [17]. We also made comparisons to two preflow-push based algorithm by Goldberg and Tarjan [13], and to an implementation of pushes along several edges without using dynamic trees.

In Sections 2. the basic notations and a short description of the Cheriyan–Hagerup algorithm are given. In Section 3. details of the implementation and the used data structures are presented. In Section 4. a summary of the experimental results and a few questions that could be addressed in the future are given.

2. Basic notions and the PLED algorithm

We assume that the reader is familiar with the generic maximum flow algorithm in [13] and refer to [13] for definitions of the terms *network*, *source* s, *sink* t, *edge capacity* $c(v,w)$, *flow*, *maximum flow*, *preflow* f, *flow excess* $e(v)$ of a vertex v, *residual graph*, *residual capacity* $rescap(v,w)$ of the edge (v,w),

---

1991 *Mathematics Subject Classification*. Primary 90B10; Secondary 68P05, 68-04.

The second author is supported in part by NSF (Grant DMS 89-06870) and AFOSR (Grants 89-0512 and 90-0008).

valid labeling d, active vertex, push, saturating push, and *nonsaturating push*. Let $G = (V, E)$ denote the digraph (assumed symmetric) corresponding to the network. Let $N = |V|$, $M = |E|$.

We call an edge (v, w) eligible if $rescap(v, w) > 0$ and $d(v) = d(w) + 1$. Each vertex $v \in V$ has an adjacency list consisting the edges $(v, w) \in E$. For each vertex v, the first eligible edge in the adjacency list of v is called its current edge and denoted by $ce(v)$ (such an edge may not exist, in which case $ce(v)$ refers to none).

Our implementation is based on the algorithm developed by Cheriyan and Hagerup (see [1]). Following [1] we shall refer to this algorithm as PLED (shorthand for Prudent Linking and Excess Diminishing). This algorithm is an instance (with one minor exception) of the generic preflow algorithm by Goldberg and Tarjan [13]. The difference (namely, the value of a push over an edge (v, w) may be less then both the excess at v and residual capacity of (v, w)) does not affect the essential properties of the generic algorithm (in particular, maximum flow is always correctly computed upon termination). PLED also uses an idea introduced by Ahuja and Orlin [6] of scaling the volume of the pushes. The scaling factor plays here, however, a slightly different role: the limits imposed on the volume of a push are not the same as in [6]. A third idea in PLED is randomization: after each relabeling of a vertex v, the edgelist of v is permuted randomly. This random permutation ensures a better theoretical running time. Noga Alon showed in [5] that this randomization can be replaced by a deterministic procedure. The worst case running time of PLED, using the randomized procedure is $O(NM + N^2(logN)^2)$ with high probability (see [2, 3, 4]). Using Alon's derandomization, the deterministic worst case bound improves to $O(NM + N^{8/3}(logN))$. The worst case bound without randomization is $O(NMlog(N))$. Randomly permuting the edgelist of the vertex after relabeling can take substantial amount of time. From the implementation point of view just one random permutation of the adjacency list of each vertex at the start of execution suffices for the best theoretical running time, though the probability of failure increases slightly, $exp(-\Theta(\sqrt{NM} + NlogN))$ versus $exp(-\Theta(N\sqrt{NM} + N^2logN))$, see [1, 2].

Three main data structures are essential for PLED.

- **Ordinary heap** that contains vertices which have big excesses and which are ordered by their distance labels. This structure supports the easy selection of a vertex for a push. (Select a vertex with the minimal distance label among the vertices having large enough excesses).
- **Fibonacci heap**(see [17]) contains the rest of vertices, ordered by their (small) excesses. This supports constant (amortized) time decrease key operation and fast update of the scaling factor.
- **Dynamic trees structure** (see [15]) to maintain a spanning forest F of G containing a subset of the current edges, where the value associated with an edge in F is its residual capacity. This structure is able to send

flow value along a path of length L in (amortized) time $O(log L)$

Sketch of the PLED algorithm:

PLED Algorithm
    initialize;
    loop(until all the excesses are eliminated)
        v := **select**;
        macropush(v);
    endloop;
end PLED

initialize;
    Perform the standard initialization of the variables f, d and e (see [**13**], pp.925) and of the dynamic trees, and Fibonacci heap data structures.
    $\Delta := max(\{e(v)|v \in V - \{s,t\}\} \cup \{0\})$;

select: vertex;
    Return an active vertex v with $e(v) \geq \Delta$ and minimum $d(v)$ or decrease Δ and then select v as before.

macropush(v: vertex);
    Send flow from v over a path in the tree to which v belongs, or if v was a root then send flow from v using edges (v, w) for which $rescap(v, w) \leq$ **limflow**(v) until v is relabeled or until $limflow(v)$ decreases to $< \Delta/2$ due to saturating pushes. If $limflow(v) > \Delta/2$ then v is linked to the other end of its current edge, and a treepush is executed.

limflow(v : vertex): real;
    return(if $e(v) \geq 2\Delta$ then $e(v)/2$ else $e(v)$);

The bottleneck factor in virtually all preflow-push type maximum-flow algorithm is the number of nonsaturating pushes. One of the properties of PLED is to try a sequence of nonsaturating pushes along several eligible edges that constitute a path in a tree of the eligible edges while avoiding saturation on that path, rather then performing a nonsaturating push along one single edge. These trees of the eligible edges are stored in a dynamic trees structure, in which one can perform a push in logarithmic time of the length of the path. Beside this property, PLED uses a scaling parameter for controlling the size of the pushes and the trees in the dynamic trees forest. A Fibonacci heap structure handles this scaling parameter efficiently. Finally for achieving the theoretically best bound, we either have to perform a random permutation or we have to generate a permutation by the deterministic scheme of [5] at the start of execution. Theoretically by permuting the adjacency lists after each relabel one could reach the best bound with higher probability, but the amount of work to do this can easily be substantial even if the number of relabel operations is just $\Theta(N)$.

3. Implementation

Since the algorithm requires the above data structures to achieve the theoretically best performance, we decided to implement all of them. In this we were encouraged by earlier experiences (see [19]) with implementations of Fibonacci heaps, and Splay trees [16]. Splay trees play a main role in the dynamic trees structure. It was found in the experiments of [19] that a particular implementation of Splay trees was quite efficient comparing to some other search trees. In [19] Splay trees were compared to Binary, Red-Black and AVL trees and the Fibonacci heap was compared to Ordinary and Pairing heaps. Their study showed, that Fibonacci heaps are better than Ordinary heaps. This suggested that the use of Fibonacci heaps can have practical advantage, and that they can be implemented efficiently. So we implemented Fibonacci heaps and the dynamic trees data structure following [15].

Beside the above data structures, we implemented a routine for randomly permuting the edge lists of the vertices. Although the deterministic permutation of Alon derandomizes the algorithm, the overhead of such a permutation generating procedure is so large that we did not expect much improvement by implementing such a deterministic procedure. Moreover since our instances were mostly randomly generated, after some preliminary experiments we used the PLED algorithm without random permutations.

For comparison reasons we coded Goldberg's simple preflow-push algorithm using simple FIFO queue for selecting the active vertices, and the dynamic trees version of this algorithm described in the same paper ([13]). Later we refer to these codes as GOLD and GOLDYN respectively. In the code GOLDYN we did not control the size of the dynamic trees. Thus we got a code which has theoretical worst case bound $O(NMlogN)$ instead of $O(NMlog(N^2/M))$. The reason for this choice was mostly lack of time and the expected overhead of such a control mechanism. Besides, in the range of examples we tested the codes, the benefit from such size control, even without the overhead, probably is not significant.

For testing the performance of the dynamic trees data structure, we implemented its operations (*find-min, add-value, find-root, link, cut, etc.*) with storing the trees explicitly and executing these operations in the obvious (linear time) way. Hence we avoided the overhead of handling Splay trees and complicated updates. Later in this paper the codes using these "non-dynamic tree" operations are called NPLED and NGOLDYN. Note that the number of elementary operations for PLED and NPLED (or GOLDYN and NGOLDYN) are the same on the same instance, only the way of handling the tree operations are different. Therefore the difference in running time shows exactly the impact of the dynamic trees structure.

In our implementations of all the codes we employed an idea, mentioned in ([13]), the so called "global-" or "big-relabeling". Our early experiments showed clearly, that in PLED just like in GOLD or GOLDYN, the running times of the variant which uses global-relabeling were much smaller (orders of magnitude) than the one which does not use it. Therefore we built in some heuristic parameters controlling the calling frequency of big-relabeling and affecting thus the running time.

A "big-relabeling" step consist of two breadth-first-searches, one starting from the sink and working on the sink side of the residual graph, and another one for the source side, starting from the source. In these breadth-first-searches the shortest distance is calculated from each vertex to the sink on the sink side, or to the source on the source side, respectively. Unfortunately breadth-first-search is a relatively expensive operation (it takes O(M) steps), so the calling frequency of big-relabel is very important and can be a subject of later studies.

The effect of a big-relabeling is to lift all the vertices as high as possible, keeping the labeling feasible. This is the point where this expensive operation can help a lot. During the steps of these "preflow-push" type algorithms the label of a vertex v (say $d(v)$) is always a lower bound on the shortest distance from v to the sink if the sink is reachable from v in the residual graph, or $(d(v) - N)$ is a lower bound on the distance from v to the source. After each saturating push along the edge (v, w), v can either have another eligible edge for later pushes, or it will be relabeled, i.e. lifted right above its neighbors. In this latter case the deletion of the edge (v, w) cuts the tree (formed by some of the current edges) to which v belonged. All the vertices in the subtree, rooted at v, will keep their old labels, and so their real distance from the sink can become larger then the corresponding label values. Hence, after some iterations, there could be many such vertices whose label is far from their real distance. The algorithm will terminate only when all the remaining excesses are sent back to the source. Therefore vertices with positive excess have to be lifted above the source (having the label value N). Since one relabel lifts a vertex only above its neighbors, meaning an increase by 1 or 2 in its label in the average case, many simple relabel operations may be necessary right before termination. In the worst case for example, if the $s-t$ minimum cut $(V_s, V - V_s)$ with V_s minimal has $|V_s| = \Theta(N)$, then we may need $O(N^2)$ simple relabels just to terminate the algorithm. In practice, we found that this so called second phase of the algorithm can take almost all the running time (because the number of relabels became significant). In contrast only one execution of big-relabel step will lift ALL the vertices as high as possible by calculating the exact distances from each vertex to the sink or to the source, respectively. Of course, if our graph is dense, one big-relabel can also take $O(M)$ time, but each of these steps are much cheaper than a simple relabeling.

We built 3 parameters to control the calling frequency of the big-relabel procedure.

(i) **Relab-freq**, a nonnegative integer. A big-relabel is called after each **Relab-freq** iterations. By iteration we mean a vertex selection. (This is a simple heuristic to call big-relabel with some constant frequency.)

(ii) **Cut-freq**, a nonnegative integer. After this many cuts in the tree structure a big-relabel is called. (This parameter estimates the total sum of the differences between the actual labels of the vertices and their shortest distances to the sink or to the source, respectively. After a cut performed at the vertex v, not only the label of v, but all the labels of its descendants in the tree can loose their exact meaning. That is the label of a vertex w in the subtree of v can be less than the shortest distance from w to the sink, since the shortest distances may increase by cutting an edge. Hence, by calculating the number of cuts only, we are not taking into account all the labels in the subtree. Some further improvement could be done by making this estimation more precise.)

(iii) **Gap**, a switch which can be on or off. If on, it checks whether the labels of the vertices, not larger than $N - 2$, form a consecutive sequence or not. If there is a gap (according to [11]), there is a cut in the residual graph, hence each vertex in the source side have to be lifted above the source, and therefore a big-relabel is called.

Our experiments showed that before the maxflow value is reached, parameters **Relab-freq** or **Cut-freq** can be helpful. Their effect seemed to be equivalent in the sense that in almost all problems we tested, the number of iterations was roughly four times more than the number of cuts. So by setting **Cut-freq** = **Relab-freq**/4 we could get about the same number of big-relabels. The use of the **Gap** parameter is extremely beneficial in this phase and its update is a very cheap operation. Therefore, in our experiments we always turned **Gap** on.

In the second phase, i.e. in the phase of sending back the excesses, **Relab-freq** or **Cut-freq** are useful, while **Gap** has no impact at all. The second phase can take quite a long time, in particular for problems with long paths from the source to the sink, and with large excess to be sent back to the source. Therefore, the use of the big-relabel is important in this final phase, too.

On the other hand, since big-relabel is a costly operation, we kept its calling frequency below a small linear function of the number of arcs, in order to keep the overall complexity at $O(NM)$.

We implemented another mechanism to achieve better running times in all three codes. Namely at initialization the algorithm calculates an upper bound U on the maximum flow value by taking the minimum of capacities of some cuts. Then it creates a new source by adding an artificial vertex S and a new arc (S, s) with capacity U to the network, where s was the old source. The new problem is obviously equivalent with the old one, and the extra cost of its implementation is negligible. The advantage of doing this is that we do not let the algorithm push too much excess into the network, reducing in this way the runtime of the second phase. We have found instances showing that without this procedure the running time was significantly bigger due to the long second phase.

4. Experimental results

For the experiments, we used the DIMACS suggested problems, and the generators GENRMF, WASHINGTON, and AC-MAX [9, 8, 10]. (See the DIMACS document "The Core Experiments"). The families of networks we report on include the ones suggested by "The Benchmark Experiments", and two classes of problems made intentionally very difficult for Goldberg's preflow-push algorithm.

The input parameters for the DIMACS benchmark problems were chosen, according to the structure of the particular networks, such that the number of nodes of the generated example increase approximately by multiples of 2 up to the memory limitations of our workstation.

4.1. Description of the problem classes.
The generator GENRMF takes 4 numbers $(a, b, c_1 and c_2)$ as input parameters. The resulting network consists of b layers of $(a \times a)$ grids, in one of the corners of the first layer is the source, and in the opposite corner of the last layer is the sink. Arcs are drawn between each neighboring grid nodes in both directions with capacities $c_2 a^2$, and two consecutive layers are connected along a randomly generated perfect matching with capacities chosen randomly in the range (c_1, c_2).

The generator WASHINGTON takes several numbers as input parameters. The first argument chooses the type of the network, and the rest are specific to the type chosen.

The generator AC-MAX generates fully dense or sparse random acyclic networks. The only parameter is n, the number of nodes in the network. (see "The Core Experiments")

The following is a description of the network families what we tested on. Exact parameters for the families GL, GW, WLM and AD are given in Table 1.

- **GL - Genrmf-Long**. Generated by the GENRMF code. They are called "long", because b is approximately a^2.
- **GW - Genrmf-Wide**. Generated by the GENRMF code. They are called "wide", because a is approximately b^2.
- **WLM - Washington-Line-Moderate**. Fairly dense networks, generated by the WASHINGTON code. The network type is "6", and the type specific parameters are n, m, deg. The resulting network has $n \times m$ nodes in a grid, plus two additional nodes for the source and sink. The source is connected to the first m nodes of the grid, and the last m nodes are connected to the sink by arcs with infinite capacities. For each node

in the grid, *deg* arcs are generated with random capacities pointing to *deg* randomly selected nodes from the next *deg* columns.
- **AD - Acyclic-Dense.** These instances are fully dense acyclic graphs with random capacities, generated by the AC-MAX code.

TABLE 1. Parameters used for generating the DIMACS Benchmark instances.

NAME	PARAMETERS	NODES	ARCS
GL1	$a=6$ $b=31$ $c_1=1$ $c_2=10000$	1116	4800
GL2	$a=7$ $b=42$ $c_1=1$ $c_2=10000$	2058	9065
GL3	$a=8$ $b=64$ $c_1=1$ $c_2=10000$	4096	18368
GL4	$a=9$ $b=100$ $c_1=1$ $c_2=10000$	8100	36819
GL5	$a=11$ $b=128$ $c_1=1$ $c_2=10000$	15488	71687
GL6	$a=13$ $b=194$ $c_1=1$ $c_2=10000$	32786	153673
GW1	$a=16$ $b=4$ $c_1=1$ $c_2=10000$	1024	4608
GW2	$a=21$ $b=5$ $c_1=1$ $c_2=10000$	2205	10164
GW3	$a=28$ $b=5$ $c_1=1$ $c_2=10000$	3920	18256
GW4	$a=37$ $b=6$ $c_1=1$ $c_2=10000$	8214	38813
GW5	$a=48$ $b=7$ $c_1=1$ $c_2=10000$	16128	76992
GW6	$a=64$ $b=8$ $c_1=1$ $c_2=10000$	32768	157696
WLM1	6 64 4 5	258	1236
WLM2	6 128 4 8	514	3975
WLM3	6 256 4 8	1026	8069
WLM4	6 512 4 11	2050	22279
WLM5	6 1024 4 16	4098	65019
WLM6	6 2048 4 22	8194	179258
AD1	fully dense	256	32640
AD2	fully dense	321	51360
AD3	fully dense	403	81003
AD4	fully dense	507	128271
AD5	fully dense	636	201930
AD6	fully dense	800	319600

- **GB - Goldberg-Bad-Case**. A deterministic family, generated by the WASHINGTON code. The type is "10", and has one parameter n. The structure is described in Figure 1. This family was designed especially difficult for Goldberg's algorithm.
  We generated 30 networks of this kind, with changed the value of n along a geometric series from 500 to 25000.

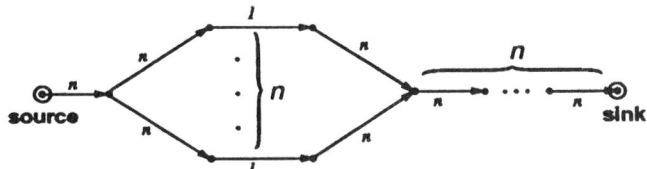

FIGURE 1. Goldberg-Bad-Case network

- **CL - Cheriyan-Graph-Long**. A deterministic family, generated by the WASHINGTON code. The type is "11", and has three parameters n, m, c. The structure is described in Figure 2. This family was designed by J. Cheriyan, and also considered as a hard case for Goldberg's algorithm.
  For CL networks, we generated 30 instances with $n = 1000$, $m = 10$ and c changed along a geometric series from 10 to 1000.
- **CW - Cheriyan-Graph-Wide**. The same family as CL, except the generated 30 instances have parameters $n = 1000$, m changed along a geometric series from 10 to 1000 and $c = 10$.

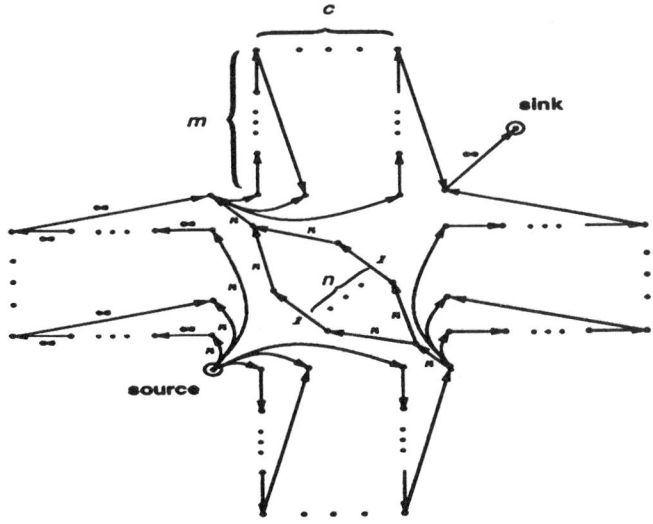

FIGURE 2. Cheriyan-Graph

The following three families were all generated by the code GENRMF, with different parameter settings. Each contains 30 instances. Capacities c_1, c_2 were set to 100 and 10000 respectively in each case.

- **GXL - Genrmf-eXtra-Long**. Extreme "long" networks, with $a = 4$, and b changed along a geometric series from 20 to 2430.
- **GM - Genrmf-Medium**. Cubic like networks, with $a = b$, and both were incremented by 1 from 5 to 34.
- **GXW - Genrmf-eXtra-Wide**. Extreme "wide" networks, with a changed along a geometric series from 20 to 94, and $b = 4$.

TABLE 2. Benchmark times (sec) on the SPARCstation 1+ used for testing.

FORTRAN	TEST 1			TEST 2		
(netflow)	real	user	sys	real	user	sys
f77	5.7	4.7	0.5	9.8	8.6	0.6
f77 -O	4.8	3.6	0.4	7.3	6.1	0.6
f77 -O1	5.4	4.6	0.5	9.7	8.5	0.6
f77 -O2	4.0	3.4	0.4	6.7	5.7	0.6
f77 -O3	4.4	3.6	0.5	7.0	6.3	0.4
f77 -O4	3.7	2.9	0.5	5.8	5.0	0.6
C	TEST 1			TEST 2		
(wmatch)	real	user	sys	real	user	sys
cc	5.5	5.2	0.2	46.2	44.7	0.5
cc -O	4.4	4.0	0.2	35.9	34.2	0.6
gcc	5.9	5.5	0.2	49.3	47.5	0.5
gcc -O	4.4	4.1	0.2	34.2	32.7	0.3

4.2. Description of the reported data. In all reported experiments we used the **Gap** switch on, **Relab-freq** $= 0.5M$, **Cut-freq** $= 0$ and the procedure to generate permutations was turned off.

Table 3. contains the mean running times of 10 random instances and the relative variance for the DIMACS Benchmark problems. All reported times are in CPU seconds on a Sun Sparc 1+ Workstation and do not include I/O time. We attempted to measure the user CPU time what the program uses during the execution of the algorithm. Since we used a multi-user workstation under Unix, our obtained running times slightly depended on the workload of the system. This can explain a part of the large variances.

Figures 3–9 contain graphical comparison of the tested algorithms on the GB, CL, CW, GXL, GM, GXW families.

The codes were written in ANSI C language, employing structures, pointers in order to handle the various operations efficiently. Memory requirements are

24 bytes/edge for all codes and 64 bytes/vertex for PLED, 20 bytes/vertex for GOLD, and 40 bytes/vertex for GOLDYN. All integers are of 4 bytes. We used the Gnu C-compiler (gcc) with -O optimization option.

Table 2. contains some benchmark results which were obtained by executing the DIMACS provided Fortran and C codes on our machine.

TABLE 3. Mean running times (secs) and relative variances (%) of the instances.

NAME	PLED		GOLD		GOLDYN	
	mean	var %	mean	var %	mean	var %
GL1	2.11	19.9	0.68	19.6	1.14	22.3
GL2	4.82	11.9	1.57	12.4	2.59	15.3
GL3	11.20	9.9	3.76	6.9	5.71	9.9
GL4	27.89	9.2	8.83	11.3	12.74	5.4
GL5	64.69	9.3	20.94	10.1	28.98	7.3
GL6	275.75	7.0	54.43	10.2	144.85	7.6
GW1	3.38	10.2	1.27	9.0	2.42	11.4
GW2	8.91	7.5	3.45	5.7	6.55	3.9
GW3	20.62	6.4	8.46	4.0	15.47	9.7
GW4	61.92	7.6	23.49	6.1	43.86	6.3
GW5	158.01	4.7	59.84	4.7	106.44	6.6
GW6	979.44	6.5	149.21	8.0	314.10	6.0
WLM1	0.20	5.3	0.07	4.2	0.13	5.0
WLM2	0.53	8.7	0.18	6.2	0.30	7.8
WLM3	1.38	7.5	0.37	11.8	0.54	8.5
WLM4	3.21	10.2	0.87	11.2	1.19	13.7
WLM5	8.05	6.5	2.28	5.8	2.79	9.5
WLM6	22.75	5.3	5.94	9.7	7.14	6.1
AD1	1.32	18.9	1.11	10.5	1.34	18.8
AD2	2.42	19.8	1.80	13.2	2.22	15.1
AD3	3.52	16.1	2.82	12.4	3.38	11.1
AD4	5.78	16.5	4.65	11.3	5.68	11.6
AD5	8.81	16.4	7.09	16.1	8.49	16.9
AD6	14.43	12.6	14.01	8.1	14.27	10.2

4.3. Evaluation of the DIMACS Benchmark problems. From Table 3. at first sight we can say that the simple preflow-push code (GOLD) performs much better on these families than our PLED or GOLDYN implementation. The only exception were the AD instances, where all codes performed about the same way. Moreover the ratio of the running times of PLED over GOLD is getting bigger in almost all the cases as the number of nodes increases. This ratio is very much depending on the structure of the network.

Having a closer look of the profile information of PLED, we found that PLED used about 45% of the execution time for heap operations and only 10-12% for dynamic trees operations. We have also found (see Table 5.) that the number of selections of the vertices for the same problem were about the same in PLED as in GOLD. Since GOLD uses only a simple FIFO policy, while PLED has a "sophisticated" minimum label selection rule, GOLD can be much faster in these cases. This means that using the global relabeling technique, in average both algorithms take approximately the same number of steps.

The running times of PLED in Table 3. corresponds to the "considered to be the best" parameter settings. From early experiments we learned that using random or deterministic permutations instead of just leaving the edgelists untouched has no practical advantage. In some cases the number of operations decreased slightly if the edgelists were permuted, but the cost of such a procedure is high, and the running times became larger. We even encountered examples, in which the number of steps taken was more with using permutations than without them. Let us remark also since all the examples were generated randomly, an additional random permutation of the adjacency lists can not add much to the randomness of these lists.

Our experiments also showed, that perhaps the best setting for the parameter **Relab-freq** (, and **Cut-freq**) is in the range $[0.4N..4N]$ ($[0.1N..N]$), however, we could not make much difference between the values in this range. For several different settings from this range we obtained almost identical running times. On the other hand, of course, too small values for these parameters result in too many big-relabel calls, while too large values do not make profit of the big-relabel procedure. As we mentioned earlier, the effect of big-relabel is so significant, that its examination could be a subject of later studies.

It is interesting that problems with short paths between the source and the sink (eg. the Acyclic-dense problems which are full graphs with diameter 1 or 2) need much less effort to solve than other problems. A reason for this perhaps, is that the maximum flow can be pushed over much sooner in these cases.

TABLE 4. Number of Saturating and Nonsaturating pushes.

NAME	SATURATING			NONSATURATING		
	GOLD	GOLDYN	PLED	GOLD	GOLDYN	PLED
WLC1	895	1036	899	2735	2172	3451
WLC2	1838	2231	1878	7444	5889	10282
WLC3	3550	4392	3775	17720	13869	26400
WLC4	7397	9629	8465	45568	36075	71340
WLC5	14001	18996	17646	104444	81626	182883
WLC6	27666	39759	38472	252319	193867	461418
GL1	1076	1217	1004	10783	8619	12343
GL2	2151	2514	2072	24849	19618	27864
GL3	4229	4890	4065	60679	45268	71501
GL4	8228	9378	8078	146039	104902	201734
GL5	17065	19465	16867	343387	245319	476582
GL6	36856	42414	38370	900646	624328	1364934
GW1	1899	3038	2598	19883	20123	15289
GW2	4858	7826	6759	51479	57841	42577
GW3	10453	17456	15319	124369	141565	95288
GW4	26700	44745	42481	342545	408457	287406
GW5	61309	102140	103667	826702	1037573	765610
GW6	138916	236455	259357	2068616	2699497	2268190
WLM1	247	251	240	652	611	841
WLM2	490	503	459	1264	1179	1972
WLM3	938	978	882	3504	2651	7603
WLM4	1773	1857	1595	6367	4747	17397
WLM5	3621	3822	3149	12077	8594	39231
WLM6	6970	7480	5978	23030	15110	91198
AD1	358	628	586	415	142	258
AD2	499	838	773	537	190	402
AD3	618	1040	936	670	235	443
AD4	759	1323	1186	858	292	534
AD5	954	1631	1483	1052	376	683
AD6	1183	1977	1902	1283	495	884

TABLE 5. Number of Vertex Selections and Relabels.

NAME	SELECT			RELABEL		
	GOLD	GOLDYN	PLED	GOLD	GOLDYN	PLED
WLC1	3374	2470	3217	713	727	736
WLC2	9108	6601	8821	1828	1940	2112
WLC3	21674	15208	21884	4264	4477	5467
WLC4	55836	40074	59103	10953	12069	15545
WLC5	126935	88635	146368	23762	26680	40123
WLC6	307239	210937	365618	57558	63994	102588
GL1	13454	8814	10152	2944	2954	3132
GL2	30803	20046	22795	6610	6844	7181
GL3	73657	43938	53240	14367	14884	16559
GL4	174621	96398	131540	31354	32443	40588
GL5	407952	218523	305694	72237	73552	95535
GL6	1054634	531145	811533	170055	178095	256119
GW1	25255	19371	17671	5773	7095	6685
GW2	65811	51368	45786	15393	19091	17494
GW3	157419	119001	106912	35319	44203	40563
GW4	435163	329006	317527	98618	123425	120234
GW5	1054655	798915	810822	241743	302175	306952
GW6	2615522	1969943	2124932	578402	752522	804951
WLM1	787	582	734	140	142	151
WLM2	1508	1143	1649	245	259	330
WLM3	4165	2511	5194	682	597	1092
WLM4	7506	4503	11088	1195	1035	2301
WLM5	14035	8046	24355	1960	1593	4975
WLM6	26470	14168	52860	3442	2572	10574
AD1	590	567	556	188	187	307
AD2	788	761	864	270	270	470
AD3	974	915	944	326	318	518
AD4	1246	1198	1152	415	418	639
AD5	1513	1465	1471	497	499	815
AD6	1864	1805	1819	576	583	1021

4.4. Testing on the GXL, GM, GXW problems.

Since the DIMACS Benchmark problems had only a limited set of networks, we decided to compare our codes on some "extreme" families too. We assumed that the length of paths from the source to the sink could be relevant for Goldberg's algorithm, so we generated the earlier described GXL, GM, GXW network families. Figures 3., 4., and 5. show graphs of running times obtained on these networks. All the scales are linear and the runtimes are shown as functions of the number of vertices.

Among these problems the GXW family seems to be the hardest to solve. Even the growth rate is the highest on GXW networks for all the codes. Again GOLD proved to be the fastest in the GM and GXW case, while PLED is the slowest in all three cases. However, on the GXL family the dynamic tree version of Goldberg's algorithm performed the best. This family has the property that the sink can be reached from the source using only relatively long paths, (compared to the size of the network). So the strategy of pushing excess along a longer path seems to have advantage in this case. GOLDYN pushes excesses as far as possible without controlling the size of this path or the amount to be pushed. On the other hand PLED does not use this "greedy" strategy (e.g. by limiting the amount to be pushed).

This experiment also suggests that the longer the relative distance between the source and the sink, the more useful the GOLDYN algorithm could be.

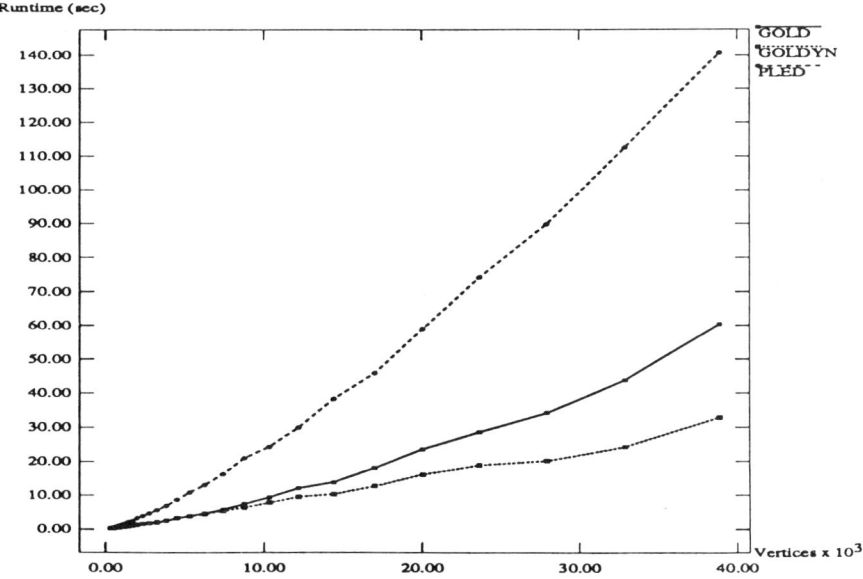

FIGURE 3. Runtimes on the Genrmf-eXtra-Long problems.

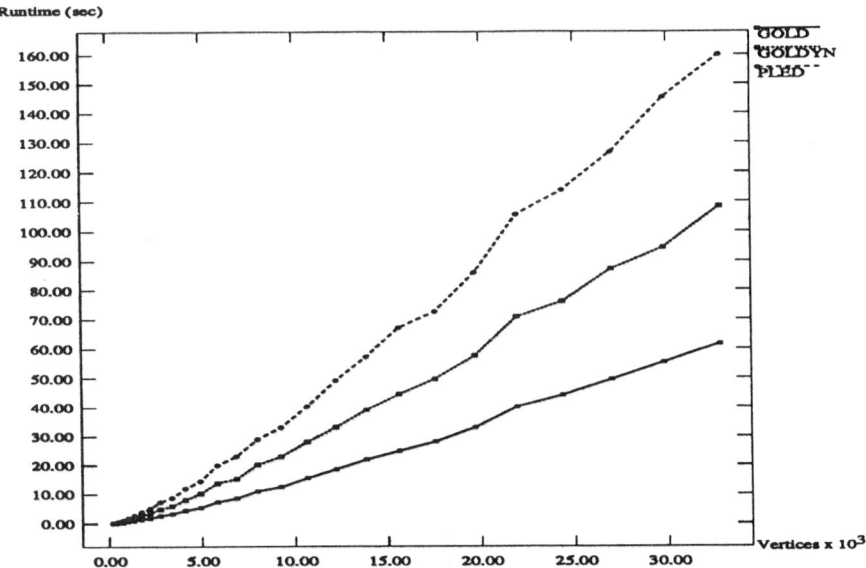

FIGURE 4. Runtimes on the Genrmf-Medium problems.

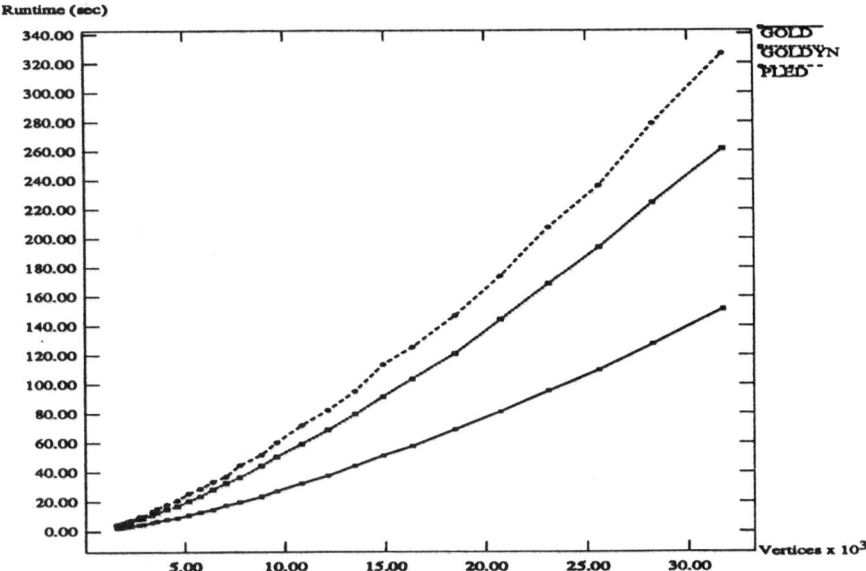

FIGURE 5. Runtimes on the Genrmf-eXtra-Wide problems.

4.5. The effect of dynamic trees.

Figures 6., 8. and 9. show the experimental results obtained on the GB, CL and CW families. To examine the effect of using the dynamic trees data structure, we also ran our NPLED and NGOLDYN code on these examples. Figure 7. contains the same runtimes as Figure 6., except the scale is different since it shows PLED and GOLDYN only.

It is clear that GOLD, even with using the big-relabel strategy, is far the slowest on the GB problems. What is more interesting is that this family was the only case where PLED and GOLDYN outperformed their "non-dynamic tree" versions. While the running time of NPLED and NGOLDYN is exponentially increasing, PLED and GOLDYN has a nice linear behavior (see Figure 7.). The difference can be as large as 16 second versus 3192 second for GOLDYN vs. NGOLDYN and 73 second vs. 1862 second for PLED vs. NPLED on the largest example. This emphasizes again that pushes along long paths can be handled by the dynamic trees structure much more efficiently.

We counted the average length of a push during the execution of PLED and GOLDYN, and found that in the case of the GB family it was significantly longer than in any other families. Significant means several hundreds or thousands for GB and less than 10 (ten) for the rest, including all the tested families. On the CL family both NPLED and NGOLDYN were faster than their counterparts, and on the CW family PLED could beat NPLED. Looking at the average push length in these cases we again found that it was very small (under 3) in the CL family for both algorithms, and slightly increasing, but still under 11 on the CW problems for GOLDYN. This length was the longest (≈ 15) on the CW family for PLED, what may explain why it could perform better then NPLED.

It is worth to mention that in all the cases PLED tends to push along longer paths as we have found that the average length of a push was always longer for PLED than for GOLDYN. This may come from the fact that PLED links trees together only if the link has enough residual capacity, while GOLDYN can link trees with a small capacity edge, and hence by pushing along a path GOLDYN may have to cut more often than PLED.

The second longest average length for pushes were found in the GXL case. This supports our conjecture that algorithms using dynamic trees like PLED or GOLDYN can have practical advantage on networks with extremely large distance between the source and the sink.

Figures 8. and 9. also show that the actual structure of the network can change completely the ranking of the execution times of these algorithms.

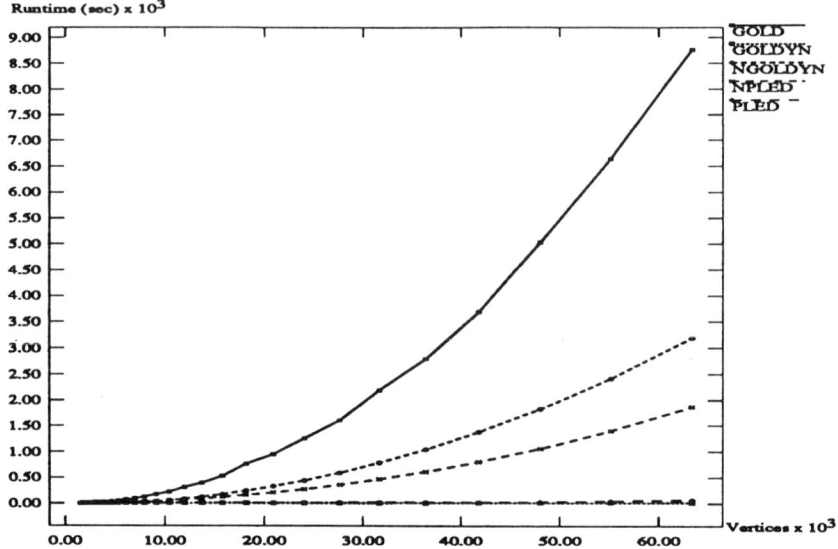

FIGURE 6. Runtimes on the Goldberg-Bad-Case problems.

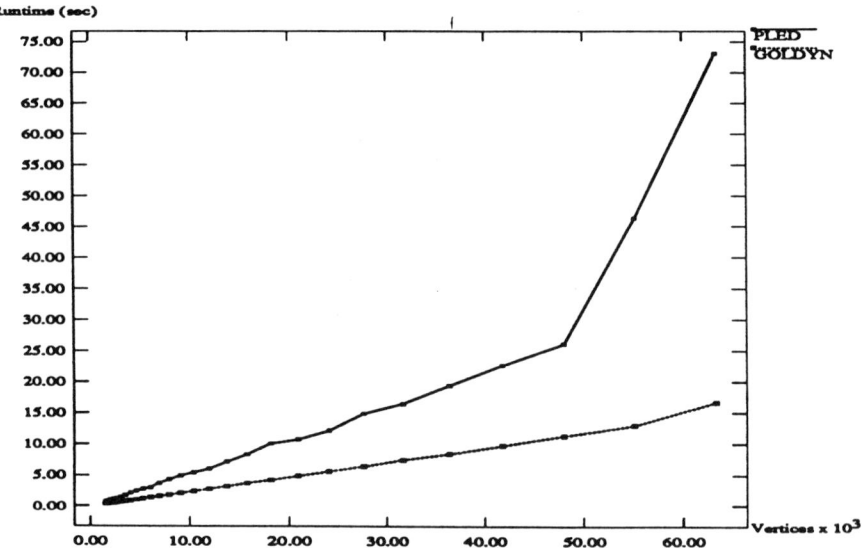

FIGURE 7. Runtimes of PLED and GOLDYN on the Goldberg-Bad-Case problems. (Note that the last to sample points for the time of PLED shows that the workstation started to swap memory at these sizes, since PLED uses the memory most extensively.)

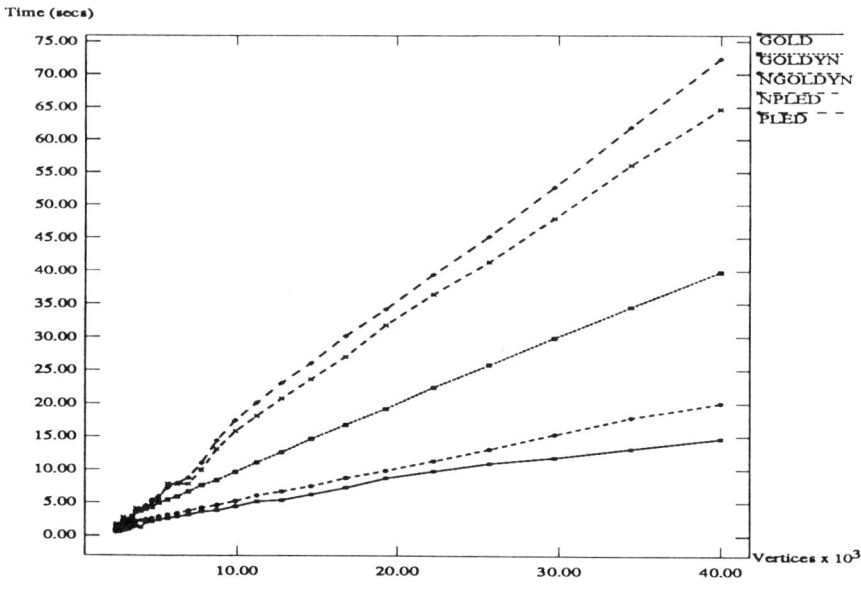

FIGURE 8. Runtimes on the Cheriyan-Long problems.

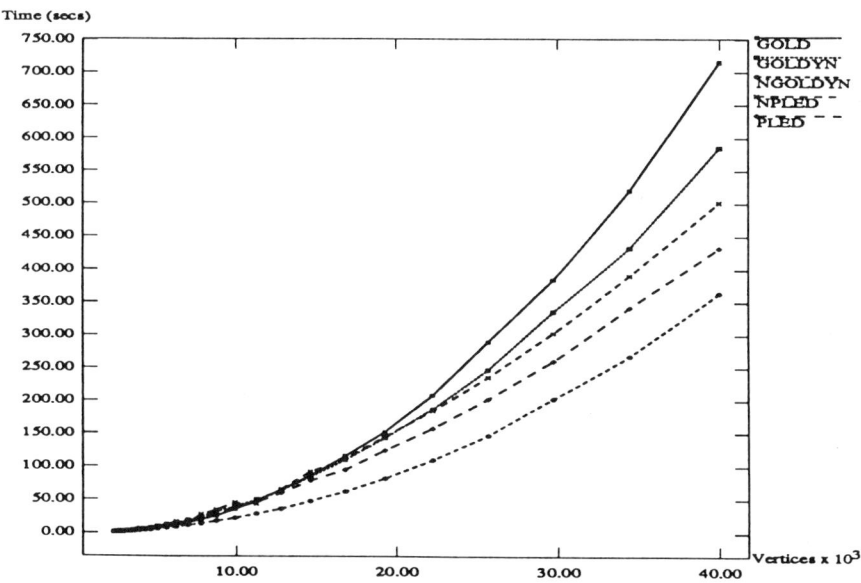

FIGURE 9. Runtimes on the Cheriyan-Wide problems.

5. Conclusions

Summarizing our work, we can conclude that although the PLED algorithm has a very good theoretical worst case bound, in practice Goldberg's simple preflow-push algorithm outperforms it on most of the examples of this study.

Our study shows that the structure of the networks is the most important factor in ranking the algorithms. (E.g. see the examples shown on Figures 8. and 9.) One such parameter to be considered, reflecting the structure of the network, could be the relative distance between the source and the sink.

In this study we were particularly interested in the effectiveness of dynamic trees. Our experiments show clearly that there are families of problems for which dynamic trees improved the performance of our code at a small cost. To determine the properties of network classes on which the algorithms GOLDYN or PLED are the best would be an interesting topic of later works.

Let us remark finally that Fibonacci heaps did not help much in these examples. They did not improve neither the running time nor the number of selection steps of PLED.

Acknowledgment: The authors are thankful to Joseph Cheriyan and to the anonymous referee for the many helpful comments and suggestions.

References

1. J. Cheriyan and T. Hagerup: "A randomized maximum-flow algorithm". in Proc.IEEE FOCS (1989), 118-123.
2. J. Cheriyan and T. Hagerup: "A randomized maximum-flow algorithm". Tech. Report 988, O.R.I.E, Cornell Univ., Ithaca, NY, Oct. 1991.
3. J. Cheriyan, T. Hagerup and K. Mehlhorn: "Can a maximum flow be computed in $o(nm)$ time?". Proc. 17th Internat. Colloquium on Automata, Languages and Programming, 1990, pp.235-248.
4. J. Cheriyan, T. Hagerup and K. Mehlhorn: "An $o(n^3)$-Time Maximum-Flow Algorithm". Tech. Report MPI-I-91-120, Max Planck Institut für Informatik, Saarbrücken, Germany, Nov. 1991.
5. N. Alon: "Generating pseudo-random permutations and maximum flow algorithms" *Information Processing Letters* **35** (1990), 201-204. North-Holland
6. R. K. Ahuja and J. B. Orlin: "A Fast and Simple Algorithm for the Maximum Flow Problem". *Operations Research* **37** (1989), 748-759.
7. R. K. Ahuja, J. B. Orlin and R. E. Tarjan: "Improved Time Bounds for the Maximum Flow Problem". *SIAM J. Computing* **18** (1989), 939-954.
8. R. Anderson, et al.: Program washington.c for creating maximum flow problem instances. Available through Dimacs.
9. T. Badics: Program genrmf.c for creating maximum flow problem instances. Available through Dimacs.
10. G. Waissi: Program ac-max.c for creating maximum flow problem instances. Available through Dimacs.
11. U. Derigs and W. Meier: "Implementing Goldberg's max-flow-algorithm — A computational investigation". *ZOR - Methods and Models of Operations Research* **33** (1989), 383-403.
12. A. V. Goldberg: "A New Max-Flow Algorithm". Tech. Rep. MIT/LCS/TM-291, Laboratory for Computer Science, Massachusetts Institute of Technology, Cambridge, MA, 1985.
13. A. V. Goldberg and R. E. Tarjan: "A New Approach to the Maximum-Flow Problem". J. ACM 35 (1988), 921-940.

14. A. V. Goldberg, É. Tardos and R. E. Tarjan: *Network Flow Algorithms*. Tech. Rep. No. 860, School of Operations Research and Industrial Engineering, Cornell University, Ithaca, NY, 1989.
15. D. D. Sleator and R. E. Tarjan: "A Data Structure for Dynamic Trees". *Journal of Computer and System Sciences* **26** (1983), 362-391.
16. R. E. Tarjan: *Data Structures and Network Algorithms*. SIAM publications, Philadelphia, PA, 1983.
17. M. L. Fredman and R. E. Tarjan: "Fibonacci Heaps and Their Uses in Improved Network Optimization Algorithms". *Journal of ACM* **34** (1987), 596-615.
18. M. L. Fredman, R. Sedgewick, D. D. Sleator and R. E. Tarjan: "The Pairing Heap: A New Form of Self-Adjusting Heap". *Algorithmica* **1** (1986), 111-129.
19. Gy. Novotny: "Comparison of Different Data Structures". *Thesis*. Eötvös Loránd University, Budapest, 1989.

RUTCOR, RUTGERS UNIVERSITY, NEW BRUNSWICK, NJ 08903-5062.
E-mail address: badics@rutcor.rutgers.edu, boros@rutcor.rutgers.edu

Implementing the Push-Relabel Method for the Maximum Flow Problem on a Connection Machine

FARID ALIZADEH AND ANDREW V. GOLDBERG

1. Introduction

In this paper we study the recent *push-relabel* [9, 13] class of algorithms for the maximum flow problem in capacitated directed networks. Several researchers [1, 5, 7, 15] concluded that their sequential implementations of push-relabel algorithms are superior to previously used codes. (For more recent results, see [3, 21].) We focus on the behavior of the push-relabel method in massively parallel environments. For a related work in the context of minimum-cost flow problem, see [20].

The push-relabel method is the first theoretically efficient algorithm for the maximum flow problem that is potentially practical; the previous algorithm of Shiloach and Vishkin [23], based on Dinitz' blocking flow method [8], requires the amount of memory quadratic in the number of nodes in the input network.

The push-relabel method uses two operations, *push* and *relabel*, to manipulate current flow and distance labeling functions. In the parallel implementation the algorithm pushes simultaneously from all nodes to which the push operation applies; we refer to this method of pushing as *parallel push*. Similarly, *parallel relabel* applies relabeling operation to all eligible nodes simultaneously.

The push-relabel algorithm works regardless of the order in which the push and relabel operations are applied. It also works with any correct labeling procedure. This robust nature of the algorithm allows us to experiment with various strategies

---

1991 *Mathematics Subject Classification*. Primary 90B10; Secondary 68Q20.

The first author was supported in part by by the Air Force Office of Scientific Research grant AFOSR-87-0127, the Minnesota Supercomputer Institute and NSF postdoctoral grant number CDA-9211106. The second author was supported in part by ONR Young Investigator Award N00014-91-J-1855, NSF Presidential Young Investigator Grant CCR-8858097 with matching funds from AT&T and DEC, Stanford University Office of Technology Licensing, a grant form Mitsubishi Electric Laboratories, and a grant from the Powell Foundation.

for ordering push and relabel operations, and with a variety of relabeling procedures. All of the push and relabel operations are suitable for parallelization. The goal is to determine the ordering and relabel procedures most suitable for a practical parallel implementation.

It is well-known that the maximum flow problem is P-complete [14]. Therefore, for any proposed parallel algorithm one may find instances of the problem which take at least linear time in the the size of input network, unless P=NC. Our goal is to find techniques that perform substantially better than linear time on most networks in a fine grain parallel environment.

This paper is a report on experiments with various strategies for the parallel push-relabel algorithm. The experiments were conducted on a Connection Machine [16] model CM-2, with 32K processors (K=1024). Each processor can directly access 1 megabits of local memory, so the entire system has 4 gigabytes of memory. The starting point for our work was the Connection Machine implementation of the algorithm described in [10]. Our current implementation, however, improves the original one in several areas, including better ordering and pipelining of operations of messages. Also, the original implementation was on CM-1 using the *LISP interpreter where as the current implementation is on CM-2 using the C* compiler.

This paper contains six sections including the introduction. In Section 2 we outline the push-relabel algorithm and discuss parallel implementation of push and various relabel operations including Derigs and Meier's gap-relabeling and parallel breadth–first search. In Section 3 we describe the mapping of networks to the Connection Machine processors. This mapping is designed to take advantage of the parallel prefix and suffix operations provided in the Connection Machine. In Section 4 we describe in more detail parallel push and relabel operations in the context of this mapping. In Section 5 we report on the running time of the program on various sample inputs. We present our conclusions in Section 6.

2. The Parallel Push-Relabel Algorithm

In this section we review some of the basic concepts of the push-relabel method and its parallel implementation. We assume that the reader is familiar with [13]. (See also [12].) Because of high-grain parallelism of the CM-2 architecture, the goal of our implementation is to get a tight inner loop of the algorithm rather then to achieve high processor utilization by careful processor scheduling (see [11, 23]).

A *flow network* is a directed graph $G = (V, E, s, t, u)$, where V and E are node set and arc set, respectively; s and t are the source and the sink, respectively; and u is a non-negative capacity function on the arcs. We define $n = |V|$ and $m = |E|$, and assume that for each arc (v, w), the arc (w, v) is also present. A flow is a function on the arcs that satisfies capacity constraints on all arcs and conservation constraints on all nodes except the source and the sink. The conservation constraint at a node v indicates that the *excess* $e_f(v)$, defined as the difference between the incoming and the outgoing flows, is equal to zero. A *preflow* [17] satisfies the capacity constraints and the relaxed version of conservation constraints that requires the excesses to be nonnegative.

An arc is *residual* if the flow on it can be increased without violating the capacity constraints, and *saturated* otherwise. The residual capacity $u_f(v,w)$ of an arc (v,w) is the amount by which the arc flow can be increased. The residual graph is induced by the residual arcs.

The *distance labeling* $d : V \to N$ satisfies the following conditions: $d(s) = n$, $d(t) = 0$, and for every residual arc (v, w), $d(v) \leq d(w) + 1$. It is easy to see that if $d(v) < n$, then $d(v)$ is a lower bound on the actual distance from v to t in the residual graph, and if $d(v) > n$, then $d(v) - n$ is a lower bound on the actual distance of v to s. A residual arc (v, w) is *admissible* if $d(v) > d(w)$.

A push-relabel algorithm maintains a preflow and a distance labeling. We say that a node v is *active* if $v \notin \{s,t\}$ and $e_f(v) > 0$. The preflow is modified using push operation, which pushes excess flow from an active node to an adjacent one that has a smaller distance label. When an active node cannot push its excess because its label is at most that of its neighbors, then the distance labeling is modified using a relabel operation, which increases the distance label of the node to the largest value allowed by the labeling constraints. Distance labels also define a $s-t$-*cut set* with respect to the current preflow. The set of nodes is partitioned into two sets by the distance labels: those nodes whose labels are larger than or equal to n, and those whose labels are less than n. When the algorithm terminates the final labeling determines the minimum cut set.

Our parallel implementation works as follows. The preflow f is initialized to $f(s,w) = u(s,w)$ for all nodes u adjacent to s, and $f(v,w) = 0$ for all $v \neq s$. The initial distance labeling is computed using breadth-first search (BFS) on the residual graph induced by the initial preflow. The preflow and distance labeling are updated in *pulses* using the push and relabel operations, respectively, until no active nodes remain. Each pulse consists of one or more parallel push operations followed by one or more parallel relabel operations.

The *parallel push* operation works as follows. For active nodes, the parallel push operation distributes their excess among admissible outgoing arcs as described in Figure 1; in this figure, "push flow from v to w" means increase the flow from v to w by the minimum of $e_f(v)$ and $u_f(w)$ and update the excesses of v and w.

For parallel relabeling we use several alternatives in our implementation. The simple parallel relabel, described in Figure 2, sets the distance label of every node

OPERATION *Parallel-Push*
For all active nodes v in parallel **do**
    push flow from v **until** $e_f(v) = 0$ **or** $\forall w$ such that $d(w) < d(v)$, $u_f(v,w) = 0$.

FIGURE 1. The parallel push operation.

OPERATION *Relabel*
For all nodes $v \notin \{s,t\}$ in parallel do
    $d'(v) \leftarrow \min\{d(w) + 1 : u_f(v,w) > 0\}$;
    If $d(v) \neq d'(v)$ then
        broadcast $d(v)$ to all neighbors of v;
    end;

FIGURE 2. The parallel relabel operation.

OPERATION *gap-relabel*
(1) Find the smallest g where $0 < g < n$ and no node has label g.
(2) For all nodes v in parallel do:
    if $g < d(v) < n$ set $d(v) \leftarrow n$.
(3) For all v which have a new label broadcast the new labels to all neighbors of v.

FIGURE 3. The gap-relabel operation.

except s and t to one plus the minimum distance label of its residual neighbors.

Another relabeling procedure is the *gap–relabel* operation of Derigs and Meier [7] based on the following observation. Let g be an integer and $0 < g < n$. Suppose at certain stage of the algorithm some nodes are labeled 0, some labeled 1, and so on, through $g - 1$, but no node is labeled g—although some other nodes may be labeled by integers between g and n. Derigs and Meier observe that the sink is not reachable from any of the nodes whose labels are strictly between g and n. Therefore, the labels of such nodes may just be increased to n. This procedure may be implemented in parallel; see Figure 3.

Note that the most accurate labeling is obtained by applying BFS backwards from the sink and forward from the source. This can be implemented by applying parallel relabel operation repeatedly until distance labels stop changing. Doing this during every pulse is too expensive. The implementation of [10] performs a breadth-first search once after a sequence of pulses so that the breadth-first search time can be amortized over the pulse time. We experimented with this approach as well as with using the gap-relabel operation instead.

No theoretical result suggests that gap-relabel and BFS operations improve the worst-case time bounds on the algorithm. In practice, both periodic BFS and gap-relabel operations reduce the number of pulses drastically. The gap-relabel operation, however, is much less expensive (usually as fast as simple parallel relabel in our implementation) and as a result gives better running times most of the time in spite of the fact that the labeling after such an operation is not exact.

Our parallel implementation can be viewed as running the parallel push, relabel, and gap-relabel (or BFS relabel) processes "simultaneously". However, due to the SIMD nature of the machine, we have to time-multiplex the operations. We can adjust relative speed of the operations by running several push operations followed by several relabel operations followed by a gap-relabel or a BFS relabel operation. (Note that it does not make sense to run two gap-relabel operations without running a simple relabel operation in between.) We can also pipeline some of the operations as described in section 4.2.

A *pulse* is a sequence of push and relabeling operations (possibly pipelined) that is repeated over and over. A general rule is that the closer the labels are to their actual distance to the sink, the fewer pushes will be performed. Therefore, the more accurate labeling may reduce the total number of pulses. However, maintaining a very accurate labeling is too expensive. Our strategy is to update labels as accurately as possible without increasing the time spent in each pulse by too much. In Section 4 we describe our implementation in more detail.

3. CM Architecture and Tools

In this section we review some aspects of the CM architecture relevant to our program. Then, we describe a set of useful operations available on the Connection Machine for accumulating in parallel partial sums, partial minimums, etc.

3.1. Connection Machine Architecture. The following is a brief outline of the CM architecture. For more details see the book by Hillis [16] and documents from the Thinking Machine Corporation, for instance [6].

The Connection Machine is a distributed memory parallel computer [19]. It consists of thousands of processors connected by a routing network. Each processor has local memory. The local memory of a processor can be accessed by other processors via the routing network. [1]

The CM is a *single instruction, multiple data* (SIMD) machine: the program is stored in a host computer which executes a sequential program containing parallel instructions. When a parallel instruction appears in the program, the host broadcasts it to all processors. Each processor, depending on its memory contents, either executes the instructions or remains idle. The operation of the machine is totally synchronous: the next instruction does not start until all processors have completed the execution of the current instruction.

Each processor can access its own memory, or it can access the memory of other processors. However, accessing the local memory is much faster than accessing other processors' memory. In general, the time required by a processor to access memory of a processor that is "close" (in the routing network) is less than the time required to access the memory of a processor that is far. If during the execution of an instruction several processors access memories of other processors, the longest memory access time determines the execution time of the instruction. A *routing cycle* is the amount of time it takes to execute such an instruction. It is important to realize that the routing cycle time varies depending on the interprocessor

---

[1] Actually 16 processors are located on a processor chip, and every pair of processor chips (32 processors) share a memory chip. The entire machine may be thought of as a hypercube whose nodes are these pairs of processor chips. For simplicity in our discussion we assume that each processor has its own memory.

communication pattern and the size of data being accessed.

The Connection Machine software also provides the notion of *virtual processors*. The user may request any number of processors he or she wishes. If this number is larger than the number of physical processors available, each physical processor simulates v virtual processors, where v is the VP ratio, that is $v = \lfloor \frac{\#\ of\ virtual\ processors}{\#\ of\ physical\ processors} \rfloor$.

We should mention that the Connection Machine processors are fairly simple *bit processors* and can operate only on one bit of information at a time. In contrast even the simplest personal computers operate on at least 8 bits of data at a time. Also, the Connection Machine clock operates at 10MHz (in contrast, even the slowest PC's have a clock rate of at least 16MHz). Therefore, in order for a program to run efficiently on the Connection Machine it must have very high grain parallelism and should keep a large number of processors busy at any given moment.

3.2. Parallel Prefix and Suffix Operations. Parallel prefix operations have been recognized as fundamental parallel operations, and their implementation and use in parallel algorithms has been widely studied; see *e.g.* [18, 22]. Given an associative binary operation "$*$" and a sequence of s numbers $a_1, a_2, \cdots, a_s$, the parallel prefix "$*$" operation maps this sequence into the sequence $a_1, a_1 * a_2, \cdots, a_1 * a_2 * \cdots * a_s$. Similarly the parallel suffix operation maps this sequence into $a_1 * a_2 * \cdots * a_s, \cdots, a_{s-1} * a_s, a_s$. In our implementation of the parallel push-relabel method, we use prefix and suffix operations with "$*$" replaced by addition, min, and copy.

An extended form of parallel prefix and suffix operations are *segmented parallel prefix* operation. In the segmented form a list of sequences $S_1, S_2, \cdots, S_l$ is mapped into $S'_1, S'_2, \cdots, S'_l$, where each sequence S'_i is obtained from the corresponding sequence S_i using the parallel prefix or suffix "$*$" operation. The lengths of sequences S_i may be different.

A sequence of numbers is stored in the Connection Machine by placing the entries of the sequence in contiguous processors. For simple operations such as addition, copying, and minimum, the time required to perform parallel prefix and suffix operations is of the same order of magnitude as the time required for one routing cycle on the data of the same size. Although the routing cycle time and the time to perform a parallel prefix operation vary depending on the communication pattern and on the type of the parallel prefix operation, we shall refer to these times as simply the routing cycle in our description of algorithms. The main point to remember is that a routing cycle is much larger than a local memory access.

4. Data Structures and Implementation Details

4.1. Parallel Implementation of the Pulse Procedure. We use a mapping of the input network to the machine as first described by Blelloch in [4]. This mapping allows us to use locality of data through the parallel prefix (suffix) operations.

Each node v is assigned a processor P_v which we call a *node processor*. Each arc (v, w) is also assigned a processor P_{vw} called an *arc processor*. Recall that for each arc (v, w) we assume that the arc (w, v) is also present in the network and therefore, in our implementation, a separate arc processor P_{wv} is assigned to this arc. We call processors P_{vw} and P_{wv} *pair processors*. In the machine, each processor P_v is followed by all the processors P_{vw} corresponding to the arcs incident on it; the positions of arc processors associated with a node are arbitrary within themselves, and so is the positions of the node processors as long as the associated arc processors follow them. Each arc processor P_{vw} stores the processor address of its pair processor P_{wv}.

The part of the program that reads in the input, allocates processors, and initializes the system is relatively simple. The main part consists of application of the *pulse* procedure until no active nodes remain. The implementation of this procedure is summarized in Figure 4. This implementation include the simple *relabel* operations as part of it because all of the versions use this operation predominantly, although occasionally other relabel operations such as BFS or *gap-relabel* is used.

Steps 1-3 and 7 implement the parallel push procedure. In Step 1, the value of excess at each node processor P_v is sent to the arc processors immediately following it. Step 2 distributes the node excess to the outgoing arcs. First, each arc processor P_{vw} determines how much excess may be sent through its arc. This is equal to the residual capacity if $d(w) < d(v)$ (that is if w is estimated to be closer to sink) and zero otherwise. Then a *seg-suffix-add* is performed on these values. Now, each arc processor P_{vw} has information about the excess to be pushed from v, the amount it can push, and the amount that can be pushed through the arcs that follow (v, w) on the arc list of v. This information is enough to compute the amount $\sigma(v, w)$ to be pushed through the arc (v, w). After an execution of the *seg-suffix-add* operation, each node processor P_v contains the information about how much excess can be pushed from the node v at this pulse. The processor sets the value of its variable $e_f(v)$ to the amount that will remain after the pushing. Finally in step 3, all arc processors P_{vw} for which the amount $\sigma(v, w) > 0$, increase $f(v, w)$ by $\sigma(v, w)$, decrease $u_f(v, w)$ by the same amount, and send the value of $\sigma(v, w)$ to their pair processor P_{wv}. Each processor P_{wv} that receives such a message decreases $f(w, v)$ by $\sigma(v, w)$ and increases $u_f(w, v)$ by the same amount. Step 7 computes the new excesses on each node by performing a *seg-suffix-add* on the amount of new flow pushed through each arc, and this amount is added to $e_f(v)$.

Steps 4-6 implement the simple *relabel* operation. In Step 4, each processor P_{vw} sets its variable *head-label* to either $d(w) + 1$ or $2n$, depending on whether the arc (v, w) is residual or not. This process involves only local memory access. Next, a *seg-suffix-min* operation is performed on the *head-label* variable, and as a result each node processor P_v contains new value of $d(v)$. In Step 5 all node processors except P_s and P_t copy this value to their corresponding arc processors using *seg-prefix-copy*. In Step 6 each arc processor P_{vw} checks if the new $d(v)$ is different

Procedure *pulse*
(1) **For all** $v \in V - \{s,t\}$ copy $e_f(v)$ to all P_{vw} using *seg-prefix-copy* operation.
(2) { distribute excess }
  For all P_{vw} use *seg-suffix-add* to compute the amount that can be pushed to lower labeled nodes through arcs that follow (v,w) on the incident list of v.
  For all P_{vw} compute the amount $\sigma(v,w)$ to be pushed from v to w.
  For all $v \in V$ compute the amount of excess that remains at v after the pushing.
(3) { Push flow }
  For all P_{vw} **do if** $\sigma(v,w > 0$ **do begin**
    $f(v,w) \leftarrow f(v,w) + \sigma(v,w); \ u_f(v,w) \leftarrow u_f(v,w) - \sigma(v,w);$
    send a message containing $\sigma(v,w)$ to processor P_{wv}.
  end.
  For all P_{wv} that received $\sigma(v,w)$ **do begin**
    $f(v,w) \leftarrow f(v,w) - \sigma(v,w); \ u_f(v,w) \leftarrow u_f(v,w) + \sigma(v.w);$
  If *simple relabel* is the relabeling operation chosen **then begin**
(4)   { Compute new distance labels }
    For all P_{vw} **do**
      if $u_f(v,w) > 0$ **then** *head-label*$(v,w) \leftarrow d(v) + 1$
      else *head-label*$(v,w) \leftarrow 2n$.
    For all $v \in V - \{s,t\}$ compute *new-d*(v) using *seg-suffix-min*.
(5)   **For all** $v \in V - \{s,t\}$ copy *new-d*(v) to all P_{vw} using *seg-prefix-copy*
(6)   { Broadcast new labels }
    For all P_{vw} such that $v \notin \{s,t\}$ **do**
      If $d(v) \neq new - d(v)$ **then**
        send a message containing the value of $d(v)$ to P_{wv}
        and set $d(v)$ to *new-d*(v) that was broadcast.
  end.
(7) { Update excess }
  For all $w \in V$ **do begin**
    Use *seg-suffix-add* to compute the amount of flow *new-e_f*(w) pushed into w,
    $e_f(w) \leftarrow e_f(w) + new\text{-}e_f(w)$
  end.

FIGURE 4. Implementation of the pulse procedure.

OPERATION *Parallel BFS*
$d'(s) \leftarrow n$, and $d'(t) \leftarrow 0$;
For all nodes $v \in V - \{s,t\}$ $d'(v) \leftarrow 2n$;
Repeat
    Run steps **(3)**, **(4)** and **(5)** of *Pulse* procedure.
Until $new\text{-}d(v) = d(v)$ for all nodes v.

FIGURE 5. Implementation of parallel breadth-first search operation.

OPERATION *Parallel gap-relabel*
(1) For each node whose label $d(v)$ satisfies $d(v) < n$ in parallel **do**:
    Broadcast a flag to the processor numbered $d(v)$;
(2) Find the smallest g where processor g did not receive any message in step 1).
(3) For all nodes v in parallel do:
    if $g < d(v) < n$ set $d(v) \leftarrow n$.
(4) For all v with new labels copy $new\text{-}d(v)$ to all P_{vw} using *seg-prefix-copy*.
(5) For all p_{vw} which received new labels send a message to P_{wv} containing $new\text{-}d(v)$.

FIGURE 6. Implementation of the parallel gap-relabel operation.

from the old one and if so, sends a message to its pair processor P_{wv} updating $d(v)$ in the pair processor.

Each step 1 through 7 contains either a segmented parallel prefix (suffix), or a communication primitive, and the running time of each step is dominated by the primitive. Therefore, the overall running time of the pulse procedure is roughly seven routing cycles of the machine. Also observe that general communications are done along paths that are fixed through entire program: each processor P_{vw} has to communicate to processor P_{wv} in steps 3 and 6 and these are the only general communication operations. Therefore, the communication path has to be computed only once for each arc processor and the same information is used throughout the program.

To implement the parallel BFS operation we simply take steps 4–6 of pulse and run them over and over until the labels do not change. The number of times the simple *relabel* is iterated in a BFS operation is at most the larger of maximum distance of a node to the sink (if sink is reachable) and maximum distance of a node to the source (if sink is not reachable) in the residual graph.

The gap-relabel procedure is also easy to implement on the Connection Machine; see Figure 6. Clearly this procedure is not much more costly than a simple *relabel* operation (roughly four routing cycles vs. three). However, it may increase the labels of many nodes by a substantial amount.

We have experimented with several variants using gap-relabel and BFS operations. In all of the variants each pulse uses push and simple relabel operations, but at certain times instead of simple relabel a BFS or a gap-relabel operation is used.

When using BFS we follow the following rule. We measure the amount of computational work done in the last call to BFS. Then we accumulate the amount of work done by the simple relabel since the last call to BFS. We also fix a parameter k. If k times the amount of work since the last call to BFS exceeds the amount of work in the last BFS then we use BFS, otherwise we use simple relabeling. The amount of work itself can be measured in several ways. One way is to simply look at the CPU time used. Another way is to count the number of "expensive" operations, in this case the number of routing cycles. Each simple relabel contributes three routing cycles (one parallel suffix copy, one parallel prefix min, and one general communication step), and the work accumulated by the simple relabeling procedure is simply three times the number of pulses since last call to BFS. The amount of work in each BFS varies as the residual graph changes with each new preflow. We have used this technique for both push–relabel and push–push–relabel methods. The latter is a variation of a the push-relabel method where we only relabel every other pulse.[2] (One may think of this method as choosing the relabeling operation that does nothing, and alternate using this operation with simple relabeling.) We also tested this technique with the pipelined variants of push–relabel and push–push-relabel techniques (to be discussed in the next section.)

Another approach is to use push and relabel operations but after each relabel to apply a gap-relabel operation as well. Our experiments show that one does not need to apply gap-relabel every time. We fix a parameter k and call gap-relabel after every k pulses. Again, we tested gap-relabel with both push–relabel and push–push-relabel variants of pulse and their pipelined versions. All of the timings are reported in Section 4.

These algorithms are expected to work well if at each pulse a large proportion of nodes have positive excess and can distribute their excess to the lower labeled nodes. In the worst case a network made up of a simple long path will have only one node active at each pulse; see Figure 7(a). On the other hand the best case happens in the network shown in Figure 7(b) where only one pulse will find the the maximum flow (in both cases we assume that all arc capacities are equal, say to one.) It is easy to see that the number of pulses is at least as large as the the length of the shortest path from source to sink. In general, our parallel implementation works best if the diameter of the network is relatively small in comparison to the number of nodes. But in networks with very long source-sink distance, there is little parallelism to exploit, and the performance is expected to be worse than serial implementations. (Recall that CM processors are very slow even compared to the slowest PC's, and if only a few of them are computing at a time, not much performance can be expected.)

4.2. Pipelining Independent Operations. In the Connection Machine, if two segmented parallel prefix or suffix operations work on exactly the same sequences and perform the same binary operations, it is possible to *pipeline* them so that the pipelined operation, performed on the two sequences at once, is faster than two operations, each performed on one of the sequences at a time. For instance, steps 1 and 5 of the pulse could be pipelined, and so could steps 2 and 7. Steps 3

---

[2]In general a pulse can have x push operations followed by y relabel operations.

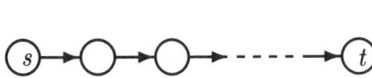

(a) worst case network,
requires at least n pulses

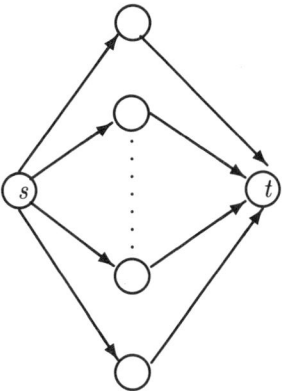

(b) best case network,
needs only one pulse

FIGURE 7. Best and worst case networks for the parallel push-relabel algorithms.

and 6 also do the same kind of operation, except that they involve message routing instructions. In theory we should be able to get better performance by pipelining the message routing steps, but on the Connection Machine we have not seen significant improvement. Pipelining the segmented prefix and suffix operations, however, results in about 10 to 20 percent improvement. (For example, instead of two *seg-suffix-copy* operations on 32 bit parallel integer variables, we may concatenate the two integers into a 64 bit long integer and apply *seg-suffix-copy* on this longer parallel variable. Then, the time required to do the latter operation is generally 10 to 20 percent faster than twice the time required to do a *seg-suffix-copy* operation on a 32 bit parallel variable.)

There are seven routing cycles in the pulse procedure (including cycles in simple relabeling). Some steps need to be performed after others (for instance, Step 6 must follow Step 5, and Step 5 must follow Step 4), whereas others are independent of each other (for example, steps 1 and 2 do not depend on each other and either may follow the other.) Figure 8 shows the dependency relationship among these steps. The arrows in the downward direction (say between steps 1 and 3) indicate dependency within a pulse; those in the upward direction (*e.g.* between steps 6 and 2) indicate dependency from one pulse to the next, (therefore, step two of the current pulse must start only after step 6 of the previous pulse is completed.)

The problem with pipelining steps 1 and 5, 2 and 7, and 3 and 6 is that they are sequentially related in the dependency graph. In order to curb this problem we take advantage of the robustness of the method. The dependency of Step 4 on Step 3 exists so that in the relabel step we have the updated residual capacities. Also, the dependency of Step 2 on Step 6 (of the previous pulse) exists so that the

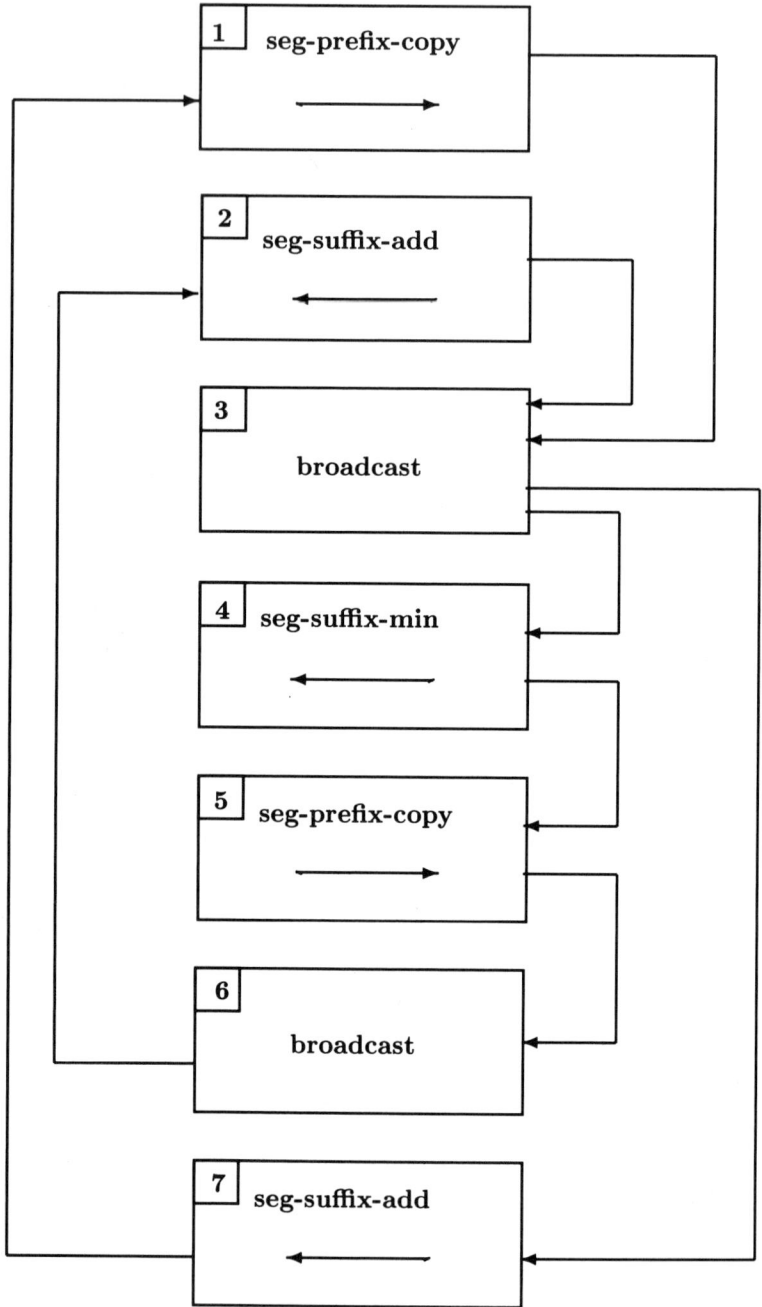

FIGURE 8. The dependency graph in the pulse procedure.

push operation is done based on new labels. These dependencies may be loosened somewhat so that steps 1 and 5, steps 2 and 7, and steps 3 and 6 can be pipelined. The disadvantage is that now the labeling becomes less accurate, and this translates into more iterations of the pulse procedure. The advantage is that now each pulse has only four routing cycles, albeit three of which are pipelined (and thus operate on longer data), see Figure 9.

The question is whether the time saved in each pulse more than compensates the time lost due to increased number of pulses. We report on the experiments in the next section. Notice that the SIMD nature of the Connection Machine does not allow us to exploit the independence of other operations such as, for example, steps 2 an 4 (one is a *seg-suffix-min* operation and the other a *seg-suffix-add*.)

We also used pipelining on the push-push-relabel implementation. (Recall that in this implementation we call the relabel operation only every other pulse.) The dependency graph for this version of the algorithm is unfolded in Figure 10. Notice that every stage in Figure 10 is equivalent to two pulses, only one of which has a relabel stage. Thus in this form the total number of routing cycles for two pulses is seven, whereas in Figure 9 there are eight routing cycles per two pulses.

5. Experimental Results

In this section we report on the running times of our program on several classes of medium and large networks. These experiments were conducted on two similar Connection Machines, one located at the Thinking Machine Corporation in Cambridge, Massachusetts, and the other one at the Army High-Performance Computing Research Center (AHPCRC) at the University of Minnesota in Minneapolis. Both machines have 32K processors. The timings reported here are based on test runs on the machine in AHPCRC. The program was coded using the new C* programming language, an extension of standard C for data parallel programming.

For each class of networks we have run four basic algorithms: push-relabel without pipelining (PR0), push-relabel with pipelining (PR1), push-push-relabel without pipelining (PPR0) and push-push-relabel with pipelining (PPR1). In all cases the simple relabel is used in the pulse procedure, except at certain pulses gap-relabel or BFS operation is used instead, as discussed in the previous section. Gap-relabel is used once in every k pulses for some parameter k. We experimented with values of k at 6, 8, 10, 12, and 14. The optimal value of k depends on the structure of the input network, and it is somewhat of an art to find it without excessive computation. Our experiments show that in the range $6 \leq k \leq 14$ the running times are reasonably good over all input classes.

For BFS, as mentioned earlier, we compare the amount of "work" (in our experiments the number of routing cycles) done by the last BFS, and if it exceeds k times the accumulated "work" done by simple relabels then we use BFS. We experimented with values of k ranging over $\{1.5, 1.7, 1.9. 2.1, 2.3, 2.5\}$. Although there

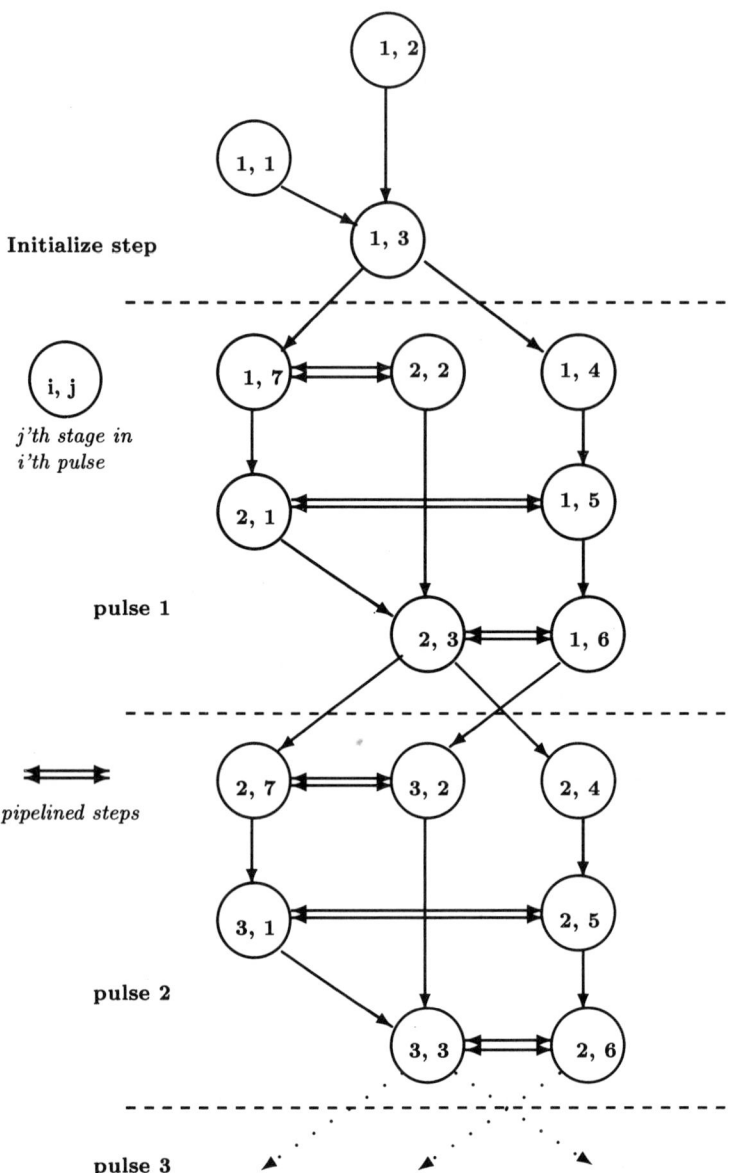

FIGURE 9. Unfolded dependency graph of the pulse procedure with pipelined steps.

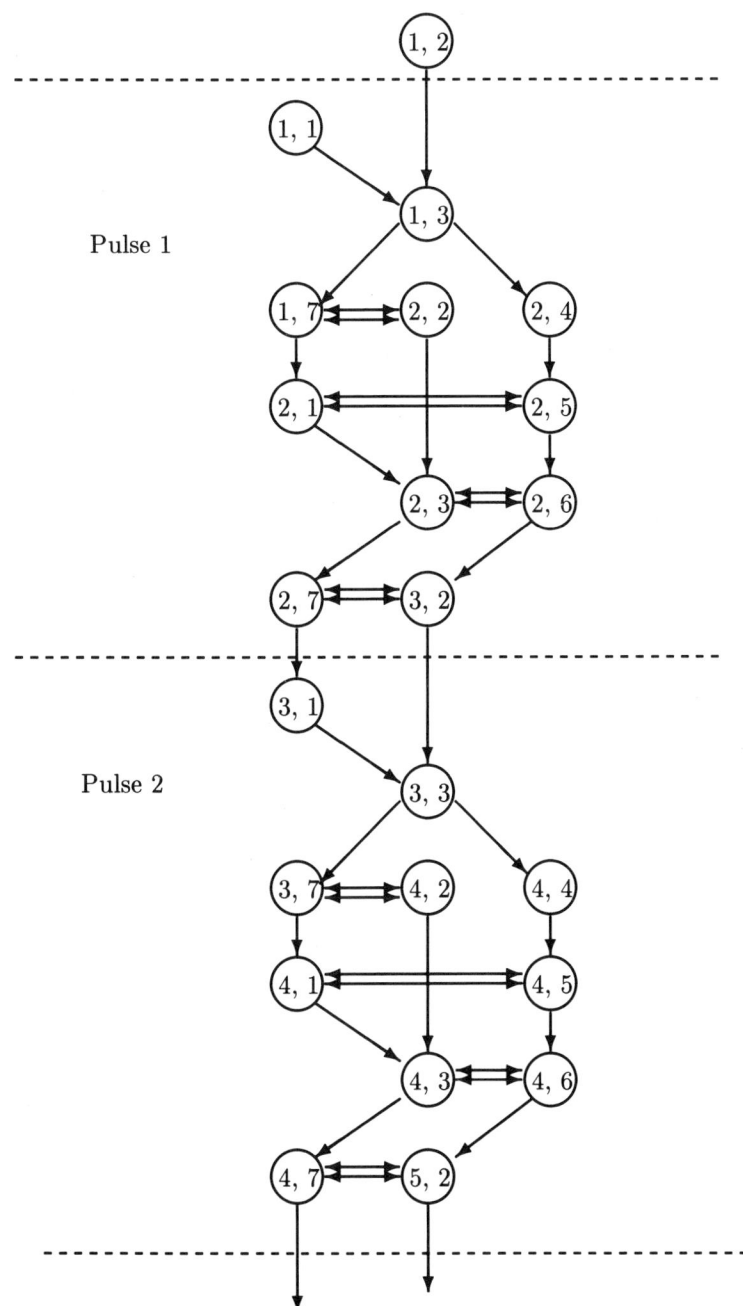

FIGURE 10. Unfolded dependency graph of push–push–relabel procedure with pipelined steps.

seems to be a large variation in the running time of the algorithms with varying k this range seems to work well overall.

The programs were run using either 8K, 16K, or 32K processors as appropriate. We experimented with a number of networks recommended by DIMACS. The timing and number of pulses for each run are reported in the sequel. In reporting the timings, we used the Connection Machines' CM-timer library. These timings report only the cpu time used by the Connection Machine and exclude computations done by the front end. In our implementation, the front end has minimal computational role. Aside from broadcasting the parallel instructions and simple computations (such as comparing two numbers to decide whether to do simple relabel or gap/BFS relabel) virtually all computing is done by the Connection Machine. Before we report on our computational results we should also mention that for the push–push–relabel heuristic with pipelining, the number of pulses should be multiplied by two. This is because every pulse in this variant corresponds to two pulses in other algorithms.

The first group are the fully dense acyclic networks constructed on a complete graph with an acyclic orientation. The characteristics of these networks are summarized in Table 11. For all variants of push–relabel algorithm, the fully dense acyclic networks seem to be particularly easy. The number of pulses for all of the variants are very small, and even the largest one, with over two million arcs is solved in a matter of seconds.

The second and third group consist of networks created by the "genrmf" generator. This generator takes as input two parameters a and b, and creates a network made of b parallel grids of size $a \times a$ (for a total of $a^2 b$ nodes in the network altogether). Nodes in each grid are connected to nodes to their right, left, up and down. In addition, ach node of a grid is connected to exactly one node of the next grid determined by a random permutation. We experimented with two classes of such networks: wide and long. Each network has approximately 2^x nodes. For wide networks we set $a \approx 2^{2x/5}$ and $b \approx 2^{x/5}$. For long networks we set $a \approx 2^{x/4}$ and $b \approx 2^{x/2}$. We experimented with values of x set at 10, 12, 14, and 16. The characteristics of these networks are summarized in Tables 12 and 13. For this class of networks the timings and pulse counts are shown in Figures 18, 19, 20, and 21. It seems that the gap-labeling heuristic outperforms BFS-labeling for RMF-Long class. However, between push–relabel and push–push–relabel variants there is no clear winner. Also, notice the sharp fluctuation for the BFS labeling as k varies. For instance, in the pipelined version of push-push-relabel with BFS labeling the running time for $k = 1.7$ is almost twice slower than for $k = 1.9$.

The timing results for RMF–Wide networks are summarized in figures 22, 23, 24 and 25. Among all networks recommended by DIMACS, this class by far requires the largest number of pulses. As it can be seen from these tables the number of pulses for each member of this class was almost in the same order of magnitude as the number of nodes. The large number of pulses for this class is reflected in the unusually large cpu time.

Finally, we experimented with networks known as "basic line mesh" class. These networks are made of 2^{x-2} rows and only 4 columns. Nodes in each column are connected to d random nodes in the next column. The parameter d is set to $\sqrt{n}$. The characteristics of these networks are summarized in Table 14. In figures 26, 27, 28, and 29 the timings and pulse counts for this class of networks are reported. It seems that for this group the gap-relabel heuristic is more efficient than the BFS labeling. Also the results in Table 28 indicate that the parallel speed up is linear. In addition, it seems that for this class push–push–relabel algorithm outperforms push–relabel, even though generally the number of pulses in the former exceeds that of the latter variation. (Remember that push–push–relabel applies relabeling only every other pulse.) For the BFS labeling heuristic there is virtually no variation in the number of pulses and in the running time of the program as the parameter k changes, except for BLM1. This insensitivity is rather atypical as the other classes often show sharp fluctuations with even the smallest change in k. Our version of push–relabel with pipelining failed to halt for this class of graphs when gap–relabeling were used. Even for BFS relabeling, when the program took unusually long time and had a much larger than expected pulse count. This may be due to an error in the code which manifests itself only for this class of problems.

The success of parallel implementation of the push–relabel algorithm rests largely on the number of pulses required to find the solution. The number of pulses in turn depends on the amount of parallel computation taking place in each pulse. If in each pulse a large number of nodes push their excess or a large number of nodes are relabeled, then we expect that fewer pulses are required to find the maximum flow. In fact, BFS and gap–relabeling operations are meant to increase the average amount of relabeling in each pulse. For the push operation no such heuristics are known to us. Each node may participate in a push only if it has positive excess. Therefore, it is reasonable to assume that if on average a large proportion of nodes have positive excess, then more computation is performed in each pulse and therefore, fewer pulses may be required. For each run of our program we calculated the number of nodes and arcs with positive excess in each pulse. The results are shown in Figure 30. It appears that only a very small fraction of nodes have positive excess on average. However, the number of physical processors we used is at most 32K and the size of networks most of the time far exceeds this number. Therefore, on average, a much larger proportion of physical processors participate in computation of each pulse.

A comparison between RMF-Long and RMF-Wide networks with approximately equal size is instructive. The number of active nodes and arcs in RMF-Long is almost one order of magnitude larger than that of RMF-Wide class; the converse is true for their running time.

6. Conclusions

We described several parallel implementations of the push-relabel method for the maximum flow problem and evaluated several heuristics that improve the practical performance of the method. The most encouraging aspect of our results is that the number of pulses, which is essentially the main bottleneck for parallel implementation of push–relabel algorithm, is usually only a fraction of the number of nodes in the networks. The small number of pulses is a direct consequence of occasional use of more accurate relabeling procedures such as BFS and gap–relabel. This shows that our approach will result in much better timings as more powerful processors are employed in future generations of Connection Machine or other parallel machines. In fact, since the speed up from 8K to 16K to 32K is almost linear, we expect the 64k Connection Machine would be almost twice as fast as the 32K machine on large problems.

Out of sequential implementations for maximum flow codes for the DIMACS Challenge, test examples of Nguyen and Venkateswaran [21] are most similar to ours, and their sequential implementation seems quite good. The sequential implementation is done on SUN SPARC-2 workstation. For the biggest problem sizes for which the data is available, the 32K parallel implementation and sequential implementation compare as follows. On the acyclic dense problems with 1024 nodes, the parallel implementation is about 40 to 50 times faster. On the RMF-long problems with 65536 nodes, the parallel implementation is about 3 to 4 times slower. On the RMF-wide problems with 63504 nodes, the parallel implementation is about 4 to 7 times slower. On Basic Line Mesh problems with 4098 nodes, the parallel implementation is about 3 to 5 times slower.

For the problem families we tried, one can conclude that the parallel implementation is competitive with the sequential one. On one hand, this is disappointing since CM-2 is much more expensive than SPARC-2. On the other hand, this is encouraging since the SPARC chip technology is superior to that of CM-2.

CM-2 is a SIMD machine. In a MIMD environment we expect to take advantage of independence of some parallel instructions more effectively. In particular the pipelining operation should result in much larger savings in time, (though not necessarily in the number of pulses.) Also, in the presence of wider communication bandwidth between processors the pipelining operations should become a lot faster. (In the CM-2 the bandwidth between two processors is only 1 bit.)

Another interesting research direction is towards obtaining an efficient implementation of the maximum flow method on parallel machines with powerful processors. On a small parallel machine, this problem was addressed in [2].

Acknowledgment

We would like to thank the Thinking Machine Corporation and the Army High Performance Computing Research Center at the University of Minnesota for making their Connection Machines available to us. The first author would also like to thank J. B. Rosen for his support during the course of this work, and Gene Golub and the Stanford University Computer Science department for their hospitality.

Graph	nodes	arcs	size
AC1	512	130816	262144
AC2	1024	523776	1048576
AC3	2048	2096128	4194304

FIGURE 11. Complete acyclic networks.

Graph	a	b	nodes	arcs	size
rmf-L1	6	32	1152	4956	7224
rmf-L2	8	64	4096	18368	26496
rmf-L3	11	128	15488	71687	102542
rmf-L4	16	256	65536	311040	441856

FIGURE 12. Rmf-Long graphs: $a \approx 2^y$, $b \approx 2^{2y}$.

Graph	a	b	nodes	arcs	size
rmf-W1	16	4	1024	4608	6400
rmf-W2	28	5	3920	18256	25312
rmf-W3	49	7	16807	80262	111475
rmf-W4	84	9	63504	307440	427392

FIGURE 13. Rmf-Wide graphs: $a \approx 2^{2y}$, $b \approx 2^y$.

Graph	nodes	arcs	size
BLM1	1026	8071	17168
BLM2	4098	65035	134168
BLM3	16386	522259	1060904
BLM4	65538	4186035	8437608

FIGURE 14. Moderately sparse Basic Line Mesh networks

GR	n-PRC	LBL	k	PR0		PR1		PPR0		PPR1	
				pls	tm	pls	tm	pls	tm	pls	tm
AC1	16K	gap	6	21	0.83	28	1.29	25	0.82	16	1.03
			8	22	0.86	28	1.30	25	0.81	16	1.03
			10	22	0.86	28	1.31	25	0.82	16	1.02
			12	21	0.84	28	1.33	27	0.88	16	1.02
			14	22	0.88	32	1.51	31	0.99	16	1.03
		bfs	1.5	21	1.17	19	1.22	22	1.25	14	1.24
			1.7	16	0.88	19	1.24	25	1.22	14	1.12
			1.9	21	1.05	26	1.52	22	1.15	15	1.18
			2.1	21	1.05	25	1.39	22	1.03	15	1.18
			2.3	21	1.07	26	1.44	22	1.03	16	1.24
			2.5	21	1.07	26	1.44	25	1.10	17	1.30
	32K	gap	6	21	0.46	28	0.71	25	0.46	16	0.58
			8	22	0.48	28	0.72	25	0.46	16	0.57
			10	22	0.48	28	0.72	25	0.46	16	0.57
			12	21	0.47	28	0.74	27	0.49	16	0.57
			14	22	0.49	32	0.83	31	0.56	16	0.58
		bfs	1.5	21	0.67	19	0.69	22	0.73	14	0.71
			1.7	16	0.50	19	0.70	25	0.71	14	0.64
			1.9	21	0.60	26	0.86	22	0.67	15	0.67
			2.1	21	0.60	25	0.78	22	0.59	15	0.67
			2.3	21	0.61	26	0.81	22	0.59	16	0.71
			2.5	21	0.61	26	0.81	25	0.63	17	0.74

FIGURE 15. computational results for AC1 network, $|V| = 512$, $|E| = 130816$.

GR	n-PRC	LBL	k	PR0 pls	PR0 tm	PR1 pls	PR1 tm	PPR0 pls	PPR0 tm	PPR1 pls	PPR1 tm
AC2	16K	gap	6	7	1.44	10	2.13	7	1.26	6	1.85
			8	7	1.45	10	2.13	7	1.26	6	1.81
			10	7	1.45	10	2.13	7	1.26	6	1.81
			12	7	1.45	10	2.13	7	1.26	6	1.81
			14	7	1.45	10	2.13	7	1.26	6	1.81
		bfs	1.5	7	1.97	8	2.34	7	1.78	5	2.14
			1.7	7	1.97	8	2.34	7	1.78	5	2.14
			1.9	7	1.97	8	2.34	7	1.78	5	2.14
			2.1	7	1.95	8	2.34	7	1.78	5	2.14
			2.3	7	1.97	8	2.34	7	1.78	5	2.14
			2.5	7	1.97	8	2.34	7	1.78	5	2.14
	32K	gap	6	7	0.72	10	1.09	7	0.63	6	0.95
			8	7	0.73	10	1.09	7	0.63	6	0.93
			10	7	0.73	10	1.09	7	0.63	6	0.93
			12	7	0.73	10	1.09	7	0.63	6	0.93
			14	7	0.73	10	1.09	7	0.63	6	0.93
		bfs	1.5	7	0.99	8	1.18	7	0.89	5	1.08
			1.7	7	0.99	8	1.18	7	0.89	5	1.08
			1.9	7	0.99	8	1.18	7	0.89	5	1.08
			2.1	7	0.99	8	1.18	7	0.89	5	1.08
			2.3	7	0.99	8	1.18	7	0.89	5	1.08
			2.5	7	0.99	8	1.18	7	0.89	5	1.08

FIGURE 16. computational results for AC2 network, $|V| = 1024$, $|E| = 523776$.

GR	n-PRC	LBL	k	PR0		PR1		PPR0		PPR1	
				pls	tm	pls	tm	pls	tm	pls	tm
AC3	16K	gap	6	8	7.33	11	10.15	9	6.94	7	9.29
			8	8	7.33	11	10.15	9	6.94	7	9.13
			10	8	7.36	11	10.15	9	6.94	7	9.13
			12	8	7.36	11	10.15	9	6.94	7	9.13
			14	8	7.36	11	10.15	9	6.94	7	9.13
		bfs	1.5	8	10.60	10	12.75	9	13.64	6	11.59
			1.7	8	10.60	10	12.75	9	13.64	6	11.59
			1.9	8	10.60	10	12.75	9	10.17	6	11.59
			2.1	8	10.60	10	12.75	9	10.17	6	11.59
			2.3	8	10.60	10	12.75	9	10.17	6	11.59
			2.5	8	10.60	10	12.75	9	10.17	6	11.59
	32K	gap	6	8	3.84	11	5.27	9	3.64	7	4.75
			8	8	3.84	11	5.27	9	3.64	7	4.75
			10	8	3.86	11	5.27	9	3.64	7	4.75
			12	8	3.86	11	5.27	9	3.64	7	4.75
			14	8	3.86	11	5.27	9	3.64	7	4.75
		bfs	1.5	8	5.62	10	6.70	9	7.30	6	6.12
			1.7	8	5.62	10	6.70	9	7.30	6	6.12
			1.9	8	5.62	10	6.70	9	5.40	6	6.12
			2.1	8	5.62	10	6.70	9	5.40	6	6.12
			2.3	8	5.62	10	6.70	9	5.40	6	6.12
			2.5	8	5.62	10	6.70	9	5.40	6	6.12

FIGURE 17. computational results for AC3 network, $|V| = 2048$, $|E| = 2096128$.

GR	n-PRC	LBL	k	PR0		PR1		PPR0		PPR1	
				pls	tm	pls	tm	pls	tm	pls	tm
rmfl1	8K	gap	6	137	1.01	158	1.09	167	1.00	93	1.06
			8	142	0.95	158	1.18	167	0.90	92	1.05
$a=6$			10	146	1.06	158	1.37	167	0.89	101	1.02
$b=32$			12	140	0.93	158	1.08	169	0.99	101	1.02
			14	141	0.93	164	1.23	167	0.88	92	0.93
		bfs	1.5	146	1.16	221	1.89	160	1.30	142	1.69
			1.7	159	1.24	241	1.91	170	1.46	155	1.91
			1.9	173	1.43	262	2.05	186	1.46	169	1.94
			2.1	187	1.41	282	2.18	199	1.51	183	2.16
			2.3	200	1.49	303	2.32	152	1.12	196	2.18
			2.5	214	1.58	324	2.55	158	1.15	210	2.40

FIGURE 18. computational results for RMF-Long1 network, $|V|=1152$, $|E|=4956$.

GR	n-PRC	LBL	k	PR0		PR1		PPR0		PPR1	
				pls	tm	pls	tm	pls	tm	pls	tm
rmfl2	16K	gap	6	484	4.36	552	5.10	628	4.63	344	4.67
			8	482	4.31	534	4.99	621	4.51	333	4.47
$a=8$			10	507	4.49	524	4.77	625	4.49	343	4.53
$b=64$			12	499	4.51	552	4.97	643	4.59	354	4.91
			14	540	4.73	560	5.03	638	4.53	382	4.94
		bfs	1.5	492	5.12	840	8.73	485	5.35	512	7.48
			1.7	569	5.81	889	9.08	609	6.35	586	8.40
			1.9	637	6.43	543	5.58	684	6.91	337	4.85
			2.1	722	7.17	555	5.52	548	5.27	357	5.10
			2.3	792	7.80	595	5.87	599	5.67	384	5.42
			2.5	438	4.26	635	6.23	669	6.15	410	5.73
	32K	gap	6	484	3.16	552	3.79	628	3.34	344	3.51
			8	482	3.14	534	3.64	621	3.26	333	3.36
			10	507	3.29	524	3.56	625	3.26	343	3.43
			12	499	3.24	552	3.72	643	3.33	354	3.51
			14	540	3.48	560	3.76	638	3.29	382	3.76
		bfs	1.5	492	3.84	840	6.60	485	4.09	512	5.78
			1.7	569	4.34	889	6.84	609	4.84	586	6.49
			1.9	637	4.81	543	4.23	684	5.26	337	3.74
			2.1	722	5.36	555	4.16	548	3.99	357	3.92
			2.3	792	5.82	595	4.42	599	4.28	384	4.17
			2.5	438	3.18	635	4.68	669	4.63	410	4.41

FIGURE 19. computational results for RMF-Long2 network, $|V|=4096$, $|E|=18368$.

GR	n-PRC	LBL	k	PR0 pls	PR0 tm	PR1 pls	PR1 tm	PPR0 pls	PPR0 tm	PPR1 pls	PPR1 tm
rmfl3	16K	gap	6	811	21.57	864	23.62	959	19.38	533	21.63
			8	788	20.28	864	22.85	959	18.58	533	20.61
$a = 11$			10	833	20.47	864	22.38	959	18.11	533	20.08
$b = 128$			12	787	18.83	864	22.05	959	17.80	533	19.77
			14	789	18.55	864	21.86	959	17.57	533	19.53
		bfs	1.5	782	21.44	798	23.68	873	22.86	825	34.21
			1.7	923	24.45	798	24.04	870	22.02	552	23.14
			1.9	1052	27.26	836	25.15	864	20.68	565	23.72
			2.1	800	19.73	914	27.07	904	21.67	617	25.48
			2.3	809	20.06	992	28.99	910	21.98	669	27.14
			2.5	823	20.46	1071	30.92	889	19.66	721	28.80
	32K	gap	6	811	12.94	864	13.88	959	11.70	533	12.70
			8	788	12.00	864	13.44	959	11.23	533	12.18
			10	833	12.25	864	13.21	959	10.96	533	11.87
			12	787	11.40	864	13.05	959	10.77	533	11.67
			14	789	11.26	864	12.90	959	10.65	533	11.52
		bfs	1.5	782	13.07	798	14.10	873	14.03	825	20.27
			1.7	923	14.91	798	14.30	870	13.51	552	13.73
			1.9	1052	16.61	836	14.95	864	12.68	565	14.07
			2.1	800	12.07	914	16.08	904	13.28	617	15.11
			2.3	809	12.27	992	17.20	910	13.46	669	16.09
			2.5	823	12.50	1071	18.33	889	12.02	721	17.07

FIGURE 20. computational results for RMF-Long3 network, $|V| = 15488$, $|E| = 71687$.

GR	n-PRC	LBL	k	PR0 pls	PR0 tm	PR1 pls	PR1 tm	PPR0 pls	PPR0 tm	PPR1 pls	PPR1 tm
rmfl4	16K	gap	6	1946	196.11	1882	194.45	2335	174.35	1292	196.57
$a = 16$			8	1853	173.42	1882	186.74	2335	164.91	1292	185.90
$b = 256$			10	1870	167.61	1882	182.21	2335	159.35	1292	179.55
			12	1918	166.44	1882	179.05	2335	155.60	1292	175.24
			14	1876	159.21	1882	176.99	2335	152.95	1292	172.20
		bfs	1.5	1890	174.94	2675	280.78	1974	180.69	1735	249.33
			1.7	2309	206.34	2693	292.01	1992	173.47	1747	260.77
			1.9	2616	230.53	1903	194.17	2034	177.42	1229	175.43
			2.1	2446	212.68	1909	196.82	2083	172.94	1254	179.76
			2.3	2351	202.20	2003	206.25	2082	168.46	1338	190.17
			2.5	2614	222.02	2153	219.81	2164	170.94	1441	202.02
	32K	gap	6	1946	120.64	1882	108.31	2335	102.01	1292	113.43
			8	1853	103.66	1882	102.25	2335	94.50	1292	105.17
			10	1870	98.27	1882	98.68	2335	90.13	1292	99.97
			12	1918	96.14	1882	96.15	2335	87.18	1292	96.60
			14	1876	91.20	1882	94.53	2335	85.02	1292	94.17
		bfs	1.5	1890	92.84	2675	144.66	1947	95.02	1735	129.39
			1.7	2309	109.18	2693	150.62	1992	91.27	1747	135.34
			1.9	2616	122.02	1903	99.96	2034	93.34	1229	90.99
			2.1	2446	112.48	1909	101.39	2083	91.05	1254	93.24
			2.3	2351	107.00	2003	106.28	2082	88.73	1338	98.66
			2.5	2614	117.52	2153	113.28	2164	90.04	1441	104.81

FIGURE 21. computational results for RMF-Long4 network, $|V| = 65536$, $|E| = 311040$.

				PR0		PR1		PPR0		PPR1	
GR	n-PRC	LBL	k	pls	tm	pls	tm	pls	tm	pls	tm
rmfw1	8K	gap	6	862	5.13	812	5.17	1134	5.53	624	5.92
$a = 16$			8	822	4.89	812	5.63	1127	5.43	631	6.09
$b = 4$			10	795	4.80	806	5.13	1127	5.45	634	5.90
			12	806	4.83	812	5.08	1153	5.51	624	5.88
			14	807	4.81	852	5.63	1109	5.46	624	5.85
		bfs	1.5	739	5.47	871	6.52	892	6.81	641	7.11
			1.7	825	6.16	825	5.98	836	6.13	652	7.02
			1.9	729	5.28	942	6.75	926	6.49	653	6.84
			2.1	727	5.13	922	6.62	1042	7.20	589	6.12
			2.3	781	5.55	921	6.56	920	6.07	616	6.41
			2.5	699	4.97	933	6.62	1094	7.11	696	7.08

FIGURE 22. computational results for RMF-Wide1 network, $|V| = 1024$, $|E| = 4608$.

				PR0		PR1		PPR0		PPR1	
GR	n-PRC	LBL	k	pls	tm	pls	tm	pls	tm	pls	tm
rmfw2	8K	gap	6	2613	33.45	2821	38.84	3305	33.72	1770	35.46
$a = 28$			8	2354	29.60	2821	38.39	3318	33.08	1778	35.23
$b = 5$			10	2258	28.21	2821	37.78	3305	32.69	1777	34.89
			12	2401	29.78	2821	37.58	3305	32.23	1770	34.16
			14	2358	29.28	2855	37.92	3318	32.20	1766	33.71
		bfs	1.5	2386	35.37	2557	39.66	2801	42.48	1795	40.04
			1.7	2268	32.39	2428	36.85	2807	40.56	1985	42.91
			1.9	2309	32.45	2708	40.50	2965	41.20	1922	40.29
			2.1	2326	32.32	2559	37.36	2792	37.46	2046	42.53
			2.3	2672	36.54	3048	44.62	2712	35.49	1743	35.63
			2.5	2340	31.47	2872	41.24	3127	39.90	1661	34.98
	16K	gap	6	2613	22.17	2821	24.44	3305	23.07	1770	22.76
			8	2354	19.72	2821	24.13	3318	22.69	1778	22.39
			10	2258	18.80	2821	23.97	3305	22.44	1777	22.12
			12	2401	19.90	2821	23.77	3305	22.23	1770	21.86
			14	2358	19.50	2855	23.93	3318	22.20	1766	21.76
		bfs	1.5	2386	24.29	2557	25.62	2801	29.55	1795	26.05
			1.7	2268	22.51	2428	23.78	2807	28.17	1985	28.14
			1.9	2309	22.49	2708	25.97	2965	28.69	1922	26.43
			2.1	2326	22.32	2559	23.89	2792	26.20	2046	27.71
			2.3	2672	25.30	3048	28.47	2712	24.69	1743	23.37
			2.5	2340	21.79	2872	26.42	3127	27.90	1661	22.17

FIGURE 23. computational results for RMF-Wide2 network, $|V| = 3920$, $|E| = 18256$.

GR	n-PRC	LBL	k	PR0 pls	PR0 tm	PR1 pls	PR1 tm	PPR0 pls	PPR0 tm	PPR1 pls	PPR1 tm
rmfw3	16K	gap	6	10441	254.17	9551	236.69	14240	265.30	7292	270.70
$a=49$			8	10280	236.60	9551	229.20	14240	253.65	7292	259.06
$b=7$			10	10253	228.20	9551	224.71	14240	246.69	7292	252.11
			12	9112	197.76	9551	221.74	14240	242.00	7292	247.44
			14	10651	226.98	9551	219.59	14240	238.67	7292	244.14
	32K	gap	6	10441	150.05	9551	138.90	14240	157.45	7292	157.61
			8	10280	140.96	9551	134.92	14240	151.39	7292	151.46
			10	10253	136.65	9551	132.54	14240	147.76	7292	147.76
			12	9112	119.25	9551	130.95	14240	145.35	7292	145.35
			14	10651	137.12	9551	129.80	14240	143.61	7292	143.59
		bfs	1.5	8361	123.91	9101	141.40	9104	140.75	6550	144.52
			1.7	8527	122.80	10615	160.84	11202	164.36	6920	149.53
			1.9	9216	129.86	10402	155.58	11026	156.19	7670	162.15
			2.1	9164	127.51	9864	144.64	10307	141.42	6650	140.61
			2.3	8885	122.37	10582	154.42	11347	150.77	7009	145.71
			2.5	8446	115.46	10140	147.59	11344	147.34	7096	146.29

FIGURE 24. computational results for RMF-Wide3 network, $|V| = 16807$, $|E| = 80262$.

GR	n-PRC	LBL	k	PR0 pls	PR0 tm	PR1 pls	PR1 tm	PPR0 pls	PPR0 tm	PPR1 pls	PPR1 tm
rmfw4	32K	gap	6	22608	1103.01	22926	1103.61	+	+	+	+
$a=84$			8	23863	1086.00	22926	1057.43	+	+	+	+
$b=9$			10	21753	938.90	22926	1029.70	+	+	+	+
			12	21771	908.77	22926	1011.25	+	+	+	+
			14	22835	933.58	22926	998.03	+	+	+	+
		bfs	1.5	21044	914.96	23099	1088.31	+	+	+	+
			1.7	21142	901.94	21599	1009.38	+	+	+	+
			1.9	20751	870.86	22478	1027.95	+	+	+	+
			2.1	19885	820.40	20746	944.18	+	+	+	+
			2.3	20004	811.68	21837	977.08	+	+	+	+
			2.5	20230	813.74	23194	1029.25	+	+	+	+

FIGURE 25. computational results for RMF-Wide4 network, $|V| = 63504$, $|E| = 307440$. (No experiments were conducted where a + appears.)

GR	n-PRC	LBL	k	PR0		PR1		PPR0		PPR1	
				pls	tm	pls	tm	pls	tm	pls	tm
blm1	8K	gap	6	155	2.08	*	*	190	2.07	104	2.22
			8	155	2.09	*	*	196	2.22	104	2.42
			10	154	2.07	*	*	190	2.15	105	2.31
			12	152	2.09	*	*	195	2.09	104	2.18
			14	146	2.07	*	*	207	2.31	107	2.24
		bfs	1.5	167	2.74	223	4.22	147	2.50	143	3.68
			1.7	182	2.93	245	4.46	126	1.99	158	3.97
			1.9	196	3.12	267	4.79	132	2.05	172	4.24
			2.1	211	3.21	289	5.03	140	2.13	187	4.72
			2.3	226	3.61	311	5.27	148	2.20	202	4.91
			2.5	241	3.80	333	5.70	158	2.40	217	5.10

FIGURE 26. computational results for Basic Line Mesh network BLM1, $|V| = 1026$, $|E| = 8071$. (The program exceeded its imposed time limit wherethere is a *.)

GR	n-PRC	LBL	k	PR0		PR1		PPR0		PPR1	
				pls	tm	pls	tm	pls	tm	pls	tm
blm2	8K	gap	6	90	9.70	*	*	97	8.92	53	9.53
			8	87	9.37	*	*	97	8.83	53	9.37
			10	88	9.49	*	*	97	8.78	53	9.54
			12	91	9.68	*	*	97	8.58	53	9.29
			14	91	9.54	*	*	97	8.64	53	9.54
		bfs	1.5	92	12.29	313	37.96	97	11.43	53	12.14
			1.7	92	12.19	354	41.15	97	11.31	53	12.17
			1.9	92	12.40	395	44.79	97	11.42	53	12.06
			2.1	92	12.30	437	47.82	97	11.34	53	12.04
			2.3	92	12.31	478	50.88	97	11.15	53	12.06
			2.5	92	12.60	519	54.12	97	11.43	53	12.10
	16K	gap	6	90	4.72	*	*	97	4.31	53	4.58
			8	87	4.56	*	*	97	4.26	53	4.54
			10	88	4.60	*	*	97	4.21	53	4.51
			12	91	4.69	*	*	97	4.21	53	4.49
			14	91	4.67	*	*	97	4.18	53	4.47
		bfs	1.5	92	5.92	313	17.91	97	5.36	53	5.64
			1.7	92	5.92	354	19.50	97	5.36	53	5.64
			1.9	92	5.92	395	21.09	97	5.36	53	5.64
			2.1	92	5.92	437	22.72	97	5.37	53	5.64
			2.3	92	5.92	478	24.31	97	5.37	53	5.64
			2.5	92	5.92	519	25.89	97	5.37	53	5.64
	32K	gap	6	90	2.63	*	*	97	2.37	53	2.52
			8	87	2.53	*	*	97	2.34	53	2.48
			10	88	2.54	*	*	97	2.30	53	2.47
			12	91	2.59	*	*	97	2.30	53	2.45
			14	91	2.58	*	*	97	2.28	53	2.43
		bfs	1.5	92	3.24	*	*	97	2.91	53	3.08
			1.7	92	3.24	*	*	97	2.91	53	3.08
			1.9	92	3.24	*	*	97	2.91	53	3.08
			2.1	92	3.24	*	*	97	2.91	53	3.08
			2.3	92	3.24	*	*	97	2.91	53	3.08
			2.5	92	3.24	*	*	97	2.91	53	3.08

FIGURE 27. computational results for Basic Line Mesh network BLM2, $|V| = 4098$, $|E| = 65035$.

GR	n-PRC	LBL	k	PR0		PR1		PPR0		PPR1	
				pls	tm	pls	tm	pls	tm	pls	tm
blm3	16K	gap	6	168	69.38	*	*	172	61.46	92	64.72
			8	166	68.64	*	*	172	60.86	92	64.29
			10	168	69.06	*	*	172	60.55	92	63.82
			12	164	67.80	*	*	172	60.40	92	63.52
			14	166	68.52	*	*	172	60.25	92	63.37
		bfs	1.5	165	89.01	601	272.47	172	80.46	91	83.25
			1.7	165	89.01	680	295.52	172	80.46	91	83.25
			1.9	165	89.01	760	318.87	172	80.46	91	83.25
			2.1	165	89.01	840	342.22	172	80.46	91	83.25
			2.3	165	89.01	920	365.57	172	80.46	91	83.25
			2.5	165	89.01	999	388.63	172	80.46	91	83.25
	32K	gap	6	168	36.57	*	*	172	31.96	92	33.54
			8	166	35.98	*	*	172	31.53	92	33.11
			10	168	36.08	*	*	172	31.32	92	32.90
			12	164	35.34	*	*	172	31.22	92	32.69
			14	166	35.67	*	*	172	31.11	92	32.58
		bfs	1.5	165	45.93	*	*	172	41.32	91	42.63
			1.7	165	45.93	*	*	172	41.32	91	42.63
			1.9	165	45.93	*	*	172	41.32	91	42.63
			2.1	165	45.93	*	*	172	41.32	91	42.63
			2.3	165	45.93	*	*	172	41.32	91	42.63
			2.5	165	45.93	*	*	172	41.32	91	42.63

FIGURE 28. computational results for Basic Line Mesh network BLM3, $|V| = 16386$, $|E| = 522259$.

GR	n-PRC	LBL	k	PR0		PR1		PPR0		PPR1	
				pls	tm	pls	tm	pls	tm	pls	tm
blm4	32K	gap	6	301	515.07	*	*	312	462.82	166	482.97
			8	303	516.02	*	*	312	458.43	166	478.58
			10	299	510.52	*	*	312	455.92	166	476.07
			12	300	510.66	*	*	312	454.66	166	474.18
			14	300	510.34	*	*	312	453.40	166	472.93
		bfs	1.5	298	675.46	*	*	312	616.96	158	622.07
			1.7	298	675.46	*	*	312	616.96	158	622.07
			1.9	298	675.46	*	*	312	616.96	158	622.07
			2.1	298	675.46	*	*	312	616.96	158	622.07
			2.3	298	675.46	*	*	312	616.96	158	622.07
			2.5	298	675.46	*	*	312	616.96	158	622.07

FIGURE 29. computational results for Basic Line Mesh network BLM4, $|V| = 65538$, $|E| = 4186035$.

Network	size (× 1K)	PR0		PR1		PP R0		PPR1	
		gap	BFS	gap	BFS	gap	BFS	gap	BFS
AC1	262.0	3.0	3.6	3.8	3.6	4.0	3.0	3.5	3.9
AC2	1048.0	4.8	4.8	4.7	4.4	4.8	4.4	5.1	4.7
AC3	4194.0	5.6	5.6	5.2	5.3	5.5	5.5	5.9	6.1
rmfl1	7.2	0.4	0.4	0.3	0.2	0.4	0.4	0.3	0.2
rmfl2	26.0	0.8	0.7	0.7	0.6	0.8	0.8	0.7	0.7
rmfl3	102.0	2.5	3.0	2.0	3.0	2.4	2.5	2.3	2.1
rmfl4	441.0	7.5	6.7	6.2	6.7	7.2	7.3	6.3	6.74
rmfw1	6.4	0.1	10.3	0.1	0.1	0.1	0.1	0.1	0.1
rmfw2	25.0	0.3	0.2	0.2	0.2	0.2	0.2	0.2	0.2
rmfw3	111.0	0.4	0.4	0.4	0.43	0.3	0.4	0.3	0.4
rmfw4	427.0	1.1	1.2	0.9	1.0	+	+	+	+
BLM1	17.0	0.7	0.5	*	0.5	0.6	0.7	0.6	0.5
BLM2	134.0	3.8	3.3	*	0.9	4.1	4.1	4.2	4.1
BLM3	1060.0	18.0	14.6	*	3.8	20.1	20.1	20.3	20.4
BLM4	84.0	78.0	59.5	*	+	85.7	85.7	88.1	92.4

FIGURE 30. The average number of active nodes and arcs in units of 1K.

REFERENCES

1. R. K. Ahuja and J. B. Orlin. Personal communication. 1987.
2. R. J. Anderson and J. C. Setubal. On the Parallel Implementation of Goldberg's Maximum Flow Algorithm. In *4th Annual ACM Symp. on Parallel Algo. and Arch.*, pages 168–177. ACM, 1992.
3. R. J. Anderson and J. C. Setubal. Goldberg's Algorithm for the Maximum Flow in Prespective: a Computational Study. In D. S. Johnson and C. C. McGeoch, editors, *DIMACS Implementation Challenge Workshop: Algorithms for Network Flows and Matching*. AMS and ACM, to appear.
4. G. Blelloch. Parallel Prefix vs. Concurrent Memory Access. Technical report, Thinking Machines, Inc., 1986.
5. B. V. Cherkassky. Personal communication. 1991.
6. Thinking Machine Corporation. *Connection Machine Model CM2 Technical Summary*. 1990.
7. U. Derigs and W. Meier. Implementing Goldberg's Max-Flow Algorithm — A Computational Investigation. *ZOR — Methods and Models of Operations Research*, 33:383–403, 1989.
8. E. A. Dinic. Algorithm for Solution of a Problem of Maximum Flow in Networks with Power Estimation. *Soviet Math. Dokl.*, 11:1277–1280, 1970.
9. A. V. Goldberg. A New Max-Flow Algorithm. Technical Report MIT/LCS/TM-291, Laboratory for Computer Science, M.I.T., 1985.
10. A. V. Goldberg. *Efficient Graph Algorithms for Sequential and Parallel Computers*. PhD thesis, M.I.T., January 1987. (Also available as Technical Report TR-374, Lab. for Computer Science, M.I.T., 1987).
11. A. V. Goldberg. Processor-Efficient Implementation of a Maximum Flow Algorithm. *Information Processing Let.*, 38:179–185, 1991.
12. A. V. Goldberg, É. Tardos, and R. E. Tarjan. Network Flow Algorithms. In B. Korte, L. Lovász, H. J. Prömel, and A. Schrijver, editors, *Flows, Paths, and VLSI Layout*, pages 101–164. Springer Verlag, 1990.
13. A. V. Goldberg and R. E. Tarjan. A New Approach to the Maximum Flow Problem. *J. Assoc. Comput. Mach.*, 35:921–940, 1988.

14. L. M. Goldschlager, R. A. Shaw, and J. Staples. The Maximum Flow Problem is Log Space Complete for P. *Theoretical Computer Sci.*, 21:105–111, 1982.
15. M. D. Grigoriadis. Personal communication. 1988.
16. W. D. Hillis. *The Connection Machine.* MIT Press, 1985.
17. A. V. Karzanov. Determining the Maximal Flow in a Network by the Method of Preflows. *Soviet Math. Dok.*, 15:434–437, 1974.
18. R. E. Ladner and M. J. Fischer. Parallel Prefix Computation. *J. Assoc. Comput. Mach.*, 27:831–838, 1980.
19. C. E. Leiserson and B. M. Maggs. Communication-Efficient Parallel Graph Algorithms. In *Proc. of International Conference on Parallel Processing*, pages 861–868, 1986.
20. X. Li and S. A. Zenios. Data-level Parallel Solution of Min-Cost Network Flow Problems Using ϵ-Relaxations. Technical Report 91-05-04, Decision Sciences Department, The Wharton School, University of Pennsylvania, Philadelphia, PA 19104, 1991.
21. Q. C. Nguyen and V. Venkateswaran. Implementations of Goldberg-Tarjan Maximum Flow Algorithm. In D. S. Johnson and C. C. McGeoch, editors, *DIMACS Implementation Challenge Workshop: Algorithms for Network Flows and Matching.* AMS and ACM, to appear.
22. J. T. Schwartz. Ultracomputers. *ACM Trans. Prog. Lang. and Syst.*, 2:484–521, 1980.
23. Y. Shiloach and U. Vishkin. An $O(n^2 \log n)$ Parallel Max-Flow Algorithm. *J. Algorithms*, 3:128–146, 1982.

ANDREW V. GOLDBERG, COMPUTER SCIENCE DEPARTMENT, STANFORD UNIVERSITY, STANFORD, CA 94305.
E-mail address: goldberg@cs.stanford.edu

FARID ALIZADEH, INTERNATIONAL COMPUTER SCIENCE INSTITUTE, BERKELEY, CA 94706
E-mail address: alizadeh@icsi.berkeley.edu

A Case Study in Algorithm Animation: Maximum Flow Algorithms

GREGORY E. SHANNON, JOHN MACCUISH,

AND ELIZABETH JOHNSON

December 1, 1992

ABSTRACT. This paper reports our efforts to capitalize on current software and hardware capabilities for developing algorithm animations and using them in algorithms research. Our animations make a number of previously claimed insights about Goldberg push-relabel maximum flow algorithms readily apparent. A previously unclaimed observation that we discovered using our animations is that the final flow values on some edges in push-relabel algorithms can be a factor of $O(n)$ larger than the initial saturating flow from the source. We also found that the gap relabeling strategy appears to performs better (in terms of pushes and relabels) than the global relabeling strategy on long, narrow networks, while the global relabeling strategy does better than the gap on short, wide networks. Overall, we found that there are critical needs for integrated software and animation design methods. Though sophisticated animations are still difficult to create and use, our work corroborates assertions that algorithm animation has broad potential to facilitate algorithm research and design.

1. Introduction

This paper reports on our efforts to design and use algorithm animations in the analysis of maximum flow algorithms. While we have the ambitious goal of using animations to identify behavioral properties of state-of-the-art maximum flow algorithms, and thereby develop more efficient algorithms, most of our efforts to date have been spent on designing and implementing the animations. In

---

1991 *Mathematics Subject Classification.* Primary 68Q25; Secondary 68R10, 68Q20, 05-04, 05C85.

This research was supported in part by NSF Grants #CCR 89–09535 and #CCR 92–01005.

Technical support was provided by the Center for Innovative Computer Applications at Indiana University.

This paper is in final form and no version of it will be submitted for publication elsewhere.

particular, we have focused on animations of various maximum flows algorithms of the push-relabel type – also known as algorithms using the Goldberg-Tarjan method [16]. Though limited analyses of maximum-flow algorithms through animation have been possible thus far, we have been able to make interesting observations. Therefore, this paper is mainly a case study on designing and implementing animations for push-relabel type maximum-flow algorithms and a collection of our observations that are relevant to research on maximum flow algorithms.

Algorithm animation and visualization[1] provide two basic functions. The first is to display literal events and properties of an algorithm as a shorthand for their rapid viewing and assimilation. This aspect is often very beneficial for verifying an algorithm's correctness and providing greater intuitive understanding of the overall workings of the algorithm. The second function is to enable the researcher to deduce relationships, correlations, and patterns from time-sequenced images of basic events. In this way, the researcher can generate ideas about an algorithm's performance, new algorithms, or more interesting data sets on which to animate the algorithm. Such inspiration is difficult to derive from numerical and textual traces of a program that implements a particular algorithm and is run on various large data sets.

In designing an animation, two aspects must be addressed. First, the salient events in an algorithm and properties of the input data must be identified. In a graph algorithm, properties of the input data might include the vertex interconnections and their weights, while events in an algorithm might include specific edge traversal or the building of a spanning tree. Second, the ultimate use of the animation must be considered. While it would appear that the algorithm and data are the primary influence on the animation design, we found that how it is used is equally important. There are often subtle, at times major, differences in effective animations for different uses.

In pedagogical animations, minute details of the algorithm are portrayed so that the novice viewer can gain understanding of the algorithmic process. Since these details usually are trivial to the experienced researcher, our interest is in examining the behavior of algorithms at a higher level. We concentrated on developing animations for exploratory use. In these animations, the viewer wants to examine algorithm behavior on a variety of data sets, typically large data sets or pathological cases. Only those details of interest need to be portrayed. Such features can often be displayed more abstractly than in an animation created for pedagogical purposes. In addition to information about individual operations, summary data for a particular data set may also be needed so that the user can quickly compare performance between data sets. These qualities, abstraction and summary, are essential for using animations in experimental research.

For our animations, we developed 3-dimensional graph-based interpretations

---

[1]Visualization refers to any graphical interpretation and includes static images of algorithms.

of various push-relabel maximum flow algorithms [10, 16]. Vertices, edges, flow-value information, and algorithm graph operations are directly or indirectly displayed. We used several software packages to create and instrument the algorithms and implement the animations, including NETPAD [21], SchemeGraphs [4, 5, 25], and AVS [29]. We relied heavily on AVS's ability to provide depth cues in the animations through perspective projections, motion parallax, and directional lighting. We implemented more than two dozen different push-relabel algorithms (from a generic implementation to several periodic relabeling strategies, using various vertex selection procedures). Each implementation was instrumented to generate trace data for building animations.

Our basic approach in this project differed from other participants in the DIMACS Challenge. Rather than implementing algorithms with the aim of testing their run-time efficiency, we set out to implement and animate algorithms in order to obtain visual images of how they operated on various data sets. We did compare combinatorial complexity across algorithms and data sets, but this mainly served to identify potentially interesting data and algorithm variations for the animations.

Our initial work has lead to three observations about push-relabel type algorithms. First, animations of the generic push-relabel algorithm show that local and global bottlenecks and cycles are a major reason for poor performance. Observations about the location of excess flow with respect to the cut make it clear that periodic-relabeling strategies (such as global [16] and gap relabeling [11]) significantly improve performance. Second, animation of the gap and global relabelings suggest that gap works well (with respect to the number of pushes and relabels) in long, narrow networks and global relabeling works better in short, wide networks. Third, animations for a variety of algorithms using a GENRMF network showed that a cycle can occur in the final flow network where the assigned values of the flow on individual edges in the cycle may be more than that of the maximum-flow value. Indeed, the assigned flows on the cycle edges can be a factor of $O(n)$ larger than the sum of the edge capacities at the source (or in the minimum cut). These cycles can occur on either side of the minimum cut. While these cycles do not contradict the theory, their occurrence is non-intuitive and may reflect computational inefficiency in the current algorithms. Only in [15] do Goldberg and Tarjan mention that cycles may appear, and they only discuss cycles with respect to intermediate flows. They discuss neither the possible flow magnitudes nor cycles in the final flow network.

We are not aware of any previous work on animating maximum flow algorithms. There has been previous work in using animations to analyze sorting algorithms [7, 2] and more recently to understand computational geometry algorithms [7, 24]. Most of the work in algorithm animation has focused on building animation development environments that are easy to use and efficient [28, 6, 8, 3]. This is because until reasonable tools are available, generating sophisticated animations to help with research is quite difficult. Our work demonstrates, however, that technology for animation has reached the point where non-trivial animations of discrete algorithms can be used in algorithms research.

In the next section, we provide some preliminaries for the maximum flow problem. In § 3, we discuss our approach to designing animations. In § 4, we describe the development process. In § 5, we explain how we used the animations and with what research results. In § 6, we present our conclusions and outline future work.

2. Problem and Algorithm Descriptions

The maximum flow problem can be defined informally as follows: given a network, with n vertices, m edges with real-valued capacities (c), one vertex designated as the source (s), and one as the sink (t), compute the maximum amount of flow that can be transported from the source to the sink. The flow is a function mapping a real number onto each edge. An edge's residual capacity (r) is that edge's capacity minus the current (or pre-) flow (f) on that edge, $r(u,v) = c(u,v) - f(u,v)$. The flow or preflow in a network induces a residual network consisting of edges with residual capacities. We define an edge in the initial network as an *original edge*. An edge in the residual network which is not an original edge is a *virtual edge*.

Flow in a network must meet certain constraints:

- Flow on a particular edge must be no greater than each edge's capacity.

$$f(u,v) \leq c(u,v) \qquad \text{capacity constraint}$$

- Total flow entering a vertex u, other than s or t, must equal total flow leaving the vertex.

$$\sum_v f(u,v) = \sum_v f(v,u) \qquad \text{flow conservation constraint}$$

- Flow on edge (u,v) must be equal to the negative of the flow on edge (v,u).

$$f(u,v) = -f(v,u) \qquad \text{antisymmetry constraint}$$

When a maximum flow algorithm ends, a minimum cut can be found which partitions the vertices into two sets in the network. The source is in one set; the sink is in the other. Vertices in the source set have no paths to the sink in the residual network. The maximum flow value is the sum of the flow on the edges that comprise the cut, i.e. the edges (u,v) such that u is in the source set and v is in the sink set.

Push-relabel algorithms compute maximum flow by incrementally pushing flow across single edges based on vertex distance labels, vertex excess capacities, and residual edge capacities. For these algorithms to work, the flow conservation constraint is relaxed for intermediate flows (also known as preflows).

$$\sum_v f(u,v) \leq \sum_v f(v,u) \qquad \textit{intermediate flow conservation constraint}$$

However, at the end of the algorithm, the original flow conservation constraint must be satisfied by all vertices. This idea is not unique to the push-relabel method; Karzanov first used this concept in a version of the Dinics algorithm [19]. *Active vertices* have excess flow, i.e. the total flow on the vertex's incoming edges can be greater than the total flow on its outgoing edges. The vertex distance labels, $d(u)$, correspond to a lower bound on the distance from the vertex u to the sink in the residual network (or to the source if no path exists from the vertex to the sink). *Admissible edges* are edges (u,v) for which $r(u,v) > 0$ and $d(u) = d(v) + 1$.

A *push* operation is used to move flow along edges. During this operation, all or part of the excess flow from one vertex is transferred to an adjacent vertex along an admissible edge. If a vertex u has excess flow, but none of its edges are admissible, the vertex is relabeled. The new label $d(u)$ is equal to $min(d(v))+1$, over all vertices v such that edge (u,v) is in the residual network. Applying push or relabel operations to a vertex until the vertex has no excess is known as *discharging* the vertex.

The push-relabel algorithm can be understood and implemented as a two-stage process. The algorithm begins by saturating all of the outgoing edges from the source. This puts a large amount of flow into the network. In the first stage, this flow is pushed towards the sink. Eventually the minimum cut is found and no more flow can be pushed to the sink. This marks the end of the first stage. In the second stage, excess flow on the source side of the minimum cut must be pushed back to the source. Depending on the implementation, these phases might not be distinct sequentially, i.e. flow may be pushed back to the source before all possible flow has been pushed to the sink.

Basic variations for the push-relabel algorithms involve varying strategies for edge and vertex selection, and initial vertex labeling. There are two common choices for initial labeling: naive and exact. In naive labeling, the source is given label n, and all other vertices are given label 0. In exact labeling, the source is also labeled n, but other vertices are labeled with their distance from the sink, determined in a backwards breadth-first search. Edge selection determines which admissible edge will be selected for the next push operation from the current vertex. Edge selection strategies include choosing the edge with highest or lowest residual capacity.

Vertex selection strategies determine the order in which vertices will be discharged. The generic push-relabel algorithm does not specify which vertex will be relabeled or pushed next, but arbitrarily chooses one from the set of active vertices. We implemented the basic Goldberg push-relabel algorithm with four different vertex selection schemes: 1) first-in, first-out (FIFO/queue), 2) last-in, first-out (LIFO/stack), 3) maximum-excess vertex, and 4) highest-label vertex [15, 16]. The best asymptotic complexity among the variations we implemented is $O(n^2\sqrt{m})$ for the highest label algorithm [9].

Other variations involve some type of periodic relabeling of some or all vertices. This relabeling is in addition to the local relabeling operation described above. In *global relabeling* algorithms, after every k discharge operations, breadth-first searches from the source and the sink in the residual network are performed. All vertices (except the source and sink) are relabeled simultaneously with the minimum of the distance from the sink and $n+$ the distance from the source, determined during the breadth-first searches. This has the effect of resetting the vertex labels to the exact distance to the source ($+ n$) or sink. The constant k is a parameter, often related to the number of discharge operations performed. The only restriction imposed is that it must have no effect on the asymptotic complexity of the algorithm. In our work, we used $k = m/2$, based on a suggestion in [1]. Goldberg and Tarjan describe several other possibilities, including after every n local relabels, or just after an edge to the sink becomes saturated or all flow on a source edge has been pushed back [16].

Another type of periodic relabeling is *gap relabeling*, as described in [11]. Under this strategy, the number of vertices in the network with each label is maintained. Whenever a relabel is done on the only remaining vertex with label i, all vertices v with $i < d(v) < n$ are relabeled to n. This is done because the vertices being relabeled no longer have a path to the sink in the residual network. Relabeling them immediately upon discovering the gap reduces the amount of work needed to push the excess flow back to the source in the second phase. We also divided the work in the implementation into two distinct phases so that only vertices with height less than n were discharged during the first phase. At the end of the first phase, the minimum cut has been found. At the beginning of the second phase, a backwards breadth-first search is done from the source to give the vertices exact distance labels. Processing continues normally until all excess flow has been pushed back to the source.

Vertex selection techniques have been shown to affect the asymptotic complexity of the algorithms, but periodic relabeling and the choice of initial labeling have no known asymptotic effect. Nevertheless, these methods appear to affect the run-time efficiency of these algorithms [1, 11]. Obviously, additional operations and data structures are needed to maintain information for the relabeling, but these extra costs do not appear to have a significant impact on the overall efficiency of the algorithms.

3. Designing Animations

In designing animations, the first task is to identify important characteristics of the data and events in the algorithm. In the maximum flow problem, interesting data characteristics include the specific edges between vertices, edge capacities, and edge flow values. Significant events in push-relabel algorithms include individual pushes and relabels.

The relevant characteristics and events are further defined by considering how the resulting animations will be used. Pedagogical animations include details about the algorithm presented in a step-by-step fashion so that the viewer can grasp the processing and results of the algorithm. Exploratory animations may skip minute details in favor of a more abstract view of the algorithm which focuses on the results of each operation rather than the operations themselves. A global view of the data is provided so that patterns within and between data sets can be observed. When exploratory animations are created for use as tools in experimental analysis, they must also provide the ability to examine large data sets and to compare results over many data sets.

Another issue in animation design is that of maximizing information presented while not overloading the viewer's visual comprehension abilities. Abstraction aids in this task, but care must be taken to create feature representations which are intuitive, correct, and provide the information needed.

3.1. Push-relabel animations. A natural intuition behind push-relabel algorithms is to equate the vertex labels with heights, and to imagine that flow moves downhill from vertices with excess flow [10, 22]. In conjunction with this view, a local relabel can be thought of as a *lift* of the vertex in 3D-space. In order to design a 3D animation of this intuitive model, one must start with a convenient 2D initial embedding of the network that will remain visually informative as it changes to 3D. Vertices, conventionally represented as small circles in 2D, can be represented as small spheres. Showing the properties of vertices (are they active vertices? can they push flow?) can be done easily with color and relative size. Edges can retain their 2D line representation.

A more difficult issue is representing properties of the edges. Exact flow capacities and magnitudes are simply too cumbersome (say, with text labels or cylinders with variable diameters for edges) to visualize them all in one 3D image. Also, displaying bi-directional edges with simple straight lines (or arcs) with arrows is largely trivial in 2D, but is much more difficult in 3D for networks of any considerable size or density. For instance, implementing cones for arrows has two drawbacks. The first is that it would slow considerably the running time to generate each frame of the animation. The second is that visual clarity would be difficult to maintain if the networks were larger than just a few vertices and edges. In short, it is best to try to reduce the amount of visual clutter created by transforming a network from 2D to 3D, while striving to maximize, or at least maintain, the amount of information shown. To do this, we found that a degree of abstraction with the goal of simplicity is essential.

As part of the animation design, it was clear that simple, static properties of the network need to be shown. The source and the sink need to be clearly discernible throughout. Since the theoretical bound for the height of any node in the network is $2n-1$, the network should be contained in a cube with $2n$-unit length sides. This aids in comparing the relative heights of the vertices.

For the significant algorithm events that dynamically affect the data to be displayed, we chose pushes, lifts, and global and gap relabelings. For pushes, we distinguish between pushes that saturate edges and those that do not. Active vertices are distinguished by having at least one admissible edge (overwhich flow can be correctly pushed) or none. The pushing vertex may change to a non-active vertex, and the receiving vertex may change to an active vertex. A lift changes an active vertex that cannot push flow to one that can, and also changes the admissible network. Global and gap relabeling (and exact relabeling performed at the beginning of the second phase) can change some of the active vertices with no pushable edges to ones with pushable edges. As a result the admissible network can also change. The admissible network reveals where there is potential for the flow to be pushed in the residual network. Saturated edges that remain in the network help to define cuts, and eventually, show the minimum cut. Unsaturated edges that remain in the network over a number of states, can be used to distinguish current edges and maximal excess vertices.

With these events and properties displayed during the animation, we could begin to analyze such algorithm variations as vertex and edge selection techniques, and global and gap relabeling. We realized that we could not reasonably show more in a single view and maintain scalability to larger networks. Since much of the pertinent information is contained in the states of the residual network, we chose to display the vertices and edges so that the structure of the residual network was readily apparent. However, distinguishing an original edge from a virtual edge in the residual network is difficult if some properties have priority over others and edge directions are not specified. If the direction of edges is shown, and original and virtual edges are distinguished in an abstract fashion (not necessarily lines and arrows), the conflicting priorities largely disappear. Though we obtain an increase in the amount of information that the animation can show us, it is at the cost of cluttering the intuitive visual model.

4. Developing the Animations

Because visualization is a new approach in the analysis of network flow algorithms, much of our time was devoted to developing the tools necessary for our research. As described in this section, we now have a useful set of tools for continued work.

4.1. Software. In order to produce network flow animations, we first needed to implement algorithms and produce trace data to drive our animations. We utilized three graph software systems: GraphView and SchemeGraphs by Shannon *et. al* [**25, 4, 5**], NETPAD by Dean *et. al* [**21**], and Combinatorica by Skiena [**30, 26**]. While each software system can be used for the above tasks, they differ in their approaches and capabilities for graph representation, manipulation, and animation.

All three graph manipulation systems have the ability to display graphs, but only NETPAD and GraphView allow users to create and manipulate displayed graphs using mouse-driven commands. Graphs can be generated using either programs from the built-in libraries (NETPAD and Combinatorica) or user-written programs (all three). In NETPAD, these programs are written off-line in C. Programs to access GraphView graphs are written in Chez Scheme, a Lisp-like interactive functional language. Combinatorica uses the interactive Mathematica programming language. All three systems also have explicit data structures for graphs. This simplified the task of implementing algorithms from pseudocode in the literature.

In order to produce animations with minimal development time, we also needed to utilize animation software. NETPAD and Mathematica have simple animation capabilities built in and GraphView can use animation services available with the NeXT computer. However, none of these animation capabilities proved adequate for our needs. They lacked the flexibility and sophistication that our applications required.

We did produce prototype animations of the generic push-relabel algorithm using Mathematica. Unfortunately, to create even a small network and to change it for each cell was laboriously slow. In addition to this, Mathematica (v1.4) has a problem maintaining consistent 3D viewpoints and scalings. Based on our experiences with Mathematica, we realized that we needed to look for software designed specifically for visualization. We settled the Application Visualization System (AVS) system by Stardent Corporation [**29**].

AVS is a visualization system with sophisticated animation capabilities. The system provides the ability to animate 3D objects and to rotate, scale, and translate them (in either orthographic or perspective projections) so that different parts and views of the objects can be studied. AVS also incorporates the ability to implement graph properties as colors and has real-time interactive facilities so that 3D transformations of the network can be done interactively during the animation by using either a dial box or a mouse. In addition to the depth cues provided by perspective projections, AVS enables us to provide additional cues with motion parallax (using the dialbox or mouse) and directional lighting. Another feature which aided us in our experiments was the ability to display multiple animations on the screen running concurrently. This allowed us to compare the performance of various algorithms on the same data set.

4.2. Hardware. We used three platforms to carry out our research. NETPAD was run on the Sun SPARCstation IPC. SchemeGraphs, GraphLab and Mathematica/Combinatorica were all run on a NeXT Cube with a 68040 board. The Stardent Titan P3000 with a G3 graphics board provided the AVS platform.

4.3. Tool development. In this project, three types of tools were necessary: graph generators/embedders, algorithm implementations, and animations. In this section, we describe the development process, difficulties we encountered, and the tools themselves.

4.3.1. *Data generators/embedders.* In our work for this paper, we used graphs generated by the programs described in the Benchmark Experiments document [18]. The DIMACS data generators provide network characteristics, but do not include embedding information. We required an embedding in order to produce the network objects for our animations, so graph embedders were another area of software development. We needed embeddings that would aid, or at least not hinder, the visual intuitions provided by the animations.

The GENRMF graphs had a natural embedding that consisted of a series of square frames connected by low capacity edges. We wrote a program using SchemeGraphs that created this embedding from the DIMACS data. We also wrote software to create a layered embedding from a DIMACS data file and used this to embed Washington Line graphs. The layered embedding is the most desirable in terms of the intuitions behind the push-relabel algorithm, but not all graphs can be effectively embedded in layers. In particular, the acyclic dense graphs were converted to NETPAD format and embedded using a built-in circle embedder in NETPAD. The latter embedding was not as satisfactory as layered embeddings, but the density of acyclic dense graphs made other approaches largely useless.

4.3.2. *Algorithm implementations.* In order to generate data for the animations, we first had to implement the algorithms. Issues in this process include the level of efficiency desired, choice of the algorithms to be implemented, and whether animations would utilize trace data or be displayed as the algorithms run.

We were not particularly concerned with run-time efficiency in this project, but our experiences do raise questions about using the current software for more traditional experimental work. We found that while the use of the data structures from NETPAD and SchemeGraphs sped up the implementation and debugging process, they significantly slowed down the program execution time. We used NETPAD for the push-relabel algorithms. This choice was initially a matter of programmer preference, but the processing of large networks for the push-relabel animations made NETPAD's comparative efficiency a necessity.

In choosing the algorithms to implement, we narrowed our efforts to the best-known variations. We implemented the generic push-relabel algorithm (vertices are selected in an arbitrary, but fixed, order). In addition, we implemented the following vertex selection variations of push-relabel algorithms: maximum-excess vertex, FIFO, LIFO, and highest-label vertex. There are three versions of each push-relabel variation. One has no periodic relabeling, one has gap relabeling, and one has global relabeling. All 12 of the push-relabel variations utilize exact

	priority	b&w	color	
vertices		gray	gray	without excess – not active
		dark	blue	active without admissible edges
		dark	green	active with admissible edges
edges	3	white	white	no flow
	2	light	green	admissible
	2	light	blue	an unsaturated push
	1	dark	red	saturated

TABLE 1. Key for residual network properties.

labeling at the beginning of both the first and second phases. Though we implemented several variations of edge selection strategies, we did not explore them using animations.

In order to provide the data for the animations, our approach was to generate trace data during algorithm execution which could be saved to a file and read by the animation program. This resulted in significantly faster animations and made debugging the algorithm implementation and animation separate and more manageable.

4.3.3. *Animations.* Our push-relabel AVS animation begins with a view of the entire graph, seen from above in its original embedding. This 2D view can be changed to 3D and back again interactively at any point during the animation. Initially, the vertices are gray and the edges white. If exact labeling is used in determining the initial label, or height, of each vertex, all the vertices except the sink are raised above the base 2D plane. The animation then shows the initialization process of the algorithm. The source rises to the height n, and the edges from the source turn red, indicating that they have been saturated. The vertices adjacent to the source turn green and enlarge, signifying that they have excess flow and also have admissible edges.

While the animation of the algorithm is proceeding, the vertices and residual edges have properties that are represented by color, as outlined in Table 1. All active vertices, in addition to being colored, are enlarged slightly to help to highlight them in large networks. During each operation that we chose to animate, properties of the vertices and edges change accordingly, as described in Table 2.

Figures 1 and 2 are black and white stills of an animation. To a limited extent, they show how information about the state and operation of an algorithm can be conveyed visually. Table 1 shows the corresponding black and white key values for the two figures. Unfortunately, because of various difficulties, these stills barely reflect the detail and clarity of the information presented in the animations. In particular, these static pictures cannot portray the ability to interactively create motion parallax with the Ardent's mouse or dial-board. Such parallax significantly amplifies the perspective and lighting depth cues used.

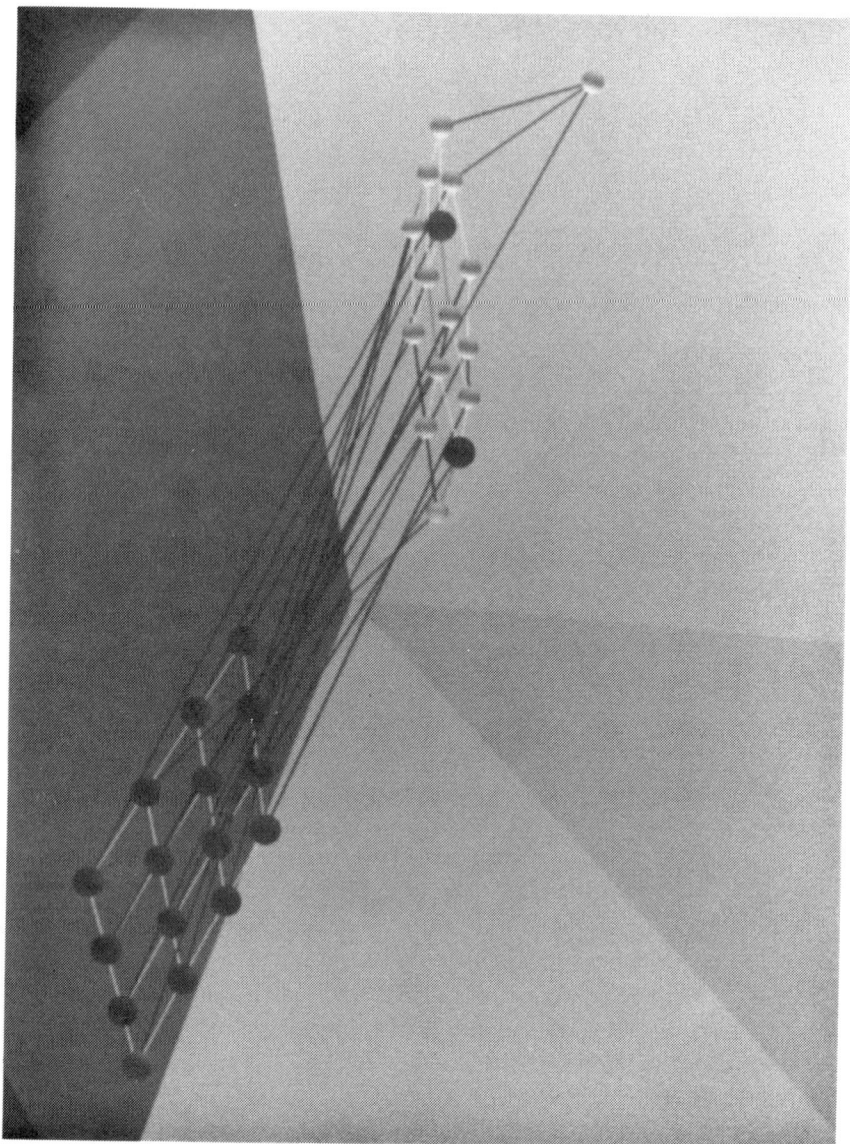

FIGURE 1. Above is a GENRMF 4X4X2 network (rotated 90 degrees) about half way through a push-relabel algorithm without periodic relabeling. The source is raised at the left and the sink is the rightmost sphere. The dark, enlarged spheres are active vertices, and the dark lines are the red, saturated edges. The light lines are the green, admissible edges or the blue, nonsaturated edges.

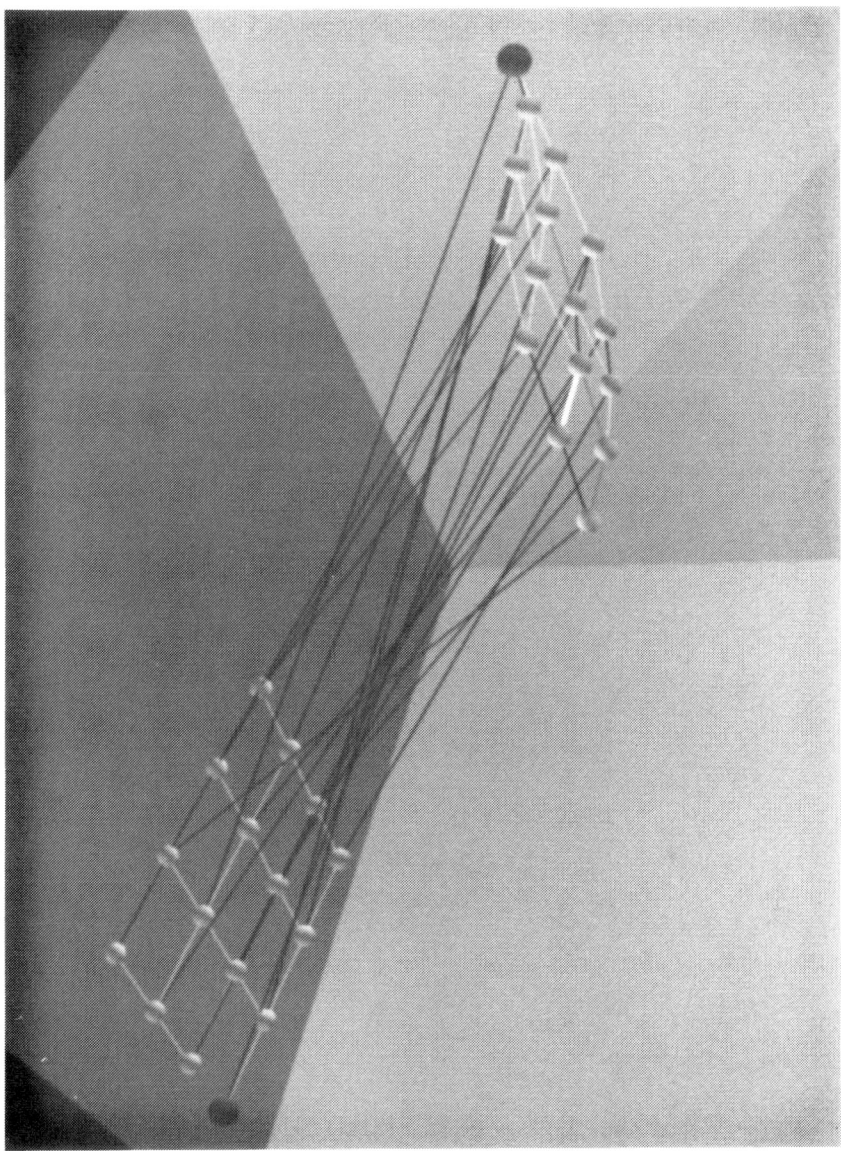

FIGURE 2. In the above picture (rotated 90 degrees), the animation and algorithm from Figure 1 has completed, and the source and sink vertices are the only vertices with excess flow. All the vertices in the frame that includes the source are raised above the source. All the edges between the two frames are saturated, thus denoting the cut. The one white edge is the only edge with a flow value of zero.

Push	vertex	Remains green if still active
		Otherwise, turns gray
	edge pushed	Red, if saturating push
		Blue, if unsaturating push
Lift	vertex u	Raises and turns green
	adjacent edges, (u,v)	If white or blue, and $d(u) = d(v) + 1$, turns green
Global relabeling	vertices	Heights of all vertices other than source and sink readjusted
		Some active vertices may change colors (blue becomes green)
	edges	If white or blue, and $d(u) = d(v) + 1$, turns green
Gap relabeling	vertices	Vertices above the gap, but below height n are readjusted to n
		Some active vertices may change colors (green becomes blue)
	edges	If white or blue, and $d(u) = d(v) + 1$, turns green

TABLE 2. Key for animations of algorithm events.

Because edge directions are not shown, a single line represents edges (u,v) and (v,u). Our primary solution prioritizes the edge properties in order to determine which color each edge should have at each step in the animation. For example, if a non-saturating push is done on edge (u,v), the line turns blue. If the edge (v,u) subsequently becomes admissible due to a relabel of v, the line becomes green and the information (blueness) about the flow on edge (u,v) is no longer displayed. If (v,u) is saturated and (u,v) becomes admissible, then the undirected edge would be colored green. The color/property priorities for the display are then (1) green/admissible, (2) blue/pushed and red/saturated (since (u,v) colored blue and (v,u) colored red is not possible), and (3) white. Though our displays do show a great deal of information, much is left out. For instance, they clearly do not show all the information about the residual network.

A second solution we implemented for dealing with edge directions consists of positioning a very small sphere along the line that represents an original edge, (u,v). This sphere is placed close to the vertex u and marks the existence of a virtual edge, (v,u), that belongs to the residual network. The spheres are colored using the same color legend as the original edges. This solution however still has several drawbacks. First, it does not reveal the original network if there are bi-directional edges in the network. This problem could be overcome by engineering the animation so that those bi-directional edges would be represented by small cubes, thus distinguishing them from the virtual edges. Second, since the residual network contains both original and virtual edges, the visual makeup of the network edges consists of green lines and small green spheres. The mix of spheres and lines is not consistent with the conventional model of a network composed of lines, and is more difficult to readily comprehend.

Another inconsistency arises due to the prioritizing of the edge properties. This difficulty is not related to edge direction and therefore is not addressed by our second solution. There is a conflict in representing the properties of edge (u,v) itself. For example, if a non-saturating push is done on (u,v) the line turns blue. If the edge later becomes admissible, it turns green and the information that a partial flow existed on the edge (blueness) is not displayed.

5. Using the Animations

In this section we discuss the input datasets for the algorithms we chose to use and our resulting observations from the animations about particular algorithms and datasets. The algorithms we focused on are the 12 described in Subsection 4.3.2. As mentioned earlier, our use of animations was for exploratory analysis rather than a more thorough experimental analysis.

5.1. Data sets used. Initially, in order to develop and test our algorithm implementations and animations, we started with a small collection of simple, planar networks gleaned from texts [10, 17]. For more extensive testing, we added three slightly larger (32 to 128 vertices) planar networks that contained several pathological properties such as strategically placed bottlenecks and cycles. These were derived from analysis contained in Galil [13] (for Dinics algorithm), and Anderson [1] (for the push-relabel algorithm). We chose planar networks because they are easier to embed in 2D, and we found that they tended to retain their readability in 3D. From this base collection of networks we were able to debug our implementations and clarify visualization needs and problems. The pathological instances also gave us our first insights into the interaction of the algorithm variations with bottlenecks and cycles in the network.

Once we had refined our algorithm variation implementations, we generated a collection of networks from the DIMACS generators. Specifically, we used GENRMF (Wide and Long), the Washington Line Moderate, and Acyclic Dense generators. Choices for the sizes of the networks were often dictated by what we could effectively see in the animation, and the running time of the animation. (Animations of GENRMF networks with 256 nodes could run as long as an hour if one wanted to watch each push and lift operation.) The number of samples from each type of generator was chosen for exploratory purposes and was not designed for statistical analysis (see Table 3).

5.2. Observations. While running a generic push-relabel algorithm, it is readily apparent from watching the animations that cycles and local cuts (or bottlenecks) in the residual network are the major reason for poor performance. Local cuts form in the network. Cycles of vertices on the source side of these

Generators	Vertices	Edges	Samples	Other parameters
Acyclic dense	16	120	10	none
	32	496	10	none
GENRMF Long	128	492	10	Frame size: 4×4 No. of frames: 8
	256	1004	3	Frame size: 4×4 No. of frames: 16
GENRMF Wide	32	108	2	Frame size 4×4 No. of frames: 2
	128	508	5	Frame size: 8×8 No. of frames: 2
Washington line moderate	64	108-120	10	none
			50 Total	

TABLE 3. Data sets

cuts, no longer having any paths to the sink, tend to climb incrementally (via relabels) while flow gets pushed around the cycle until the vertices in the cycle finally have labels greater than the source (or a local source). The number of push and relabel operations are increased substantially by such behavior. This was most noticeable in the Washington and GENRMF networks.

When we introduced the various vertex selection procedures into the algorithm, we noticed that there was a considerable reduction in this behavior, as expected. Far fewer cycles climbed the full distance from the initial exact labeling to the height of the source. The periodic relabeling strategies that we introduced into the algorithm helped curtail this behavior in an even more dramatic way. When a global relabeling is executed, cycles no longer connected by paths to the sink automatically are raised to a height greater than n, thus greatly reducing the amount of climbing. We found that gap relabeling had fewer relabels and pushes on some networks than global relabeling while global relabeling performed better on others. This was especially apparent on the GENRMF graphs. Gap relabeling was more effective in terms of combinatorial operations on the GENRMF Long graphs while global relabeling worked better on the GENRMF Wide graphs.

Recent performance data confirms the improved run-time performance of gap relabeling on long vs. wide networks [12]. After viewing the animations, we concluded that this performance difference occurs because the long graphs are likely to have more bottlenecks, enabling gaps to be found immediately after a cut is formed, but in wide graphs the cut edges are saturated long before the gap relabeling is triggered. Global relabeling, however, occurs on a periodic basis that does not rely on the graph characteristics, and depending on the period, more readily recognizes and relabels to heights above n those vertices that no longer have paths to the sink. The savings in push and relabel operations may be offset, however, by the added cost of the breadth-first searches required in order to determine exact global labeling. The gap method does not perform breadth-first searches, but merely identifies the vertices in the gap and relabels these to a fixed height, n.

Our most significant observation is about the occurrence of cycles in the final flow network, where the assigned flow on some of the edges in the cycle is greater than the sum of the capacity of the edges adjacent to the source. We call these *excess cycles*. The occurrence of excess cycles was first observed in the GENRMF networks, which have cycles and bi-directional edges in the initial network. By constructing additional test cases, we confirmed that the excess cycles can occur in any graph that contains a cycle in the original network, independent of the presence of bi-directional edges. The excess cycles also occur independent of any particular push-relabel variation that we used.

We observed these cycles on both the source and sink sides of the minimum cut. In the cycles on the source side, edge assignments may be $O(n)$ greater than the cut at the source, while those on the sink side may be $O(n)$ greater than the minimum cut. In Figure 3, we show how such cycles can be created. Vertex labels appear next to each vertex and vertices with excess flow are represented by dark circles. The current flow on each edge is also indicated, with bold type for saturated edges.

The initial network is displayed in Figure 3A. Vertices (except source and sink) are labeled with the distance to the sink. Flow has already been pushed from the source to its neighbor. In Figure 3B, 3 pushes have been done, resulting in saturation of the edge adjacent to the sink. No further flow can be pushed to the sink at this point.

The upper-right vertex is relabeled to 3 in Figure 3C so that flow can be pushed to its neighbor, but the neighbor is immediately relabeled to 4 and the flow pushed back to the upper-right vertex (Figure 3D). This vertex is then relabeled to 5 and pushes flow to its other neighbor, completing the first trip around the cycle.

In Figure 3E, relabels and pushes have resulted in another 1150 units of flow being pushed around the cycle. The upper-left vertex is relabeled to 6 and flow is pushed back to the upper-right vertex (Figure 3F). This vertex is relabeled to 7 and the flow is returned to the upper-left vertex (Figure 3G). With the help of several pushes and relabels, the flow continues around the cycle once more. When it completes the circuit, the label of the upper-left vertex is high enough to push flow back to the source. The 1150 excess units are pushed back to the source (Figure 3H), resulting in a total flow of 50 units. Note that the edges in the cycle itself have flow greatly in excess of this final flow.

The flow assignments in the above example were calculated using a push-relabel algorithm with exact initial labeling and maximum excess vertex selection. While the gap and global relabeling strategies can diminish these assign-

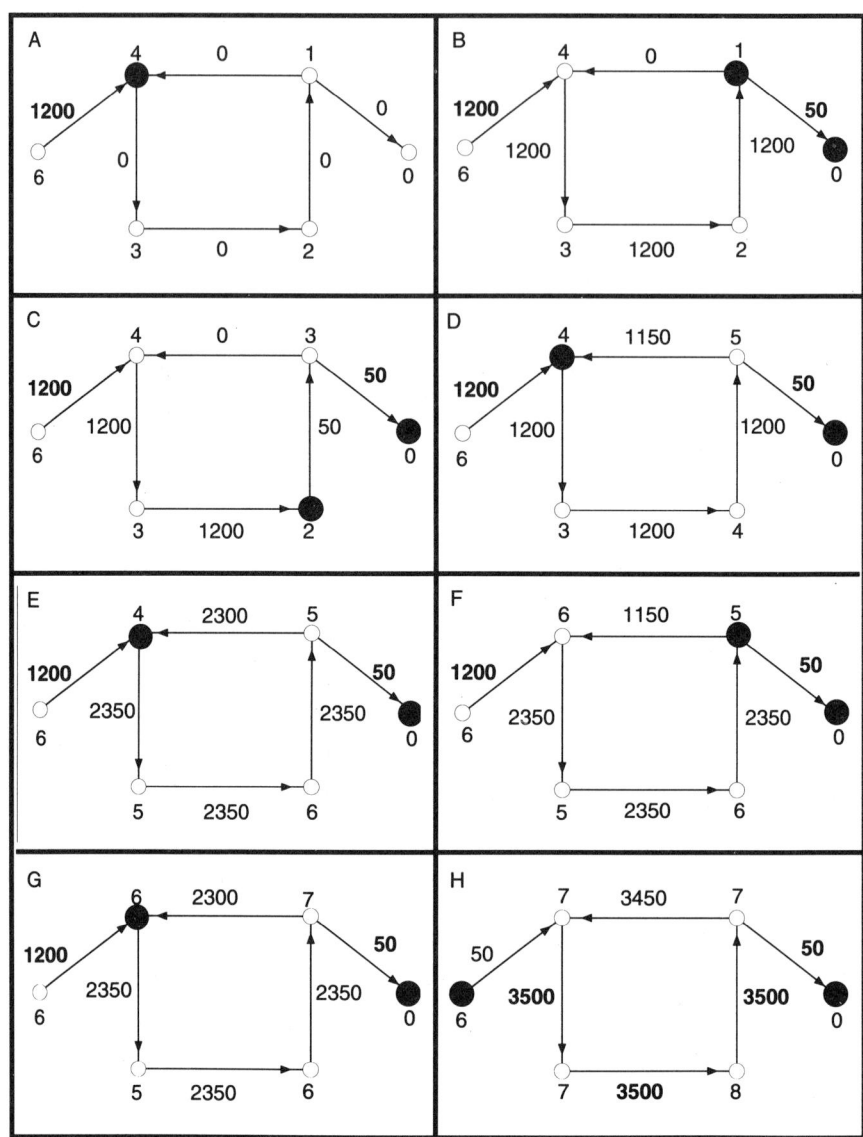

FIGURE 3. An example of how excess cycles can be created.

ments in some instances, in the worst case the size of the excess flow is unaffected by any variation of the algorithm.

Excess cycles raise two concerns. First, if the final edge assignments are to be used in an application, cycles, in general, are inappropriate because they represent an unneeded detour on the path from source to sink. Excess cycles are particularly disturbing because edge assignments greater than the total network maximum-flow value may be impossible or meaningless at best in practice. Eliminating any cycle in the final flow network is therefore important. A traditional approach might be to perform a flow decomposition [14] on the final flow network and thereby identify cycles in order to eliminate them and reduce assignments on the excess cycle edges. But there is also a second concern that may affect decisions about when to remove the cycles. If flow is pushed around a cycle, it is not making progress towards the source or sink. The pushes and relabels involved in creating cycles are therefore superfluous and represent wasted effort. This means that the cycles and the circumstances that generate them should be eliminated as soon as possible in the algorithm.

As a way of improving second stage processing, Goldberg and Tarjan [15] propose that the cycles be eliminated between the first and second stages of the algorithm, i.e. after the minimum cut has been found and before flow has begun to be pushed back towards the source. They suggest using the cycle-canceling algorithm of Sleator and Tarjan [27] to convert the residual network on the source side of the cut to an acyclic graph. The vertices are then processed in reverse topological order during the second stage. This approach is intended to eliminate cycles initially and help to promote efficient processing of excess vertices in the second stage. While this will certainly improve performance, it does not address the impact of the cycles in the first stage nor the occurrence of cycles in the second stage as edges become saturated and flow is diverted to alternative paths — possibly creating excess cycles in the final flow.

The relationship between the excess cycles and combinatorial complexity is not clear, but these cycles appear to be pervasive in worst-case situations. Clearly, the impact of excess cycles has not been addressed in the push-relabel variations we explored and therefore will be a subject of further study by our group.

6. Conclusions

Based on our work, we see great promise in the current tools available for animation development, and in using animations in algorithms research. Unfortunately, most of the previous work has concentrated on animations for pedagogical purposes. While pedagogy is an important and fruitful area for animation use, the algorithm research community would benefit greatly if general or domain-specific methods were established for the design of exploratory animations. We also found the task of defining clear, simple displays while still providing maximum information to be a difficult one. If the benefits of animations are to be fully realized in algorithm research, techniques must be developed to facilitate the creation of abstract, yet intuitive, visual representations of algorithm events and data properties.

There is also a critical need for a systematic methodology to apply animations in algorithms analysis. As discussed in § 5, our use of animations was not very systematic or thorough. Aside from the time demands of implementing the algorithms and animations, we now realize that our uneven approach is actually reflective of the lack of general methods for experimentally analyzing algorithms (as opposed to programs). Only with the development of systematic approaches to experimental analysis and the use of animations in analysis can the full research potential of algorithm animation be realized.

Even with the current weaknesses in our experimental approach, our work in exploring maximum-flow algorithms has shown that animations have great potential as a tool in experimental research. We see the role of animation as being very similar to the role that statistical analysis has in the sciences — it enables the researcher to recognize patterns and trends and to identify anomalies in the data. We conclude that use of animations will lead to greater understanding of current algorithms as well as the development of more efficient ones.

6.1. Future animation work. It would be very useful to have multiple views of the animation that run concurrently on the same network and algorithm, whereby individual properties are displayed in each. For example, we would like to include views of the admissible edge network and the saturated and unsaturated edge network. A view that highlights cycles in the residual network could also be created by augmenting the algorithm code with a cycle-finding algorithm. Having the direction of the edges clearly marked in all views would also be an important improvement.

Currently, the algorithm trace files can comfortably accommodate 256 node graphs and roughly 10,000 relabel and push operations, without running into memory problems. Virtual display capabilities with 3D scaling however enable one to easily show considerably larger networks. Thus, we would like to optimize the space needs of the trace files in order to examine networks of between 1,000 and 10,000 nodes.

The animation of a 128 node graph with the faster push-relabel variations can take roughly 10 minutes. We would like, especially for larger graphs, to be able to control which animation frames we wish to view, so that we could skip around, backwards and forwards, in an animation that might take up to one hour to run. This feature can be added relatively easily to the AVS module that performs the frame sequencing.

Though we have been able to introduce motion parallax with an AVS keyframe animator module, we would also like to explore viewing the 3D animations with stereoscopic displays. The addition of stereo parallax for depth cueing might improve the overall 3D quality of the animations, allowing the viewers, for example, to easily see that various vertices had been lifted while the network is still in its 2D embedding. This hardware technology would allow us the benefit of parallax depth cueing, without degrading the speed of the real-time animations with motion parallax.

6.2. Future work on maximum-flow algorithms. With additional improvements to the animation software, we would like to use our tools more extensively for their intended purpose: experimental research in maximum flow algorithms. We hope to perform better-defined experiments, where we couple combinatorial complexity and running-time performance data with what we see from the animations. The expectation is that this will deepen our insight into the algorithms and the problem. A good place to start this work is to better differentiate the global and gap periodic relabeling strategies on the GENRMF networks. Not only should we study the performance impact of network length and width, but also the proximity of the mincut to the source or sink.

To perform such experiments, we envision implementing the most recent algorithms [20], [16], with their attendant data structures (e.g. dynamic trees), and widening the test bed. By examining different relabeling strategies in light of our observations regarding cycles and bottlenecks in the networks, further progress could be made in understanding the nature of these pathologies. How, for instance, could relabeling strategies in combination with vertex and edge selection be improved in such a way that the cut could be found more efficiently during the first phase? What, if any, methods could be used to reduce or eliminate the occurrence of (excess) cycles in the residual network? We plan to explore these issues using our animations in well-defined experimental designs.

REFERENCES

1. R. Anderson and J. Setubal. Parallel and sequential implementations of maximum-flow algorithms. In *DIMACS Implementation Challenge Workshop: Algorithms for Network Flows and Matching, DIMACS Technical Report 92-4*, pages 17–41, 1992.
2. R. Bayer and E. McCreight. Sorting out sorting, 1991. 16mm color sound film, shown at the 1991 ACM SIGGRAPH conference.
3. J. Bentley and B. Kernighan. A system for algorithm animation. *Computing Systems*, 4(1):5–30, 1991.
4. B. Birgisson and G. Shannon. GraphView: A workstation-based environment for viewing graphs and animating graph algorithms. Technical Report 295, Computer Science Department, Indiana University, December 1989.
5. B. Birgisson and G. Shannon. GraphView documentation. Technical Report 299, Computer Science Department, Indiana University, January 1990.
6. M. Brown. *Algorithm Animation*. ACM Distinguished Dissertations. MIT Press, Cambridge, MA, 1987.
7. Marc H. Brown. Exploring algorithms using BALSA–II. *IEEE Computer*, 21(5):14–36, 1988.
8. Marc H. Brown. Zeus: A system for algorithm animation. In Proc. 1991 Workshop on Visual Languages, 1991.
9. J. Cheriyan and S. Maheshwari. Analysis of preflow push algorithms for the maximum network flow. *SIAM Journal on Computing*, 18(6):1057–1086, 1989.

10. T. Cormen, C. Leiserson, and R. Rivest. *Introduction to Algorithms*. McGraw-Hill, 1990.
11. U. Derigs and W. Meier. Implementing Goldberg's max-flow algorithm – a computational investigation. *Methods and Models of Operations Research*, 33:383–403, 1989.
12. U. Derigs and W. Meier. An evaluation of algorithmic refinements and proper data-structures for the preflow-push-approach for maximum flow. In *Extended Abstract for DIMACS Implementation Challenge Workshop*, October 1991.
13. Z. Galil. On the theoretical efficiency of various network flow algorithms. *Theoretical Computer Science*, 14:103–111, 1981.
14. A. Goldberg, E. Tardos, and R. Tarjan. Network flow algorithms. In *Paths, Flows, and VLSI-Layout*, pages 101–164, 1990. Algorithms and Combinatorics #9, Springer-Verlag, New York.
15. A. Goldberg and R. Tarjan. A new approach to the maximum flow problem. In *Proceedings of the 18th Annual ACM Symposium on Theory of Computing*, pages 136–146, 1986.
16. A. Goldberg and R. Tarjan. A new approach to the maximum-flow problem. *Journal of the ACM*, 35(4):921–940, 1988.
17. F. S. Hillier and G. J. Lieberman. *Introduction to Operations Research*. Holden-Day, 1986.
18. D. Johnson and C. McGeoch. DIMACS implementation challenge workshop: Algorithms for network flows and matching. Technical Report 92-4, DIMACS NSF Science and Technology Center, New Brunswick, New Jersey, January 1992.
19. A. V. Karzanov. Determining the maximal flow in a network by the method of preflows. *Sov. Math. Dokl.*, 15:434–437, 1974.
20. V. King, S. Rao, and R. Tarjan. A faster deterministic maximum flow algorithm. In *Proceedings of the 23rd Annual ACM Symposium on Theory of Computing*, pages 157–164, 1991.
21. M. Mevenkamp, N. Dean, and C. Monma. NETPAD user's guide and reference guide, 1990.
22. B. M. E. Moret and H. D. Shapiro. *Algorithms from P to NP*, volume I: Design and Efficiency. The Benjamin/Cummings Publishing Company, 1991.
23. C. Papadimitriou and K. Steiglitz. *Combinatorial Optimization: Algorithms and Complexity*. Prentice-Hall, 1982.
24. Peter Schorn. The XYZ GeoBench for the experimental evaluation of geometric algorithms. Manuscript., 1991.
25. G. Shannon, L. Meeden, and D. Friedman. SchemeGraphs: An object-oriented environment for manipulating graphs, 1990. Software and documentation.
26. S. Skiena. *Implementing Discrete Mathematics: Combinatorics and Graph Theory with Mathematica*. Addison-Wesley, 1990.
27. D. Sleator and R. Tarjan. A data structure for dynamic trees. *Journal of Computer and System Sciences*, 24:362–391, 1983.
28. J. Stasko. *TANGO: A Framework and System for Algorithm Animation*. PhD thesis, Brown University, May 1989.
29. C. Upson, T. Faulhaber, D. Kamins, D. Laidlaw, D. Schlegel, J. Vroom, R. Gurwitz, and A. van Dam. The application visualization system: A computational environment for scientific visualization. *IEEE Computer Graphics and Applications*, 9(4):30–42, 1989.
30. S. Wolfram. *Mathematica: A System for Doing Mathematics by Computer*. Addison-Wesley, 1988.

DEPARTMENT OF COMPUTER SCIENCE, INDIANA UNIVERSITY, BLOOMINGTON, INDIANA 47405
E-mail address: {shannon,jmaccuis,ejohnson}@cs.indiana.edu

An Empirical Study of Min Cost Flow Algorithms

R. G. BLAND, J. CHERIYAN, D. L. JENSEN AND L. LADÁNYI

ABSTRACT. Extensive computational testing of four minimum cost network flow codes over several distributions of sparse networks is conducted. Regression analyses yield models of mean execution times that predict well over the specified distributions.

1. Introduction

There is considerable practical and theoretical interest in algorithms for network flows. Several new algorithms with attractive theoretical properties have been presented in recent years (see the surveys [1, 15]). Little is known about the empirical behavior of the recent algorithms; only a few detailed computational studies have appeared, e.g., [6, 8, 11]. This study focuses on the computational behavior of codes implementing four min cost flow algorithms based on cost scaling (see [8]), successive approximations (see [17]), the network simplex method (see [18]) and relaxation (see [6]). Our main goal is to model mean execution time as a function of the input parameters over several populations (distributions) of network instances. A second goal, related to the first, is to study the inherent variability of individual execution times. Techniques from regression analysis are applied [12].

The analysis of the main experiment concerns estimation of the responses of four fixed codes to variations in certain key problem characteristics over specific input distributions. Since run-times can be sensitive to both variations in the codes and in the population from which inputs are drawn, one should resist broad interpretations of the experiment that reach for conclusions on the relative

---

1991 *Mathematics Subject Classification*. Primary 90B10; Secondary 68Q20.

The authors were supported in part by AFOSR, NSF and ONR through NSF grant #DMS-8920550.

The second author was supported in part by the Lucille and David Packard Fellowship of Éva Tardos.

efficacy of the four algorithms, as opposed to the fixed codes, or conclusions concerning wider classes of inputs. Indeed, the data offer ample evidence of large variations in the behavior of a fixed code across different populations, and, consequently, of the need to limit interpretations of empirical data to the realm in which the computation was performed.

This study does show that it is possible to deduce accurate models of average computational behavior over an entire (carefully circumscribed) population from tests on a modest-sized sample. Such models provide insights that may be useful in improving codes, and may serve as a first step toward both more robust models and more robust codes.

The populations of networks studied here are each associated with a network generator and a random mechanism for selecting the generator inputs. The generator accepts a list of inputs, and outputs a directed graph together with a cost and a capacity for each arc, and associated demands and supplies for some of the nodes. In our experiments the generator inputs are determined by a smaller set of control variables, some of which coincide with generator inputs (e.g., the seed, the number of nodes); each of the remaining generator inputs is a fixed function of the control variables. Treating the control variables as random variables with a fixed joint distribution implicitly determines a distribution (population) of network instances. We are specifically interested in modeling computational behavior as a function of the control variables over the populations that arise from independent uniform distributions on each control variable. If accurate models can be determined, this will facilitate examination of several secondary issues: comparisons of different codes on fixed input distributions; comparisons of the behavior of a fixed code across different distributions; analysis of the inherent variability of execution time about the mean as a function of the control variables.

Section 2 discusses briefly the four codes (SCALE, MIN3, RNET 3.61 [18], and RELAXT-III [6]) and the two populations of networks on which the codes were tested in the main experiment. These populations derive from the network generators CAPT [8] and NETGEN [19]. Section 3 presents the main experiment and statistical analysis. Some additional experimentation on populations from the GRIDGEN [20] and GOTO [14] generators is outlined in Section 4. Subsequent to the completion of the main experiments we conducted additional ad hoc runs trying different versions of SCALE, MIN3 and RNET 3.61; Section 5 reports on these runs, which offer indications of how some of the codes can be speeded up on some classes of inputs. Section 5 also comments on ad hoc runs on networks of larger size or different density from those considered in the main experiment. Brief conclusions appear in Section 6.

Additional data concerning these experiments are stored in the directory pub/netflow/instances/mincost/bcjl at dimacs.rutgers.edu; the file README in that directory has additional information. Those who are interested in reproducing parts of the experiments should contact the first author for access to executable SCALE and MIN3 codes for the SUN Sparc-2 workstation.

2. The Generators and The Algorithms

The main experiments used two generators to construct instances of min-cost flow problems: the NETGEN generator of Klingman, Napier and Stutz [19]; and a new version of the CAPT generator of Bland and Jensen [8]. (See Section 4 for a brief summary of additional experiments based on the GRIDGEN generator of Lee and Orlin [20] and the GOTO generator of Goldberg [14].) Each generator accepts a list of parameter values as inputs, and outputs a directed graph together with a cost and a capacity for each arc, and associated demands and supplies for some of the nodes.

The version of NETGEN we used (supplied by DIMACS and dated 9/6/90) has fifteen input parameters. Of these, the following are of interest here: *seed*, an integer used by the pseudo random number generator; *nodes*, the number of nodes; *arcs*, the (approximate) number of arcs; *nsorc*, the number of sources; *nsink*, the number of sinks; *supply*, the amount of flow to be sent from the sources to the sinks; a range on the arc costs, *locost* ... *hicost*; a range on the arc capacities, *locap* ... *hicap*; and the percentage of capacitated arcs, *ipcap*. (Our networks had no transshipment sources or sinks, and all of the "skeleton" arcs were specified to have the maximum cost.)

In order to ensure that the network will be feasible, i.e., the given node demands and supplies and arc capacities will not be inconsistent, NETGEN first constructs a "skeleton" which is an extremely sparse subnetwork. The skeleton arcs are capacitated in such a way that the specified amount of flow may be sent from the sources to the sinks using only these arcs. After generating the skeleton, NETGEN throws in additional "random" arcs until the total number of arcs roughly equals the requested number, by visiting each non-sink node and randomly adding arcs leaving that node; no duplicate arcs are created. (It was necessary to correct one line of the NETGEN code in order to allow additional arcs entering the highest-numbered sink node.)

In the NETGEN networks of the main experiment the number of nodes was in the range 2^9 to 2^{13}, the number of arcs was in the range $2 \cdot nodes$ to $32 \cdot nodes$, the minimum arc cost, *locost*, was equal to zero, the maximum arc cost, *hicost*, was in the range 2^8 to 2^{17}, the number of sources was in the range $\frac{nodes}{2^7}$ to $\frac{nodes}{2}$, *nsink* was equal to *nsorc*, and the amount of flow, *supply*, was in the range $2^6 \cdot nsorc$ to $2^{13} \cdot nsorc$. The percentage of capacitated arcs was either zero (for about one-seventh of all the networks) or 100. For the capacitated networks, the maximum arc capacity, *hicap*, was taken to be $k \cdot \frac{supply}{nsorc} \cdot \frac{nodes}{arcs}$, where k was one of the constants 1, 2, 3, 4, 5 or 6, and for the uncapacitated networks *hicap* was set to *supply*. We chose these values of k after preliminary tests in which we computed an optimal min-cost flow and counted the number of saturated arcs (i.e., arcs whose flow was at the upper bound); for the values of k above, between 5% and 39% of the arcs were saturated in our tests. Our motivation was that if the percentage is too high, then the skeleton arcs may play a dominant role, absorbing most of the flow; if the percentage is too low, then the network becomes effectively uncapacitated. For example, in the "benchmark" NETGEN networks, even those that are capacitated have very few saturated arcs in the optimal solutions:

benchmark problem no.	capacitated arcs in the network	saturated arcs in the optimal solution
16–19	20%	0.05%
20–23	40%	0.09%
24–27	80%	0.66%

We used a version of CAPT (available from DIMACS) with eleven input parameters, of which the following are of interest here: *seed*, a real number between zero and one used by the pseudo random number generator; *nsorc*, the number of sources; *nsink*, the number of sinks; *arcs*, the (approximate) number of arcs; *avgsup*, the average supply at each source; *intlen*, which controls how tightly capacitated the network will be; and *bcost*, the maximum bit-length for the arc costs. Arc costs are independent and uniform on the integers between 0 and $2^{bcost} - 1$.

CAPT generates only transportation problems. A network generated by CAPT essentially consists of a bipartite graph, one of whose node partitions is the set of sources while the other is the set of sinks. The bipartite graph is chosen uniformly from the set of *nsorc* × *nsink* bipartite graphs having the requested number of arcs; no duplicate arcs are present. CAPT next constructs what will be a feasible flow. Consider the unit simplex in the Euclidean space whose dimension is equal to the number of arcs in the bipartite graph, (i.e., each arc has a corresponding coordinate). CAPT creates a preliminary flow by choosing a point uniformly in the unit simplex, and using the coordinates of the point to allocate the proportion of the total flow (*nsorc* · *avgsup*) on each arc. The preliminary supply or demand at a source or sink node is given by the sum of the preliminary flows of the bipartite graph arcs incident with it. The capacity on each arc is set to the preliminary flow value on the arc plus an additive perturbation chosen uniformly from zero to *intlen*. Supplies (demands) are obtained by perturbing the preliminary values up (down) similarly.

The ranges for the parameters in our experiments with CAPT were similar to the NETGEN parameters, except that *nsorc* and *nsink* were each equal to half the total number of nodes. Roughly four-fifths of the networks were capacitated, and the maximum additive perturbation, *intlen*, was taken to be $k \cdot avgsup \cdot \frac{nodes}{arcs}$, where k was one of the constants 1.5, 2.5, 3.5, or 4.5. For the remaining one-fifth of the networks *intlen* was taken to be $10 \cdot avgsup$, and these networks were regarded to be uncapacitated.

Four codes were tested in our computational experiments: SCALE, MIN3, RNET 3.61 and RELAXT-III. The code SCALE was implemented by us; it is based on a cost scaling algorithm of Bland and Jensen (see [8]), which is related to earlier work by Bland and Jensen [7] and by Röck [23]. We implemented the code MIN3 mostly following the successive approximation algorithm of Goldberg and Tarjan [17], which combines and extends: the notion of cost scaling due, independently, to Bland and Jensen [7] and Röck [23], and motivated by Edmonds and Karp [13]; the push-relabel approach to maximum flow due to Goldberg and Tarjan (see [16]); and the relaxation method of Bertsekas (see [4]), which is rooted in Bertsekas's auction algorithm for the assignment problem [3]. Tardos's use of scaling and approximation to achieve the first strongly polynomial min cost flow algorithm [24] also motivated [17]. The interested reader is referred to [17, 5, 15, 1] and their references for comments on the development of these min cost flow algorithms and their antecedents. RNET (version 3.61–12/79) of Grigoriadis and Hsu [18] is a subroutine library to implement the primal network simplex algorithm (see [10]). We denote by RNET 3.61 the network simplex code used in the main experiments in which our driver fixes the RNET 3.61 parameters as follows: frq=0.5 (default=1.0, ... ,8.0), p_0=200 (default=1), and p_1=1.5 (default). RNET 3.61's execution times can be quite sensitive to these parameters, especially frq. The code RELAXT-III (April 1990 – no version number) of Bertsekas and Tseng is based on their relaxation algorithm [6]. This code has no user-settable parameters. (Henceforth, we refer to this code as RELAX.)

Bland and Jensen (see [8]) presented a cost-scaling algorithm for the minimum cost flow problem in which the computational engine is a maximum flow routine. After some preliminary comparisons among implementations of several maximum-flow algorithms, we chose the algorithm of [16] that uses the FIFO policy and does periodic computations of exact distance labels (but does not use the dynamic trees data structure). On several families of networks, this max-flow implementation runs as fast or faster than other max-flow implementations that we have tested; for example, even on "long and thin" (e.g., $10,000 \times 10$) networks the execution time of this max-flow implementation compares favorably with that of the code of [2]. An earlier version of SCALE implemented by Bland and Jensen [8] used the MPM maximum flow algorithm [21]. Perhaps because of the MPM-based maximum flow routine, the earlier version of SCALE performed poorly on sparse networks, and even on dense networks the MPM routine was problematic, since the maximum flow subproblems encountered in SCALE become very sparse as the iterations progress.

We have implemented the successive approximation algorithm of [17] with several parameters that specify different variants of the algorithm. The main variant described in [17] maintains a parameter ϵ which is iteratively decreased by a scaling factor of two; each iteration executes the *refine* procedure during which the current solution, which is guaranteed to be ϵ-optimal, is manipulated to yield a solution which is (ϵ/fac)-optimal ($fac = 2$ is the scaling factor). When ϵ decreases from its initial value by a factor of nC, where n is the number of nodes and C is the maximum absolute cost, then the algorithm terminates with an optimal solution. The basic step in the refine procedure is to select any *active* node and apply a *discharge* step to it. Several different policies for selecting the node to discharge have been proposed and analyzed (in the worst case) e.g., queue (FIFO), wave, and first active [17]. We have implemented five policies, namely, stack, queue, q-max (a hybrid of queue and stack), wave, and first-active. We found the stack and q-max policies to be the fastest among these five policies on all networks in our preliminary tests. MIN3 uses the stack policy.

As suggested by Goldberg and Tarjan, one variant of the implemented algorithm periodically executes Bellman-Ford computations to find an ϵ within a factor of two of minimum such that the current solution is ϵ-optimal; when possible, the Bellman-Ford computation also updates the node prices (i.e., the dual solution). We have tried several heuristics for invoking the Bellman-Ford computation, including one variant that invokes it after every i refines ($i = 1, 2, 4, \ldots$), and another one that attempts to balance the overall number of operations executed by the refine computation and the Bellman-Ford computation.

On most networks we have tested, the successive approximation algorithm finds an optimal flow after executing significantly fewer refines than the worst-case bound of $\log nC$. Unfortunately, the algorithm has no direct way to check when an optimal solution has been found; this is because the node prices are only guaranteed to be ϵ-optimal with respect to the current flow. This raises the issue of detecting an optimal flow and terminating the algorithm. In each refine, a quick estimate of the total cost of the current flow is computed using 32-bit arithmetic for the computation. When two consecutive refines have the same value for the estimated total cost, we may have an optimal flow, so a Bellman-Ford computation is run to find zero-optimal node prices; if the Bellman-Ford computation succeeds, then we have an optimal flow. Also, a successful Bellman-Ford computation is rather fast. If the Bellman-Ford computation is invoked before optimality, then it runs slower and returns no useful information. MIN3 employs the variant above, which worked well on all networks tested, running faster than both (1) the variant that executes only refines (but no Bellman-Ford computations), and (2) the variant of [17] that periodically executes Bellman-Ford computations.

At the conclusion of each refine, MIN3 fixes the flow at its current value on all arcs on which the absolute value of the current reduced cost exceeds $n\epsilon$, thereby reducing the effective number of arcs in the remaining computation.

We considered the use of several scaling factors in our preliminary tests, and found that a factor of four usually gave the fastest execution times, and almost always gave faster execution times than a factor of two. The reported times for MIN3 use a scaling factor of four.

In addition to the preliminary testing of SCALE and MIN3 described above, some preliminary tests were performed running RNET 3.61 on NETGEN networks to determine the settings for the parameters frq and p_0. All tuning of the SCALE and MIN3 codes and the choice of the parameter settings for RNET 3.61 were fixed prior to the design of the main experiments, which took place prior to the DIMACS Workshop in October 1991.

The compilers used are: Gnu C (for MIN3), and SUN Fortran (for SCALE, RELAXT and RNET 3.61), all with the standard optimization level, "-O" (i.e., "big oh"). The execution times reported are the "elapsed user times" in seconds on a SUN Sparc-2 with 16 megabytes of memory, measured to an accuracy of $\frac{1}{60}$ seconds; the times for input and initialization are never included.

3. The Main Experiments

The primary goal of the experimental analysis is to model mean execution time as a function of problem characteristics for each of the several codes and input distributions discussed previously. The problem characteristics, or *control variables*, identified for study here are: n (the number of nodes); m (the number of arcs); C (the maximum absolute value of any cost); f (the total flow through the network); s (the number of source nodes); and d (a measure of tightness of the capacity constraints on the arcs – for CAPT, d denotes *intlen*, while for the other generators it denotes *hicap*.

3.1. Experiment design. Based on earlier work on statistical analysis of network flow algorithms [8], models of execution time of the form:

$$(3.1) \qquad T = \varepsilon e^{\beta_0} n^{\beta_1} m^{\beta_2} (\ln C)^{\beta_3} f^{\beta_4} d^{\beta_5} s^{\beta_6}$$

seemed a plausible first choice. Here $\beta_0, \ldots, \beta_6$ are the parameters of the model to be estimated by regression analysis, e is the base of the natural logarithm, and the value of β_0 reflects the units of measurement of the dependent random variable T, the execution time, so that the random variable ε has mean one. Both T and ε must be regarded to be functions $T(n, m, \ln C, f, d, s)$ and $\varepsilon(n, m, \ln C, f, d, s)$ of the control variables. An important part of the analysis to follow is the question of whether the random variables $\varepsilon(n, m, \ln C, f, d, s)$ are independent and identically distributed. We will address an even more stringent *lognormality assumption*:

$$(3.2) \qquad \varepsilon(n, m, \ln C, f, d, s) \text{ are i. i. d. lognormal with mean 1.}$$

For the CAPT distribution examined here s is always $n/2$, and so s is dropped from the model (3.1). It is natural to use $\ln C$ rather than C in the model, since the number of iterations executed by SCALE is linear in $\ln C$ for all networks, and MIN3 has similar behavior. Section 4 has more discussion on the model, including consideration of different choices of some of the control variables and the issue of how to measure d for uncapacitated networks.

We did not expect a model as simple as (3.1) to be sufficiently robust to portray accurately algorithmic behavior over the entire class of networks constructible by any one of the generators. Even an unabashed optimist could not reasonably hope for an accurate model much simpler than a sum of terms each of the form (3.1):

$$(3.3) \quad T = \sum_{i=1}^{k} \varepsilon_i e^{\beta_{0,i}} n^{\beta_{1,i}} m^{\beta_{2,i}} (\ln C)^{\beta_{3,i}} f^{\beta_{4,i}} d^{\beta_{5,i}} s^{\beta_{6,i}}.$$

However, (3.1) is plausible for limited ranges of the control variables (it picks out a dominant term from (3.3) over those ranges) and much easier to analyze. Moreover, statistical analysis based on (3.1) might be a productive first step toward analysis of more complicated models like (3.3).

The main experiment was designed to have four stages:
(1) preliminary screening;
(2) collection of large data sets for detailed analysis;
(3) estimation – regression analysis of the data from (2);
(4) validation – evaluation of the predictive power of the models developed in (3).

The regression analysis of the data in both stages (1) and (3) is facilitated by a logarithmic transformation of the model (3.1) to the equivalent model:

$$(3.4)$$
$$Y = \ln \varepsilon + \beta_0 + \beta_1 \ln n + \beta_2 \ln m + \beta_3 \ln \ln C + \beta_4 \ln f + \beta_5 \ln d + \beta_6 \ln s$$

of Y, the natural logarithm of execution time. Standard techniques of linear regression apply to (3.4). The package SAS was used for this purpose.

The screening tests were based on a half-factorial design. They indicated that when the control variables were chosen from the intervals of values previously selected (see Section 2), then the model (3.4) was sufficiently promising to continue with large-scale tests. Consequently we designed the experiment of stage (2) to generate 100 observations where each of n, m, and C was chosen uniformly from the respectives values $n \in \{$ 512, 1024, 1536, 2048, 2560, 3072, 3584, 4096, 4608, 5120, 5632, 6144, 6656, 7168, 7680, 8192 $\}$, $m/n \in \{$ 2, 4, 6, 8, 10, 12, 14, 16, 18, 20, 22, 24, 26, 28, 30, 32 $\}$, $C \in \{2^8, 2^9, 2^{10}, 2^{11}, 2^{12}, 2^{13}, 2^{14}, 2^{15}, 2^{16}, 2^{17}\}$. With NETGEN s is determined by choosing uniformly $n/s \in \{2, 2^2, 2^3, 2^4, 2^5, 2^6, 2^7\}$; with CAPT $s = n/2$. In both cases f is set by choosing uniformly $f/s \in \{2^6, 2^7, 2^8, 2^9, 2^{10}, 2^{11}, 2^{12}, 2^{13}\}$, and d is chosen to be $k \cdot (f/s) \cdot (n/m)$ with k selected uniformly from $\{1, 2, 3, 4, 5, 6, sm/n\}$ for NETGEN and $\{1.5, 2.5, 3.5, 4.5, 10m/n\}$ for CAPT. Each observation in stage (2) consisted of a min cost flow instance solved to optimality by each of the four codes. The values of the control variables, seed, other generator input parameters, the optimal solution value, and the execution time (in seconds) for each code were recorded, along with information about the number of saturated arcs in the optimal solution, the step times and counts, etc.

3.2. Regression analysis. The regressions on the data from stage (2) based on (3.4) are summarized in Table 1 which shows the least-squares estimators b_i of β_i, $(0 \leq i \leq 6)$, and their standard errors, as well as the R^2 values for each of the eight samples (4 codes × 2 generators) of 100 instances.

Examining the third column of Table 1(b), for example, reveals that the model estimates the mean of the natural logarithm of execution time for RNET 3.61 on the NETGEN networks by

$$(3.5) \quad -12.87 + 0.58 \ln n + 0.92 \ln m + 0.08 \ln \ln C + 0.27 \ln f - 0.20 \ln d - 0.07 \ln s$$

which translates to an estimate

$$(3.6) \quad e^{-12.87} n^{0.58} m^{0.92} (\ln C)^{0.08} f^{0.27} d^{-0.20} s^{-0.07}$$

of the median execution time (in seconds). Point estimates of mean execution time (based on the lognormality assumption) exceed the estimates of median execution time by 0.1% to 1.6%, for the ranges of values of the control variables noted above; the mean exceeds the median by no more than 1.9% for RELAX and 0.4% for MIN3 and SCALE. On the CAPT population these percentages are smaller than for NETGEN by a factor of 2 to 4.

Figures 1–4 (a) illustrate the growth in predicted execution times with the logarithm of the number of nodes when $m = 16 \cdot n$ and the other control variables are held fixed, and Figures 1–4 (b) illustrate the growth in predicted execution times with the logarithm of the number of arcs when all other control variables are held fixed. The plots may look different if we choose different values for the control variables that are held fixed; each choice specifies a different slice of a multi-dimensional surface.

The correlation coefficient R between the vector of observations and the vector of predictions is the cosine of the angle formed by the vector of differences between the observations $Y = \ln T$ and their mean $\overline{Y}$ and the vector of differences between the predicted values

$$\widehat{Y} = b_0 + b_1 \ln n + b_2 \ln m + b_3 \ln \ln C + b_4 \ln f + b_5 \ln d + b_6 \ln s$$

and $\overline{Y}$. R^2 can be regarded to be the fraction of the variation about $\overline{Y}$ that is explained by the model. High R^2 values indicate that the model has fitted the data well. Seven of the eight R^2 values are between 0.962 and 0.993; R^2 for RELAX on the NETGEN sample is 0.914.

Because of the large number of control variables, it is difficult to portray the goodness of fit by a conventional plot of observed values versus a single control variable together with the fitted curve. Figures 5–12 (a) offer that sort of information in a slightly different form. Here, we have accumulated the effect of the entire set of control variables by plotting for each of the observations

$i = 1, \ldots, 100$ the predicted value $\widehat{Y}_i$ versus the observed value Y_i. The clustering of the pairs $(Y_i, \widehat{Y}_i)$ about the 45°-line is a clear indication of good fit for each of the samples.

The standard errors given in Table 1 allow the reader to ascertain the precision of the estimators. An estimator that is small in magnitude (relative to its standard error), e.g. the estimator b_3 associated with $\ln C$ in the RNET 3.61 model for the NETGEN population, suggests that the associated control variable may not be helpful in explaining variations in the response variable, Y. The ratio of an estimator b_i of β_i to its standard error is t_i, the t-value of the estimator. Here most of the t-values are large and have very small significance probabilities (for the test of the hypothesis that $\beta_i = 0$). The estimator b_6 associated with s tends to have small t-values, suggesting that the control variable s may be dropped from the model.

All of the analysis reported in this section is based on regressions with all of the control variables present in the model; however for any one code and distribution several of those parameters may be insignificant. Consideration of the information in Table 1 together with other data gathered in the regression analysis yields insights about which control variables are the most critical in explaining variations in execution times of a given code on a particular population. For example regressions with only n and m are sufficient to give R^2 values of better than 0.924 for MIN3 on the CAPT sample and 0.948 on the NETGEN sample. Similarly for SCALE n and $\ln C$ are enough to get $R^2 = 0.948$ on the NETGEN sample.

It is interesting to consider by how much R^2 drops when a single control variable is excluded from the model — this gives a measure of the sensitivity of the fit to the inclusion of that variable in the model. (Note that this is not the same issue as whether the estimate b_i of β_i is large in magnitude, since there may be correlations among the variables.) For example on the NETGEN sample SCALE is sensitive to n, but not to m: SCALE's R^2 value falls by 0.332 if n is dropped from the model, but by only 0.010 when m is dropped. On the CAPT sample the respective numbers are 0.093 and 0.005. SCALE is not as sensitive as one might expect to $\ln C$ — for both samples R^2 remains above 0.900 when $\ln C$ is dropped from the model.

All four codes are insensitive to s; R^2 never drops by more than 0.007 when s is excluded from the NETGEN models. (Recall that s does not appear in the CAPT models.)

MIN3 is also relatively insensitive to $\ln C$, d and f, with R^2 never decreasing by more than 0.018 on the NETGEN sample or 0.055 on the CAPT sample with the deletion from the model of any one of these three parameters; dropping all of them simultaneously still gives R^2 of 0.948 on NETGEN and 0.924 on CAPT.

On the CAPT sample, RELAX is extremely sensitive to each of d (R^2 decreases by 0.431) and f (0.237) and a little sensitive to n (0.080), but not $\ln C$ (0.000) or m (0.001). However, on the NETGEN sample it is somewhat sensitive to m (0.131) and not at all sensitive to d (0.000).

RNET 3.61 is quite sensitive to d on the CAPT sample (0.262), less so on the NETGEN sample (0.070), where it is sensitive to m (0.184). In both cases it is impervious to $\ln C$ (0.000).

For each of the four codes the estimate of the exponent of d is negative. All four codes seem to run faster, on average, as the arc capacities are loosened. This effect is less pronounced for the NETGEN population, where RNET 3.61 had the largest negative exponent on d, -0.203.

In order to evaluate further the effects of variations in the tightness of capacities, each of the 100 instances in each of the CAPT and NETGEN base tests was re-run twice – once with very loose capacities and once with the parameter k fixed at 2.5 on the CAPT instances, where it determines *intlen*, and fixed at 2 on the NETGEN instances, where it determines *hicap*. All input parameters other than k were identical in the two replicates. Figures 13–16 show clearly that all four codes ran much slower on the CAPT instances with tight capacities and all but MIN3 ran slower on the NETGEN instances with tight capacities. The effects on RNET 3.61 times are especially noteworthy – in both sets RNET 3.61 ran slower on all 100 instances with tight capacities. The percentage of saturated arcs at optimality for the instances with tight capacities ranged between 9.2% and 28.8% and averaged 24.0% on the CAPT set, and ranged between 2.3% and 32.2% and averaged 15.4% on the NETGEN set.

3.3. Analysis of residuals. Some of the statistical interpretations of the regressions depend on the lognormality assumption, which is equivalent to the assumption that the random variables $\ln \varepsilon(n, m, \ln C, f, d, s)$ are independent normals with mean zero and constant variance. In order to evaluate the accuracy of that assumption, we produced scatter plots of residuals $(Y - \widehat{Y})$ versus the different control variables; these plots do not show marked deviations from what one might expect under the lognormality assumption. More revealing are the normal probability plots of Figures 17–20. If the lognormality assumption were valid and the model were correct, the studentized residuals, i.e., the residuals divided by their standard errors, should be independent standard normals. The normal probability plots are produced by ordering the hundred studentized residuals from smallest R_{j_1} to largest $R_{j_{100}}$ and plotting each pair (R_{j_i}, Z_i), where Z_i is the inverse value of the ith percentile of the standard normal cumulative distribution function. Clustering of these pairs about the 45°-line indicates that the lognormality assumption is approximately correct. For comparison, Figures 21–22 show normal probability plots from two sets of 100 pseudorandom standard normal variates generated by SAS using sums of twelve uniform variates, and the Box-Mueller method, respectively. The probability plots of studentized residuals in Figures 21–22 do not differ appreciably from the plots of Figures 17–20.

3.4. Tests with independent data sets. In the fourth stage of the experiment we generated eight new sets of 100 observations each, as in stage (2), but the new *validation sets* are independent of the earlier *base sets* on which the regressions were performed and predictions are based. Figures 5–12 (b) show the plots of predicted values versus observed values of the logarithm of execution times for the new validation sets. These plots indicate clearly that the predictions are good. Variations about the predictions appear to be symmetric and not sensitive to the magnitude of the predicted value. The plots for the validation set do not differ substantially from those for the base set on which the estimation was carried out. The mean errors from prediction on the CAPT validation set are almost as small as on the CAPT base set. They are relatively larger for the NETGEN validation set.

One can compute confidence intervals on individual observations based on the lognormality assumption and the assumption that the model is correct. Table 2 presents counts on the number of observations out of 100 in the validation sets that fall below, within, and above specified confidence intervals. These counts provide further evidence that the model (3.1) is approximately correct over the distributions of networks from which the base and validation sets were drawn and that the lognormality assumption is approximately correct.

3.5. Lognormality assumption. The lognormality assumption underlies some of the statistical evaluations, interpretations, and applications of the regressions, for example, the evaluation of precision based on t-statistics, and the construction of confidence intervals for prediction of individual observations. Our use of confidence intervals above was in an inverse role, to validate the model and the lognormality assumption. If correct, the lognormality assumption would indicate that factors contributing to the inherent variability of execution times accumulate in a multiplicative way. This seems to be interesting in its own right, and suggests further study of other optimization problems and codes for similar behavior.

The estimates from the main experiment appear to lead to acceptable models of median execution time. However, one should be cautious in accepting the lognormality assumption as a close enough approximation to the truth to invoke formal statistical interpretations that derive from it, or to pretend that the inherent variability of execution times (separate from responses to the control variables) is understood well. In order to examine this issue further, we collected additional observations with large numbers of replicates. (In replicates all generator input parameters except the seed are held fixed.)

For each of CAPT and NETGEN we generated 2500 new observations in five sets of 500 replicates, each set with a different choice of values of the control variables. Figures 23–24 show the normal probability plots for these observations, where residuals are measured from the sample mean of the 500 replicates. These probability plots are quite striking. The lognormality assumption appears to be accurate for all four codes on the CAPT networks, and on the NETGEN networks except for RELAX, where deviation from lognormality is apparent outside of the center of the distribution. The probability plots suggest that under the distributions of networks considered here, although (3.1) may not be quite right, there are accurate models of median and mean execution times with the same choice of control (independent) variables. The large numbers of replicates in these sets revealed some significant bias in the current model (with the estimators from our base tests). It appears that this may be due largely to inaccuracy in the way in which the model reflects the effects of capacitation; see the next section. Further study is needed. It appears from these replicate sets that the greater variability in residuals from predictions for RNET 3.61 and RELAX over SCALE and MIN3 is due to greater inaccuracy in the RNET 3.61 and RELAX models; the residuals from the sample means for each set of 500 replicates were relatively smaller for RNET 3.61 and RELAX. Indeed, the combination of faster run-times and smaller variability necessitated running RNET 3.61 and RELAX on larger (harder) replicated inputs than SCALE and MIN3 – the small variations in RNET 3.61 and RELAX runtimes on the smaller networks tended to result in large numbers of duplicate values among the 500 replicates (due to the 1/60 second accuracy in the timer).

3.6. More on the model. One should not attach too much importance to the estimates b_i of the individual coefficients β_i; it is the aggregation of their effects that is interesting. Not only are there strong (positive and negative) correlations between pairs of the b_i's, but their values depend on certain arbitrary choices on how the model is represented. Such substitutions as m/n for m or dms/nf for d do not really change the model. They result in exactly the same fit, merely transforming its representation in a way that alters the coefficients b_i in accordance with the substitution.

A variation on (3.1) in which $\ln u$ and $\ln f$ replace u and f, respectively, gives a symmetry in the treatment of u and f with C, which plays a role dual to u and f in the minimum cost flow problem. For some algorithms that are naturally symmetric in their treatment of primal and dual variables this symmetry in the model might be desirable. Regressions with $\ln u$ and $\ln f$ in place of u and f did not change the R^2 values appreciably – in some cases they increased slightly, and in others they decreased slightly. Perhaps if the experimental ranges on u and f were wider, the effects would be more noticeable.

It has been suggested that the diameter of a network may be influential on the run-times of certain minimum cost flow codes [5]. It might be interesting to study this over some network distributions where diameters are easy to estimate.

The inclusion in these tests of uncapacitated networks along with capacitated networks causes a difficulty in our attempt to evaluate responses of run-times to variations in capacitation. In a model of the basic form (3.1) applied to capacitated networks there are simple and natural ways of defining a parameter d that measures strictness of capacities, for example, the maximum capacity. When uncapacitated networks are also included, some convention is needed to extend the definition of d. Regarding uncapacitated networks to be capacitated with large, effectively infinite capacities leads to an assignment to the d above of a value representing effectively infinite capacity. The effect on the regression of arbitrary choices that may differ by orders of magnitude can be substantial. It seems plausible that one should avoid, if possible, assignment of a value much larger than necessary to guarantee, or, at least, make it likely that no edge with that capacity will be saturated in an optimal solution. It appears that the accuracy of the estimation based on our main experiment is improvable by a more careful treatment of this issue. Capacitation does seem to be contributing more to bias in the model than any other control variable. One could use the fraction of arcs that are saturated in an optimal solution as a good measure of capacitation. (Let *bfrac* denote this fraction.) This might be useful in evaluating how run-times respond to changes in capacitation, but not of much practical use for prediction, since it requires solving the problem (perhaps with a different code) in order to evaluate one of the independent variables. Some regressions were run using *bfrac* in place of d. It did not improve the R^2 values as much as we expected on the CAPT and NETGEN samples.

4. Some Additional Experiments

Subsequent to the main experiment we ran additional tests along the same lines using two other generators: GRIDGEN [20] and GOTO [14]. The model (3.1) fit the GRIDGEN data well with R^2 values between .95 and .99. See Table 3a and Figure 3. For the GOTO sample the model (3.1), without the control variables f and s (which are not controllable inputs with GOTO), did not fit as well here as in the samples from the three other generators, with R^2 values between 0.711 and 0.909 (see Table 3). This generator seemed to produce more difficult instances than the other three, and there is some indication that the difficulty of the GOTO networks is related to the fraction of the arcs that are saturated at optimality. When *bfrac* (which should not be regarded as a generator input) is included to extend the model (3.1) by multiplying the right hand side by $(bfrac)^\gamma$, the R^2 values go up dramatically (see the last line of Table 3). The relative behavior of the codes on the GOTO sample was rather different from what we saw in the main experiment (see Figure 4).

5. Ad Hoc Runs

While the analysis of the previous sections leads to statistically significant interpretations, the comments below are impressions based on small numbers of additional ad-hoc tests. They raise some points for further consideration.

5.1. DIMACS benchmarks. In order to facilitate run-time comparisons with other computing environments, Table 4 gives run-times of the four codes on some benchmark networks described in the DIMACS document [9]. In each case the seed was set to 270001.

5.2. Large sparse networks. In order to evaluate responses to density, it was necessary to design the main experiment so that the sample included a wider range of densities than one often sees in network flow experiments; this limited the number of nodes in those networks to the range 512–8,192. Later we ran the codes on sparse networks with 16,384–55,056 nodes and 32,768–165,167 arcs. There were twenty networks generated by each of CAPT and NETGEN. The total run-times for each code are summarized below.

Generator	RNET 3.61	RELAX	MIN3	SCALE
CAPT	773.79	9666.43	2505.48	5262.06
NETGEN	9210.97	1342.92	8932.87	19681.8

RNET 3.61 was the fastest on 19 of the 20 CAPT networks and RELAX was the slowest on 13 of the 20. RELAX was the fastest on all 20 NETGEN networks and SCALE was the slowest on 19 of the 20. Because these networks have far more nodes than those tested in the main experiment, there is reason to expect greater deviations here from predictions based on the main experiment. Indeed the mean errors are much larger here, except for the combinations of RNET 3.61 with CAPT and RELAX with NETGEN.

5.3. A few large, dense networks. We tested a small number of dense networks larger (in m) than those included in the main experiment. The models that fit well earlier did not predict so well on these new networks outside the population on which the model was developed. These few large, dense networks also offer further evidence that relative performance of codes may be highly sensitive to the population of test problems. We solved four networks from each of CAPT and NETGEN, with 520,000–530,000 arcs and 8,000–33,000 nodes. Cumulative times show that on the four CAPT networks RNET 3.61 ran faster than the next fastest code by a factor of four and RELAX ran slower than the next slowest by a factor of two. On the four NETGEN networks RELAX ran faster than the next fastest by a factor of 2.5 and RNET 3.61 ran slower than the next slowest by a factor 2.5. We also ran some very dense NETGEN networks on 2048 nodes, two with approximately 524,000 arcs and two with more than one million arcs. On these networks and the sparser $m > 500,000$ networks, uncapacitated or very loosely capacitated examples were much easier for RNET 3.61 and RELAX, and to a lesser extent, SCALE, than the more tightly capacitated examples.

5.4. Variations on the codes. Interpretation of these experiments should be viewed in terms of the specific implementations of the four algorithms. Clearly there are many modifications that might improve performance.

The cost-scaling approach has not reached a mature implementation. The version of SCALE employed here differs from the first version [8] primarily in the substitution of a max flow algorithm based on [16] for one based on [21].

This appears to yield substantial speed-ups on sparse networks. The choice of a scaling factor of four in SCALE was based on testing with the old max flow routine; this should be re-examined. It would also be interesting to examine arc-fixing, early termination, and pre-processing in SCALE.

With preprocessing of the cost data to reduce the maximum absolute cost one might hope to reduce the number of iterations required by SCALE and MIN3. In a small number of trials on CAPT and NETGEN networks the costs were transformed, but with no appreciable benefit for any of the codes; for relatively dense networks the preprocessing time was substantial and outweighed any improvements in run-times. There were surprising outcomes in some runs of SCALE on GOTO networks. The initial node numbers were changed from zero to minimize the absolute value of the costs incident with a given node; the nodes were processed in the order in which they were given and several passes were done. We were not able to reduce the number of scaling iterations, but the SCALE execution times (including preprocessing) dropped by factors of 6–8 on some of the larger GOTO networks ($n = 8192$, $m = 60{,}000$–$80{,}000$) run. It is not yet clear how consistently such speed-ups occur on GOTO networks.

Some ad hoc testing of variations on the MIN3 code offer indications of potential for general improvements, as well as some changes that could make the code run faster on specific input classes. MIN3 only fixes an arc when the magnitude of its reduced cost exceeds $n\epsilon$, the theoretical worst-case criterion. Fixing at lower values, with provision for correcting for mistakes at the end, led to improved running times on CAPT and NETGEN networks. A scaling factor of two was found to work better on GOTO networks, whereas four seems to work better with the other generators examined here. More robust price updates than the single node update, as described in [**17**] and employed in MIN3, look promising, but we have not yet tested this. There seems to be a significant potential for improved implementations of the relatively new successive approximation algorithm.

Version 3.61 of RNET (1979) as used here was not designed for networks of the size and density of some of those in our tests. It is likely that substantial speedups could be achieved on large networks by better choices of the RNET 3.61 parameters, and by modification of the pricing routine, e.g., employing a candidate queue.

We have done some experimentation with the RNET 3.61 parameters *frq* and p_0. The frequency *frq* controls the partitioning of the edge set for partial pricing, and p_0 controls the initial costs on the artificial edges. For each of two CAPT and two NETGEN networks from the base sets we ran RNET 3.61 with several different values of *frq* and each choice of $p_0 \in \{1, 2, 3, 4, 10, 200\}$.

On the CAPT networks the choice of p_0 made little difference; for every choice of *frq* only one pass was required to drive out the artificial variables, and the variations in run-times over the six choices of p_0 never exceeded 8.5%. Run-times were sensitive to the choice of *frq*, sometimes exhibiting abrupt responses to small variations. For example, on one of the two networks the six run-times for each of frq=2.50, 2.22, and 2.00 ranged between 20.93-21.28, 45.61-48.15, and 22.9-23.0 seconds, respectively. General responses to frq were somewhat unpredictable, but in both tests the minimum run-time (among all 144 combinations of 24 different values of frq between .06 and 20.0 and the six choices of p_0) occurred with frq=2.5 and p_0=3:

Network	n	m	Base test time	Min. observed time
CAPT3-115	8,096	194,304	76.1	23.35 at frq=2.5 p_0=3
CAPT3-146	7,680	215,040	54.1	20.93 at frq=2.5 p_0=3

This prompted us to re-run all 100 networks from the CAPT validation set with frq=2.5 and p_0=3. It ran faster on all 100 networks by a minimum of 17.5%. The average improvement of 59.2% is almost a 2.5-fold speedup. (Approximately the same speedup was also achieved on the large sparse CAPT networks reported on above.)

On the NETGEN networks the choice of p_0 as well as frq made a substantial difference. Here we examined 14 values of frq between .06 and 4.

Network	n	m	Base test time	Min. observed time
NETG4-121	5,632	123,904	191.6	188.7 at frq=0.50 p_0=10
NETG4-140	8,192	180,224	348.0	291.7 at frq=0.25 p_0=3

In contrast with the CAPT networks, here the parameter values used in the main experiment were not far from the best observed. This is not surprising, since the choices of the RNET 3.61 parameter settings for the main experiment were based on preliminary tests on NETGEN networks. Note that our choice of $p_0 = 200$ for the main experiments essentially forces RNET 3.61 into a Big–M scheme, taking a single pass to force all of the artificial edges out of the basis. This is contrary to the intentions of RNET 3.61's authors, and it can have a disastrous effect when combined with an unfortunate choice of frq. However in our (limited) testing with fixed values of frq and varying values of p_0, the unusually high value of p_0 did not cause run-times much greater than values in the ranges they suggest (and sometimes gave faster times), as long as frq was not too large.

Four modifications of the RNET 3.61 pricing strategy (other than the choice of frq) were tested on a set of 100 large, sparse NETGEN networks and a few

GOTO networks. The four strategies are:
  (i) the original RNET strategy – the arc list is partitioned into approximately $\frac{m \cdot frq}{100}$ blocks consisting of arcs whose indices are congruent modulo $\frac{100}{frq}$;
  (ii) the original RNET strategy, but with the blocks constructed by taking sets of $\frac{100}{frq}$ arcs with consecutive indices, beginning at the top of the arc list;
  (iii) the original RNET strategy, but with the blocks constructed by taking sets of $\frac{100}{frq}$ arcs with consecutive indices, beginning at the top of a random permutation of the arc list;
  (iv) a candidate queue strategy (see [22]). Each time the queue is constructed it is filled with basic arcs and negative transformed cost arcs until it has $\frac{m}{20}$ arcs in addition to the basic arcs. The queue is refreshed after pricing for at most $\frac{m}{200}$ iterations. The number of arcs priced is set to $\frac{100}{frq}$, as above.

A variation on these strategies is to topologically sort the arcs, and scan them in groups incident with a node[22].

The table below displays the total runtimes with each of the four strategies for each of two settings of frq, 0.5 and 0.1. All times are on an IBM RS/6000 Model 550 with 256M of memory and running AIX 3.2, and with $p_0 = 2$.

frq	i	ii	iii	iv
0.5	22042.8	16857.6	16438.7	10871.2
0.1	23032.4	12292.3	12512.5	16486.0

Strategy (iii) was faster than (i) on all 200 runs. Strategy (ii) exhibited somewhat more variability than (iii). Strategy (iv) combined with $frq = .5$ appears to be superior to the other seven combinations tested on these NETGEN networks; it achieved the minimum run-time among the eight on 69 of the 100 networks, and was only about 10% slower than RELAX in total runtime on these observations (excluding one on which we were unable to get RELAX to terminate within specified limits). It would be interesting to examine all of the strategies further, particularly with a wider range of values of frq. In contrast with the NETGEN runs, in runs on a small number of GOTO networks strategy (iv) was the slowest by far and (iii) was the fastest.

6. Conclusions

These experiments yielded good models of mean execution time over two distributions of networks. They offer strong evidence that over those populations, for three of the four codes the inherent variability of execution times is very accurately approximated by a lognormal distribution with constant variance. It would be interesting to extend the general model (3.1) to the form (3.3) and attempt to get accurate fits to data from distributions arising from NETGEN, CAPT, etc., with a broader domain of generator inputs.

None of the four codes studied here can be regarded to be dominant over the entire set of distributions examined.

An accurate model of the average behavior of a particular code over a population of networks based on empirical data presents very different information from an estimate of worst-case complexity, and may appear to be of greater practical interest. However there are inherent limitations to such studies as these. In principle, the worst-case complexity of an algorithm over any permissible population of inputs can be estimated, and for the answer to be nontrivial the population will have to have infinite variation in the parameters of interest. However, for a sufficiently large diverse population there may not exist any simple model that accurately captures the average behavior of a particular code. Furthermore, even if an accurate and simple model exists, limitations on computational resources and on our ingenuity at generating test instances may not allow for adequate sampling of a large diverse population.

This study, like all empirical studies we know of in mathematical programming, samples from rather narrow classes of inputs. Thus we may find that what emerges is a good model of empirical behavior over a very limited population, and that may be the best we can hope for initially. Such models certainly extend our understanding of the codes under examination, and, with appropriate qualifications, the algorithms that the codes implement. If well designed, they may offer insights that can lead to more robust models and codes that are also more robust. However performance can be quite sensitive to variations in the codes and in the population of inputs. Certainly the realm of commercial applications of the minimum cost flow model does not have a sufficiently close connection to the populations that we test in this study, or any other comprehensive study of network flow algorithms that we know of, to draw fine conclusions about computation in that realm. In particular, we can construct from the generators different populations that have much more in common with one another than with the "real-world", but over which the relative merits of codes studied here shift dramatically.

When we observe consistent behavior of a code across diverse populations of networks, we may be on firmer ground in offering broader assessments, and even prognosticating on its merits for practical computation. Our success at obtaining good models for a few codes over a few (admittedly narrow) populations offers some hope in that direction.

Acknowledgments. We acknowledge with pleasure the frequent advice of David Ruppert on the statistical analysis.

REFERENCES

1. R. K. Ahuja, T. L. Magnanti, and J. B. Orlin, *Network flows*, in: Optimization (G. L. Nemhauser, A. H. G. Rinnooy Kan, and M. J. Todd,eds.), Handbooks in Operations Research and Management Science, vol. 1, North-Holland, Amsterdam, 1989, pp. 211–369.
2. T. Badics, *Maximum flow code PLED*, 1991, First DIMACS International Algorithm Implementation Challenge.
3. D. P. Bertsekas, *A distributed algorithm for the assignment problem*, unpublished LIDS working paper, March 1979.
4. D. P. Bertsekas, *Distributed asynchronous relaxation methods for linear network flow problems*, Tech. Report LIDS-P-1606, Dept. of EE&CS, M. I. T., September, 1986, Revised November 1986.
5. D. P. Bertsekas, *Linear Network Optimization*, MIT Press, Cambridge, MA, 1991.
6. D. P. Bertsekas and P. Tseng, *Relaxation methods for minimum cost ordinary and generalized network flow problems*, Operations Research **36** (1988), 93–114.
7. R. G. Bland and D. L. Jensen, *On the generality of network flow theory*, U. California-Berkeley, lecture, May 1979.
8. R. G. Bland and D. L. Jensen, *On the computational behavior of a polynomial-time network flow algorithm*, Mathematical Programming **54** (1992), 1-39.
9. DIMACS, *The First DIMACS International Algorithm Implementation Challenge: The Benchmark Experiments*, December 1991, New Brunswick, NJ.
10. G. B. Dantzig, *Linear programming and extensions*, Princeton University Press, Princeton, NJ, 1963.
11. U. Derigs and W. Meier, *Implementing Goldberg's max-flow-algorithm — a computational investigation*, ZOR - Methods and Models of Operations Research **33** (1989), 383–403.
12. N. R. Draper and H. Smith, *Applied regression analysis*, second ed., Wiley, New York, 1981.
13. J. Edmonds and R. M. Karp, *Theoretical improvements in algorithmic efficiency for network flow problems*, J. ACM **19** (1972), 248–264.
14. A. V. Goldberg, *The Grid-On-TOrus (GOTO) generator*, 1991, First DIMACS International Algorithm Implementation Challenge.
15. A. V. Goldberg, É. Tardos, and R. E. Tarjan, *Network flow algorithms*, in: Paths, Flows and VLSI-Design (B. Korte, L. Lovász, H.J. Proemel, and A. Schrijver, eds.) Springer Verlag, 1990, pp. 101-164.
16. A. V. Goldberg and R. E. Tarjan, *A new approach to the maximum-flow problem*, J. ACM **35** (1988), 921–940.
17. A. V. Goldberg and R. E. Tarjan, *Finding minimum-cost circulations by successive approximation*, Mathematics of Operations Research **15** (1990), 430–466.
18. M. D. Grigoriadis, *An efficient implementation of the network simplex method*, Mathematical Programming Study **26** (1986), 83–111.
19. D. Klingman, A. Napier, and J. Stutz, *Netgen: A program for generating large scale capacitated assignment, transportation, and minimum cost flow network problems*, Management Science **20** (1974), 814–821.
20. Y. Lee and J. B. Orlin, *Computational testing of a network simplex algorithm*, 1991, First DIMACS International Algorithm Implementation Challenge.
21. V. M. Malhotra, M. P. Kumar, and S. N. Maheshwari, *An $O(|V|^3)$ algorithm for finding maximum flows in networks*, Information Processing Letters **7** (1978), 277–278.
22. J. Mulvey, *Pivot strategies for primal-simplex network codes*, Journal of the ACM **25** (1978), 266–270.
23. H. Röck, *Scaling techniques for minimal cost network flows*, Discrete Structures and Algorithms (V. Page, ed.), Carl Hansen, Munich, 1980.
24. É. Tardos, *A strongly polynomial minimum cost circulation algorithm*, Combinatorica **5** (1985), 247–255.

TABLE 1. Parameter estimates for (a) CAPT (b) NETGEN. Standard errors are shown in parentheses.

CAPT	SCALE	MIN3	RNET	RELAX
R^2	0.990	0.993	0.963	0.963
b_0	-8.940	-9.728	-8.413	-12.104
	(0.165)	(0.142)	(0.341)	(0.524)
b_1: $\ln n$	0.683	0.332	0.268	1.029
	(0.023)	(0.020)	(0.047)	(0.073)
b_2: $\ln m$	0.127	0.699	0.489	-0.070
	(0.019)	(0.016)	(0.038)	(0.059)
b_3: $\ln \ln C$	1.144	0.433	0.009	0.040
	(0.046)	(0.040)	(0.095)	(0.146)
b_4: $\ln f$	0.356	0.213	0.375	0.740
	(0.010)	(0.008)	(0.020)	(0.030)
b_5: $\ln d$	-0.331	-0.159	-0.356	-0.697
	(0.007)	(0.006)	(0.014)	(0.021)

NETGEN	SCALE	MIN3	RNET	RELAX
R^2	0.976	0.979	0.962	0.914
b_0	-9.173	-9.293	-12.875	-12.442
	(0.248)	(0.250)	(0.508)	(0.555)
b_1: $\ln n$	1.126	0.771	0.584	0.593
	(0.032)	(0.032)	(0.065)	(0.071)
b_2: $\ln m$	0.130	0.519	0.929	0.591
	(0.021)	(0.021)	(0.044)	(0.048)
b_3: $\ln \ln C$	1.124	0.214	0.084	1.065
	(0.067)	(0.067)	(0.137)	(0.150)
b_4: $\ln f$	0.086	0.115	0.271	0.171
	(0.013)	(0.013)	(0.026)	(0.028)
b_5: $\ln d$	-0.032	-0.006	-0.203	-0.097
	(0.008)	(0.008)	(0.015)	(0.017)
b_6: $\ln s$	-0.079	-0.092	-0.079	-0.047
	(0.017)	(0.017)	(0.035)	(0.038)

TABLE 2. Confidence interval tallies for (a) CAPT (b) NETGEN.

CAPT Confidence Interval Results of Validation Test									
	SCALE	MIN3	RNET	RELAXT		SCALE	MIN3	RNET	RELAXT
LO	4	3	0	3	LO	18	21	24	22
95% IN	95	94	97	97	60% IN	64	59	61	58
HI	1	3	3	0	HI	18	20	15	20
LO	7	5	2	3	LO	25	32	37	30
90% IN	92	90	94	94	40% IN	45	37	35	39
HI	1	5	4	3	HI	30	31	28	31
LO	13	14	9	9	LO	32	41	46	39
80% IN	79	77	82	81	20% IN	25	19	18	23
HI	8	9	9	10	HI	43	40	36	38

NETGEN Confidence Interval Results of Validation Test									
	SCALE	MIN3	RNET	RELAXT		SCALE	MIN3	RNET	RELAXT
LO	2	4	2	2	LO	14	11	13	17
95% IN	93	90	93	94	60% IN	63	66	67	59
HI	5	6	5	4	HI	23	23	20	24
LO	5	4	2	2	LO	22	17	19	25
90% IN	89	88	90	94	40% IN	47	53	51	36
HI	6	8	8	4	HI	31	30	30	39
LO	10	7	5	5	LO	35	31	33	38
80% IN	79	78	82	84	20% IN	24	29	24	17
HI	11	15	13	11	HI	41	40	43	45

TABLE 3. (a) Parameter estimates for GRIDGEN. (b) Parameter estimates for GOTO. (c) R^2 values, when *bfrac* is included in the model. Standard errors are shown in parentheses.

GRIDGEN	SCALE	MIN3	RNET	RELAX
R^2	0.965	0.983	0.978	0.951
b_0	-9.414	-9.634	-12.003	-12.971
	(0.303)	(0.234)	(0.377)	(0.508)
b_1: $\ln n$	0.780	0.467	0.346	0.459
	(0.040)	(0.031)	(0.050)	(0.068)
b_2: $\ln m$	0.428	0.821	1.125	0.714
	(0.033)	(0.025)	(0.041)	(0.055)
b_3: $\ln \ln C$	1.142	0.109	-0.201	0.921
	(0.090)	(0.070)	(0.113)	(0.152)
b_4: $\ln f$	-0.003	-0.014	0.131	0.261
	(0.017)	(0.013)	(0.022)	(0.029)
b_5: $\ln d$	0.056	0.136	-0.107	-0.163
	(0.019)	(0.015)	(0.024)	(0.032)

GOTO	SCALE	MIN3	RNET	RELAX
R^2	0.839	0.878	0.909	0.711
b_0	-16.131	-13.862	-17.965	-25.688
	(1.426)	(1.088)	(1.027)	(2.378)
b_1: $\ln n$	0.847	0.989	0.191	-1.143
	(0.335)	(0.256)	(0.241)	(0.559)
b_2: $\ln m$	1.443	1.068	2.033	3.318
	(0.313)	(0.238)	(0.225)	(0.521)
b_3: $\ln \ln C$	0.155	0.255	0.074	3.679
	(0.365)	(0.279)	(0.263)	(0.609)
b_5: $\ln d$	0.103	0.066	0.077	0.078
	(0.049)	(0.037)	(0.035)	(0.082)
R^2 incl. *bfrac*	0.978	0.980	0.978	0.963

TABLE 4. DIMACS benchmark tests with the generators (a) GGRAPH (b) GOTO (increasing density) (c) NETGEN. For other network parameters, see [**9**, Section 4].

GGRAPH	SCALE	MIN3	RNET	RELAXT
$N = 2^{12}$ (long)	485.500	113.470	8.050	13.400
$N = 2^{13}$ (long)	1959.100	907.450	13.850	36.167
$N = 2^{14}$ (long)	6950.183	3563.580	44.700	132.050
$N = 2^{15}$ (long)	29553.200	12596.330	151.967	182.867
$N = 2^{10}$ (square)	27.200	9.270	1.167	3.083
$N = 2^{12}$ (square)	258.983	69.180	7.100	76.117
$N = 2^{14}$ (square)	2401.650	791.300	90.500	1119.900
$N = 2^{12}$ (wide)	171.583	56.320	5.933	99.233
$N = 2^{13}$ (wide)	410.250	199.030	11.967	502.900
$N = 2^{14}$ (wide)	962.133	256.980	32.217	3397.717
$N = 2^{15}$ (wide)	2324.033	691.330	70.683	16644.583

GOTO	SCALE	MIN3	RNET	RELAXT
$N = 2^8$	11.850	5.180	3.100	78.850
$N = 2^9$	47.750	31.730	15.217	513.567
$N = 2^{10}$	246.067	166.070	66.300	2441.633
$N = 2^{11}$	768.550	748.420	355.033	11735.250

NETGEN	SCALE	MIN3	RNET	RELAXT
Prob. 11	1.483	0.350	0.200	0.150
Prob. 20	3.367	0.750	0.167	0.167
Prob. 27	2.517	0.850	0.267	0.100
Prob. 28	8.867	3.670	0.500	0.333
Prob. 32	16.483	7.980	1.067	0.700
Prob. 36	206.417	64.170	12.150	14.450
Prob. 38	56.250	36.630	9.000	6.133
Prob. 40	53.117	28.470	6.617	3.650

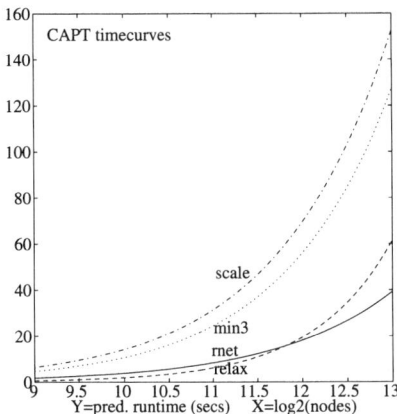
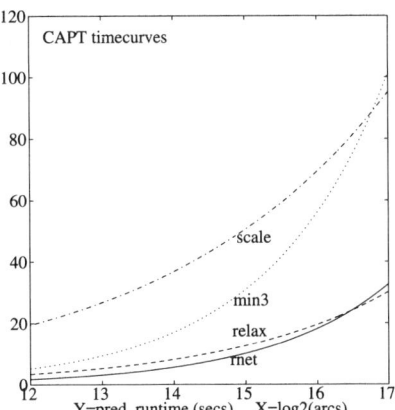

FIGURE 1. Predicted execution times (seconds) of SCALE, MIN3, RNET, and RELAX on CAPT networks: $hicost = 2^{14}$, $supply = 10^6$, $intlen = 5 \cdot supply/arcs$. (a) $nodes = 2^9, \ldots, 2^{13}$, $arcs = 16 \cdot nodes$, (b) $nodes = 2^{12}$, $arcs = 2^{12}, \ldots, 2^{17}$.

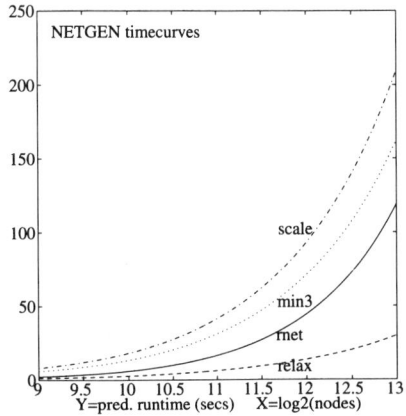
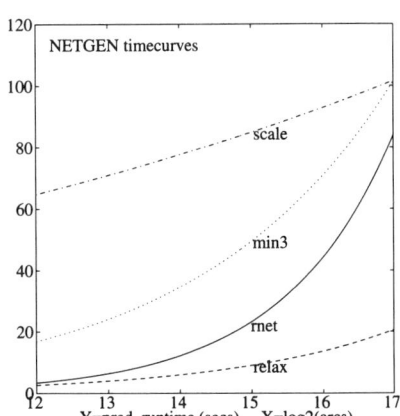

FIGURE 2. Predicted execution times (seconds) of SCALE, MIN3, RNET, and RELAX on NETGEN networks: $hicost = 2^{14}$, $hicap = 10^4$, $supply = 10^6$. (a) $nodes = 2^9, \ldots, 2^{13}$, $arcs = 16 \cdot nodes$, (b) $nodes = 2^{12}$, $arcs = 2^{12}, \ldots, 2^{17}$.

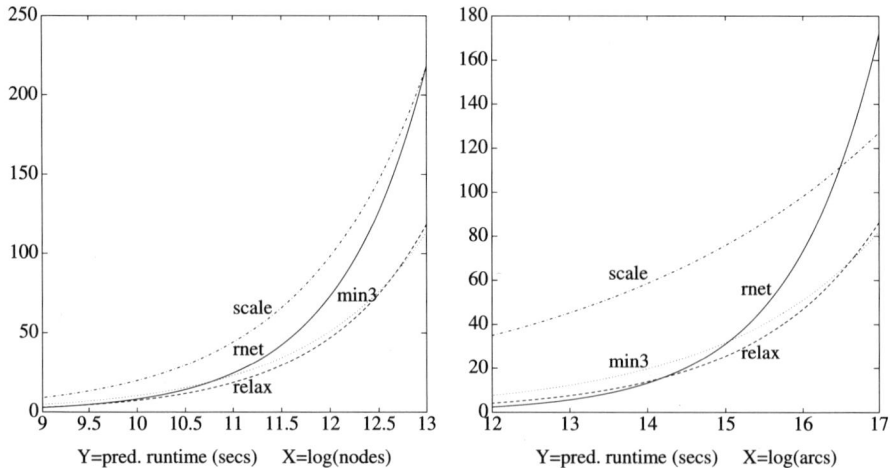

FIGURE 3. Predicted execution times (seconds) of SCALE, MIN3, RNET, and RELAX on GRIDGEN networks: $hicost = 2^{14}$, $supply = 10^6$, $hicap = 3 \cdot supply/arcs$, $width = \sqrt{nodes}$. (a) $nodes = 2^9, \ldots, 2^{13}$, $arcs = 16 \cdot nodes$, (b) $nodes = 2^{12}$, $arcs = 2^{12}, \ldots, 2^{17}$.

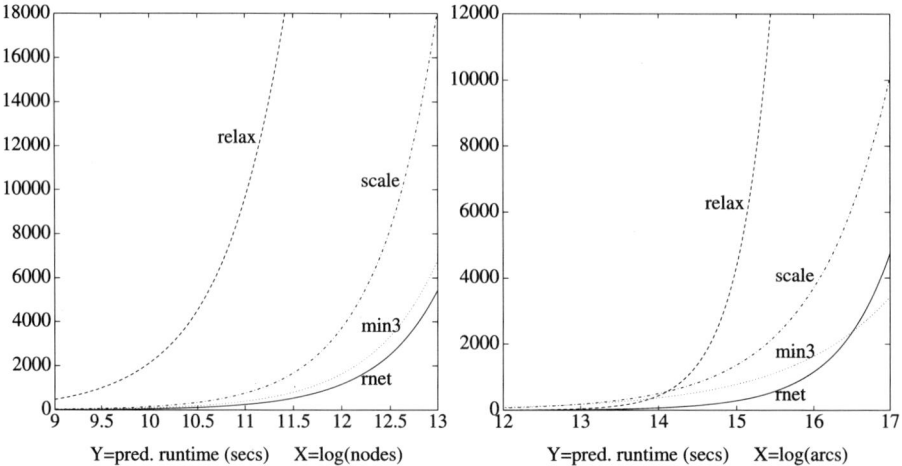

FIGURE 4. Predicted execution times (seconds) of SCALE, MIN3, RNET, and RELAX on GOTO networks: $hicost = 2^{14}$, $hicap = 10^4$. (a) $nodes = 2^9, \ldots, 2^{13}$, $arcs = 16 \cdot nodes$, (b) $nodes = 2^{12}$, $arcs = 2^{12}, \ldots, 2^{17}$.

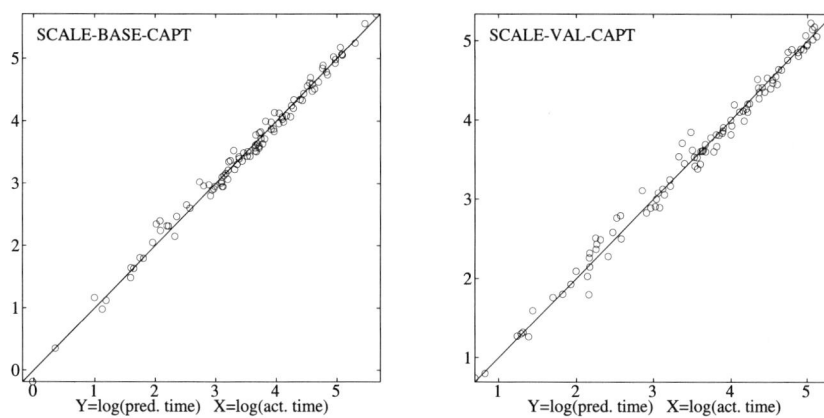

FIGURE 5. SCALE execution times on CAPT networks: Log of predicted versus log of observed execution time. (a) Base dataset. (b) Validation dataset.

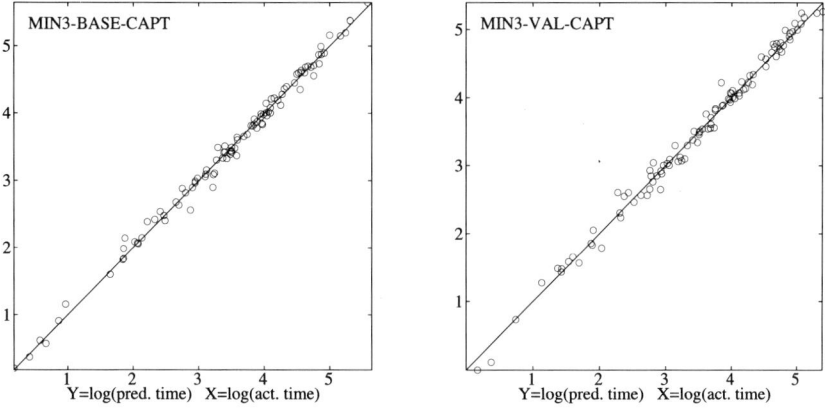

FIGURE 6. MIN3 execution times on CAPT networks: Log of predicted versus log of observed execution time. (a) Base dataset. (b) Validation dataset.

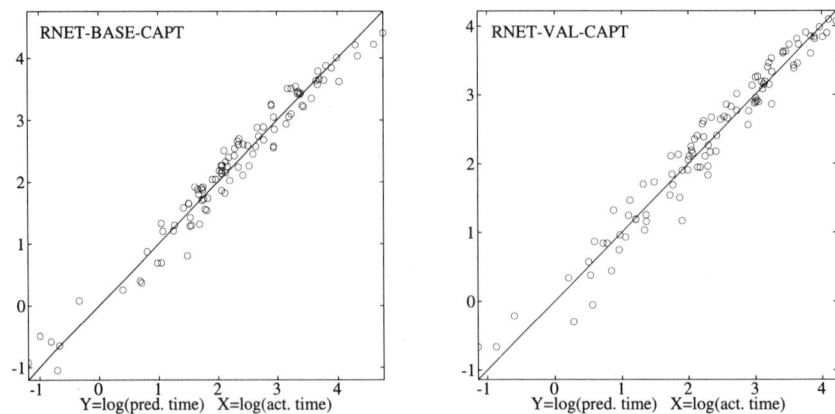

FIGURE 7. RNET execution times on CAPT networks: Log of predicted versus log of observed execution time. (a) Base dataset. (b) Validation dataset.

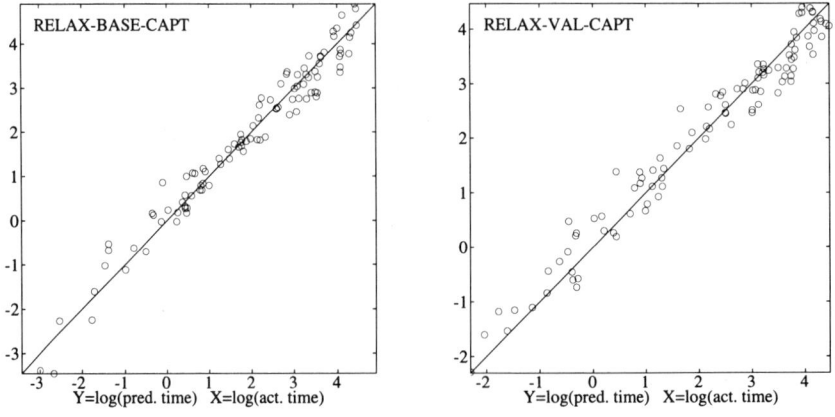

FIGURE 8. RELAX execution times on CAPT networks: Log of predicted versus log of observed execution time. (a) Base dataset. (b) Validation dataset.

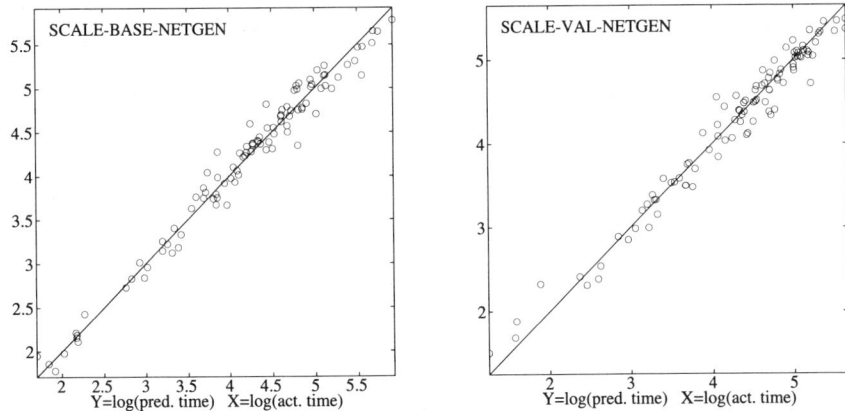

FIGURE 9. SCALE execution times on NETGEN networks: Log of predicted versus log of observed execution time. (a) Base dataset. (b) Validation dataset.

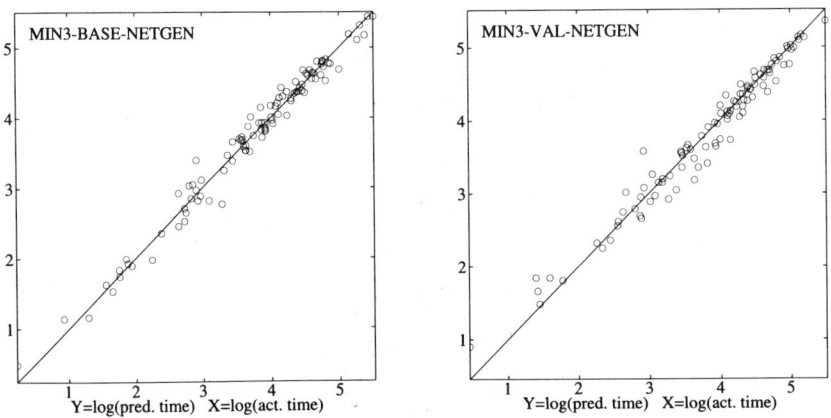

FIGURE 10. MIN3 execution times on NETGEN networks: Log of predicted versus log of observed execution time. (a) Base dataset. (b) Validation dataset.

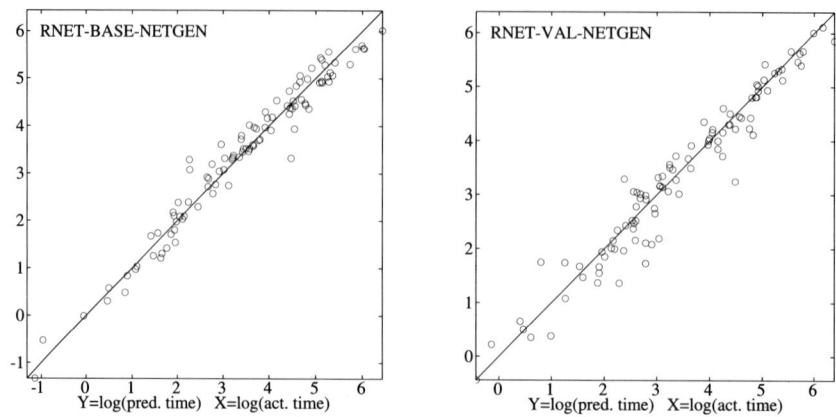

FIGURE 11. RNET execution times on NETGEN networks: Log of predicted versus log of observed execution time. (a) Base dataset. (b) Validation dataset.

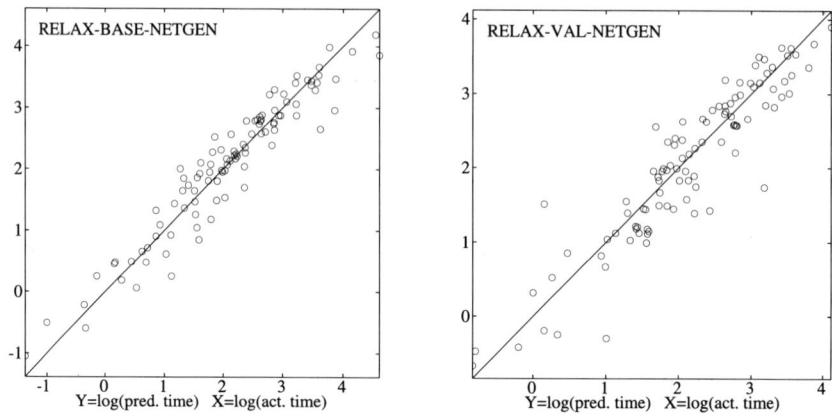

FIGURE 12. RELAX execution times on NETGEN networks: Log of predicted versus log of observed execution time. (a) Base dataset. (b) Validation dataset.

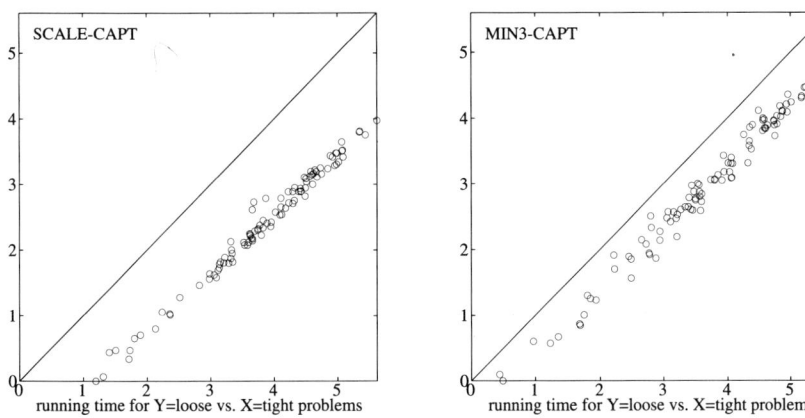

FIGURE 13. Time plots for tight and loose CAPT networks: Running times with tight capacities versus running times with loose capacities. (a) SCALE. (b) MIN3.

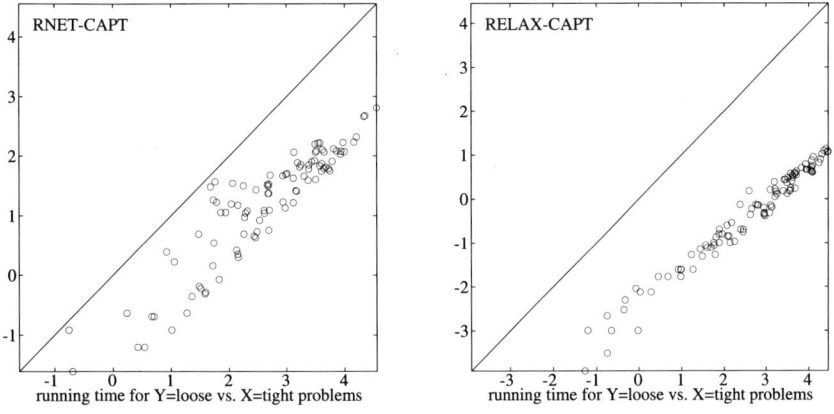

FIGURE 14. Time plots for tight and loose CAPT networks: Running times with tight capacities versus running times with loose capacities. (a) RNET. (b) RELAX.

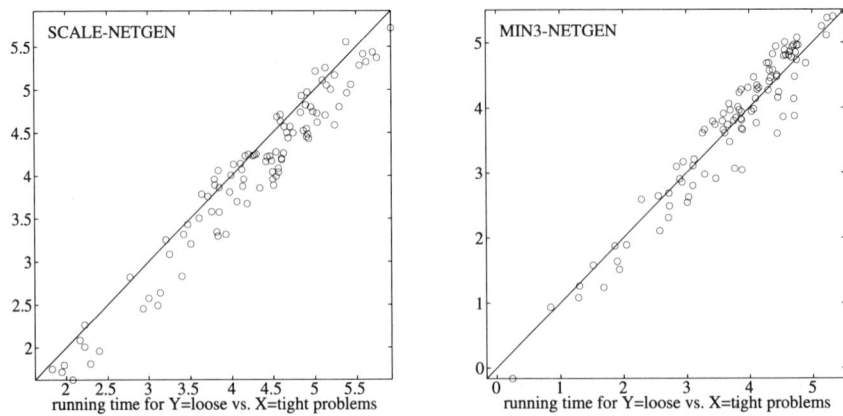

FIGURE 15. Time plots for tight and loose NETGEN networks: Running times with tight capacities versus running times with loose capacities. (a) SCALE. (b) MIN3.

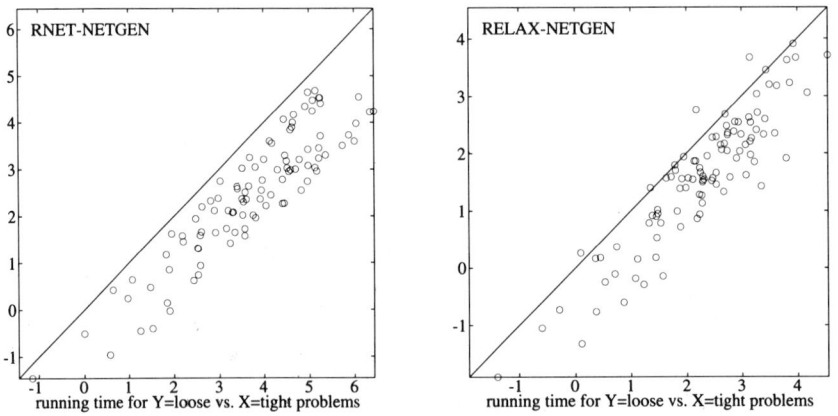

FIGURE 16. Time plots for tight and loose NETGEN networks: Running times with tight capacities versus running times with loose capacities. (a) RNET. (b) RELAX.

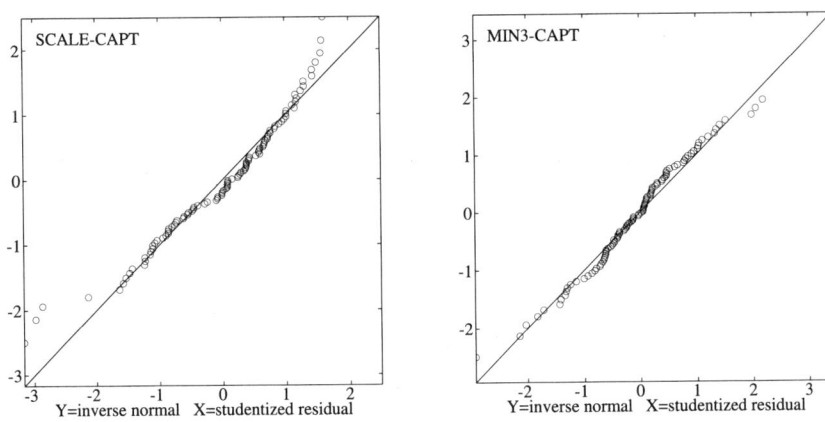

FIGURE 17. Probability plots for CAPT networks: Inverse of normal distribution versus studentized residuals. (a) SCALE. (b) MIN3.

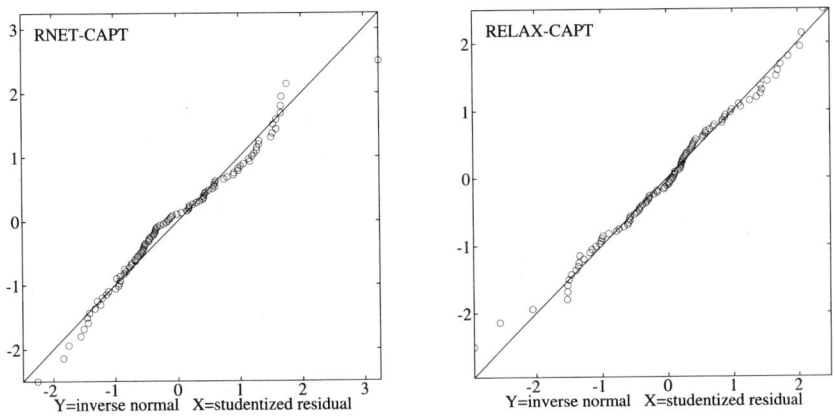

FIGURE 18. Probability plots for CAPT networks: Inverse of normal distribution versus studentized residuals. (a) RNET. (b) RELAX.

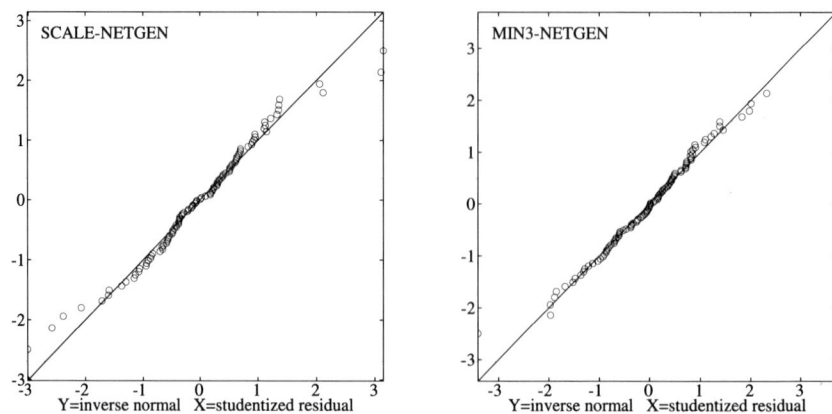

FIGURE 19. Probability plots for NETGEN networks: Inverse of normal distribution versus studentized residuals. (a) SCALE. (b) MIN3.

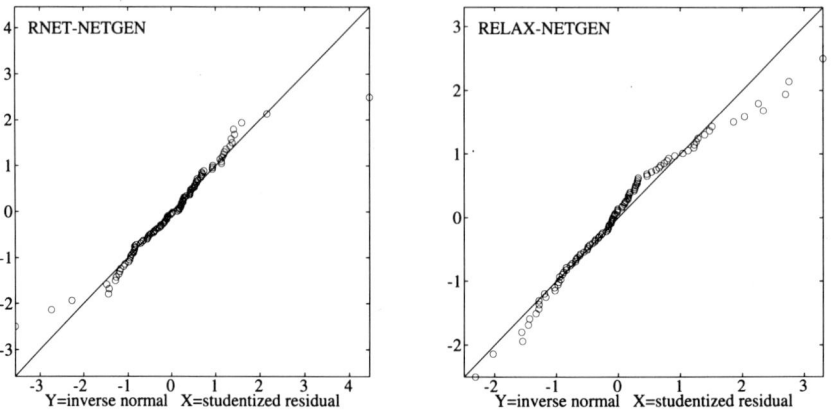

FIGURE 20. Probability plots for NETGEN networks: Inverse of normal distribution versus studentized residuals. (a) RNET. (b) RELAX.

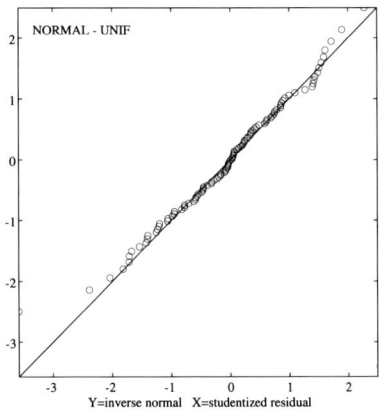

FIGURE 21. Probability plots for randomly generated normal variates created using uniform random variables. Inverse of normal distribution versus standard normal variates.

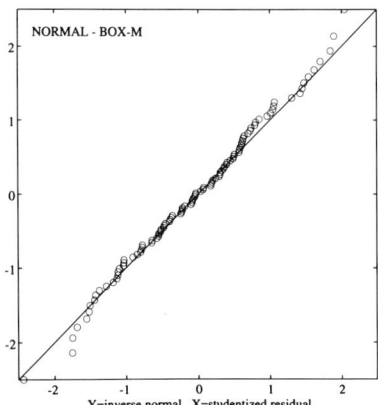

FIGURE 22. Probability plots for randomly generated normal variates created using Box-Mueller method. Inverse of normal distribution versus standard normal variates.

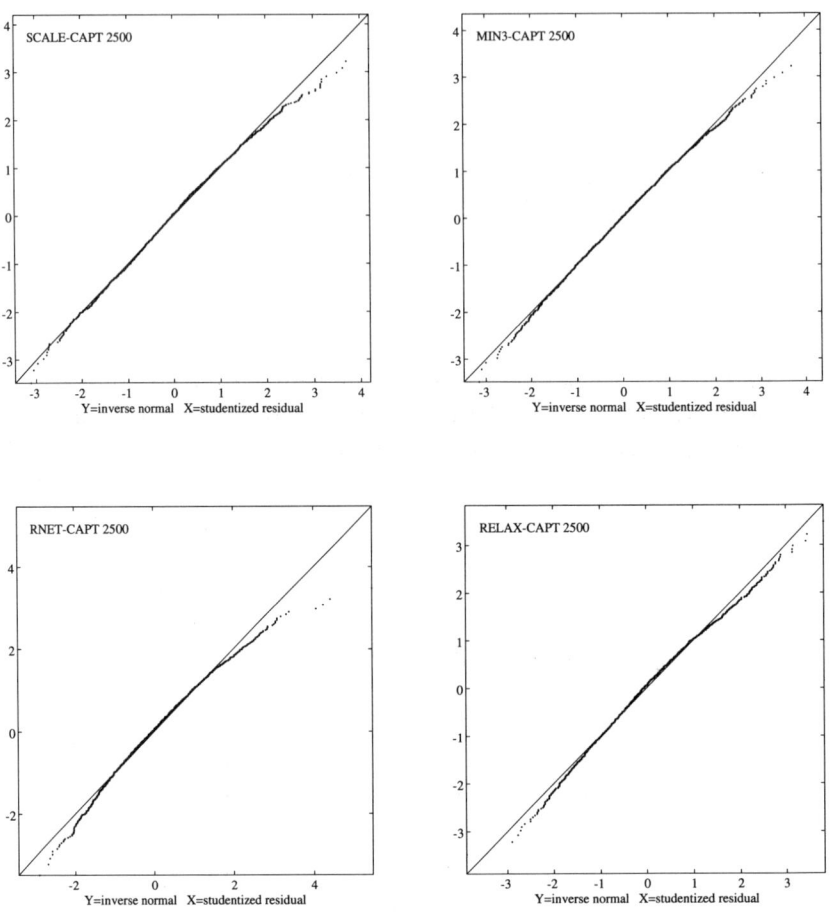

FIGURE 23. Probability plots for 2500 CAPT networks: Inverse of normal distribution versus studentized residuals. (a) SCALE (b) MIN3 (c) RNET (d) RELAX.

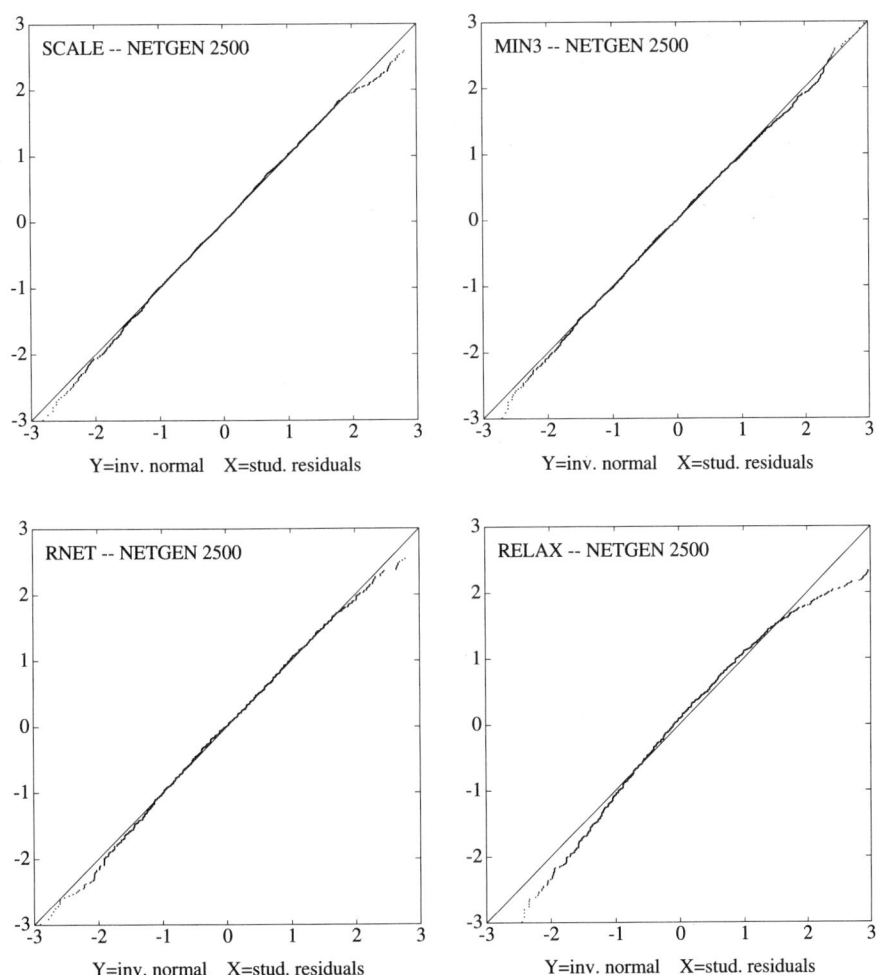

FIGURE 24. Probability plots for 2500 NETGEN networks: Inverse of normal distribution versus studentized residuals. (a) SCALE (b) MIN3 (c) RNET (d) RELAX.

School of O. R. and I. E., Cornell University, Ithaca, NY 14853-3801.
E-mail address: bland@orie.cornell.edu

Dept. of Combinatorics and Optimization, University of Waterloo, Waterloo, Ontario, Canada N2L 3G1.
E-mail address: jcheriyan@watdragon.waterloo.edu

I. B. M., T. J. Watson Research Center, Yorktown Heights, NY 10598.
E-mail address: djensen@ibm.com

School of O. R. and I. E., Cornell University, Ithaca, NY 14853-3801.
E-mail address: ladanyi@cs.cornell.edu

On Implementing Scaling Push-Relabel Algorithms for the Minimum-Cost Flow Problem

ANDREW V. GOLDBERG AND MICHAEL KHARITONOV

ABSTRACT. The invention of the scaling push-relabel method is an important theoretical development in the area of minimum-cost flow algorithms. In this paper we study implementations of this method. We are especially interested in heuristics that improve the performance of the algorithms in practice. Our results show that the technique works well on several problem classes with very different network structure. We also find a problem class on which the method does not perform as well and suggest directions for further improvement.

1. Introduction.

Significant theoretical progress has been made recently in the area of minimum-cost flow algorithms (see [1, 15] for surveys). Little is known, however, about the practical behavior of recent methods. Detailed studies of somewhat older methods include investigations of network simplex [18], cost-scaling [6], and relaxation [4].

The goal of this paper is to study practical performance of one recent technique, the *successive approximation push-relabel method* of Goldberg and Tarjan [12, 17]. This method combines and extends the ideas of cost-scaling, developed independently by Röck [28] and Bland and Jensen [6], the push-relabel maximum flow framework of Goldberg and Tarjan [11, 16], and the relaxation method of Bertsekas [3]. The experimental study of this new method is promising for two reasons. First, the inner loop of the method is based on the push-relabel algorithm for the maximum flow problem, which in that context has been shown superior to previous

---

1991 *Mathematics Subject Classification.* Primary 90B10; Secondary 68Q20.

The authors were supported in part by ONR Young Investigator Award N00014-91-J-1855, NSF Presidential Young Investigator Grant CCR-8858097 with matching funds from AT&T and DEC, Stanford University Office of Technology Licensing, and a grant form Mitsubishi Electric Laboratories.

The second author was supported in part by a Fannie and John Hertz Foundation Fellowship

codes in several experimental studies [**2, 7, 8, 19**]. Second, the successive approximation technique used in the method requires fewer iterations of the inner loop compared to the closely related cost-scaling.

One of the main advantages of the push-relabel framework is its flexibility: it runs in polynomial time regardless of the order in which its basic operations, *push* and *relabel*, are applied. Both theoretical and practical performance, however, depend on the operation selection strategy. This is true in the contexts of both the maximum flow and the minimum-cost flow problems; in the latter context, cost scaling adds another dimension of flexibility.

We study *heuristics* that improve the method's performance on a wide range of problems without necessarily changing its worst-case asymptotic behavior. Such heuristics proved crucial in the maximum flow context; a good example is gap-relabel of Derigs and Meier [**8**]. In some cases, heuristics apply in both maximum flow and minimum-cost flow contexts, but often the maximum flow heuristics do not apply or do not help in the minimum-cost flow context and vice versa. For example, gap-relabel does not apply to the minimum-cost flow problem.

Because of the variety of available heuristics and their interrelation, finding a combination of heuristics that works best in practice is a difficult task. We describe the heuristics we found useful in our experiments and those that we believe are worth further investigation. We implement several heuristics whose combination significantly improves the algorithm's performance.

Our experimental results show that the method works very well on a wide class of problems. We also identify a class of problems on which our implementation does not work as well, suggesting a direction for further improvement.

Recently, the first author developed another code based on the lessons learned from the work described in this paper and the ideas from [**14**]. The performance improvement of the new code over SPUR is considerable. These results are briefly discussed in Section 8, and will be described in detail in an upcoming paper.

This paper is organized as follows. Section 2 gives the relevant definitions and outlines the method. Section 3 talks about variations of the algorithm and comments on some implementation aspects. Section 4 discusses heuristics that can be used in the method. Section 5 describes our experimental setup and Section 6 gives the experimental results. Section 7 examines how much heuristics help. In Section 8, we give our conclusions and suggest directions for further research.

2. Background

In this section we briefly review the push-relabel method. For more detail, see [**17**].

2.1. Definitions and Notation. Our implementation works with the *capacitated transshipment* version of the minimum-cost flow problem defined as follows. A *network* is a directed graph $G = (V, E)$ with a real-valued *capacity* $u(a)$ and a real-valued *cost* $c(a)$ associated with each arc a, and a real-valued *demand* $d(v)$ associated with each node v.[1] For the rest of this paper, we assume that all costs, capacities, and demands are integers, as is the case in our implementation. We assume that G is *symmetric*, i.e., $a \in E$ implies that the reverse arc $a^R \in E$. (We add the reverse arcs during parsing.) The cost function satisfies $c(a) = -c(a^R)$ for each $a \in E$ and the total demand is zero, i.e., $\sum_V d(v) = 0$. We denote the size of V by n, the size of E by m, and the biggest input cost by C.

A *pseudoflow* is a function $f : E \to R$ satisfying the following *capacity* and *antisymmetry* constraints for each $a \in E$: $f(a) \leq u(a)$, $f(a) = -f(a^R)$.

For a pseudoflow f and a node v, the *excess flow into* v, $e_f(v)$, is defined by $e_f(v) = \sum_{(u,v) \in E} f(u,v) - d(v)$. A node v with $e_f(v) > 0$ is called *active*. Note that $\sum_{v \in V} e_f(v) = 0$.

A *(feasible) flow* is a pseudoflow f such that, for each node v, the demand at v is met, i.e., $e_f(v) = 0$. Observe that a pseudoflow f is a flow if and only if there are no active nodes. The *cost* of a pseudoflow f is given by $cost(f) = \frac{1}{2} \sum_{a \in E} c(a) f(a)$. The minimum-cost flow problem is to find a flow of minimum cost *(optimal flow)*.

For a given pseudoflow f, the *residual capacity* of an arc a is $u_f(a) = u(a) - f(a)$. An arc a is *saturated* if $u_f(v,w) = 0$, and *residual* if $u_f(a) > 0$. The *residual graph* $G_f = (V, E_f)$ is the graph induced by the residual arcs.

A *price function* is a function $p : V \to R$. For a given price function p, the *reduced cost* of an arc (v,w) is $c_p(v,w) = c(v,w) + p(v) - p(w)$.

For a given f and p, an arc a is *admissible* if it is a residual arc of negative reduced cost. The *admissible graph* $G_A = (V, E_A)$ is the graph induced by the admissible arcs. A flow f is optimal if and only if there exists a price function p such that no arc is admissible with respect to f and p [9].

For a constant $\epsilon \geq 0$, a pseudoflow f is said to be ϵ-*optimal with respect to a price function* p if, for every residual arc a, we have $c_p(a) \geq -\epsilon$. A pseudoflow f is ϵ-*optimal* if f is ϵ-optimal with respect to some price function p. If the arc costs are integers and $\epsilon < 1/n$, any ϵ-optimal flow is optimal [3].

2.2. The Method. First we give a high-level description of the successive approximation algorithm (see Figure 1). The algorithm maintains a flow f and a price function p, such that f is ϵ-optimal with respect to p. The algorithm starts with $\epsilon = C$, with $p(v) = 0$ for all $v \in V$, and with any feasible flow. A feasible flow can be found using one invocation of any maximum flow algorithm. Any flow is C-optimal with respect to the zero price function. The main loop of the algorithm repeatedly reduces ϵ by a constant factor α, the choice of which is discussed later. When $\epsilon < 1/n$, the algorithm terminates.

Reducing ϵ is the task of the subroutine *refine*. The input to *refine* is ϵ, f, and

---

[1] Sometimes we refer to an arc a by its endpoints, e.g., (v, w). This is ambiguous if there are several arcs from v to w. An alternative if to refer to v as the tail of a and to w as the head of a, which is precise but inconvenient.

```
procedure min-cost(V, E, u, c);
    [initialization]
    ε ← C;
    ∀v,  p(v) ← 0;
    if ∃ a flow then f ← a flow else return(null);
    [loop]
    while ε ≥ 1/n do
        (ε, f, p) ← refine(ε, f, p);
    return(f);
end.
```

FIGURE 1. The successive approximation algorithm.

```
procedure refine(ε, f, p);
    [initialization]
    ε ← ε/α;
    ∀(v, w) ∈ E do if c_p(v, w) < 0 then f(v, w) ← u(v, w);
    [loop]
    while ∃ a push or a relabel operation that applies do
        select such an operation and apply it;
    return(ε, f, p);
end.
```

FIGURE 2. The generic refine subroutine.

p such that f is ϵ-optimal with respect to p. The output from *refine* is ϵ reduced by a factor of α, a new f, and a new p such that f is ϵ-optimal with respect to p.

The algorithm terminates in $\lceil \log_\alpha(nC) \rceil$ iterations. After an optimal flow is found, our implementation uses a Bellman-Ford shortest path computation to obtain an optimal price function.

The generic *refine* subroutine (described on Figure 2) begins by decreasing the value of ϵ and saturating every arc with negative reduced cost. This action converts the flow f into an ϵ-optimal pseudoflow (indeed, into a 0-optimal pseudoflow). Then the subroutine converts the ϵ-optimal pseudoflow into an ϵ-optimal flow by applying a sequence of *push* and *relabel* operations, each of which preserves ϵ-optimality. The generic algorithm does not specify the order in which these operations are applied.

A *push* operation applies to a residual arc (v, w) of negative reduced cost whose tail node v is active. It consists of pushing $\delta = \min\{e_f(v), u_f(v, w)\}$ units of flow from v to w, thereby decreasing $e_f(v)$ and $f(w, v)$ by δ and increasing $e_f(w)$ and $f(v, w)$ by δ.

A *relabel* operation applies to an active node v that has no exiting residual arcs with negative reduced cost. It consists of decreasing $p(v)$ to the smallest value allowed by the ϵ-optimality constraints, namely $\max_{(v,w) \in E_f}\{p(w) - c(v, w) - \epsilon\}$. (If such a maximum is taken over the empty set, the problem is infeasible.)

The generic implementation of the algorithm needs one additional data structure,

---

push(v, w).
Applicability: v is active, $u_f(v, w) > 0$, and $c_p(v, w) < 0$.
Action: send $\delta = \min(e_f(v), u_f(v, w))$ units of flow from v to w.

relabel(v).
Applicability: v is active and $\forall w \in V\ u_f(v, w) > 0 \Rightarrow c_p(v, w) \geq 0$.
Action: replace $p(v)$ by $\max_{(v,w) \in E_f} \{p(w) - c(v, w) - \epsilon\}$.

FIGURE 3. The *push* and *relabel* operations described in the figure are somewhat restrictive. Some heuristics use different versions of these operations as discussed later.

Discharge.
Applicability: v is active.
Action: apply *push/relabel* operations to v until v becomes inactive.

FIGURE 4. The *discharge* operation.

a set S containing all active nodes. Initially S contains all nodes whose excess becomes positive during the initialization step of *refine*. Updating S takes only $O(1)$ time per *push* or *relabel* operation. (Such an operation requires possibly deleting one node from S and adding one node to S.)

At a low level, the *push* and *relabel* operations are combined in the *discharge* operation, described in Figure 4. A *discharge* operation applies *push* and *relabel* operations to an active node until the node becomes inactive, *i.e.*, its excess drops to zero. We assume the adjacency list representation of the graph and maintain a current arc pointer for every node v. The current arc of a node is set to its first arc initially and after each relabeling of the node. The *discharge(v)* operation attempts to push flow along the current arc of v. If the current arc is not eligible for pushing, *discharge* advances the current arc pointer to the next arc on the edge list of v unless the current arc is the last arc on the list, in which case v is relabeled.

There remains the issue of the order in which to discharge active nodes. The *first-in first-out* (FIFO) algorithm consists of maintaining the set of active nodes as a queue, repeatedly discharging the front node on the queue and adding newly active nodes to the rear of the queue. Another possibility is to maintain the set of active nodes as a stack, popping a node to be discharged from the stack and pushing newly activated nodes on the stack, as the LIFO algorithm does.

Two other ways of ordering the *discharge* operations use the fact that the *admissible graph remains acyclic* during *refine*. These two methods, called *first-active* and *wave*, are described next.

The first-active method maintains a list L of all the nodes of G in topological order with respect to the current admissible graph G_A, *i.e.*, if (v, w) is an arc of

```
procedure first-active;
    let L be a list of all nodes;
    let v be the first node on L;
    while ∃ an active node do begin
        if v is active then begin
            discharge(v);
            if the discharge has relabeled v then
                move v to the front of L;
        end;
        else replace v by the node after v on L,
    end;
end.
```

FIGURE 5. The *first-active* method.

G_A, v appears before w on L. Initially L contains the nodes of G in any order. The method consists of repeating the following step until no active nodes remain: Find the first active node on L, say v, apply a *discharge* operation to v, and move v to the front of L if the *discharge* operation has relabeled v.

In order to implement this method, we maintain a *current node* v of L, which is the next candidate for discharging. Initially v is the first node on L. The implementation, described in Figure 5, repeats the following step until there are no active nodes left: If v is active, apply a *discharge* operation to it, and if this operation relabels v, move v to the front of L; otherwise (*i.e.*, v is inactive), replace v by the node currently after it on L. The reordering of L maintains a topological ordering with respect to G_A. This implies that the implementation is correct.

The *wave* method differs from the first-active method in the following way: after moving a node v to the front of L, set the current node to the node after the *old* position of v; if v is the last node on L, set the current node to the first node on L.

The worst-case theoretical bounds on the number of basic operations invoked during an execution of *refine* are as follows:
- The number of *relabel* operations is $O(n^2)$.
- The number of *push* operations is $O(n^2 m)$ in any implementation of the generic method.
- The number of *push* operations in the first-active and wave methods is $O(n^3)$.

A dynamic tree data structure can be used to do several *push* operations at once, as described in [17]. Our experience suggests that in practice the *relabel* operations are the bottleneck, and the dynamic trees are not likely to help. We did not experiment with the dynamic tree version of the algorithm.

3. Method Variations and Implementation Notes

3.1. Scaling Factor Selection. Recall that the goal of *refine* is to reduce the error parameter ϵ by a factor of α. The running time of the algorithm depends on the value of α. This dependence, however, is not uniform: a value of α that is the best for one problem instance may not be the best for another instance.

We found that scaling factors of 4, 5, and 6 work well on the classes of problems we encountered. We used $\alpha = 5$ in our experiments. The scaling factor of 2 and large scaling factors (20 and above) are usually a bad choice.

As one would expect, if two scaling factors result in the same number of executions of *refine*, the smaller scaling factor almost always gives a better running time. Unfortunately, our implementation cannot take advantage of this observation because it uses early termination detection (described below) and cannot predict the number of executions of *refine*.

3.2. A Simpler Relabeling. The push-relabel method can use a simpler version of the *relabel* operation which replaces $p(v)$ by $p(v) - \epsilon$ instead of the maximum computed as described in Figure 3. This version of *relabel* is more efficient but results in a higher number of *relabel* operations because it often does not reduce $p(v)$ as much as the variant described in the Figure 3. We experimented with the two versions of *relabel* and found them to be very close in performance. Our final implementation uses *relabel* given in Figure 3.

We would like to note that the relative performance of the two variants of the method depends on the implementation details and on the problem structure.

3.3. A Variation of Discharge. The *discharge* procedure, as described in Figure 4, terminates when v is no longer active. An alternative, with the same worst-case asymptotic time bounds, is to terminate *discharge* when v is no longer active or when v has been relabeled by the *discharge*. We found the former version to be superior for most implementations.

3.4. Implementing First-Active and Wave Methods. Recall that the first-active algorithm selects an active node with no active predecessors. Implemented as described in Section 2.2, this algorithm maintains a topologically ordered (with respect to the admissible graph) list of nodes and finds the next active node by scanning the list.

Several researchers [29, 31] have observed that in practice a small percentage of nodes is active during an execution of *refine*. We also observed that the number of active nodes, although large at first, decreases quickly and stays low during most of the execution. The fraction of active nodes seems to go down with problem size; even for a moderate problem size (a thousand nodes) it is already below one percent.

Because of the small number of active nodes, the above mentioned implementation spends most of its time scanning the list in search of the next active node. In theoretical work, this problem came up in the context of an implementation that uses dynamic trees to reduce the number of *push* operations [17]. The proposed solution was to use finger search trees. The theoretical motivation for using this data structure, which is quite sophisticated, is to avoid introducing a logarithmic factor into the running time bound. In the context of the implementation discussed here (which does not use dynamic trees), this theoretical objective does not apply, and of course in practice one uses the data structure that works best.

One alternative is to use balanced (or self-adjusting) search trees to maintain the list of nodes. However, the logarithmic and constant factors introduced by this data structure are relatively large. We obtain a better implementation of the first-active method by using instead a priority queue implemented as a linear list. We refer to this implementation as the PFA implementation throughout the rest of the paper.

The PFA implementation maintains topological numbers of nodes which are updated during relabels. The priority queue maintains the list of active nodes ordered by their topological numbers. The active node with the largest topological number is removed from the priority queue and discharged, and newly activated nodes are added to the priority queue. Note that the PFA implementation is very much like the FIFO implementation, except that the topological numbers are maintained and a priority queue is used instead of a fifo queue. The PFA implementation works fairly well because priority queues are quite efficient in practice and the number of active nodes is small on the average.

The corresponding implementation of the wave algorithm is slightly more complicated and uses two priority queues, *current* and *next*. A node v to be discharged is removed from the current queue, and its topological number l is saved. A newly activated node w is added to the current queue if its topological number is less than l and to the next queue otherwise (the latter case can occur if *discharge* relabels v). When the current queue is empty, the current and the next queues are exchanged.

3.5. Comparison of Selection Strategies. As mentioned in Section 2.2, the efficiency of the method depends on the strategy used to select the next node to be discharged. In addition to the FIFO, LIFO, first-active, and wave selection strategies, we experimented with several other strategies, including selecting a node with maximum excess and selecting an active node with the minimum topological number.

The FIFO and LIFO strategies give similar results, both in terms of running times and operation count. The first-active and wave implementations seem to result in a slightly larger number of operations and add to the computational costs of *relabel* and node selection operations. The maximum excess strategy results in a significantly higher number of operations and these operations require more computation as well.

Although our first-active and wave implementations are not quite as fast as the FIFO and LIFO implementations, they are not drastically slower. With addition of some heuristics, the former implementation ran faster than the latter ones on some instances. We used the FIFO strategy in our final tests. According to preliminary experiments on moderate sized problems, the LIFO strategy would give similar results.

The node selection strategy performance depends on heuristics used by the implementation. Although FIFO and LIFO strategies are easier to implement, have lower computational cost, and worked faster in out study, the first-active and wave strategies cannot be dismissed; it is possible that one of these strategies will win when combined with appropriate heuristics.

4. Heuristic Improvements

4.1. Price Refinement and Termination Detection. As suggested in [17], refine may produce a solution which is not only ϵ-optimal, but also (ϵ/α)-optimal. This can be tested by adding ϵ/α to the reduced costs and performing a Bellman-Ford shortest path computation on the residual graph with respect to the resulting length function. If no negative cycles are detected, f is (ϵ/α)-optimal and the price function showing this can be obtained as a by-product. In this case we say that the heuristic succeeds. We call this heuristic *price refinement*.

As observed by Bland et. al. [5], price refinement rarely succeeds except at the end of the computation when the current solution is optimal even though ϵ is greater than $1/n$. This observation suggests using the Bellman-Ford computation (without changing the reduced costs) to test for optimality.

We check for optimality after every iteration of refine when ϵ is 1 or less, and call this *early termination detection*.

Price refinement can be also done using a minimum-mean cycle computation [22, 23]; see [17] for details. We did not evaluate this way of doing price refinement.

4.2. Arc Fixing. The arc fixing heuristic involves "removing" some arcs from the graph, thus reducing the number of times the algorithm examines an arc.

The theoretical justification of this technique is as follows [30], [17]: if the current circulation is ϵ-optimal and the absolute value of an arc cost exceeds $2n\epsilon$, the push-relabel method will not change the flow on this arc. Thus the arc does not need to be examined until the optimal flow value computation. Arc fixing can be done after every execution of *refine*.

In the dual context, arc fixing corresponds to edge contraction. Fujishige et. al. [10] propose contracting edges earlier than the theory suggests. We use this idea in the primal context and call the resulting heuristic *speculative arc fixing*. This heuristic fixes all arcs with the absolute value of reduced cost greater than β, where β is a parameter that depends on the input. Fixed arcs are not examined by *refine*, and the flow on these arcs does not change.

Speculative arc fixing is done after every execution of *refine*. During this process, we first check if all fixed arcs with current reduced cost below $-\epsilon$ are saturated. If this is not the case, the current flow is not ϵ-optimal; we say that the speculative arc fixing has failed. In this case we use the regular arc fixing strategy and invoke *refine* again without decreasing ϵ; the resulting solution is ϵ-optimal. Then we decrease ϵ

and examine all the arcs again, fixing arcs speculatively.

A proper choice of β is important. The smaller β is, the fewer arcs *refine* has to deal with. If β is too small, however, speculative arc fixing fails often and the number of invocations of *refine* increases. We used $\beta = 5 + 0.7n^{5/8}$ in our experiments.

Our implementation uses the speculative arc fixing heuristic.

4.3. Price Updates. The push-relabel method modifies prices locally, one node at a time. *Price update* heuristics modify prices in a more global way. In the maximum flow context, price updates, implemented using breadth-first search, have been shown to significantly improve practical performance of the push-relabel method.

When price update heuristics are used, the following conditions must be satisfied for the theoretical analysis of the method to apply:
  (i) ϵ-optimality must be preserved,
  (ii) acyclicity of the admissible graph must be preserved,
  (iii) prices must be monotonically decreasing,
  (iv) prices of nodes with negative excess must remain unchanged.

One way to update prices is at the beginning of *refine,* when all negative reduced cost arcs are saturated. This can be done using a shortest path computation on the residual graph with respect to the length function given by reduced costs. For each node, the shortest distance to the set of nodes with negative excesses is computed and subtracted from the node's price. Preliminary experiments suggest that this price update heuristic decreases running times a little on some classes of problems and does not help on other classes. Furthermore, this heuristic can be used at most once per an execution of *refine*.

Suppose the algorithm rounds costs to the nearest multiple of ϵ every time ϵ decreases. Then prices and reduced costs are always in units of ϵ. To update prices at any point during an execution of *refine* we can add ϵ to the reduced costs and use the resulting function as a length function. This length function is nonnegative on residual arcs. We can update prices by computing distances from all nodes to the set of nodes with negative excesses and subtracting these distances from the node prices to obtain new prices.

The rounding of costs is needed to assure that the admissible graph remains acyclic. When the costs are rounded, the only possible negative value of reduced cost of a residual arc is $-\epsilon$ by ϵ-optimality. Since the admissible graph is acyclic before a price update, the minimum mean cycle cost in the residual graph is greater than $-\epsilon$. A cycle in the admissible graph must have a mean cost of $-\epsilon$, so the admissible graph cannot have cycles after the update.

Although this kind of price update, when done often enough, reduces the number of *push* and *relabel* operations, it is expensive and did not produce consistent running time improvements in our experiments.

One of the drawbacks of the price update technique described above is that it is too global in a sense that it may change prices of most nodes. Next we describe a price update strategy that does not have this drawback and does not require the rounding of costs. This update strategy is a modification of Dijkstra's shortest path algorithm. It also can be viewed as a relaxed version of positive cut canceling [4, 20].

We define a *set-relabel* operation as follows. Let S be a set of nodes that contains all nodes with negative excess and $\bar{S}$, the complement of S, contains at least one node with positive excess. Suppose that no admissible arc goes from a node in $\bar{S}$ to a node in S. The *set-relabel* operation reduces the price of every node in $\bar{S}$ by

$$\epsilon + \min\{c_p(v,w) | (v,w) \in E_f, \, v \in \bar{S}, \, w \in S\}.$$

(Alternatively, the prices can be reduced by ϵ.) The $O(n^2)$ bound on the number of relabels per *refine* applies to set-relabels as well. If fact, if set-relabels are used as described below, we can make a stronger claim that each node participates in $O(n^2)$ set-relabels per *refine*.

The *set-relabel* operation can be applied in the following way. Initially, the set S contains a set of all nodes with negative excess. At each iteration, the set S is extended to include all nodes from which a node in S is reachable in the admissible graph. If all nodes with positive excess are in S, the computation terminates. If not, *set-relabel* is applied to S and the next iteration begins. This computation can be implemented using a priority queue in a way similar to that of Dijkstra's shortest path algorithm. Note that this implementation of price update works correctly even if it is stopped after an arbitrary iteration.

Although SPUR does not use any price update heuristics, subsequent work shows that price updates based on the *set-relabel* operation substantially improves performance of the method. See Section 8 for discussion.

4.4. Lookahead and Generalizations.

The following scenario seems to be common in practice. Consider two nodes, v and w, such that $e_f(v) > 0$ and $e_f(w) \geq 0$. Suppose v pushes flow to w, and the first time flow is pushed from w afterwards, this flow is pushed back to v. Observe that this can happen only if w does not have any outgoing admissible arcs just before the flow is pushed into it. Intuitively, the work done during the two pushes is wasted.

Such a situation can be avoided using the *lookahead* heuristic: before pushing flow to a node w, check whether w has an outgoing admissible arc or whether $e_f(w) < 0$. If this is so, do the pushing; if not, relabel w. A technical difficulty is that w may be inactive and the *relabel* operation, as described in Section 2.2, may not apply. In this case, either a node with negative excess is reachable from w or no such node is reachable. In the former case, the method remains correct if *relabel* is applied to w by the same argument as the one presented in [17] for active nodes. In the latter case, one can show that *relabel* still can be applied to w except when w has no outgoing residual arcs. In this case, however, the price of w can be decreased by an arbitrary amount without violating ϵ-optimality. For example, we can decrease the price of w by ϵ. Alternatively, we can decrease the price by a large enough amount so that all arcs adjacent to w will be fixed.

We use the lookahead heuristic in our implementation. This heuristic reduces the number of pushes significantly; in many cases, the number of pushes falls below the number of relabels. See Section 7 for experimental data.

The lookahead heuristic can be generalized in the following way. Define the *admissible capacity* of a node x to be $a_f(x) = -e_f(x) + \sum_{(x,y) \in E_A} u_f(x,y)$. (The sum taken over the empty set is defined to be zero.) The admissible capacities can be maintained by the *push* and *relabel* operations.

The *relabel* operation can be extended to apply to nodes v with nonnegative excess and no outgoing admissible arcs. The *push* operation can be modified so that the amount pushed from v to w is bounded by $a_f(w)$ as well as by e_f and $u_f(v,w)$. If $a_f(w)$ provides the tightest constraint on the amount pushed, (v,w) remains admissible and v remains active. In this case v cannot be relabeled until w is relabeled. We call such a node w *hyperactive*. We assume that discharging of v stops if it makes a node w hyperactive, and v is not processed again until w is relabeled.

Note that while a hyperactive node v is being discharged (and before it is relabeled), another node w can become hyperactive, and w must be relabeled before v is. Note, however, that the arc (v,w) remains admissible until w is relabeled. Since the admissible graph is acyclic, we can process the hyperactive nodes in the proper order. For example, we can keep such nodes on a stack. Alternatively, we can select a hyperactive node with the smallest topological number with respect to the admissible graph.

This generalization of the lookahead heuristic adds another dimension to the flexibility of the push-relabel method. Preliminary experiments show that this heuristic reduces the number of *push* operations, but additional work is needed to determine if some version of this heuristic reduces the number of *relabel* operations and if such a version yields running time improvements.

4.5. Good Initial Solution.
The method could benefit from a good initial flow and price function. For example, after finding an initial flow, one can use a minimum cycle mean computation to find a price function that minimizes the error parameter ϵ (see [17]). Better initial prices may reduce the number of iterations of *refine*, especially on networks that have a small number of arcs with very large costs (such arcs are supposed to be used only if the problem would be infeasible without them). Other ways of finding good initial solutions may be possible.

4.6. Heuristics Used.
Our implementation uses early termination detection, lookahead, and speculative arc fixing heuristics.

5. Experimental Setup

We evaluated our code on eight network families produced by three generators, all obtained from DIMACS: GOTO, NETGEN, and GRIDGRAPH.

The GOTO generator is described in the appendix. The generator takes five parameters: number of nodes, number of edges, maximum capacity, maximum cost, and a seed for the random number generator. We use GOTO to produce three example families. In all of these families, the maximum capacity parameter is set to 2^{14} (16384) and the maximum cost to 2^{12} (4096). The families are parameterized by the number of nodes n and differ in graph density as follows:

- GOTO-8 family examples have density 8;
- GOTO-16 family examples have density 16;
- GOTO-I family examples have density $\lfloor \sqrt{n} \rfloor$.

NETGEN is a "classical" generator developed by Klingman, Napier, and Stutz [25]. We used a version of NETGEN (obtained from DIMACS and fixed by Bland et. al. [5]) to generate two example families, NETGEN-HI and NETGEN-LO. The families are identical except for maximum capacity value.[2] The assignments to the 15 parameters of NETGEN are as follows:

- NETGEN-HI:

1	seed	Random number seed (a large integer).
2	problem	Problem number (for output documentation).
3	nodes	Number of nodes: $N = 2^n$.
4	sources	Number of sources: 2^{n-2}.
5	sinks	Number of sinks: 2^{n-2}.
6	density	Number of (requested) arcs: 2^{n+3}.
7	mincost	Minimum arc cost: 0.
8	maxcost	Maximum arc cost: 4096.
9	supply	Total supply: $2^{2\times(n-2)}$.
10	tsources	Transshipment sources: 0.
11	tsinks	Transshipment sinks: 0.
12	hicost	Percent of skeleton arcs given max cost: 100%.
13	capacitated	Percent of arcs to be capacitated: 100%.
14	mincap	Minimum capacity of capacitated arcs: 1.
15	maxcap	Maximum capacity of capacitated arcs: 16384.

- NETGEN-LO:
  same as NETGEN-HI except maximum capacity (the last parameter) is 16.

The GRIDGRAPH generator, written by Resende [27] and based on a generator proposed by Karmarkar and Remakrishnan [21], produces networks that form rectangular grids with a source and a sink. This generator has five parameters:

---

[2] This difference, however, may result in a substantial difference in the generated examples.

grid height X, grid width Y, maximum capacity, maximum cost, and a seed for the random number generator. We use this generator to produce three example families. In all these families, the maximum capacity and cost parameters are set to 10^4 (10000). The values of X and Y are set as follows:

- GRID-SQUARE family, $X = Y$;
- GRID-WIDE family, $Y = 16$ and X increases;
- GRID-LONG family, $X = 16$ and Y increases.

We also experimented with GTE instances. These instances were obtained by Leong, Shor, and Stein as subproblems in a multicommodity flow algorithm [26] applied to a real-life problem. The GTE networks all have the same graph structure but different demands and costs. These instances are quite small; each has 49 nodes and 520 edges.

Our code is written in C and compiled using SUN C compiler with the optimization option "–O4".

As a benchmark to compare our code against, we use the code RELAX of Bertsekas and Tseng [4]. The version we used was RELAXT, April 1990, compiled using SUN Fortran compiler with the optimization option "–O".

Our experiments were conducted on SUN Sparc-2 workstations with 40 MHz CPUs. The running times we measure are user execution times (measured with 1/60 seconds precision). The input-output time is not included.

A machine calibration experiment with the "wmatch" program provided by DIMACS was performed. The C version of the program was compiled without optimization and with level 4 optimization (used to compile our code). In each case the "run.me" shell script provided by DIMACS was run four times and results were averaged over these runs. For the wmatch test 1 running times were 2.8 real, 2.7 user, and 0.1 system (not optimized) and 2.2 real, 2.0 user, and 0.1 system (optimized). For the wmatch test 2 running times were 24.6 real, 24.2 user, and 0.2 system (not optimized) and 19.2 real, 18.1 user, and 0.3 system (optimized).

6. Experimental Results

We describe our experimental results below. For every family and every problem size, we give average running times of our code, SPUR (Scaling **PU**sh-**R**elabel), on instances whose size starts at 2^8 nodes and doubles at every step. (If a generator cannot produce a problem of exactly the desired size, we use a close size that the generator can produce.) We also give the average running times of RELAX on the same sets of instances. We take averages over four instances of each size.[3] For each problem family, we try to go to the biggest possible size, subject to the limits imposed by computation times and generator restrictions.

---

[3]Because of the large variation in running times, a bigger sample size would have given more uniform results, but would limit experiments on large problem sizes.

6.1. GOTO Families.
We measured performance of SPUR on three families of GOTO problems with densities 8, 16, and $\sqrt{n}$. The results are summarized in Figure 6. These experiments show that the algorithm's performance depends on both the number of nodes and the number of edges. The running times seem to depend more on the number of nodes than on the number of edges.

Figures 7, 8, and 9 compare the running times of SPUR on the GOTO families with those of RELAX. SPUR is faster by an order of magnitude. The running times ratio depends on the problem density. There is no noticeable asymptotic growth of the ratio as the number of nodes increases.

6.2. NETGEN Families.
Figure 10 gives running time of SPUR on the NETGEN families. Recall that these families differ only by the capacity upper bound. SPUR's performance is only slightly worse for tightly capacitated problems.

Figure 11 gives the running times of SPUR and RELAX on the NETGEN-HI family. RELAX is faster by a factor of two to three most of the time.

The NETGEN-LO family differs from the NETGEN-HI family in the capacity upper bound, which is set to 16. RELAX is a little faster on small instances, and SPUR is faster on larger instances. The RELAX/SPUR running time ratio grows with problem size, and on networks with 2^{16} nodes SPUR is faster by more than a factor of four.

6.3. GRIDGRAPH Families.
Figure 13 summarizes our experiments with the GRIDGRAPH families. The data indicates that SPUR runs much faster on shorter networks than it does on longer ones.

RELAX running times depend on the grid geometry in the opposite way. As the result, SPUR is asymptotically faster on wide grids and RELAX on long grids. On square grids, SPUR is asymptotically faster. See Figures 14 – 16 for running time comparisons.

Observe that the running times of SPUR on wide grids are very close to the running times of RELAX on long grids of the same size. However, the running times of SPUR on long grids are asymptotically better than those of RELAX on wide grids. Thus SPUR has a better "worst-case" performance on GRIDGRAPH examples.

6.4. GTE Instances.
Figure 17 gives running times of SPUR and RELAX on GTE instances. Recall that these instances are subproblems in a multicommodity flow algorithm and have the same underlying graph, but different demands and costs.

The graph is small and SPUR solves each instance in a fraction of a second. Although some problems are harder for SPUR than others, its running times are all within a factor of three of each other.

The behavior of RELAX on these instances is different. On some of them it runs very fast, and on others extremely slow. The ratio between the slowest and fastest

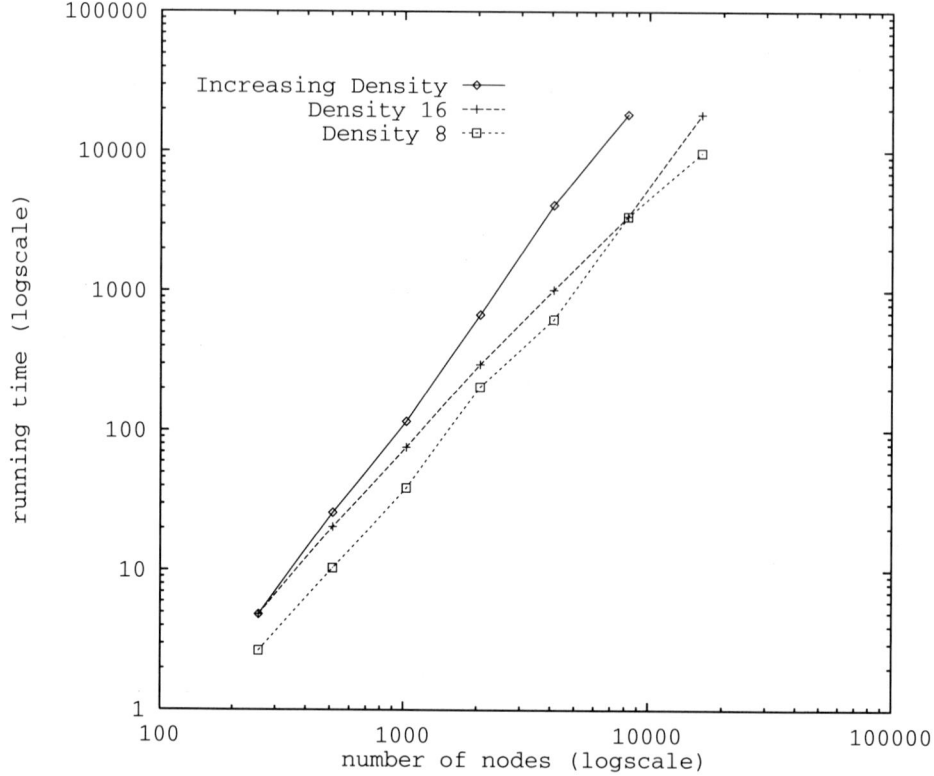

Problem size	SPUR running time			
(nodes)	Density 8	Density 16	Density $\sqrt{n}$	$\sqrt{n}$
256	2.662	4.842	4.842	16
512	10.358	20.525	25.904	23
1024	38.904	76.037	116.033	32
2048	205.767	297.875	676.196	45
4096	626.221	1018.833	4131.342	64
8192	3420.925	3444.529	18410.209	91
16384	9728.400	18331.559	timeout	128

FIGURE 6. Average running times (in seconds) of SPUR on GOTO families.

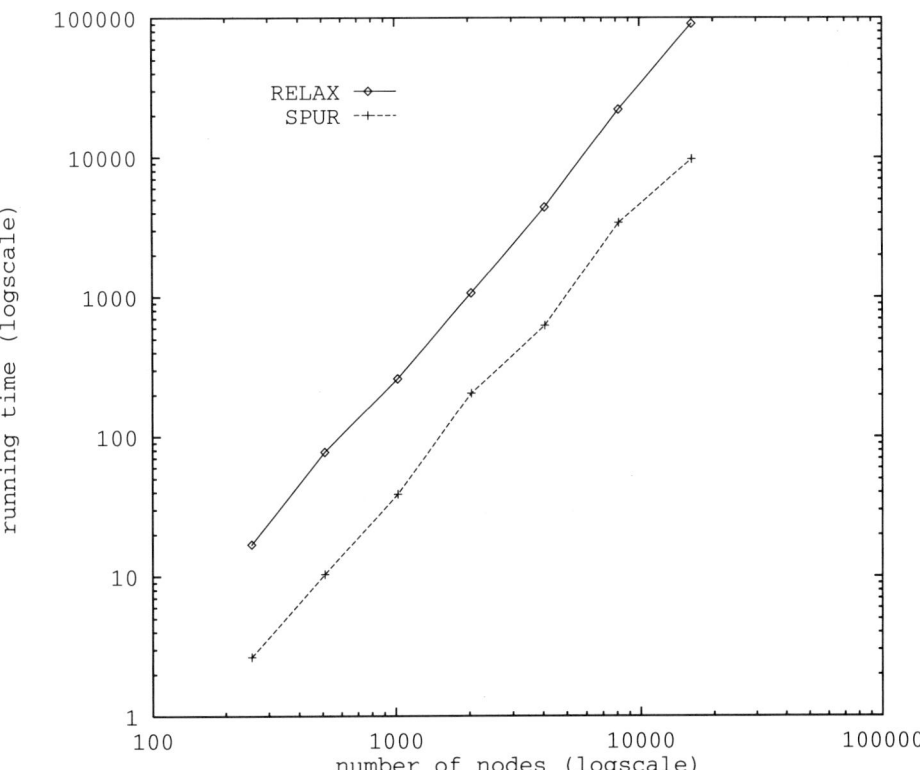

Problem size		Running time		Normalized time	
nodes	edges	SPUR	RELAX	SPUR	RELAX
256	2048	2.662	16.992	1	6.38
512	4096	10.358	77.783	1	7.51
1024	8192	38.904	260.958	1	6.71
2048	16384	205.767	1069.992	1	5.20
4096	32768	626.221	4404.150	1	7.03
8192	65536	3420.925	22173.180	1	6.48
16384	131072	9728.400	90664.312	1	9.32

FIGURE 7. Average running times (in seconds) of SPUR and RELAX on the GOTO-8 family.

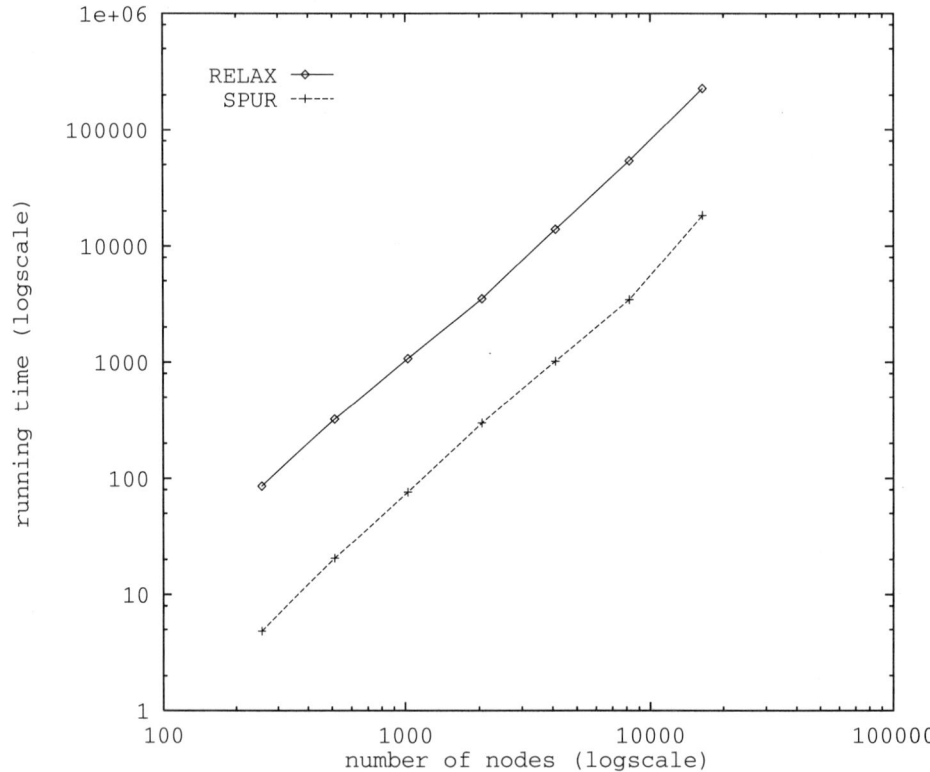

Problem size		Running time		Normalized time	
nodes	edges	SPUR	RELAX	SPUR	RELAX
256	4096	4.842	85.750	1	17.71
512	8192	20.525	322.671	1	15.72
1024	16384	76.037	1071.154	1	14.09
2048	32768	297.875	3501.650	1	11.76
4096	65536	1018.833	13950.066	1	13.69
8192	131072	3444.529	54075.504	1	15.70
16384	262144	18331.559	227889.172	1	12.43

FIGURE 8. Average running times (in seconds) of SPUR and RELAX on the GOTO-16 family.

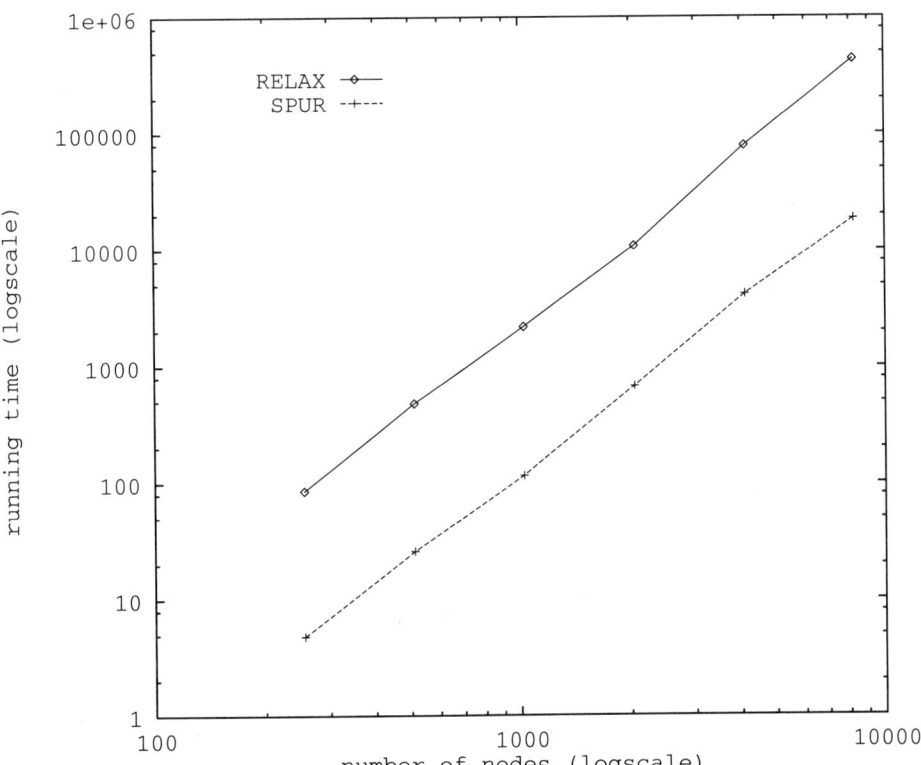

Problem size		Running time		Normalized time	
nodes	edges	SPUR	RELAX	SPUR	RELAX
256	4096	4.842	85.750	1	17.71
512	11585	25.904	479.763	1	18.52
1024	32768	116.033	2192.212	1	18.89
2048	92682	676.196	10766.146	1	15.92
4096	262144	4131.342	77956.250	1	18.87
8192	741455	18410.209	431195.250	1	23.42

FIGURE 9. Average running times (in seconds) of SPUR and RELAX on the GOTO-I family.

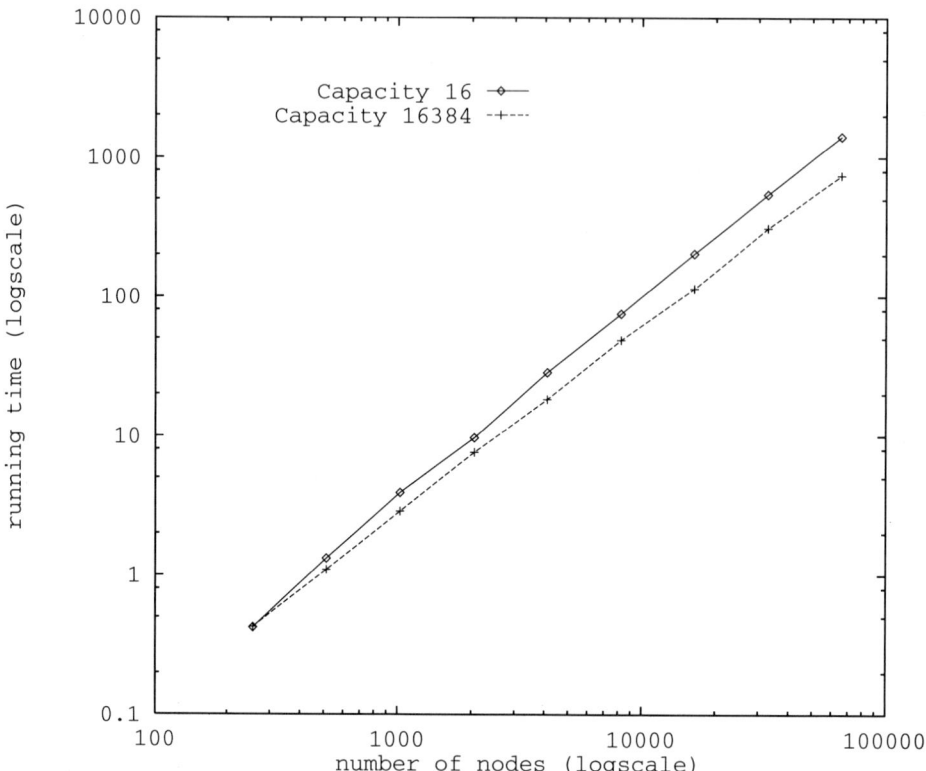

Problem size		SPUR running time	
nodes	edges	High Capacity	Low Capacity
256	2048	0.425	0.421
512	4104	1.092	1.316
1024	8200	2.875	3.904
2048	16418	7.613	9.683
4096	32854	18.300	28.508
8192	65714	48.750	75.188
16384	131458	114.096	204.054
32768	262892	310.858	543.808
65536	525830	743.075	1415.009

FIGURE 10. Average running times (in seconds) of SPUR on NETGEN families.

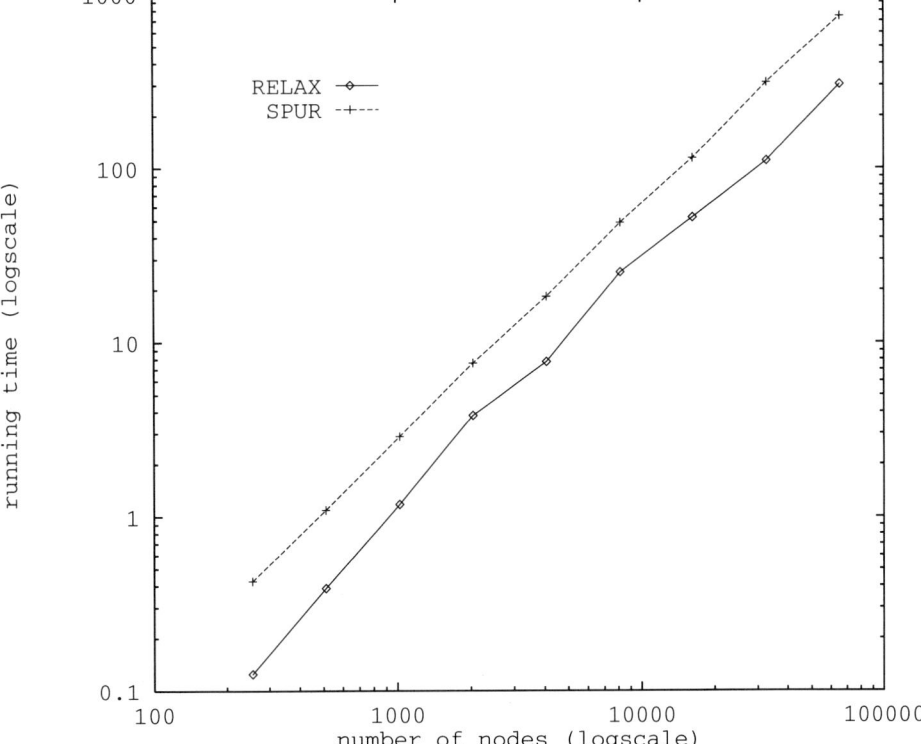

| Problem size || Running time || Normalized time ||
nodes	edges	SPUR	RELAX	SPUR	RELAX
256	2048	0.425	0.125	3.40	1
512	4104	1.092	0.387	2.82	1
1024	8200	2.875	1.175	2.45	1
2048	16418	7.613	3.808	2.00	1
4096	32854	18.300	7.738	2.37	1
8192	65714	48.750	25.312	1.93	1
16384	131458	114.096	52.167	2.19	1
32768	262892	310.858	110.292	2.82	1
65536	525830	743.075	302.992	2.45	1

FIGURE 11. Average running times (in seconds) of SPUR and RELAX on the NETGEN-HI family (maximum capacity 16384).

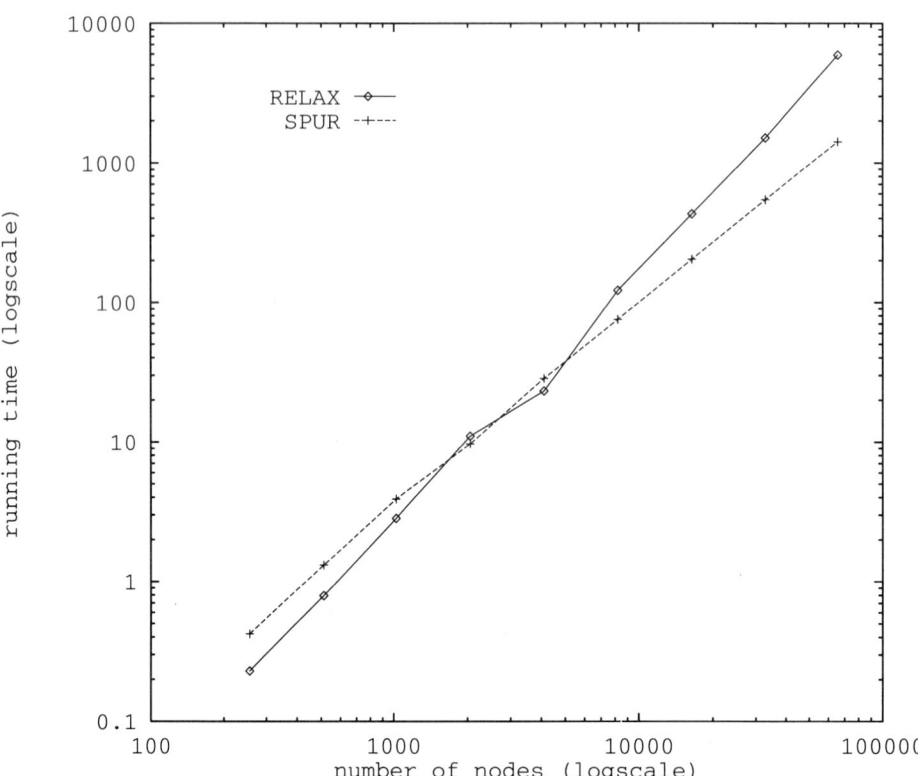

Problem size		Running time		Normalized time	
nodes	edges	SPUR	RELAX	SPUR	RELAX
256	2048	0.421	0.229	1.84	1
512	4104	1.316	0.792	1.66	1
1024	8200	3.904	2.842	1.37	1
2048	16418	9.683	11.050	1	1.14
4096	32854	28.508	23.183	1.23	1
8192	65714	75.188	122.146	1	1.62
16384	131458	204.054	429.888	1	2.11
32768	262892	543.808	1507.762	1	2.77
65536	525830	1415.009	5895.587	1	4.17

FIGURE 12. Average running times (in seconds) of SPUR and RELAX on the NETGEN-LO family (maximum capacity 16).

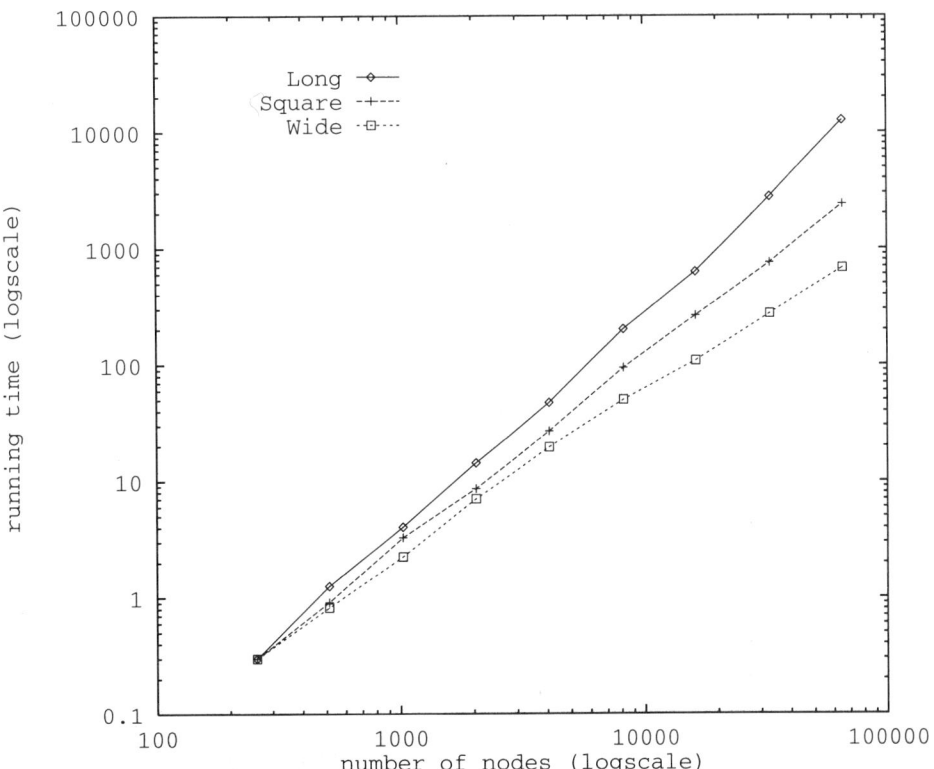

Problem size	SPUR running time		
(nodes)	Wide grid	Square grid	Long grid
258	0.300	0.300	0.300
514	0.821	0.913	1.254
1026	2.238	3.271	4.025
2050	7.008	8.575	14.254
4098	19.471	26.771	47.079
8194	49.938	93.729	201.288
16386	108.096	263.550	627.804
32770	274.504	751.346	2764.283
65538	677.912	2385.329	12577.959

FIGURE 13. Average running times (in seconds) of SPUR on GRIDGRAPH families.

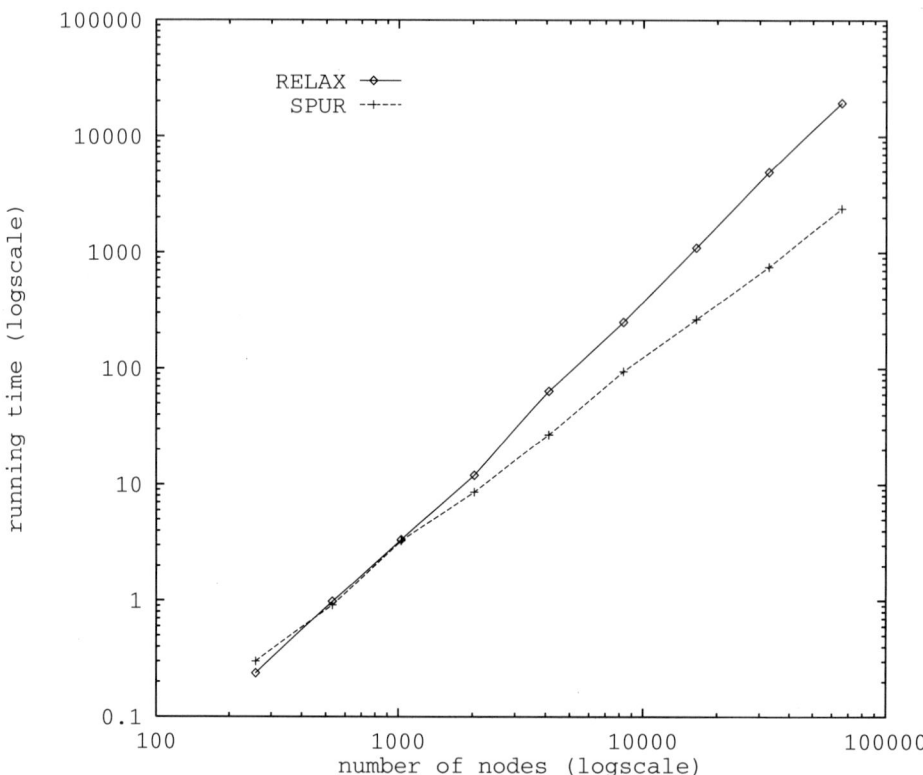

Problem size		Running time		Normalized time	
nodes	edges	SPUR	RELAX	SPUR	RELAX
258	512	0.300	0.237	1.26	1
531	1058	0.913	0.979	1	1.07
1026	2048	3.271	3.350	1	1.02
2027	4050	8.575	11.971	1	1.40
4098	8192	26.771	63.529	1	2.37
8283	16562	93.729	249.617	1	2.66
16386	32768	263.550	1094.171	1	4.15
32763	65522	751.346	4915.750	1	6.54
65538	131072	2385.329	19387.441	1	8.13

FIGURE 14. Average running times (in seconds) of SPUR and RELAX on the GRID-SQUARE family.

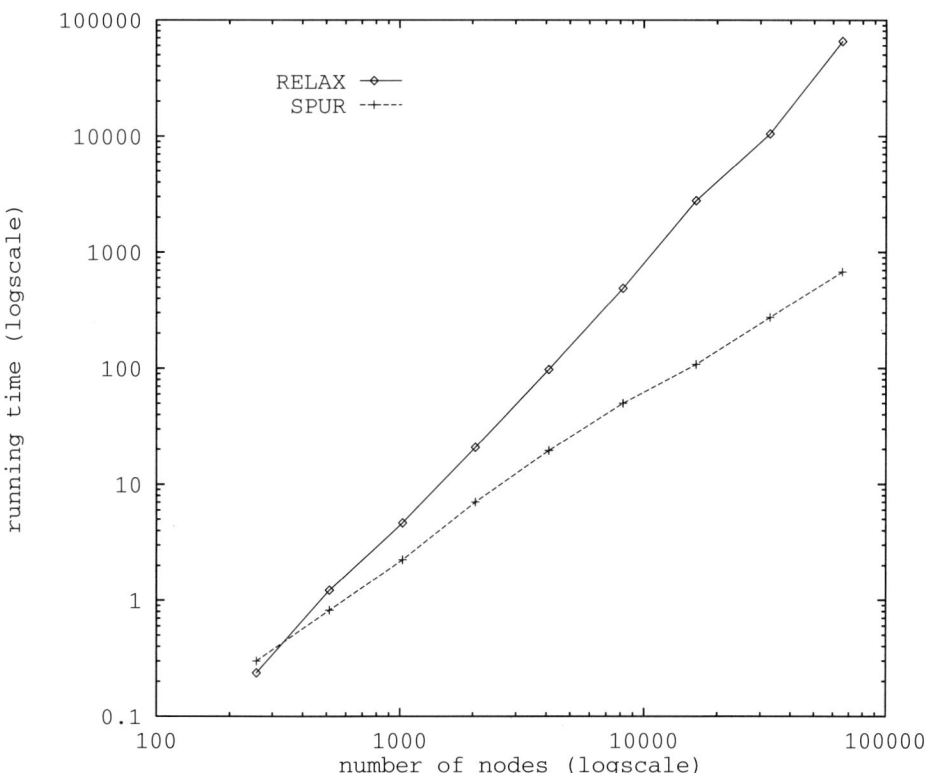

| Problem size || Running time || Normalized time ||
nodes	edges	SPUR	RELAX	SPUR	RELAX
258	512	0.300	0.237	1.26	1
514	1040	0.821	1.221	1	1.49
1026	2096	2.238	4.654	1	2.08
2050	4208	7.008	20.721	1	2.96
4098	8432	19.471	97.338	1	5.00
8194	16880	49.938	489.737	1	9.81
16386	33776	108.096	2772.962	1	25.65
32770	67568	274.504	10487.862	1	38.21
65538	135152	677.912	65298.406	1	96.32

FIGURE 15. Average running times (in seconds) of SPUR and RELAX on the GRID-WIDE family.

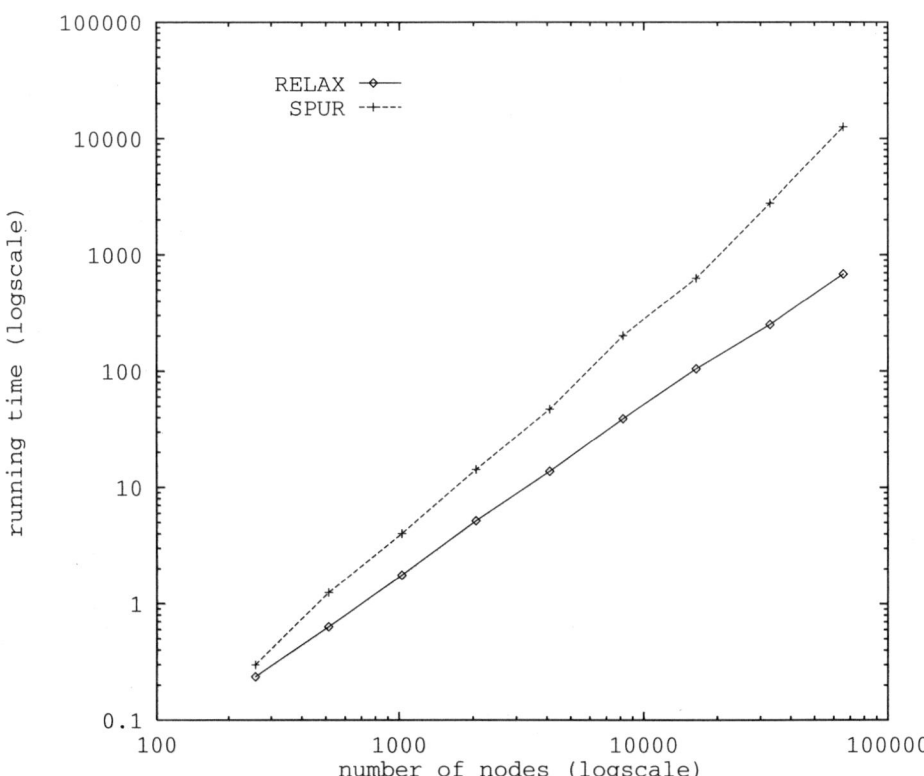

FIGURE 16. Average running times (in seconds) of SPUR and RELAX on the GRID-LONG family.

Problem	Running time		Normalized time	
	SPUR	RELAX	SPUR	RELAX
gte.20	0.050	0.017	2.99	1
gte.40	0.067	0.017	4.01	1
gte.60	0.067	0.017	4.01	1
gte.160	0.150	0.067	2.25	1
gte.200	0.100	0.233	1	2.33
gte.460	0.100	0.600	1	6.00
gte.510	0.100	0.817	1	8.17
gte.1160	0.100	1.333	1	13.33
gte.1700	0.133	0.450	1	3.38
gte.6410	0.150	4.000	1	26.67
gte.6830	0.100	7.333	1	73.33
gte.15100	0.133	0.033	3.99	1
gte.15710	0.150	0.067	2.25	1
gte.35620	0.133	33.400	1	251.13
gte.49320	0.150	22.850	1	152.33
gte.60090	0.150	58.600	1	390.67
gte.65330	0.150	0.767	1	5.11
gte.176280	0.133	0.050	2.66	1
gte.298300	0.100	0.450	1	4.50
gte.451760	0.150	67.750	1	451.67
gte.469010	0.133	70.717	1	531.70
gte.508829	0.133	97.017	1	729.45
Total	2.632	366.583	1	139.28

FIGURE 17. Running times (in seconds) of SPUR and RELAX on GTE instances. Each network consists of 49 nodes and 520 arcs.

RELAX running times is over 5700.

The total running time of SPUR on the GTE instances is better than the total running time of RELAX by two orders of magnitude.

6.5. Discussion. Figure 18 compares the execution times of SPUR on GOTO-8 and NETGEN-HI families, which have the same maximum cost and maximum capacity values. The difference in running times for the same problem size is quite drastic. GOTO problems are much harder for SPUR than NETGEN problems. (For RELAX, the difference is even bigger.) NETGEN seems to produce relatively easy instances of the minimum-cost flow problem.

In general, practical running times vary from generator to generator. By observing the behavior of a program on different families of problems, one may gain insight into frequently occurring network structures that are bad for a particular implementation, suggesting directions for improvement.

For example, the behavior of SPUR on the GRID-LONG family shows an area where the algorithm needs to be improved. We discuss a possible reason for the relatively poor performance of our code on this family in Section 8.

7. Effectiveness of Heuristics

In this section we examine the effect of heuristics on the algorithm's running time.

Our implementation spends most of its time doing *refine*. The percentage of the computation time spent in *refine* increases with the problem size, exceeding 95% on the larger problems we experimented with. The running time of *refine* depends on the number of *push* and *relabel* operations performed as well as on the speed of selecting and applying the operations. Heuristic improvements decrease both the number of operations and time per operation.

Figure 19 gives running times as well as operation counts for SPUR on problems from several classes we studied. We count the number of *refines* and the number of *pushes* and *relabels* per refine. One interesting observation here is that most of the time the number of *push* operations is smaller than the number of *relabel* operations. Since the latter operation is computationally more expensive, the number of relabels is the bottleneck for our implementation. In contrast, the theoretical worst-case bottleneck of the method is the number of pushes.

Figure 19 also gives the data for SPUR with all the heuristics turned off. Comparison of the two sets of data shows that the speedup due to heuristics is different for different problem types. For the problems in the Figure 19 the speedup varies from the factor of 2.88 on the GOTO-I instances to the factor of 6.5 on the NETGEN-HI instances. Reduction in the number of refines, reduction in the number of pushes and relabels per refine, and faster operation times all contribute to the speedup. The only exception is the number of relabels per refine on the GOTO-I instances, where the number of relabels per refine in the full implementation is somewhat

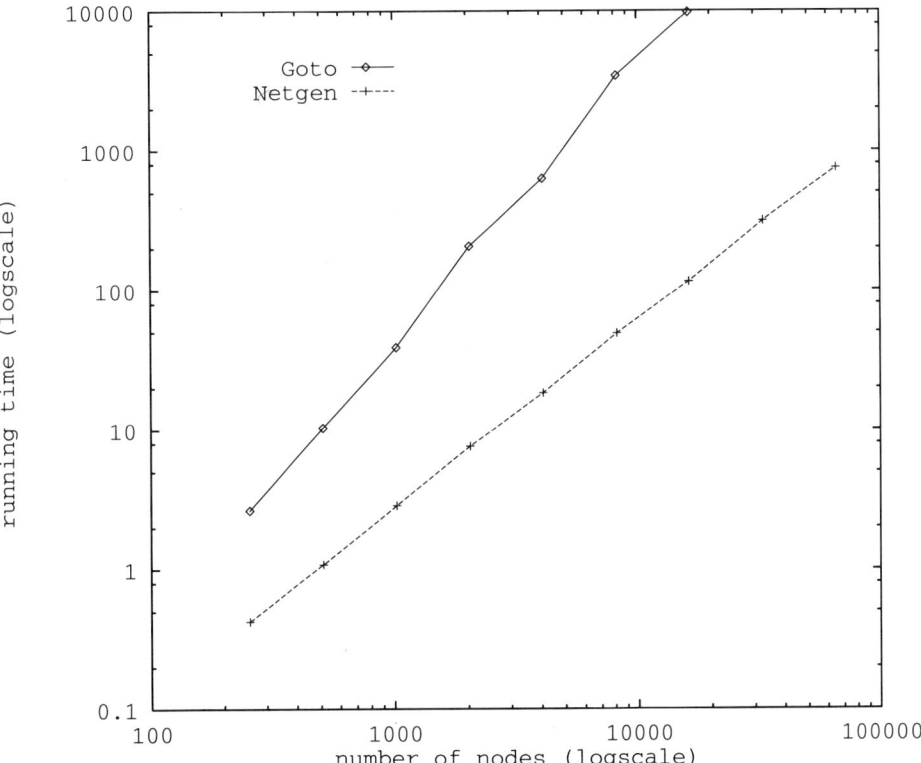

| Problem size | SPUR running time | | |
(nodes)	GOTO-8 family	NETGEN-HI family	Ratio
256	2.662	0.425	6.26
512	10.358	1.092	9.49
1024	38.904	2.875	13.53
2048	205.767	7.613	27.03
4096	626.221	18.300	34.22
8192	3420.925	48.750	70.17
16384	9728.400	114.096	85.27
32768	timeout	310.858	?
65536	timeout	743.075	?

FIGURE 18. Average running times (in seconds) of SPUR on GOTO-8 and NETGEN-HI families.

Category	with heuristics	w/out heuristics	gain factor
Time (seconds)	626.221	1964.104	3.14
Number of Refines	7.750	12.000	1.55
Relabelings/Refine	1995966.125	2129125.000	1.07
Pushes/Refine	1724047.000	4268074.500	2.48

GOTO-8 examples with 4096 nodes and 32768 edges.

Category	with heuristics	w/out heuristics	gain factor
Time (seconds)	676.196	1947.979	2.88
Number of Refines	7.750	11.000	1.42
Relabelings/Refine	534075.062	482254.875	0.90
Pushes/Refine	535869.750	1111533.125	2.07

GOTO-I examples with 2048 nodes and 92682 edges.

Category	with heuristics	w/out heuristics	gain factor
Time (seconds)	743.075	4833.667	6.50
Number of Refines	9.000	13.000	1.44
Relabelings/Refine	3143826.500	4828418.500	1.54
Pushes/Refine	1883396.875	8488667.000	4.51

NETGEN-HI examples with 65536 nodes and 525830 edges.

Category	with heuristics	w/out heuristics	gain factor
Time (seconds)	751.346	4064.438	5.41
Number of Refines	7.000	13.000	1.86
Relabelings/Refine	6707662.500	9974990.000	1.49
Pushes/Refine	4905041.500	19180974.000	3.91

GRID-SQUARE examples with 32763 nodes and 65522 edges.

FIGURE 19. Effects of heuristics on SPUR's average running times and operation counts.

greater than in the implementation without heuristics.

Figures 20 – 22 summarize the effects of individual heuristics on the algorithm's performance. Note that the heuristics are interdependent: the speedup for the code that uses all the heuristics is less then the product of the speedups for individual heuristics.

As Figure 20 shows, early termination detection reduces the number of iterations of the algorithm. The corresponding improvement in the running time, however, is not directly proportional to the reduction in the number of iterations: sometimes it is noticeably more, sometimes noticeably less. This is probably due to the fact that the network structure at the end of the execution of the implementation without heuristics (when the flow is already optimal) is different from the structure at the beginning of the execution.

Arc fixing is aimed at reducing time per operation. As Figure 21 shows, the heuristic succeeds in this respect. An unexpected side-effect of this heuristic is reduction in the number of *push* and *relabel* operations on GRID-SQUARE examples. (On other examples in the table, the number of these operations does not change much.)

The lookahead heuristic is designed to reduce the number of *push* operations. As Figure 22 suggests, this heuristic also reduces the number of *relabel* operations, although by a smaller factor.

Figure 23 compares running times of the implementations with and without heuristics on the GRID-SQUARE family. This includes the implementation with no heuristics, the implementation with all heuristics, and three implementations wich a single heuristic each.

Figures 24, 25, and 26 compare the number of refines and the number of relabels and pushes per refine for the implementation with all the heuristics and the implementation with no heuristics. There is no significant asymptotic decrease in the number of relabels and pushes per refine due to heuristics. The speedup due to heuristics also does not grow significantly.

8. Conclusions and Further Research

Our implementation works well on a wide class of problems. The heuristics we use are important for the efficiency of our implementation. Although we studied several heuristics in the context of several variants of the push-relabel method, many more remain to be studied. Several directions for further study are suggested in Section 4.

The following example illustrates why SPUR is relatively slow on some classes of networks such as the GRID-LONG family. Consider a path of k nodes going from left to right. A unit of flow excess is located at the left end of the path and a unit of deficit at the right end. Suppose all arcs are directed from left to right and have capacities of one and reduced costs of 2ϵ. The number of relabels that occur during

Category	early termination	no heuristics	gain factor
Time (seconds)	1373.833	1964.104	1.43
Number of Refines	7.750	12.000	1.55
Relabelings/Refine	2275481.250	2129125.000	0.94
Pushes/Refine	4830547.000	4268074.500	0.88

GOTO-8 examples with 4096 nodes and 32768 edges.

Category	early termination	no heuristics	gain factor
Time (seconds)	1620.292	1916.387	1.18
Number of Refines	8.000	11.000	1.38
Relabelings/Refine	556442.625	482254.875	0.87
Pushes/Refine	1364173.500	1111533.125	0.81

GOTO-I examples with 2048 nodes and 92682 edges.

Category	early termination	no heuristics	gain factor
Time (seconds)	2266.725	4833.667	2.13
Number of Refines	9.000	13.000	1.44
Relabelings/Refine	3139502.750	4828418.500	1.54
Pushes/Refine	5626424.000	8488667.000	1.51

NETGEN-HI examples with 65536 nodes and 525830 edges.

Category	early termination	no heuristics	gain factor
Time (seconds)	2451.208	4064.438	1.66
Number of Refines	7.500	13.000	1.73
Relabelings/Refine	10064268.000	9974990.000	0.99
Pushes/Refine	19821288.000	19180974.000	0.97

GRID-SQUARE examples with 32763 nodes and 65522 edges.

FIGURE 20. Effects of early termination on SPUR's average running times and operation counts.

Category	arc fixing	no heuristics	gain factor
Time (seconds)	985.912	1964.104	1.99
Number of Refines	12.000	12.000	1.00
Relabelings/Refine	2222590.750	2129125.000	0.96
Pushes/Refine	4429898.000	4268074.500	0.96

GOTO-8 examples with 4096 nodes and 32768 edges.

Category	arc fixing	no heuristics	gain factor
Time (seconds)	759.421	1916.387	2.52
Number of Refines	11.000	11.000	1.00
Relabelings/Refine	485542.094	482254.875	0.99
Pushes/Refine	1130211.625	1111533.125	0.98

GOTO-I examples with 2048 nodes and 92682 edges.

Category	arc fixing	no heuristics	gain factor
Time (seconds)	1665.158	4833.667	2.90
Number of Refines	13.000	13.000	1.00
Relabelings/Refine	4980536.000	4828418.500	0.97
Pushes/Refine	8754921.000	8488667.000	0.97

NETGEN-HI examples with 65536 nodes and 525830 edges.

Category	arc fixing	no heuristics	gain factor
Time (seconds)	2103.892	4064.438	1.93
Number of Refines	13.000	13.000	1.00
Relabelings/Refine	7264765.500	9974990.000	1.37
Pushes/Refine	14034406.000	19180974.000	1.37

GRID-SQUARE examples with 32763 nodes and 65522 edges.

FIGURE 21. Effects of arc fixing on SPUR's average running times and operation counts.

Category	lookahead	no heuristics	gain factor
Time (seconds)	1235.654	1964.104	1.59
Number of Refines	12.000	12.000	1.00
Relabelings/Refine	1539412.375	2129125.000	1.38
Pushes/Refine	1271935.500	4268074.500	3.36

GOTO-8 examples with 4096 nodes and 32768 edges.

Category	lookahead	no heuristics	gain factor
Time (seconds)	1684.221	1916.387	1.14
Number of Refines	11.000	11.000	1.00
Relabelings/Refine	439438.469	482254.875	1.10
Pushes/Refine	405540.375	1111533.125	2.74

GOTO-I examples with 2048 nodes and 92682 edges.

Category	lookahead	no heuristics	gain factor
Time (seconds)	3836.388	4833.667	1.26
Number of Refines	13.000	13.000	1.00
Relabelings/Refine	4615173.000	4828418.500	1.05
Pushes/Refine	2675020.250	8488667.000	3.17

NETGEN-HI examples with 65536 nodes and 525830 edges.

Category	lookahead	no heuristics	gain factor
Time (seconds)	1975.516	4064.438	2.06
Number of Refines	13.000	13.000	1.00
Relabelings/Refine	7177376.000	9974990.000	1.39
Pushes/Refine	5123041.000	19180974.000	3.74

GRID-SQUARE examples with 32763 nodes and 65522 edges.

FIGURE 22. Effects of lookahead on SPUR's average running times and operation counts.

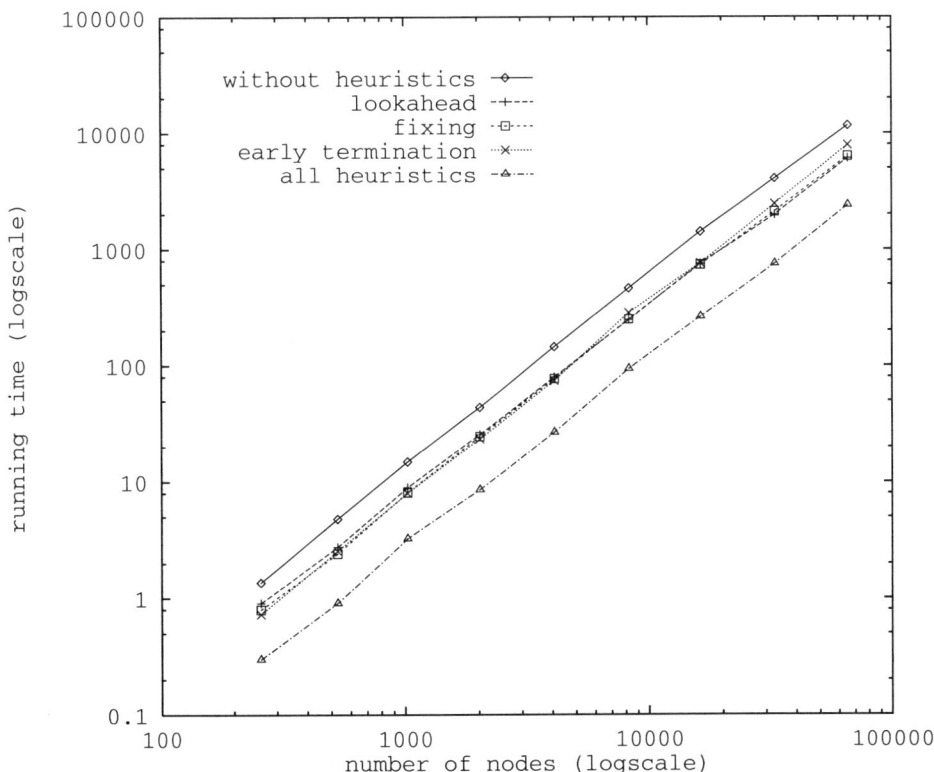

Problem size		Running Time					
		early	arc		all	without	gain
nodes	edges	term.	fixing	lookahead	heuristics	heuristics	factor
258	512	0.73	0.80	0.91	0.30	1.37	4.55
531	1058	2.56	2.43	2.77	0.91	4.81	5.27
1026	2048	8.00	8.10	8.94	3.27	14.90	4.56
2027	4050	23.44	24.58	25.58	8.57	43.75	5.10
4098	8192	73.92	76.89	79.39	26.77	144.82	5.41
8283	16562	284.52	250.29	247.05	93.73	462.63	4.94
16386	32768	760.00	738.65	751.90	263.55	1414.87	5.37
32763	65522	2451.21	2103.89	1975.52	751.35	4064.44	5.41
65538	131072	7898.13	6344.48	6070.79	2385.33	11621.89	4.87

FIGURE 23. Effects of individual and combined heuristics on SPUR's running time on the Grid-Square family.

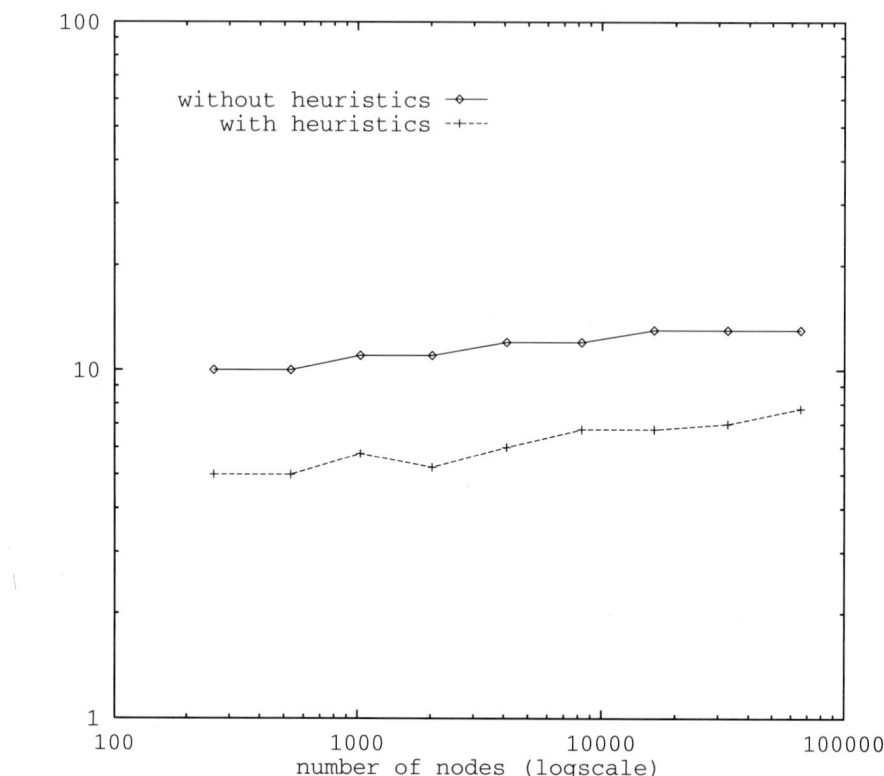

Problem size		Number of Refines		
		with	without	gain
nodes	edges	heuristics	heuristics	factor
258	512	5.000	10.000	2.00
531	1058	5.000	10.000	2.00
1026	2048	5.750	11.000	1.91
2027	4050	5.250	11.000	2.10
4098	8192	6.000	12.000	2.00
8283	16562	6.750	12.000	1.78
16386	32768	6.750	13.000	1.93
32763	65522	7.000	13.000	1.86
65538	131072	7.750	13.000	1.68

FIGURE 24. Effects of heuristics on the average number of refines done by SPUR on the Grid-Square family.

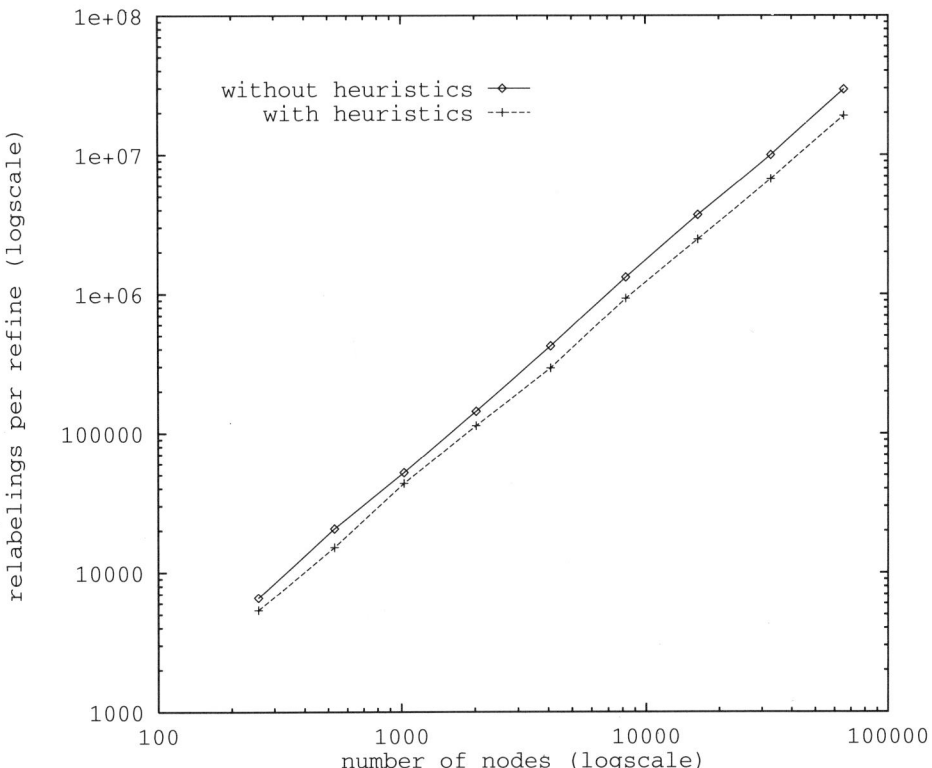

Problem size		Relabelings per Refine		
nodes	edges	with heuristics	without heuristics	gain factor
258	512	5366.500	6595.500	1.23
531	1058	15213.500	20686.775	1.36
1026	2048	43678.522	52273.727	1.20
2027	4050	113666.524	144165.636	1.27
4098	8192	294052.458	422782.938	1.44
8283	16562	931970.481	1319870.875	1.42
16386	32768	2474598.148	3700223.077	1.50
32763	65522	6707662.357	9974990.173	1.49
65538	131072	19079652.419	29456771.904	1.54

FIGURE 25. Effects of heuristics on the average number of relabels per each refine done by SPUR on the Grid-Square family.

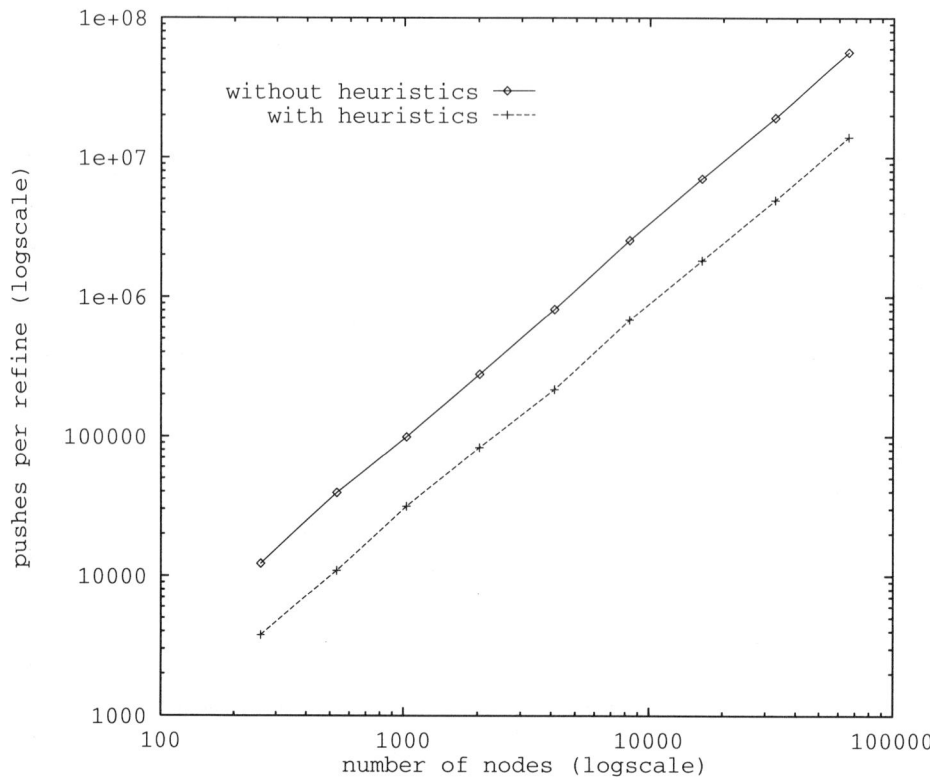

| Problem size || Pushes per Refine |||
nodes	edges	with heuristics	without heuristics	gain factor
258	512	3766.100	12258.800	3.26
531	1058	10931.800	39324.400	3.60
1026	2048	31478.304	99049.045	3.15
2027	4050	83044.333	278346.864	3.35
4098	8192	216022.292	810585.167	3.75
8283	16562	679902.630	2535172.292	3.73
16386	32768	1815986.481	7017152.000	3.86
32763	65522	4905041.643	19180973.788	3.91
6 5538	131072	13972798.419	56414415.288	4.04

FIGURE 26. Effects of heuristics on the average number of pushes per each refine done by SPUR on the Grid-Square family.

an execution of *refine* on such a network is $\Omega(k^2)$. Based on this example, one can imagine a variety of subnetworks that cause quadratic behavior of the algorithm. We call such kind of structure a *hill*.

Note that price update heuristics discussed in Section 4.3 are very effective in eliminating hills if applied at the right time. Some of these heuristics, however, may create potential hills.

Recently, the first author developed another implementation of the cost-scaling push-relabel method [**13**]. This implementation uses price update and price refinement heuristics based on the *set-relabel* operation described in Section 4 and the scaling shortest path algorithm of [**14**]. The new implementation is superior to SPUR, and on many problem families the improvement is asymptotic. This implementation does not seem to suffer from the hill phenomena.

Performance of the new implementation on GRIDGRAPH examples depends very little on the grid shape. The implementation is faster then SPUR even on GRID-WIDE examples, on which SPUR works best. The improvement on the GRID-LONG examples is dramatic. On GRID-LONG examples with 2^{16} nodes, the average running time of the current code is 215 seconds, compared to 12578 seconds for SPUR — a factor of 58.5 improvement. The new code is 3.2 times faster then RELAX on these examples.

The new implementation is also substantially faster on the GOTO examples. On GOTO-8 examples with 2^{14} nodes, the average running time is 405 seconds, compared to 9728 seconds for SPUR — a factor of 24 improvement. (See Figure 7.) The new code is 224 times faster then RELAX on these examples.

The performance of the new implementation may improve even further. The results of the work related to this implementation will be reported in an upcoming paper.

Our comparison of SPUR and RELAX shows that neither code dominates. Limited experiments with a network simplex codes NETFLO [**24**] and RNET [**19**] indicate that the same can be said for all of these codes. The new implementation of the scaling method mentioned above, however, is superior to the other codes on most families of problems. Further improvement may be possible using some of the ideas suggested in this paper.

APPENDIX: the GOTO Generator

In this section we discuss the GOTO generator, which was developed by the first author. GOTO stands for **G**rid **O**n **TO**rus. We would like to note that although the initial stage of the example generation does use the concept of a grid embeded on torus, the output of the generator is much more complicated. The goal of this generator is to produce hard instances of the capacitated transshipment problem, and in fact the problems produced by the generator are harder then those produced by other generators we experimented with.

#	name	notation	restrictions
1	number of nodes	n	$n \geq 15$
2	number of arcs	m	$6n \leq m \leq n^{\frac{5}{3}}$
3	maximum capacity	$max\text{-}cap$	$max\text{-}cap \geq 8$
4	maximum cost	$max\text{-}cost$	$max\text{-}cost \geq 8$
5	seed		

FIGURE 27. GOTO generator parameters. All parameters are integers.

The generator works as follows. Imagine an X by Y torus with a grid embedded on it. Each node of the torus with coordinates (a, b) is connected to nodes $(a + 1 \bmod X, b), \ldots, (a + x\text{-}degree \bmod X, b)$ in x direction and to nodes $(a, b + 1 \bmod Y), \ldots, (a, b + y\text{-}degree \bmod Y)$ in y direction.

The capacity of an x direction arc is chosen uniformly at random from the interval $[1, \lceil \frac{max\text{-}cap}{\alpha^{i-1}} \rceil]$, where i is the x direction distance from the tail to the head of the arc. Here $max\text{-}cap$ is an input parameter and $alpha = max\text{-}cap^{\frac{1}{x\text{-}degree+2}}$. The cost of an x direction arc is chosen uniformly at random from the interval $[0, max\text{-}cost]$, where $max\text{-}cost$ is an input parameter.

The capacity of an y direction arc is chosen uniformly at random from the interval $[1, max\text{-}cap]$. The cost of an y direction arc is chosen uniformly at random from the interval $[0, 8]$.

Then the arcs between nodes with x coordinates X and 1 are deleted, and the demand D is set to be equal to the sum of capacities of the deleted arcs. Node demands are set to zero, except for the *source node* $(1, 1)$ and the *sink node* (X, Y); the former node has a demand of $-D$ (supply of D), and the latter has a demand of D.

To assure feasibility, a hamiltonian path is added from the source to the sink. This path traverses nodes in the lexicographical order of their coordinates. The capacity of the arcs on this path is set to D, and the cost of the arcs is set to $\frac{max\text{-}cost}{Y}$. Note that capacity of the arcs on the path can exceed $max\text{-}cap$.

To make the generator easy to use, we reduced the number of input parameters as much as possible. The generator takes five parameters, described in Figure 27. The generator computes the values of X, Y, $x\text{-}degree$, and $y\text{-}degree$ from the number of nodes n and the number of arcs m so that X is approximately $n^{\frac{2}{3}}$, Y is approximately $n^{\frac{1}{3}}$, We $x\text{-}degree$ and $y\text{-}degree$ so that $y\text{-}degree^2$ is approximately $y\text{-}degree$, and the sum of the two degrees is as large as the number of available arcs allows.

Because of the integrality, we usually have $X \times Y < n$, so some of the nodes are not used in the construction described above. We call these nodes *extra nodes*. Similarly, we have *extra* arcs. We set $x\text{-}degree$ and $y\text{-}degree$ to make sure that there are enough extra arcs for the construction below. The extra nodes are connected in a path from the source to the sink by arcs of small capacity and large cost. The remaining extra arcs are used to connect extra nodes to the regular nodes. The direction, head node, and destination node of these arcs are chosen uniformly at random from the appropriate set.

The examples produced by GOTO have the following properties. On one hand, the range of costs and capacities of paths from the source to the sink is large. On the other hand, for any given path, there are usually many paths whose costs and capacities are close to those of the given path. A similar statement holds for the cuts between the source and the sink. We believe that these properties account for the difficulty of the problems generated by GOTO.

Acknowledgment

This work was done as a part of the First DIMACS International Algorithm Implementation Challenge, organized by Mike Grigoriadis, David Johnson, Cathy McGeoch, Clyde Monma, and Bob Tarjan. We would like to thank Bob Bland, Joseph Cheriyan, David Jensen, and Laszlo Ladanyi for helpful discussions and assistance. We are also thankful to Robert Kennedy and Andrew Purshottam for comments on a draft of this paper, and to the International Computer Science Institute for providing the second author with office and computing support.

REFERENCES

1. R. K. Ahuja, T. L. Magnanti, and J. B. Orlin. Network Flows. In G. L. Nemhauser, A. H. G. Rinnooy Kan, , and M. J. Todd, editors, *Optimization. Handbooks in Operations Research and Management Science, Vol. 1*, pages 211–369. North-Holland, Amsterdam, 1989.
2. R. K. Ahuja and J. B. Orlin. Personal communication. 1987.
3. D. P. Bertsekas. Distributed Asynchronous Relaxation Methods for Linear Network Flow Problems. Technical Report LIDS-P-1986, Lab. for Decision Systems, M.I.T., September 1986. (Revised November, 1986).
4. D. P. Bertsekas and P. Tseng. Relaxation Methods for Minimum Cost Ordinary and Generalized Network Flow Problems. *Oper. Res.*, 36:93–114, 1988.
5. R. G. Bland, J. Cheriyan, D. L. Jensen, and L. Ladanyi. Personal communication. 1991.
6. R. G. Bland and D. L. Jensen. On the Computational Behavior of a Polynomial-Time Network Flow Algorithm. *Math. Prog.*, 54:1–41, 1992.
7. B. V. Cherkassky. Personal communication. 1991.
8. U. Derigs and W. Meier. Implementing Goldberg's Max-Flow Algorithm — A Computational Investigation. *ZOR — Methods and Models of Operations Research*, 33:383–403, 1989.
9. L. R. Ford, Jr. and D. R. Fulkerson. *Flows in Networks*. Princeton Univ. Press, Princeton, NJ, 1962.
10. S. Fujishige, K. Iwano, J. Nakano, and S Tezuka. A Speculative Contraction Method for the Minimum Cost Flows: Toward a Practical Algorithm. The First DIMACS International Implementation Challenge, 1991.
11. A. V. Goldberg. A New Max-Flow Algorithm. Technical Report MIT/LCS/TM-291, Laboratory for Computer Science, M.I.T., 1985.
12. A. V. Goldberg. *Efficient Graph Algorithms for Sequential and Parallel Computers*. PhD thesis, M.I.T., January 1987. (Also available as Technical Report TR-374, Lab. for Computer Science, M.I.T., 1987).
13. A. V. Goldberg. An Efficient Implementation of a Scaling Minimum-Cost Flow Algorithm. Technical Report STAN-CS-92-1439, Department of Computer Science, Stanford University, 1992.

14. A. V. Goldberg. Scaling Algorithms for the Shortest Paths Problem. Technical Report STAN-CS-92-1429, Department of Computer Science, Stanford University, 1992.
15. A. V. Goldberg, É. Tardos, and R. E. Tarjan. Network Flow Algorithms. In B. Korte, L. Lovász, H. J. Prömel, and A. Schrijver, editors, *Flows, Paths, and VLSI Layout*, pages 101–164. Springer Verlag, 1990.
16. A. V. Goldberg and R. E. Tarjan. A New Approach to the Maximum Flow Problem. *J. Assoc. Comput. Mach.*, 35:921–940, 1988.
17. A. V. Goldberg and R. E. Tarjan. Finding Minimum-Cost Circulations by Successive Approximation. *Math. of Oper. Res.*, 15:430–466, 1990.
18. M. D. Grigoriadis. An Efficient Implementation of the Network Simplex Method. *Math. Prog. Study*, 26:83–111, 1986.
19. M. D. Grigoriadis. Personal communication. 1988.
20. R. Hassin. The Minimum-Cost Flow Problem: A Unifying Approach to Dual Algorithms and a New Tree-Search Algorithm. *Math. Prog.*, 25:228–239, 1983.
21. N. K. Karmarkar and K. G. Ramakrishnan. Computational Results of an Interior Point Algorithm for Large Scale Linear Programming. *Math. Prog.*, 52:555–586, 1991.
22. R. M. Karp. A Characterization of the Minimum Cycle Mean in a Digraph. *Discrete Math.*, 23:309–311, 1978.
23. R. M. Karp and J. B. Orlin. Parametric Shortest Path Algorithms with an Application to Cyclic Staffing. *Discrete Applied Math.*, 3:37–45, 1981.
24. J. L. Kennington and R. V. Helgason. *Algorithms for Network Programming*. John Wiley and Sons, 1980.
25. D. Klingman, A. Napier, and J. Stutz. Netgen: A Program for Generating Large Scale Capacitated Assignment, Transportation, and Minimum Cost Flow Network Problems. *Management Science*, 20:814–821, 1974.
26. T. Leong, P. Shor, and C. Stein. Implementation of a Combinatorial Multicommodity Flow Algorithm. The First DIMACS International Implementation Challenge, 1991.
27. M. G. C. Resende and G. Veiga. Computational Investigation of an Interior Point Linear Programming Algorithm for Minimum Cost Network Flows. 1991.
28. H. Röck. Scaling Techniques for Minimal Cost Network Flows. In U. Pape, editor, *Discrete Structures and Algorithms*, pages 181–191. Carl Hansen, Münich, 1980.
29. N. Schienker and R. E. Tarjan. Personal communication. 1990.
30. É. Tardos. A Strongly Polynomial Minimum Cost Circulation Algorithm. *Combinatorica*, 5(3):247–255, 1985.
31. J. Wein. Personal communication. 1990.

ANDREW V. GOLDBERG, COMPUTER SCIENCE DEPARTMENT, STANFORD UNIVERSITY, STANFORD, CA 94305.
E-mail address: goldberg@@cs.stanford.edu

MICHAEL KHARITONOV, COMPUTER SCIENCE DEPARTMENT, STANFORD UNIVERSITY, STANFORD, CA 94305.
E-mail address: misha@@theory.stanford.edu

Performance Evaluation of the MINET Minimum Cost Netflow Solver

ISTVÁN MAROS

ABSTRACT. In this paper we give account of the experiences obtained when participating in the DIMACS International Implementation Challenge with MINET simplex based minimum cost network flow (MCNF) LP solver. MINET has some distinguishing features among network simplex solvers in pricing and in selecting the leaving arc. With some fixed setting of strategic parameters of MINET, on the given set of test problems an $O(mn)$ performance in solution time was observed. MINET was also compared with NETFLO an industry standard netflow solver. The effect of two "external" conditions, namely, code optimization and loading conditions of the computer are also analyzed and some important facts are revealed.

1. Introduction

Efficient solution of a wide variety of minimal cost network flow (MCNF) problems, due to its theoretical and practical importance, has long been a challenge for researchers in the field. The efforts resulted in a number of powerful algorithms to solve the MCNF problem. One of them is MINET, developed by the author. MINET has some distinguishing features in pricing and row selection that will be presented briefly in section 2.

The purpose of this paper is to give account of the experiences obtained participating in the DIMACS International Implementation Challenge (for details of the Challenge, see [11]). The nature of experiences is manifold. First, MINET was tested, as required by [11], on a set of predefined problems (7 problem groups of problems generated by NETGEN [5], and 8 problems from the original NETGEN set of 40 problems). The test results have been evaluated by statistical methods to find some relationship between problem characteristics and solution time. This discussion is presented in section 3. Next, a series of real–life problems from a specific family of problems (TRGEN) [6] was solved. A statistical

---

1991 *Mathematics Subject Classification.* 90C35.
This research was supported in part by Hungarian Research Fund OTKA 2587.

analysis was also carried out. The TRGEN family turned out to contain very degenerate problems. Therefore, it appeared appropriate to test a new anti-degeneracy row selection strategy [10] for reducing the number of degenerate iterations. The very interesting experiences obtained with this technique are also presented in section 4.

In section 5 we give account of a further category of tests when we used MINET to solve the "gte_bad" set of real–life problems that made the — otherwise excellent — RELAX–III optimizer behave very strangely producing extreme timings between 0.020 to 508.829 sec. MINET solved all these problems consistently within less than 0.06 sec.

To get an idea about the performance of MINET relative to other MCNF solvers, we were able to do some comparative testing with NETFLO. NETFLO is an industry standard MCNF solver. It has been made available for the participants of the Challenge. For comparisons we chose test problems from the original and the final DIMACS Core problems both. The results of this testing are presented and discussed in section 6.

We paid special attention to some "external" conditions that may influence the solution time, with the purpose of getting better insight into the reliability of the reported solution times. In this respect, we investigated the effect of two factors, namely code optimization and loading conditions of the computer. The issue of loading conditions is discussed in section 7, while the effect of code optimization is evaluated in section 3 where the relevant data are presented in the table of the NETGEN problems.

2. The algorithm of MINET

Experience has shown that computationally the most critical operation in simplex based MCNF algorithms is pricing (selecting the incoming edge variable). Its logical and computational simplicity are misleading (two integer additions and some indirect memory accesses per scanned nonbasic edges). It is not unusual that 50–95 percent of the total computational effort is spent on pricing.

There is another remarkable characteristic feature of MCNF algorithms, namely, on many real life problems, they tend to make a large number of degenerate steps, thus slowing down the convergence to an optimal solution.

In designing and developing MINET, we tried to address mainly the above two issues, in addition to the most important efficiency considerations [7], [8].

MINET is a primal simplex based minimum cost network flow solver. It was originally planned to be a module in a hierarchically designed special–purpose optimization algorithm. On the basis of its successful performance it was taken out of that framework and has been further developed as a stand–alone MCNF optimizer.

Efficiency of network LP systems heavily depends on the data structure applied. As a good compromise between extreme possibilities, MINET keeps track of the basic spanning tree (T) using the following list functions:

- $p(i)$ predecessor of node i,
- $s(i)$ thread successor of node i,
- $t(i)$ cardinality of subtree headed by node i,
- $f(i)$ last node in subtree headed by node i,
- $id(i)$ directed node identifier of the ith basic arc.

MINET is equipped with a newly designed column selection (pricing) technique, called NETPRI, described in [9], that can take advantage of a recognized structure of a problem. The purpose of this technique is to reduce the overall computational effort spent on pricing. It is controlled by 3 parameters. By special setting of them, it is possible to reproduce most of the known pricing strategies.

Due to the fact that randomly generated problems usually do not have any recognizable structure, this sophisticated pricing feature of MINET was not used during the experiments. The present fixed setting of the parameters reproduces the *least recently considered (LRC)* entering rule [2].

The most important parts of the main algorithmic steps of MINET, as used during the testing, can be summarized as follows.

Step 0. *Initialization.* Create all–logical basis. Set principal pricing pointer $p = m$ (the number of edges) and set up pricing vector.

Step 1. *Pricing.* Start pricing at $p+1$ (mod m), price the nonbasic arcs and stop at the first improving arc. Set pointer p to the index of this arc. If no improving arc can be found, stop (solution optimal, or problem infeasible).

Step 2. *Pivot.* Trace the loop created by the entering arc in the T basis tree and, simultaneously, do the ratio test. Start from both endpoints of this arc and move upwards in the direction of the root of T. Move always on the smaller subtree. Stop ratio test immediately if a 0 ratio is encountered. The corresponding arc will leave the basis.

Step 3. *Updating.* Update solution, T, and pricing vector. Go to step 1.

Step 2 ensures that the "smaller subtree" will always be as small as possible, thus updating work will be reduced to a minimum.

Logically the most complicated part of the program is the updating of the basic spanning tree. In step 2, there is an effort to minimize this work. In-depth investigation has shown that, after all, updating is a less decisive factor in the running time of MINET than pricing. Table 5 and model (4.1) in section 4 show an example that, by Spearman's rank correlation, there may be a perfect positive relationship between the number of scanned variables and solution time. The same is usually not true for the relation between the number of iterations (which, in most cases, is a good estimate of the tree updates) and solution time. Similar observations have been made on a large number of other problems.

MINET has been used to solve large scale real–life problems for a couple of years. During this period it encountered a large number of very degenerate problems. Such problems caused MINET to make many degenerate steps. In an effort to try to decrease the number of them, a new anti–degeneracy row selection procedure was introduced [10] and experimentally implemented in MINET. The idea behind it can briefly be sketched as follows.

During degenerate iterations the value of the objective function does not change if we do not want to violate the feasibility bounds of the basic and the out–going variables. On the other hand, the objective can be improved if we relax the feasibility bounds of these variables in every iteration and allow some or all of them to go slightly more and more infeasible. Namely, this always allows us to make a positive step, at least to the extent of the relaxation of the feasibility bounds. By this principle, at least a *minimum forced step* can be made at each iteration. It is important that the value of the objective function will always change, therefore, a past basis cannot come up again, thus cycling will not occur. At the same time, the ever growing infeasibility may cause troubles. However, if it is small enough, it can be kept under control.

Though a minimum forced step corresponds to an otherwise degenerate iteration, experience has shown that by this technique usually less time is spent on degenerate vertices (this procedure generates different pivot sequence), therefore the overall efficiency of the algorithm increases. Moreover, since we know that the solution of the MCNF problem will be an integer point, it can be expected that the relaxed optimal solution will be in the close neighborhood of the integer optimum. It means that when the original feasibility bounds are restored after "relaxed optimum" has been achieved, very few if any additional iterations will have to be made to obtain integer optimum.

It is clear that the benefits of this procedure are only hopes based on some heuristic reasoning. At the same time, its weak points are evident. While with the traditional feasible pivot rule we can minimize the size of the "smaller" subtree that is to be updated at every change of the basis, with the new method this will not always be possible. Another point of degradation is the necessary use of floating point arithmetic for the solution vector instead of the more efficient integer arithmetic. These arguments suggest that the tradeoff between advantages and disadvantages of this procedure may be problem dependent. Problem family TRGEN analyzed in section 4 proved to belong to the profitable category.

MINET is written in portable FORTRAN–77. It was developed on an IBM PC compatible computer in DOS environment. This fact influenced the design of the data structure. Uploading MINET into a UNIX environment was formally straightforward. However, since it was done mechanically, some inherited structural weaknesses are present in MINET/UNIX. While much attention was paid to rectifying the sizes of integer arrays and variables (which is an issue of consideration on PCs to save memory and speed up computations), the basic data structure was not redesigned. MINET is an in–core program. The incidence matrix of the network is stored in a $\{from(j), to(j)\}$ pair of arrays. The basis is maintained as a set of list functions outlined above.

The computational testing of MINET was performed on a Sun Sparc Station 1+ at RUTCOR, Rutgers University. The computer is linked up with 5 other similar Sun Sparc Stations, and a larger number of terminals have access to any computer in the system. These facts will be important when we discuss the effect of the loading conditions of the system on the running time. The operating system is SunOS 4.1.1 (UNIX), the identifier of the Fortran compiler is $f77$.

3. Experiences with NETGEN problems

NETGEN is a widely accepted problem generator that has been available to the scientific community since 1974 [5]. It is able to generate different types of network problems, among them MCNF instances.

Two series of experiments with NETGEN were requested by [11]. One of them was with a newly designed set of generation parameters, the other one was with eight selected MCNF problems of the original set of 40 problems in [5].

3.1. New NETGEN problems.

The description how to carry out the first category of computational tests for the DIMACS Challenge with 7 classes of MCNF problems can be found in [11], Section 7, Minimum Cost Flows.

Unfortunately, the version of NETGEN obtained from DIMACS to generate these 7 classes of problems appeared to have some bugs. Therefore I used a revised version received directly from Cornell University, Ithaca, NY. This instance of NETGEN was able to generate most of the required test problems. In one case, however, I experienced that even this version did not work properly. Namely, in class 5, NETGEN–Lo–Supply the only requested size that was accepted by the generator was 128×1024 resulting in an actual size of 128×1039. Any further attempt with very different, but larger sizes in this class made NETGEN stop at the very beginning of the run with a message of *STOP 9000*. This is a certain error message that has many meanings.

I did not carry out testing in class 6 that is apparently identical with class 4 (NETGEN–Hi–Hi ≡ NETGEN–Med–Supply).

In each class, 7 different problem sizes were chosen. NETGEN generated problems identical or very close to the required sizes. The number of nodes was taken to be powers of 2, as requested.

In Table 1, times in each line were obtained as average solution times (in seconds) of 10 problems generated with different random seeds. Standard deviations are also provided. Solution times do not include input. Running times were observed at low load conditions (Saturday night) at RUTCOR, Rutgers University. The problem classes are identified in the same way as in [11], like NETGEN–Lo–Lo, etc. MINET was compiled by the $f77$ compiler in two different ways: without and with code optimization ($-O4$ option). The table contains solution times with both versions. Index 1 refers to data obtained without, while index 2 with code optimization. Column "Ratio of Times" shows the ratio of the performance of the optimized and the non–optimized code.

Problem	Nodes (n)	Edges (m)	Average time-1	Standard dev-1	Average time-2	Standard dev-2	Ratio of Times
NETGEN-Lo-Lo:	sources=n/8,	sinks=n/8					
N11	128	1024	0.24	0.03	0.15	0.02	0.625
N12	256	2048	0.74	0.06	0.44	0.04	0.595
N13	512	4096	2.50	0.14	1.51	0.08	0.604
N14	1024	8196	8.99	0.40	5.28	0.24	0.587
N15	2048	16388	33.18	1.47	20.50	0.90	0.618
N16	4096	32807	132.58	4.17	84.15	2.67	0.635
N17	8192	65627	536.94	12.76	335.39	7.80	0.625
NETGEN-Lo-Hi:	sources=n/8,	sinks=n/2					
N21	128	1024	0.28	0.04	0.16	0.02	0.571
N22	256	2068	0.89	0.05	0.49	0.03	0.551
N23	512	4123	3.10	0.22	1.73	0.12	0.558
N24	1024	8259	11.23	0.68	6.31	0.37	0.562
N25	2048	16541	43.31	1.69	26.14	0.95	0.604
N26	4096	33065	170.57	4.19	108.65	2.65	0.637
N27	8192	66155	681.42	7.40	424.28	4.44	0.623
NETGEN-Hi-Lo:	sources=n/2,	sinks=n/8					
N31	128	1024	0.31	0.03	0.18	0.02	0.581
N32	256	2048	1.01	0.04	0.56	0.02	0.554
N33	512	4096	3.40	0.15	1.91	0.07	0.562
N34	1024	8192	12.41	0.35	6.98	0.20	0.562
N35	2048	16384	49.57	1.02	28.54	0.60	0.576
N36	4096	32768	188.42	3.20	119.92	2.02	0.636
N37	8192	65536	736.44	11.65	460.11	7.07	0.625
NETGEN-Hi-Hi:	sources=n/2,	sinks=n/2					
N41	128	1025	0.39	0.03	0.22	0.02	0.564
N42	256	2048	1.30	0.07	0.73	0.03	0.562
N43	512	4124	4.62	0.22	2.55	0.12	0.552
N44	1024	8262	16.53	0.43	9.19	0.23	0.556
N45	2048	16524	62.18	2.43	37.48	1.43	0.603
N46	4096	33021	243.64	4.61	153.11	2.97	0.628
N47	8192	66119	902.75	11.46	556.43	7.84	0.616
NETGEN-Lo-Supply:	supply=100						
N51	128	1039	0.31	0.04	0.17	0.03	0.548
NETGEN could not generate larger problems in this class							
NETGEN-Med-Supply:	supply=10^5 ≡ NETGEN-Hi-Hi						
NETGEN-Hi-Supply:	supply=10^8						
N71	128	1025	0.39	0.02	0.22	0.01	0.564
N72	256	2048	1.31	0.09	0.73	0.05	0.555
N73	512	4124	4.63	0.19	2.52	0.11	0.579
N74	1024	8262	17.40	0.45	9.55	0.24	0.533
N75	2048	16524	72.25	2.72	40.90	1.58	0.581
N76	4096	33018	291.25	5.69	183.94	3.63	0.638
N77	8192	66112	1244.80	22.39	810.69	13.52	0.604

TABLE 1.

The above data provide good basis for statistical analysis of solution times of MINET as a function of problem characteristics. Among several possible models, we found that the simple linear regression model described solution times very accurately if it was defined as follows:

$$T = a + bx, \qquad (3.1)$$

where x is the problem size expressed as mn, i.e., the size of the incidence matrix. Coefficients a and b vary by problem class.

During the regression analysis the unit of the problem size for all 7 observations in each class was defined as $mn \times 10^{-6}$ which resulted in x_i values in the range of 0.13 to 541.9 such that $x_i \approx 4x_{i-1}$ (for $i = 2, \ldots, 7$).

For the 5 different classes we determined the correlation coefficients, standard errors, and 95 % confidence intervals of b for the "Average time–2" values. They are summarized in Table 2.

Problem	Correlation coefficient (r)	Standard error of the estimate	Confidence interval for b
NETGEN–Lo–Lo:	0.999998	0.260936	(0.622452, 0.625220)
NETGEN–Lo–Hi:	0.999978	1.060852	(0.777804, 0.788968)
NETGEN–Hi–Lo:	0.999940	2.035211	(0.846901, 0.868520)
NETGEN–Hi–Hi:	0.999686	5.689803	(0.997979, 1.057886)
NETGEN–Hi–Supply:	0.999713	52.514020	(1.222111, 1.775359)

TABLE 2.

The confidence intervals for b are included to give a better characterization of the goodness of fit. For the first four classes these intervals are narrow, while class 5 has a relatively wider confidence interval.

The above data show that the performance of MINET can well be estimated by the (3.1) model within the experimental range of $0.13 \leq x \leq 542$.

It is instructive to investigate the predicting power of this model. For this purpose we did the following. We applied the (3.1) model for only the first 5 observations and used the resulting regression line for predicting the remaining 2 data points. The findings are presented in Table 3.

The relatively small sample (number of observations is equal to 5 in each class there) appears to support the applicability of the (3.1) linear model. Consequently, we can expect that the regression model based on 7 observations will behave similarly beyond the experimental region, therefore, it can be used for at least short range predictions. And this is approximately the same what other forecasting models can promise.

Problem	i	x_i	Observed T_i	Predicted $\hat{T}_i$
Netgen–Lo–Lo	6	134	84.15	81.68
	7	538	335.39	326.35
Netgen–Lo–Hi	6	135	108.65	104.23
	7	542	424.28	417.02
Netgen–Hi–Lo	6	134	119.92	113.83
	7	537	460.11	455.13
Netgen–Hi–Hi	6	135	153.11	151.81
	7	542	556.43	607.84
Netgen–Hi–Supply	6	135	183.94	165.95
	7	542	810.69	665.02

TABLE 3.

It is important to note that the validity of the model and results are restricted to the present pricing strategy of MINET on the NETGEN problems.

3.2. Old NETGEN problems.

The second group of NETGEN experiments was carried out with 8 problems from the NETGEN test set of [5] selected by DIMACS. In Table 4 they are identified by their original sequential number. Running times were obtained at low load conditions of the computer system. The table contains solution times with two versions of MINET. Index 1 refers to data obtained without, while index 2 with code optimization. Times are given in seconds. Column "Time Ratio" shows the ratio of the performance of the optimized code to the non–optimized one.

Problem	Nodes	Edges	Iterations	Time–1	Time–2	Time Ratio
11	400	1500	5095	2.37	1.36	0.574
20	400	1416	1468	0.74	0.37	0.500
27	400	2676	2323	0.99	0.54	0.545
28	1000	2900	4503	3.32	1.85	0.557
32	1500	4342	6594	6.07	3.41	0.562
36	8000	15000	34461	118.48	76.78	0.648
38	3000	35000	54105	75.73	45.93	0.606
40	3000	23000	41595	48.46	33.86	0.699

TABLE 4.

This set of NETGEN problems consists of structurally different MCNF problems. Therefore, the (3.1) regression analysis is not applicable here since (3.1) was designed to characterize the behavior of MINET on a structurally similar set of problems.

In Tables 1 and 4 we included a column (Ratio of Times) that shows the ratio of solution times with and without code optimization. This serves to evaluate the gain that can be obtained by using the $-O4$ code optimizing option of the $f77$ compiler. The mean value of this figure is 58.8% with a small standard deviation. It means that the gain in solution time that can be obtained by code optimization is 41.2% which is very considerable.

4. Experiences with TRGEN problems

4.1. The TRGEN real–life problems.

As a second category of testing, I generated some real–life problems based on the trailer subproblems for "Traffic scheduling via Benders decomposition" [6]. These problems are based on 4 parameters, say p_1, p_2, p_3, p_4. The size of a generated problem is:

\# of nodes : $2(p_1 + p_2)$,
\# of edges : $(p_1 + p_2)^2 + p_2$.

p_3 and p_4 play role in generating the cost coefficients.

I submitted the FORTRAN version of this generator to DIMACS to make it available for other interested researchers.

The test results with MINET (optimized code, low loading conditions) can be found in Table 5 which contains two additional columns: "Degenerate iterations" and "Scanned variables". The term "Scanned variables" refers to the total number of scalar products computed in search for an incoming variable throughout the algorithm. In column "Problem" the four generation parameters are also given for reproducibility.

Here again, we can investigate if there is a significant linear relationship between problem size (measured in mn) and running time. A regression analysis based on the linear model of (3.1) for the above data gives a correlation coefficient $r = 0.950881$, while the 95% confidence interval for b is $0.102864 \leq b \leq 0.209920$. This relatively wide interval is caused by the seemingly irregular behavior of TR8. If we leave it out from the investigations then values of $r = 0.997614$ and $0.141881 \leq b \leq 0.169919$ are obtained which indicates the existence of a close linear relationship.

Table 5 suggests an additional interesting investigation. We can test whether there is a significant linear relationship between solution time (T) and the number of scanned variables (x) measured in units of 10^6, i.e.,

$$T = a + bx \qquad (4.1)$$

Problem & Gen. params	Nodes	Edges	Iterations	Degenerate iterations	Scanned variables	Sol. time seconds
TR4 50,150,200,48	400	40150	16171	15797	325348	3.33
TR5 75,150,200,48	450	50775	19633	19209	462524	4.31
TR6 100,150,200,48	500	62650	22488	22014	588475	5.32
TR7 50,250,300,48	600	90250	38484	37867	735998	8.57
TR8 100,250,300,168	700	122750	58949	58372	4119085	25.18
TR9 150,250,400,168	800	160250	68029	67305	3385990	22.71
TR10 200,300,600,168	1000	250300	101709	100871	6026781	39.07

TABLE 5.

For this model we obtain $r = 0.996183$ and $5.400 \leq b \leq 6.609$. This means that under the present setting of strategic parameters of MINET, solution time is a linear function of the number of scanned variables.

Similarly, we can compute the Spearman rank correlation coefficient for these data. Doing so, we obtain a value of $r_s = 1.00$ that suggests a perfect positive relationship between the two variables.

Unfortunately, values based on model (4.1) that cannot be computed in advance, therefore this model cannot be used to estimate solution time. However, this information can be useful for researchers who study the behavior of algorithms with the purpose of trying to improve the performance of them.

4.2. Effects of an anti–degeneracy row selection strategy.

The very large number of degenerate iterations (up to 99 % of the total number of iterations) that appears in Table 5 suggests that solution times could be reduced if some "cheap" anti–degeneracy strategy could be found. Though in the framework of simplex based MCNF algorithms degenerate steps cannot completely be avoided (for detailed analysis see [3]), there seems to be a chance for achieving some improvement. For the TRGEN problems we tried the anti–degeneracy row selection Version–1 described in [10] and obtained the results displayed in Table 6 (MINET, optimized code). Row "Old" presents data with the feasible (traditional) row selection, row "New" with Version–1. In a "New" row a degenerate iteration indicates the number of minimum forced steps. If any small step larger than this was made, it was not counted degenerate.

Problem	Method	Iterations in Phase–I	Deg. itns in Phase–I	Iterations (Total)	Deg. itns (Total)	Sol. time in *sec*
TR4	Old	12798	12598	16171	15797	3.33
	New	399	0	3675	0	4.04
TR5	Old	14398	14173	19633	19209	4.31
	New	449	0	3601	0	3.37
TR6	Old	16623	16373	22488	22014	5.32
	New	499	0	4113	0	4.81
TR7	Old	32948	32648	38484	37867	8.57
	New	599	0	3952	0	6.97
TR8	Old	36773	36423	58949	58372	25.18
	New	699	0	6521	0	11.66
TR9	Old	43098	42698	68029	67305	22.76
	New	799	0	8899	2	15.16
TR10	Old	65748	65248	101709	100871	39.07
	New	999	0	12664	4	25.59

TABLE 6.

The improvement in the number of iterations seems to be considerable, while the solution times do not show a proportional decrease. This is partly due to the rough experimental implementation of the simplest version of the new row selection procedure, and partly to the fact that with this procedure we usually get a much larger "smaller subtree" that is to be updated at every change of the basis, as outlined in section 2. This latter can be improved by some refinements and also by the implementation of versions 2 and 3 proposed in [10]. However, it is inevitable that the computational cost of an iteration will be higher than with the traditional row selection technique. This is the price to be paid for the reduction in the number of iterations. The tradeoff may be problem dependent.

5. GTE–instances

The third category of our experimentation was a set of recently discovered real–life problems identified as GTE–instances [12]. They have been obtained from a multi–commodity model. The problems are small (49 nodes, 520 edges) and differ from one another in the capacities and demands.

The remarkable feature of these problems is that they are apparently "bad" for Bertsekas–Tseng's RELAXT–III algorithm making it to produce a large variation in solution times. This is why this family of problems received the **bad** suffix. The range of the running times is 0.020 *sec* through 508.829 *sec*. Clearly, this is a challenge for the simplex based network solvers to see how efficiently and with what variance in running time they can solve this family of problems.

MINET showed a very stable performance on the gte_bad problems, as it can be seen in Table 7. Solution times never exceeded 0.06 *sec*, which is still in the magnitude of the resolution of the clock in the test environment of MINET. The extension of problem names after the period gives the solution time in milliseconds obtained by RELAXT–III. The last column contains times obtained by MINET (optimized code).

Problem	Iterations	Degenerate iterations	Scanned variables	Sol. time in *sec*
gte.20	121	117	2741	0.04
gte.60	128	44	2461	0.03
gte.200	154	48	2465	0.03
gte.510	226	116	3943	0.04
gte.1160	248	138	3571	0.06
gte.6830	211	108	3301	0.04
gte.15100	139	87	2609	0.03
gte.15710	163	71	2869	0.04
gte.35620	192	47	3129	0.04
gte.60090	290	147	4229	0.06
gte.298300	135	39	2554	0.03
gte.451760	178	68	3721	0.05
gte.469010	131	41	2852	0.05
gte.508829	127	38	3041	0.03

TABLE 7.

6. Comparison of MINET and NETFLO

NETFLO is an industry standard MCNF solver developed by R. V. Helgason and J. L. Kennington. The FORTRAN version of NETFLO has been made available for the participants of the DIMACS Challenge at DIMACS /netflow/mincost/solver-1 together with some hints how to use it for timing. (Actually, this is the only MCNF solver available there.) This version of NETFLO is dated February 1988.

To enable the comparison, one little change had to be made in NETFLO. Since NETFLO makes some preprocessing during input, it was necessary to include it in timing. MINET does not generate any detailed solution output after the optimality condition is fulfilled. To ensure comparability, we changed NETFLO to stop after optimality condition is detected, included input for timing of MINET, and used the run time system *logtime* provided in the above referenced subdirectory of DIMACS for timing. Software tool *logtime* separates *user time* and *system time* and gives them in seconds with an accuracy of one decimal for a program run under its supervision.

For the comparison, we used seven of the eight selected NETGEN problems. The smallest of them (#11) is an assignment problem that could not be converted for NETFLO by the supplied conversion routine, therefore it was not considered here.

In the other rounds of comparisons we used problem classes from the final DIMACS Core problems [12]. First, problems generated by the GGRAPH generator, producing grid graphs, were used, then several different classes generated by the GOTO generator, producing capacitated transportation problems on a Grid-On-TOrus network, were solved. The GOTO problems are acknowledgedly hard.

For the tests, both MINET and NETFLO were compiled with the recommended $-O$ optimization feature of the $f77$ compiler.

Results with the selected NETGEN problems are summarized in Table 8.

Problem	Nodes	Edges	MINET Iterations	MINET User time	NETFLO Iterations	NETFLO User time
20	400	1426	1468	1.9	1338	2.6
27	400	2676	2323	3.3	1900	5.3
28	1000	2900	4503	5.0	3565	7.7
32	1500	4342	6594	8.1	5704	15.2
36	8000	15000	34461	85.3	31246	186.0
38	3000	35000	54105	79.7	25077	333.7
40	3000	23000	41595	54.6	23267	195.7

TABLE 8.

There are a few things in Table 8 that deserve attention. First, the two codes give different sequences of problems if sorted by solution time. Second, NETFLO tends to make considerably less iterations than MINET. Third, solution times of MINET get significantly smaller than that of NETFLO as problem size increases.

The GGRAPH problems can be generated by 5 parameters: X: Grid height, Y: Grid width, MAXCAP: Capacities are uniform on [0, MAXCAP], MAXCOST: Costs are uniform on [0, MAXCOST], SEED: Seed of the random number generator.

For all of the GGRAPH problems the values MAXCAP= 10^4 and MAXCOST= 10^4 were fixed. In Table 9, under problem generation parameters, we give the values for X, Y, and SEED.

The last two GGARPH problems in Table 9 differ only in the value of the SEED. For this set we ignore regression analysis because the problems are structurally inhomogeneous. Nevertheless, it is interesting to observe that both solvers show a smooth performance in the sense that solution times increase smoothly with problem size. NETFLO is always faster by a narrow margin.

Generation parameters	Nodes	Edges	MINET Iterations	MINET User time	NETFLO Iterations	NETFLO User time
32 32 270001	1026	2048	2674	3.5	1869	3.0
16 128 270001	2050	3984	4988	8.8	2282	5.9
128 16 270001	2050	4208	5762	9.3	6260	8.9
64 64 270003	4098	8192	13645	23.3	10168	18.8
256 16 270001	4098	8432	12290	25.8	13210	22.5
80 80 270004	6402	12800	19539	42.2	21384	40.8
512 16 270001	8194	16680	23815	77.1	33306	75.2
512 16 270011	8194	16680	22811	75.9	34968	74.4

TABLE 9.

The GOTO problems are generated by following 5 parameters: N: Number of nodes, M: Number of arcs, MAXCAP: Maximum capacity (some arcs may be generated with capacities greater than this to ensure feasibility), MAXCOST: Maximum arc cost, SEED: Seed for random number generator.

We investigated three groups of GOTO problems. Each of them contained structurally similar problems on which both codes obeyed the (3.1) model. Therefore, we could use regression analysis for comparison. It turned out that the computed values of the correlation coefficients for MINET and NETFLO in the groups were quite the same that established a basis for comparison of the regression lines.

The first group of GOTO problems is referenced as Grid–Density–8, (GD–8) and has the feature that $M = 8N$. For the generated test problems the last three parameters, in all cases, were fixed: MAXCAP=16384, MAXCOST=4096, and SEED=270001. Results of the test are summarized in Table 10.

Nodes	Edges	MINET Iterations	MINET User time	NETFLO Iterations	NETFLO User time
128	1024	1409	1.6	2138	1.8
256	2048	3188	3.8	5450	4.7
512	4096	6860	11.8	16496	16.5
1024	8192	11674	41.9	30003	54.8
2048	16384	28759	108.1	67287	182.7
4096	32768	75322	329.3	342391	905.9

TABLE 10.

The second group of GOTO problems is called Grid–Density–16 (GD–16), and is characterized by the relation $M = 16N$. The value of the last three parameters, in all cases, were fixed in the same way as in Grid–Density–16. Results of the test are presented in Table 11.

Nodes	Edges	MINET Iterations	MINET User time	NETFLO Iterations	NETFLO User time
128	2048	2858	3.2	5750	4.5
256	4096	6812	8.4	14265	11.0
512	8192	14948	24.2	31305	30.4
1024	16384	29753	77.0	72282	109.3
2048	32768	70538	239.4	172723	408.3
4096	65536	143122	1042.4	460458	1759.5

TABLE 11.

The third group of GOTO problems is referenced as Grid–Increasing–Density (GDI), and has the feature that $M = \lceil N^{3/2} \rceil$. The rest of the parameters is as above. Results of the test can be found in Table 12.

Nodes	Edges	MINET Iterations	MINET User time	NETFLO Iterations	NETFLO User time
128	1449	2040	2.2	2725	2.5
256	4096	6812	8.4	14265	11.0
512	11586	19490	37.0	48493	50.7
1024	32768	60980	166.1	190027	297.4
2048	92682	199899	832.7	680137	1953.9

TABLE 12.

From Tables 10–11–12 we can compute linear regression data for the three GOTO groups by (3.1). The obtained values of r and b are summarized in Table 13.

Group	Parameter	MINET	NETFLO
GD–8	r	0.996360	0.998699
	b	2.406176	6.731219
GD–16	r	0.999618	0.999803
	b	3.856446	6.540692
GDI	r	0.999784	0.999698
	b	4.352572	10.330800

TABLE 13.

Based on the near equality of the correlation coefficients (r) of MINET and

NETFLO in Table 13, we can compare the slopes (b). This suggests that on the GOTO problems MINET is 2 to 3 times faster than NETFLO.

7. Running times versus loading conditions of the computer

The most widely accepted measure of performance of optimization algorithms is running time. Other measures fail to express the total computational effort needed to solve a problem. One such example is the number of iterations which can heavily be influenced by the work per iteration, therefore, the meaning of an iteration can be different from algorithm to algorithm.

The running time of a program was believed to be proportional to the real computational effort required to solve a problem and therefore suitable for comparing the performance of different algorithms. Standards have been set up for reporting computational experiences (c.f., [1] or [4]).

Timing on PCs is relatively simple and accurate. For comparability of results, it is more or less sufficient to specify the type of the processor and clock rate, together with the *solution time* which is usually equal to the *residence time*.

Workstations, however, run under UNIX operating system that provides multiprocessing, multitasking, and networking. Therefore, in any short time interval, a larger number of programs may run in the computer. This immediately explains why the question of timing accuracy was raised by the organizers of the First DIMACS International Algorithm Implementation Challenge [11].

Clearly, under UNIX environment, residence time can be anything depending on the nature of the concurrently running processes, therefore it cannot be used for timing. However, most of the systems offer time measuring procedures that can separate *system time* and *user time*.

We investigated with MINET the following question: How accurately can solution times be reproduced if the same program solves the same problem over and over again?

MINET uses the *etime* routine of the UNIX FORTRAN library. Its first parameter is supposed to measure user time.

First, the performance of *etime* was tested when the Spark 1+ was run in a stand alone mode. The problem to be solved was chosen from the GGRAPH family and identified as **gg2** with 2027 nodes and 4050 arcs. MINET solved **gg2** three consecutive times in 7.82, 7.73, and 7.79 seconds.

When two identical copies (say, A and B) of MINET were loaded and both worked on **gg2** concurrently, the following running times were obtained at tree different attempts:

Run	Copy-A	Copy-B
1	7.80	7.87
2	7.75	7.89
3	7.84	7.88

When the experiment was repeated with three copies of MINET running concurrently on **gg2**, solution times provided by *etime* were 7.62, 7.70, and 7.67 seconds.

These figures seem to justify the belief that *etime* really returns the user time of a program, and not residence time or something else.

However, when MINET was tested on the same **gg2** problem under everyday multiprogramming, multitasking, and networking circumstances, we observed a rather erratic behavior.

The method of testing was the following. Within a selected time interval MINET was run to solve the same problem repeatedly. The lengths of the intervals were different, and the intervals were selected on different days of the week and different hours of the day. For each interval the mean ($\bar{X}$) of the solution times is computed, and also the t_{max}/t_{min} ratio in percents. For all the three problems a "Total" row is also created displaying the total sample mean and the t_{max}/t_{min} ratio for the sample. Problem characteristics are summarized below. **gg2** and **gg3** are from the GGRAPH family, while `cap1` was generated by `capt`, also from [12].

Problem	# of nodes	# of arcs
gg2	2027	4050
gg3	4098	8192
cap1	514	33025

The results obtained with **gg2**, **gg3**, and `cap1` are summarized in Tables 13, 14, and 15, resp.

Time Interval	Observations (t_i)	mean ($\bar{x}$)	t_{max}/t_{min} %
2 min	7.91 7.96 7.95	7.940	100.63
5 min	7.59 7.77 7.66 7.62 7.68 7.58	7.650	102.51
5 min	8.26 8.08 8.27 8.07 8.19	8.174	102.48
2 min	9.68 9.82 9.60	9.700	102.29
5 min	10.96 10.95 7.57 7.58 7.88 7.59	8.757	144.78
	Total	8.357	144.78

TABLE 14.

Time Interval	Observations (t_i)	mean ($\bar{x}$)	t_{max}/t_{min} %
5 min	25.09 25.61 25.17 25.07 24.03	24.994	106.58
6 min	24.05 24.06 24.91 25.08 24.19	24.458	104.28
8 min	31.43 30.13 25.18 27.13 25.23	27.820	124.82
5 min	34.30 23.98 24.01 24.88	26.793	143.04
	Total	25.975	143.04

TABLE 15.

Time Interval	Observations (t_i)	mean ($\bar{x}$)	$t_{\max}/t_{\min}$ %
3 min	8.14 8.17 8.20 7.85	8.090	104.46
3 min	9.70 9.51 9.46 9.32 9.88	9.574	106.01
3 min	9.89 8.18 8.17 9.20	8.860	121.05
	Total	8.898	125.99

TABLE 16.

The obtained figures do not look very encouraging. Running times seem to vary too much. A bit more careful analysis reveals some important details.

Within the first four rows of gg2 (Table 14) we can see a uniform behavior of running times. The same uniform figures cannot be observed between rows. The mean values of 7.940, 7.650, 8.174, and 9.700 differ nearly 27%. Then comes row 5 with a very inhomogeneous row performance and globally extreme values that also define the $t_{\max}/t_{\min}$ ratio for the whole sample.

The same can be said if we look at the observed data for gg3 (Table 15). The only difference is that in this case the variable behavior can be seen in two rows.

Though in the third series of observations cap1 (Table 16) did not show such a more than 40% deviation, it is still too variable with its 26%.

It is our belief that if a single counterexample can be found then this is sufficient to reject the unconditional reliability of time measuring in UNIX. What we presented here is definitely more than just a single counterexample. Therefore, if we imagine a long test run including the serial solution of hundreds of problems in a blind batch mode then we really cannot be certain about the comparability of such solution times.

Having seen the above instances several questions can be asked. What is the real solution time? How can we compare algorithms on the basis of reported running times? We think that nowadays, when a 25% performance improvement of an algorithm is considered significant, these questions should be addressed correctly.

Our purpose here is to point out the above problems. Answering them may require a considerable effort. At the same time, it appears inevitable that such a study must be conducted.

8. Summary

Based on the quick overview of the above experiences, it can be concluded that MINET shows a relatively stable and more or less predictable performance with the present setting of the available strategic parameters.

In the discussion — among others — we addressed two important issues that may influence the comparability of reported running times and, therefore the relative efficiency of codes. It turned out that the optimizing capability of a compiler can improve timings by more than 40%. It also became evident that time measuring under UNIX operating system can be rather unreliable giving an at least 40% wide range of solution times. The former observation suggests that information on code optimization must always be provided, while the latter necessitates an in–depth analysis of time measuring in UNIX.

9. Acknowledgments

The author is indebted to the two anonymous referees for their helpful comments.

References

1. Crowder, Dembo, and Mulvey, *Reporting Computational Experiments in Mathematical Programming*, Mathematical Programming **15** (1978), 315–329.
2. Cunningham, W. H., *Theoretical Properties of the Network Simplex Method*, Mathematics of Operations Research **4** (1979), 196–208.
3. Gal, T., Geue, F., *A New Pivoting Rule for Solving Various Degeneracy Problems*, Operations Research Letters **11** (1992), 23–32.
4. Greenberg, H. J., *Computational Testing: Why, How, and How Much*, ORSA Journal on Computing **2** (1990), 94–97.
5. Klingman, D., Napier, A., Stutz, J., *NETGEN—A program for generating large-scale (un)capacitated assignment, transportation and minimum cost flow problems*, Management Science **20** (1974), 814–821.
6. Love, R. R., *Traffic scheduling via Benders decomposition*, Mathematical Programming Study **15** (1981), 102–124.
7. Maros, I., *MINET a Fast Network LP Solver*, IIASA Working Paper WP–87–50, International Institute for Applied Systems Analysis, A–2361 Laxenburg, Austria, 1987.
8. Maros, I., *MINET Fast Network LP Solver, Description and User's Guide for V2.00*, IIASA Working Paper WP–88–6, International Institute for Applied Systems Analysis, A–2361 Laxenburg, Austria, 1988.
9. Maros, I., *A Structure Exploiting Pricing Procedure in the Primal Network Simplex Algorithm,*, RUTCOR Research Report, RRR #18–91, RUTCOR, Rutgers University, 1991.
10. Maros, I., *A Practical Anti–Degeneracy Row Selection Technique in Network Linear Programming*, RUTCOR Research Report, RRR #4–92, 1992.
11. McGeoch, C. C., *The First DIMACS International Algorithm Implementation Challenge: The Core Experiments*, DIMACS, August, 1991.
12. McGeoch, C. C., *The First DIMACS International Algorithm Implementation Challenge: The Benchmark Experiments*, DIMACS, December, 1991.

Computer and Automation Institute, Hungarian Academy of Sciences, H–1518 Budapest, P.O.Box 63., Hungary

Current address: RUTCOR, Rutgers Center for Operations Research, Rutgers University, P.O.Box 5062, New Brunswick, NJ 08903

E-mail address: maros@rutcor.rutgers.edu and h107mar@ella.hu

A Speculative Contraction Method for Minimum Cost Flows: Toward a Practical Algorithm

SATORU FUJISHIGE, KAZUO IWANO,

JUN NAKANO, AND SHU TEZUKA

Abstract. We propose a new heuristic method, called a *speculative contraction method*, for uncapacitated minimum cost flows based on Plotkin and Tardos' algorithm, a variant of Orlin's algorithm. In a Δ scaling phase, Plotkin and Tardos' algorithm contracts arcs of flow value more than $5n\Delta$. However, the speculative contraction method contracts arcs of flow value much smaller than $5n\Delta$, and corrects the possible primal infeasibility caused by inappropriate contractions at the end. Our experiments show that the new method significantly reduces the running time of the original algorithm. In particular, for sparse graphs, it shows a speed up of 3.4. Moreover, it runs faster than our implementation of the Primal-Dual method. We hope our experimental results on speculative contractions provoke further research toward theoretically and practically faster scaling algorithms.

1. Introduction

The minimum cost flow problem is one of the most important network optimization problems and has been widely studied for the last four decades. (See surveys in [2, 13].) In particular, there have been emerging theoretical developments since Tardos invented the first strongly polynomial time algorithm for this problem [23].

Many of recent algorithms are based on the scaling methods on cost or capacity. For example, the cost scaling algorithms include Tardos [23], Galil and Tardos [10], and Goldberg and Tarjan [14], while the capacity (excess) scaling algorithms include Orlin [20], Fujishige [8], Orlin [21], and Plotkin and Tardos [22].

1991 *Mathematics Subject Classification.* Primary 90B10; Secondary 90C35, 68Q25.

Key words and phrases. min-cost flow, network algorithms, heuristics, speculative contraction.

The first author is partly supported by a grant-in-aid of the Ministry of Education, Science and Culture of Japan. A preliminary version of this paper appeared in [9].

Ahuja, Goldberg, Orlin, and Tarjan devised the currently fastest polynomial time algorithm, $O(nm \log \log U \log(nC))$ time, based on a double scaling method [1]. Among strongly polynomial algorithms Orlin's $O(m \log n(m + n \log n))$ time algorithm [21] is the fastest known algorithm. Here, n (resp. m) denotes the number of nodes (resp. arcs), and U (resp. C) is the maximum capacity (resp. cost) in an input network.

Despite recent theoretical developments, very little has been known about the practical efficiency of recent scaling algorithms. DIMACS, the Center for Discrete Mathematics and Theoretical Computer Science, thus, organized the competition of implementing recent network flow algorithms [5]. According to [2], Grigoriadis' implementation of the primal simplex method [16] and Bertsekas and Tseng's relaxation algorithms [3] seem to be the currently fastest in practice.

In the above context, we propose a new heuristic method based on Orlin's algorithm, and perform experiments on its practical efficiency. Plotkin and Tardos' variant [22] of Orlin's algorithm is simple, does not assume the strong connectivity of an input graph, and runs in the same complexity as Orlin's algorithm does. Therefore, in this paper we discuss Plotkin and Tardos' variant (hereafter, we call it PT) instead of Orlin's original algorithm. Since Orlin's algorithm (thus, Plotkin and Tardos' algorithm) is designed for transshipment networks, our experiments are also limited to transshipment networks, which are uncapacitated.

An excess scaling algorithm PT consists of Δ phases starting from $\Delta = U$ to 0, and it reduces Δ by at least a half at the end of each phase. During a Δ phase the algorithm contracts every arc which carries flow exceeding $5n\Delta$. In contrast, we propose a *speculative contraction*, which contracts every arc of flow value much less than $5n\Delta$. Since it may contract an arc inappropriately, it may result in a negative flow value on some arc. Thus, we adjust these negative flow values by using the Primal-Dual algorithm.

Our speculative contraction method significantly reduces the running time of the original algorithm (we observe a factor of about 3.4 speed up for sparse graphs). Moreover, it runs faster than our implementation of the Primal-Dual method. We also conduct comparisons with NETFLO [18] and RELAXT-III [3] for uncapacitated networks. As a result, for the netgen density/size family RELAXT-III (resp. NETFLO) is 7.6 times faster (resp. 3.5) than our method with the threshold 8Δ, which we denote by $PT(8\Delta)$. For uncapacitated networks transformed from capacitated networks in the DIMACS core set [5], $PT(8\Delta)$ runs competitively to RELAXT-III: i.e., $CPU(PT(8\Delta))/CPU(\text{RELAXT-III})$ is 1.25 for the Grid families and 0.50 for the goto families on average. NETFLO runs faster than $PT(8\Delta)$ by 3.6 (resp. 5.2) for the Grid (resp. goto) families. However, since our code is still premature and does not involve refined heuristics, we hope that our proposal of speculative contractions provoke further research toward scaling algorithms that are faster in theory and practice.

Section 2 reviews terminology and the Plotkin and Tardos algorithm. Section

3 introduces the concept of speculative contractions and describes our implementation. Section 4 introduces the specifications of our experiments. Section 5 discusses heuristics we investigate. Section 6 shows our experimental results followed by our observations. Section 7 compares our code with other minimum cost flow codes such as NETFLO [18] and RELAXT-III [3]. Section 8 investigates future research topics.

2. Preliminaries

In this section we first introduce our terminology and then briefly review Plotkin and Tardos' algorithm [22].

We basically follow Plotkin and Tardos' notations in [22]. Let $G = (V, A)$ be a directed graph with $|V| = n$ nodes and $|A| = m$ arcs. For the sake of simplicity, we assume there are no parallel or opposite arcs. We define two real-valued functions: one is an arc *cost* function $\gamma : A \longrightarrow R$, while the other is a node *demand* function $d : V \longrightarrow R$ such that $\sum_{v \in V} d(v) = 0$. For $v, w \in V$ with $(v, w) \in A$, we define $\gamma(w, v) = -\gamma(v, w)$. We also define a real-valued arc *capacity* function $c : A \longrightarrow R$. We call a tuple $\mathcal{N} = (G, \gamma, d, c)$ a *network*. In particular, when $c(a) = \infty$ for each $a \in A$, we call $\mathcal{N}$ a *transshipment network*. A non-negative real-valued function f on A is called a *pseudoflow*. For a pseudoflow f, we define the excess of each node v by $ex_f(v) = \sum_{(w,v) \in A} f(w, v) - \sum_{(v,w) \in A} f(v, w) - d(v)$. A pseudoflow f is called a *transshipment* if the excess of every node with respect to f is 0; that is, a flow conservation rule is satisfied at each node. The cost of a pseudoflow f is $\gamma(f) = \sum_{a \in A} \gamma(a) f(a)$.

Given a pseudoflow f in a transshipment network $\mathcal{N}$, we define the *residual network* with respect to f by $\mathcal{N}_f = (G_f = (V, A_f), \gamma, d, c_f)$, where $c_f(v, w) = \infty$ for all $(v, w) \in A$, $c_f(v, w) = f(w, v)$ for $(v, w) \in V \times V$ when $(w, v) \in A$, $c_f(v, w) = 0$ otherwise, and $A_f = \{(v, w) \mid c_f(v, w) > 0, v, w \in V\}$.

The *transshipment problem* is the problem of finding a minimum cost transshipment in a given transshipment network. The following theorem on the optimality of a transshipment is well known. Before introducing the theorem, we need a few more definitions. A *node price function* (or *potential function*) p is a real-valued function defined on V. The *reduced cost function* γ_p with respect to a price function p is defined by $\gamma_p(v, w) = \gamma(v, w) + p(v) - p(w)$.

THEOREM 2.1 ([7]). *A transshipment f is optimal if and only if there exists a price function p which satisfies (1) the dual feasibility constraints: for each arc $a \in A$, $c_p(a) \geq 0$ and (2) the complementary slackness constraints: for each arc $a \in A$, if $c_p(a) > 0$, then $f(a) = 0$.* □

We can now introduce Plotkin and Tardos' algorithm (PT) as shown in Figures 1, 2, and 3. Algorithm PT takes a transshipment network $\mathcal{N}$ as an input. Notice that we define $S_f(\Delta) = \{v \in V \mid ex_f(v) > \Delta\}$ and $T_f(\Delta) = \{v \in V \mid ex_f(v) < -\Delta\}$. Algorithm PT consists of excess scaling phases with Δ starting from the

> **Procedure** *Plotkin-Tardos*
> (1) $\Delta \leftarrow \max_{v \in V} |d(v)|$;
> (2)   **while** $\Delta \neq 0$ **do begin**
> (2.1)   **while** $S_f(\frac{n-1}{n}\Delta) \cup T_f(\frac{n-1}{n}\Delta) \neq \emptyset$ **do begin**
> (2.2)   **if** $S_f(\frac{n-1}{n}\Delta) \neq \emptyset$
> (2.3)   **then** $AUGMENT(S_f(\frac{n-1}{n}\Delta), T_f(\frac{1}{n}\Delta), \Delta)$;
> (2.4)   **else** $AUGMENT(S_f(\frac{1}{n}\Delta), T_f(\frac{n-1}{n}\Delta), \Delta)$;
> (2.5) $CONTRACT(5n\Delta)$;
>   **end**
>   **if** f is zero on all uncontracted arcs
> (3)   **then** $\Delta \leftarrow \max_{v \in \tilde{V}} |ex_f(v)|$, where $\tilde{V}$ denotes the set of remaining nodes;
> (4)   **else** $\Delta \leftarrow \Delta/2$;
>   **end**
> (5) $f \leftarrow UNFOLD$;

FIGURE 1. Algorithm Plotkin-Tardos

> **Procedure** $AUGMENT\ (S,T,\Delta)$
> (1) $\pi(s) \leftarrow 0$ for each $s \in S$; Find a shortest path tree starting from nodes in S with respect to a reduced cost γ_p in G_f. Let $\pi(v)$ be the obtained cost in the tree for each $v \in V$.
> (2) $p(v) \leftarrow p(v) + \pi(v)$ for each $v \in V$.
> (3)   For each $s \in S$, if a subtree of the shortest path forest containing s includes a node $t \in T$, move Δ units of flow from s to t along a tree path.

FIGURE 2. Procedure AUGMENT

> **Procedure** $CONTRACT\ (threshold)$
>   For each $e = (u,v) \in A_f$ such that $f(e) \geq threshold$, create a new node w with $\pi(w) = \pi(u)\ (= \pi(v))$ and $ex_f(w) = ex_f(u) + ex_f(v)$, delete two nodes u and v and the arc e. For each arc (x,u) (resp. (x,v), (u,x), and (v,x)) in A_f, replace this arc by (x,w) (resp. (x,w), (w,x), and (w,x)).

FIGURE 3. Procedure CONTRACT

maximum absolute value of node demands and ending at 0. In the Δ-phase, until $S_f(\frac{n-1}{n}\Delta) \cup T_f(\frac{n-1}{n}\Delta) = \emptyset$, the algorithm repeats augmentations and contractions of every arc of flow value more than $5n\Delta$ by $CONTRACT(5n\Delta)$, and finally it resets Δ by at least a half. From now on we call the factor which divides Δ at Step (5) in Figure 1 a *scaling factor*, denoted by α: that is, $\alpha = 2$ in the figure. The procedure $AUGMENT(S, T, \Delta)$ shown in Figure 2 first finds a shortest path forest starting from nodes in S with respect to a reduced cost γ_p, and then augments flow from nodes in S to nodes in T with disjoint minimum cost paths in the forest. At the end of algorithm PT, the procedure UNFOLD reverses the contraction process to unfold the network and compute the resulting arc flows. Since every node is involved in augmentations at most $O(\log n)$ times before being contracted, there are at most $O(n \log n)$ phases. Therefore, the algorithm PT attains a strongly polynomial time, $O(n \log n(m + n \log n))$. See the detailed discussion in [22].

3. Speculative Contractions

In this section we introduce our heuristic method called a *speculative contraction*, and briefly discuss our implementation.

As observed in [21] and [22], once an arc carries a flow of more than $5n\Delta$ at a Δ phase, a flow on the arc will never vanish in the succeeding scaling phases. Therefore, algorithm PT contracts such an arc, and reduces the size of a graph. In [22], $5n\Delta$ is determined by the maximum possible flow value to a single arc. That is, at each phase there are at most $2n$ augmentations of value Δ, whose summation over the succeeding phases is $4n\Delta$, and at the end of the algorithm there is at most $n\Delta$ flow coming from the contracted node. However, it is unlikely that flow of the estimated $5n\Delta$ units gathers to a single node. Therefore, we propose a *speculative contraction*, which contracts an arc of flow value much less than $5n\Delta$, and we may expect this heuristic method to work well for some value $\beta\Delta$ (e.g. $2\Delta, 4\Delta, 8\Delta, \ldots$).

One drawback of speculative contractions is that it may result in an infeasible flow f: that is, for some arc a, $f(a)$ may be negative. In order to resolve the infeasibility, we use the implementation shown in Figure 4. First, we run algorithm PT with a contraction threshold value $\beta\Delta$. Let f be the obtained flow function. Notice that there may exist negative arc flow values. We denote them by the set $A_{f_-} = \{a \in A \mid f(a) < 0\}$. We now define a new network $\mathcal{N}_g = \{G_g, \gamma, d_g, c_g\}$ where a flow g is defined as $g(a) = f(a)$ for $a \notin A_{f_-}$ and 0 for $a \in A_{f_-}$, and a new demand function d_g is defined as $d_g(v) = -\sum_{(v,w) \in A_{f_-}} f(v,w) + \sum_{(w,v) \in A_{f_-}} f(w,v)$ for each $v \in V$. Lastly, we solve new minimum cost flow problem $\mathcal{N}_g$ by the Primal-Dual method.

Notice that there is a temptation to regard a network $\mathcal{N}_g$ as a residual graph which appears in the middle of executing PT, and resume the execution of PT for $\mathcal{N}_g$ with an increased contraction threshold value to find an optimal flow.

> **Procedure** SPECULATIVE
> (1)   Run *Plotkin-Tardos* with a contraction threshold $\beta\Delta$.
>   Let f be the obtained flow, and $A_{f-} = \{a \in A \mid f(a) < 0\}$.
> (2)   **if** $A_{f-} \neq \emptyset$ **then begin**
> (2.1)   Create a network $\mathcal{N}_g$;
> (2.2)   Run Primal-Dual for $\mathcal{N}_g$;
>   **end**

FIGURE 4. Algorithm SPECULATIVE

However, this method may not work due to the following facts. In order to resume an execution of *PT* from a Δ-phase, it is required that every residual capacity should be an integral multiple of Δ. This condition may not be true in $\mathcal{N}_g$. Let Δ be the maximum absolute value of excesses in $\mathcal{N}_g$, and let c_{min} be the minimum non-zero absolute value of residual capacities. Then c_{min} may be too small to be an integral multiple of Δ. Therefore, we find an optimal solution by regarding $\mathcal{N}_g$ as a new instance of the capacitated minimum cost flow problem. Then, we use the Primal-Dual method for solving $\mathcal{N}_g$. In fact, Step (2.2) of *SPECULATIVE* can be replaced with any other minimum cost flow algorithm like the network simplex method. The Primal-Dual method we use is the same as the *successive shortest path* algorithm in [2]; that is, we successively find a shortest path from a source to a sink and augment flow along this path.

Hereafter, we denote *SPECULATIVE* (resp. *Plotkin-Tardos*) with a contraction threshold $\beta\Delta$ by *SPECULATIVE*($\beta\Delta$) (resp. *PT*($\beta\Delta$)). We denote the Primal-Dual method by *PD*.

We will investigate the following issues on the speculative contraction method: (1) How practical and effective is this new method? (2) How can we determine an appropriate contraction threshold? (3) What are the theoretical issues regarding this method? We consider these questions in the succeeding sections.

4. Specification of Experiments

In this section we describe specifications of our experiments.

Language, OS, CPU: All programs are written in *C*-programming language. The main program, *SPECULATIVE*, consists of about 1,500 statements. We use an IBM POWERserver model 540 workstation under the UNIX environment.

Performance measurement: We use the following measurements: CPU times of the whole execution, the execution of PT, and major subroutines. We use the **-pg** option of the C compiler in order to obtain the CPU time of each subroutine call.

Number of random trials: In order to obtain reliable figures, we follow the test procedure suggested by [5]. That is, we run $t = 15$ random trials for each sampling point and take a mean value as a representative value. We ran $PT(2\Delta)$ based on different random seeds for a network generated by *netgen* of size $n = 400$ and $m = 1600$. We chose $t = 15$ because four independently obtained means of 15 random trials differ by at most 7% from each other. On the other hand when $t = 10$, we observed at most 27% difference among four independent means. Therefore, we chose $t = 15$ instead of 10.

Data structure: We use a binary heap for implementing Dijkstra's shortest path algorithm.

4.1. Instance generators. We employ two instance generators, *netgen* [19] and *washington* [5] for creating transshipment and transportation networks based on random seeds. Hereafter, we use n (*resp. m*) to denote the number of nodes (*resp.* arcs) of a generated network.

The netgen family: We prepare the following families of transshipment networks generated by *netgen*:

The density/size family: This family consists of (n, m) transshipment networks such that $n = 200, 400, 800, 1600$, and $m = 4n, n\log_2 n, n^2/16, n^2/4$. Other parameters are set as follows: $source = sink = n/4$, $mincost = 1$, $maxcost = 1000$, and $total\text{-}supply = 10^6, 2 \cdot 10^6, 4 \cdot 10^6$, and 10^7 for $n = 200, 400, 800$, and 1600, respectively.

The supply/demand ratio family [5]: This family consists of (n, m, s, t) transshipment networks where (s, t) represents the numbers of source and sink nodes. We use $m = 8n$, and vary (s, t) as described later for $n = 200, 400$. Other parameters are set as follows: $source = sink = n/2$, $mincost = 1$, and $maxcost = 1000$.

The total-supply family [5]: This consists of $(n, m, total)$ transshipment networks where *total* represents the number of the amount of total supply. We use $m = 8n$, $total = 10^3, 10^5, 10^7, 10^9$ for $n = 200, 400$. Other parameters are set as follows: $mincost = 1$, $maxcost = 1000$, and $total\text{-}supply = 10^5$.

The washington family: We use *washington* to produce the following network families, some of which are suggested by [5]. Since the Plotkin-Tardos algorithm, on which our code is based, assumes a transshipment network as an input, we post-process a transportation network generated by *washington* to create a feasible transshipment network while preserving the original network topology. We first change all arc capacities infinite, and then we assign appropriate node excesses which make the network feasible. Therefore, notice that a newly generated transshipment network is not equivalent to the original transportation network, but preserves the original network topology.

The network density family [5]: A network in this family is either (a) a *square mesh* ($function = 5$), denoted by (*square*,

s, l, d), which has a square grid of s nodes with l nodes on a side and each node has d outgoing arcs, or (b) a basic line network ($function = 6$), denoted by ($line, s, d$), which has s nodes on a line and each node has d outgoing arcs. Networks we tested are in the following four categories with $s = 64, 256,$ and 1024. (1) Square-Sparse: ($square, s, \sqrt{s}, 4$), (2) Square-Moderate: ($square, s, \sqrt{s}, \sqrt{s}/4$), (3) Line-Sparse: ($line, s, 4$), and (4) Line-Moderate: ($line, s, \sqrt{s}/4$).

The network-width family: This family consists of *random leveled graphs* ($function = 2$), denoted by (RLG, s, l, w, d), which has a rectangle shape of length l and width w with s nodes and each node has d outgoing arcs, Networks we tested are in the following categories with $x = 256, 512, 1024,$ and 2048. (1) RLG-Wide [5]: ($RLG, x, x/64, 64, 3$), (2) RLG-Long [5]: ($RLG, x, 64, x/64, 3$), and (3) RLG-Lean: ($RLG, x, 4, x/4, 3$).

5. Heuristics

Contraction threshold: For each network, we run *Primal-Dual*, and also run *SPECULATIVE* for contraction thresholds of $5n\Delta, 16\Delta, 8\Delta, 4\Delta,$ and 2Δ.

Scaling factor: We use scaling factors $\alpha = 2, 4, 6, 8, 10$ for networks of the *netgen* density/size family with $n = 400$. As shown in Table 1, the scaling factor of 2 gives us the fastest running time. Therefore, we adopt $\alpha = 2$ for our further experiments. It is interesting that for the cost scaling method of Goldberg and Tarjan [14], the scaling factor of 2 is reported as a bad choice in [4] and [11].

TABLE 1. Scaling factors vs. CPU time (sec)

(n, m)	$\beta\Delta$	Scaling factor (α)				
		2	4	6	8	10
(400, 1600)	4Δ	1.78	2.63	3.71	4.13	2.90
	8Δ	1.83	2.37	3.68	3.22	3.17
(400, 2397)	4Δ	2.01	3.30	4.56	4.32	3.76
	8Δ	2.06	3.08	4.45	3.58	3.50
(400, 10000)	4Δ	5.67	9.51	11.90	12.0	10.2
	8Δ	6.04	7.47	11.50	9.75	10.51
(400, 40000)	4Δ	21.92	35.09	46.57	44.92	38.16
	8Δ	22.58	33.66	46.99	35.32	36.99

d-heap: Table 2 shows results of $PT(4\Delta)$ applied to the *netgen* density family when we used various d-heap. From the table, appropriate choice seems to be 2 or 4, and thus we adopt $d = 2$ for our further experiments.

TABLE 2. CPU time of $PT(4\Delta)$ with various d-heap ($n = 400$)

m	$d = 2$	4	8	16	32	64	128
1600	1.8	1.7	1.8	2.0	2.7	3.3	3.7
2397	2.0	2.0	2.1	2.3	2.9	3.6	4.3
10000	5.7	5.3	5.5	5.4	6.2	7.0	7.5
40000	21.9	21.7	22.1	21.3	22.6	23.6	25.1

TABLE 3. Effectiveness of heuristics

(n, m)	$CPU(PT)$			$\#SP$		
	FO	MER	ratio	FO	MER	ratio
(800, 3200)	8.05	5.71	1.41	175	128	1.36
(800, 5347)	10.35	7.34	1.41	175	127	1.38
(800, 40000)	52.92	39.06	1.35	172	130	1.32

Choice of a source-sink pair: At Step (3) of *AUGMENT* in Figure 2, we can arbitrarily choose a sink node. Thus, we investigate the following two strategies:

The fixed ordering rule (FO): For a source node $s \in S$, we pick up a sink node $t \in T$ with the lowest *index* which is in a shortest subtree of s. The index is with respect to a pre-fixed ordering of nodes.

The maximum excess reduction rule (MER): For a source node $s \in S$, we pick up the sink node $t \in T$ such that t is in a shortest subtree of s and the absolute value of excess at t is maximum. With this rule, we may expect fast decrease of total supplies, which implies a smaller number of shortest path calls.

Table 3 illustrates the CPU time of $PT(8\Delta)$ in *SPECULATIVE*(8Δ) and the number of shortest path calls (indicated by $\#SP$) when we applied the fixed ordering rule (FO) and the maximum excess reduction rule (MER). It also indicates ratios of performance with FO and that with MER. As shown in the table, *PT* with MER runs faster by 1.4 than *PT* with FO. This speed up seems to come from the reduction of the number of shortest path calls. In fact, the number of shortest path calls in *PT* with MER is reduced by a factor of 1.35. Therefore, we adopt the maximum excess reduction rule in the succeeding sections.

6. Experimental Results

The density/size family by netgen: Tables 4 and 5 show results of a test family generated by *netgen* for $n = 400$ and 800 and $m = 4n$, $n \log_2 n$, $n^2/16$. (See Appendix A for results for other networks with $n = 200$ and 1600.) Each entry consists of the total CPU time without I/O in seconds, a percentage of the CPU time spent for post-processing PD, the number of negative flow edges after $PT(\beta\Delta)$, denoted by #(neg. edges), and a speed up factor of SPECULATIVE($\beta\Delta$), (that is, CPU-time(SPECULATIVE($\beta\Delta$)) divided by CPU-time($PT(5n\Delta)$)). Figure 6 illustrates the CPU times for networks of size $n = 400$ for different contraction thresholds. As Figure 6 and Tables 4 and 5 show, our speculative contraction method outperforms both of the original Plotkin-Tardos' algorithm and the Primal-Dual method. The average speed up of SPECULATIVE(8Δ) relative to the Plotkin-Tardos (resp. Primal-Dual) is about 3.4 (resp. 2.1) for sparse graphs and about 2.5 (resp. 1.9) for dense graphs.

TABLE 4. The density family for $n = 400$

m	Threshold	$5n\Delta$	16Δ	8Δ	4Δ	2Δ	PD
1600	CPU(total)	5.68	2.18	1.83	1.78	2.93	3.27
	CPU(PD)%	0%	.8%	3.2%	22.8%	59.5%	100%
	#(neg. edges)	0	.5	1.5	8.3	26.3	-
	Speed up	1.00	.38	.32	.31	.52	.58
2397	CPU(total)	6.96	2.55	2.06	2.01	3.41	4.67
	CPU(PD)%	0%	.4%	1.4%	16.4%	55.4%	100%
	#(neg. edges)	0	.3	.9	6.0	27.9	-
	Speed up	1.00	.37	.30	.29	.49	.67
10000	CPU(total)	17.61	7.10	6.04	5.67	8.06	13.38
	CPU(PD)%	0%	.3%	2.9%	14.8%	45.6%	100%
	#(neg. edges)	0	.5	1.8	7.6	27.1	-
	Speed up	1.00	.40	.34	.32	.46	.76
40000	CPU(total)	63.47	26.84	22.58	21.92	30.76	47.82
	CPU(PD)%	0%	.1%	1.6%	14.6%	41.8%	100%
	#(neg. edges)	0	.1	1.1	6.4	22.7	-
	Speed up	1.00	.42	.36	.35	.48	.75

Notice that as $CPU(PD)\%$ in Tables 4 and 5 show the above speed up comes mainly from the speed up of $PT(\beta\Delta)$, not from the post-processing of PD. In particular, for some small $\beta\Delta$, $PT(\beta\Delta)$ can obtain an optimal solution without involving a post-processing of PD. For example, $PT(16\Delta)$ can find optimal solutions for $n = 200$ with $m = 800, 2500, 10000$ as Appendix A shows. Figure 7 illustrates the total CPU time, the CPU time of PD, and the number of negative flow edges after $PT(\beta\Delta)$ when we run $SPECULATIVE$ for the netgen density family with $(n, m) = (800, 40000)$. As the figure shows, $CPU(PD)$ decreases as the contraction threshold increases. Moreover, $CPU(PD)$ decreases as almost the same rate as the number of negative edges decreases. Notice that the number of negative edges after the first stage of PT can be regarded as a measure of how far the current solution is apart from the optimal. Table 6 shows average speed up ratios obtained from complete test families generated by netgen for $n = 200, 400, 800$, and 1600.

Table 7 shows the number of shortest path calls and a percentage of the CPU time spent by shortest path calls. As the table shows, when a contraction threshold is 4Δ or 8Δ, the number of shortest path calls and a percentage for shortest path calls, a dominant CPU usage, is minimized.

TABLE 5. The density family for $n = 800$

m	Threshold	$5n\Delta$	16Δ	8Δ	4Δ	2Δ	PD
3200	CPU(total)	22.80	7.10	6.05	6.59	15.00	14.76
	CPU(PD)%	0%	.5%	5.5%	28.4%	68.4%	100%
	#(neg. edges)	0	.5	3.2	13.9	59.6	-
	Speed up	1.00	.31	.27	.29	.66	.65
5347	CPU(total)	27.47	8.95	7.87	8.08	14.33	21.11
	CPU(PD)%	0%	.9%	6.7%	.26%	59.9%	100%
	#(neg. edges)	0	.5	4.4	15.8	59.0	-
	Speed up	1.00	.33	.29	.29	.52	.77
40000	CPU(total)	128.76	46.44	41.62	42.70	72.29	112.84
	CPU(PD)%	0%	1.2%	6.2%	23.1%	51.0%	100%
	#(neg. edges)	0	1.0	5.3	17.4	55.7	-
	Speed up	1.00	.36	.32	.33	.56	.88

TABLE 6. Average speed up ratios for the density/size family

m	$5n\Delta$	16Δ	8Δ	4Δ	2Δ	PD
$4n$	1.00	.35	.29	.30	.55	.61
$n \log_2 n$	1.00	.36	.30	.29	.49	.70
$n^2/16$	1.00	.41	.35	.33	.47	.77
$n^2/4$	1.00	.47	.39	.36	.49	.74

Surprisingly enough, the *optimal contraction threshold value* $\beta\Delta$, which makes our speculative contraction method fastest, seems to be a constant between 4Δ and 8Δ regardless of network density, size, supply/demand ratio, total supply, and network generators. We hope further theoretical research may explain this phenomenon.

Figure 5 shows the growth rate of the CPU time when we increase the network size from $n = 200$ to 1600 while keeping the density at $m = 4n$. We obtained almost straight lines with logarithmic axes. This means that the CPU time is proportional to some fixed power of n (say, $CPU = cn^{f(\beta\Delta)}$). According to Figure 6, $f(8\Delta) = 1.77$, and it seems to be slightly smaller than $f(5n\Delta) = 1.98$ and $f(PD) = 2.09$.

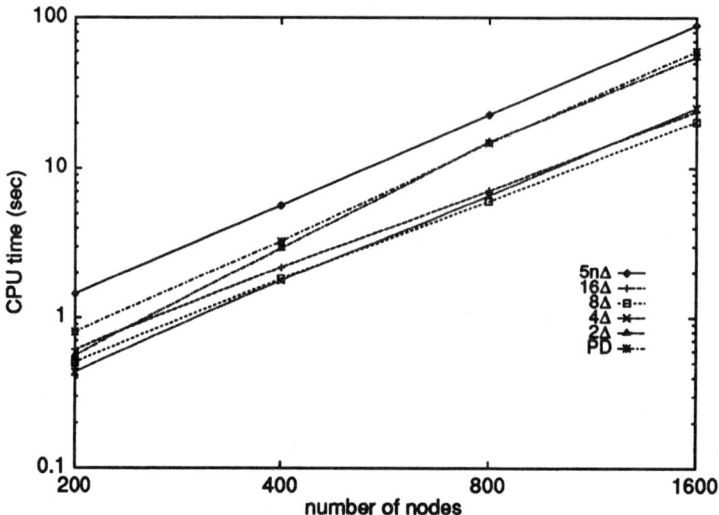

FIGURE 5. The CPU time growth (netgen, $m = 4n$)

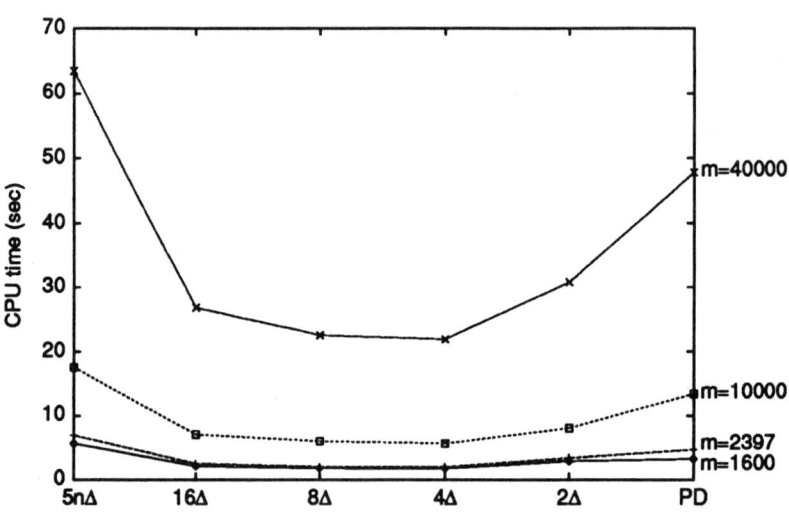

FIGURE 6. The density family (netgen, $n = 400$)

TABLE 7. The number of shortest path calls vs. speed up

$n = 400$

m	Threshold	$5n\Delta$	16Δ	8Δ	4Δ	2Δ	PD
1600	Speed up	1.00	.38	.32	.31	.52	.58
	CPU(SP)%	80.0%	72.3%	70.8%	69.4%	74.4%	84.1%
	#(SP)	239	105	87	81	124	115
2397	Speed up	1.00	.37	.30	.29	.49	.67
	CPU(SP)%	79.4%	70.9%	69.5%	68.3%	72.8%	83.7%
	#(SP)	245	100	82	77	119	140
10000	Speed up	1.00	.40	.34	.32	.46	.76
	CPU(SP)%	72.5%	66.1%	63.6%	61.0%	62.4%	79.6%
	#(SP)	245	103	86	78	106	174
40000	Speed up	1.00	.42	.36	.35	.48	.75
	CPU(SP)%	70.0%	63.7%	60.4%	57.9%	59.3%	78.8%
	#(SP)	245	101	83	78	109	185

$n = 800$

m	Threshold	$5n\Delta$	16Δ	8Δ	4Δ	2Δ	PD
3200	Speed up	1.00	.31	.27	.29	.66	.65
	CPU(SP)%	82.7%	72.4%	70.5%	70.6%	76.7%	85.5%
	#(SP)	417	156	134	135	276	228
5347	Speed up	1.00	.33	.29	.29	.52	.77
	CPU(SP)%	80.6%	70.9%	68.4%	67.4%	72.2%	84.3%
	#(SP)	415	156	134	133	218	273
40000	Speed up	1.00	.36	.32	.33	.56	.88
	CPU(SP)%	74.8%	65.0%	61.4%	59.7%	62.4%	80.0%
	#(SP)	457	160	139	140	235	382

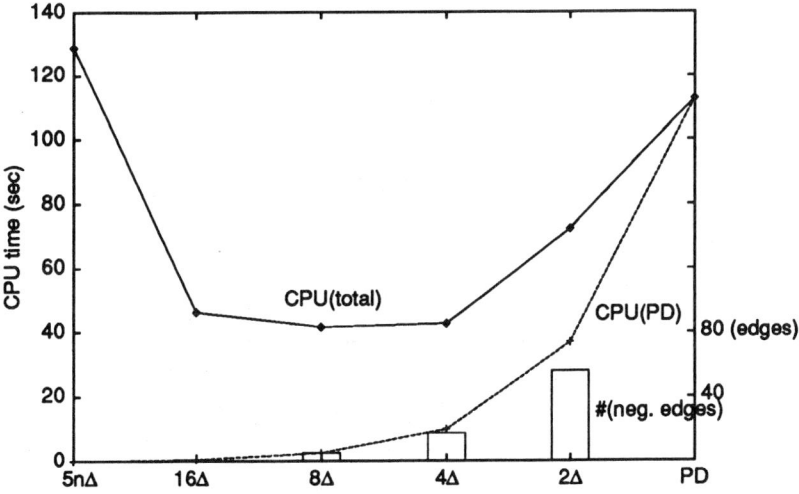

FIGURE 7. The CPU(total), CPU(PD), and the number of negative edges

The total supply family: Figure 8 shows the growth rate of the total CPU time when we fix a network configuration (that is, $n = 400$, the supply/demand ratio is $(n/2, n/2)$) and change the total supply as 10^3, 10^5, 10^7, and 10^9. Since the 4Δ curve almost matches the 8Δ curve, we omit it from the figure. As a result, CPU times of $SPECULATIVE$ for 4Δ, 8Δ, and 16Δ grow much slower and saturate earlier than those of $PT(5n\Delta)$ and PD.

TABLE 8. Speed up ratios for the *washington* network density family

	$5n\Delta$	16Δ	8Δ	4Δ	2Δ	PD
Square-Sparse						
$(64, 8, 4)$	1.00	.58	.47	.36	.35	.68
$(256, 16, 4)$	1.00	.38	.32	.28	.32	.57
$(1024, 32, 4)$	1.00	.27	.22	.21	.24	.51
Square-Moderate						
$(64, 8, 2)$	1.00	.49	.44	.36	.29	.38
$(256, 16, 4)$	1.00	.38	.32	.28	.32	.57
$(1024, 32, 8)$	1.00	.29	.23	.20	.24	.54
Line-Sparse						
$(64, 4)$	1.00	.32	.25	.23	.29	.48
$(256, 4)$	1.00	.21	.19	.19	.26	.39
$(1024, 4)$	1.00	.13	.14	.16	.19	.29
Line-Moderate						
$(64, 2)$	1.00	.35	.28	.24	.24	.31
$(256, 4)$†	1.00	.21	.19	.19	.26	.39
$(1024, 8)$	1.00	.20	.17	.18	.22	.37

†This network is the same as Square-Sparse $(256, 4)$.

TABLE 9. Speed up ratios for the network width family

(w,l)	$5n\Delta$	16Δ	8Δ	4Δ	2Δ	PD
RLG-Wide						
$(4, 64)$	1.00	.35	.29	.25	.25	.33
$(8, 64)$	1.00	.26	.25	.24	.28	.41
$(16, 64)$	1.00	.26	.25	.24	.28	.41
$(32, 64)$	1.00	.26	.25	.24	.28	.51
$(64, 64)$	1.00	.50	.42	.40	.50	1.08
RLG-Long						
$(64, 4)$	1.00	.37	.29	.26	.28	.50
$(64, 8)$	1.00	.32	.25	.23	.29	.51
$(64, 16)$	1.00	.25	.20	.19	.26	.48
$(64, 32)$	1.00	.24	.19	.18	.24	.51
RLG-Lean						
$(4, 64)$†	1.00	.35	.29	.25	.25	.33
$(4, 128)$	1.00	.30	.26	.24	.25	.29
$(4, 256)$	1.00	.20	.19	.19	.23	.20
$(4, 512)$	1.00	.16	.17	.19	.20	.14
$(4, 1024)$	1.00	.19	.24	.29	.27	.13

†This network is the same as $(4, 64)$ of RLG-Wide.

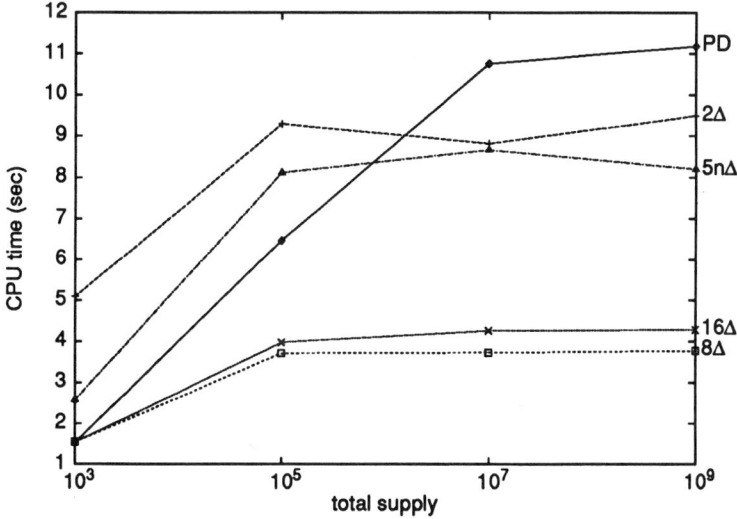

FIGURE 8. The total supply family (netgen, $n = 400$)

The supply/demand ratio family: We first choose $(s,t) = (n/8, n/8)$, $(n/8, n/2)$, $(n/2, n/8)$, $(n/2, n/2)$. The contraction thresholds 4Δ and 8Δ work well, whose speed up ratios range from 1.6 to 3.0 to the Plotkin-Tardos and do not vary very much by changing supply/demand ratio. We also vary the number of sources (*resp.* sinks) from 50, to 250 by 50 while keeping the number of sinks (*resp.* sources) the same as 150 for networks of $n = 400$. We observe that the CPU-time increases up to 1.6 times as the number of sinks increases, while it remains roughly the same within 1.1 times when the number of sources increases. This difference seems to be caused by the asymmetry of the implementation of our algorithm with regard to sources and sinks.

The Washington family: Tables 8 and 9 show lists of speed up ratios for the network families generated by *washington*. For the network density family, as Table 8 shows, *SPECULATIVE*(4Δ) runs faster than the Plotkin-Tardos by 2.8 to 4.9 and average by 3.7 for square networks, and by 4.2 to 5.6 and average by 5.0 for line networks. Its speed up over the Primal-Dual is about 2.0 on average. For *RLG-Wide* and *RLG-Long* networks, as Table 9 shows, our method with the contraction threshold of 4Δ or 8Δ also runs faster than $PT(5n\Delta)$ by 4.1 and the Primal-Dual by 2.1. For *RLG-Lean* networks, a similar phenomenon appears except for $(4, 512)$ and $(4, 1024)$, in which the Primal-Dual runs faster than our method with any chosen contraction threshold. This is probably because the Primal-Dual needs only small number of shortest path calls for lean networks. In fact, we observed that the Primal-Dual needs 249 shortest path calls for the $(4, 1024)$ network, while *SPECULATIVE*(4Δ) requires 727 calls for *PT* and 62 calls for *PD*. See details in Appendix A.

7. Comparisons with Edmonds-Karp, NETFLO, and RELAXT-III

TABLE 10. Comparison with the Edmonds-Karp algorithm ($n = 400$)

m	Threshold	$5n\Delta$	8Δ	4Δ	EK
1600	Speed up	1.00	.32	.31	1.08
	CPU	5.68	1.83	1.78	6.11
	#(SP)	239	87	81	287
2397	Speed up	1.00	.30	.29	1.02
	CPU	6.96	2.06	2.01	7.13
	#(SP)	245	82	77	278
10000	Speed up	1.00	.34	.32	.97
	CPU	17.61	6.04	5.67	17.10
	#(SP)	245	86	78	285
40000	Speed up	1.00	.36	.35	1.01
	CPU	63.47	22.58	21.92	64.11
	#(SP)	245	83	78	289

First, we compared our code with an Orlin's variation of the Edmonds-Karp scaling algorithm [21], which scales excesses instead of capacities and does not use any contraction [6, 21]. Table 10 illustrates the CPU time in seconds and the number of shortest path calls for each of $PT(5n\Delta)$, $PT(8\Delta)$, $PT(4\Delta)$, and EK (which stands for the Edmonds-Karp algorithm) when applying to the netgen density/size family. From the table, the original Plotkin-Tardos algorithm ($PT(5n\Delta)$) runs slightly faster than the Edmonds-Karp algorithm by 1.02. On the contrary, $PT(4\Delta)$ (resp. $PT(8\Delta)$) attains speed up of 3.2 (resp. 3.1) over the Edmonds-Karp algorithm.

Second, we compared our method with two other minimum cost flow codes, NETFLO [18] and RELAXT-III [3]. Since our implementation of the Plotkin-Tardos algorithm is for transshipment networks as the original algorithm does, our comparison study has been done mainly for transshipment networks. They are the netgen density/size family with $n = 800$ in Section 6 and three transshipment networks in the DIMACS core set [5] whose parameters are shown in Table 11.

As shown in Table 12, RELAXT-III (resp. NETFLO) is 7.6 times faster (resp. 3.5) than our method with the threshold 8Δ.

TABLE 11. The transshipment networks by netgen in the DIMACS core set

Case 28	28	1000	50	50	2900	1	10000	1000000	0	0	0	0	0
Case 32	32	1500	75	75	4342	1	10000	1500000	0	0	0	0	0
Case 38	38	3000	125	500	35000	1	10000	2000000	25	50	0	0	0

Each of above cases uses 13502460 as a random seed.

According to a detailed run-time analysis of $SPECULATIVE(8\Delta)$ for an $(800, 40000)$ network, 90 percent of the total CPU time (46.7 seconds) consists of shortest path calls (50.7%), augmentation routine calls other than shortest path

TABLE 12. Comparison in CPU time (sec)

(n,m)	8Δ	4Δ	NETFLO	RELAXT-III
(800,3200)	6.05	6.59	0.85	3.00
ratio	7.12	7.75	1.00	3.53
(800,5347)	7.87	8.08	1.04	2.51
ratio	7.57	7.77	1.00	2.41
(800,40000)	41.62	42.70	5.86	6.22
ratio	7.10	7.29	1.00	1.06
case-28	3.16	2.88	0.50	0.95
ratio	6.32	5.76	1.00	1.90
case-32	7.21	5.92	0.99	1.99
ratio	7.28	5.98	1.00	2.01
case-38	100.29	106.32	9.43	19.84
ratio	10.64	11.27	1.00	2.10
avg. ratio	7.67	7.64	1.00	2.17

call (18.7%), and the contraction overhead (20.0% — 14% for the contraction routine and 6% for the unfold routine). In particular, only 14% of the total shortest path calls are devoted to heap operations; the other 86% are devoted to scanning nodes and edges and maintaining node labels and pointers. In this 86% of the shortest path calls (which take up 43.6% of the total CPU time) and in the 20% contraction overhead, we believe that our implementation could be improved by adopting an appropriate storage management method other than pointer-based implementation. A much more effective way of improving our code, however, would be to reduce the number of shortest path calls itself. This will require further investigations to obtain new heuristics and theoretical results.

For a capacitated network $\mathcal{N}$ with n nodes and m arcs, we can transform $\mathcal{N}$ to an equivalent uncapacitated network $\mathcal{N}'$ with $n+m$ nodes and $2m$ arcs by introducing a new node for each arc (See details in [2]). We have conducted comparison study for uncapacitated networks obtained by the above transformation of capacitated networks in the DIMACS core set [5]. Although testing for capacitated networks is out of scope of our experiments, we show our detail comparison results in Appendix B for a reference. Table 13 shows summary of comparison results for networks in the DIMACS core set [5] including the GTE instances, the Grid families, and the goto families. For transformed uncapacitated networks, $PT(8\Delta)$ runs competitively to RELAXT-III: that is, $CPU(PT(8\Delta))/CPU(\text{RELAXT-III})$ is 1.25 for the Grid families and 0.50 for the goto families on average. NETFLO runs faster than $PT(8\Delta)$ by 3.6 (resp. 5.2) for the Grid (resp. goto) families. Comparing to running times of NETFLO and RELAXT-III for the original capacitated networks of size (n,m), $PT(8\Delta)$ for the transformed uncapacitated networks of size $(n+m, 2m)$, $CPU(PT(8\Delta))/CPU(\text{RELAXT-III})$ (resp. $CPU(PT(8\Delta))/CPU(\text{NETFLO})$) is 0.5, 5.9, and 2.0 (resp. 66.0, 40.4, and 32.9) for the GTE instances, the Grid families, and the goto families.

TABLE 13. Comparison table among $PT(8\Delta)$, RELAXT-III, and NETFLO

	Transformed			Original	
	$PT(8\Delta)$	RELAXT-III	NETFLO	RELAXT-III	NETFLO
gte-bad	66.04	-	-	138.13	1.0
Grid-square	4.47	4.79	1.00		
	29.45	32.97	6.62	6.48	1.00
Grid-long	3.53	1.91	1.00		
	59.67	31.33	21.22	6.39	1.00
Grid-wide	5.17	4.80	1.00		
	34.47	32.53	6.68	7.86	1.00
Grid-average	4.40	3.89	1.00		
	40.43	32.32	11.20	6.88	1.00
Grid-density-8	3.71	7.46	1.00		
	19.66	39.83	5.31	11.96	1.00
Grid-density-16	6.21	12.07	1.00		
	46.12	91.54	7.44	19.86	1.00
goto-average	4.96	9.77	1.00		
	32.89	65.69	6.38	15.91	1.00

8. Concluding Remarks

We have proposed a speculative contraction method for minimum cost flows based on the Plotkin-Tardos algorithm, a variant of Orlin's algorithm. According to our experiments, very small contraction thresholds like 4Δ and 8Δ outperform the original Plotkin and Tardos' algorithm with a $5n\Delta$ contraction threshold and the Primal-Dual method. We observed an average speed up factor to the Plotkin-Tardos algorithm (*resp.* the Primal-Dual method) of 3.4 (*resp.* 2.1) for sparse graphs and 2.5 (*resp.* 1.9) for dense graphs. One surprising thing is that this optimal contraction threshold seems to be invariant with respect to the density, size, supply/demand ratio, total supply, and types of networks.

After introducing our idea of the speculative contraction in [9], Goldberg and Kharitonov adopted the dual of our concept to their implementation of scaling push-relabel algorithms as an *aggressive pricing out* [12]. We hope our experimental results on speculative contractions will provoke further theoretical and practical research toward practical scaling algorithms.

Acknowledgements

We would like to thank two anonymous referees for their suggestions which improve the quality of this paper and Dr. David Johnson and Dr. Catherine C. McGeoch for their efforts on organizing the DIMACS implementation challenge [5].

REFERENCES

1. R.K. Ahuja, A.V. Goldberg, J.B. Orlin, and R.E. Tarjan, Finding Minimum-cost Flows by Double Scaling. Technical Report CS-TR-164-88, Department of Computer Science, Princeton University, 1988.

2. R.K. Ahuja, T.L. Magnanti, and J.B. Orlin, Network Flows. In G.L. Nemhauser et al., Eds., *Handbooks in OR & MS, Vol. 1*, Elsevier Science Publishers B.V. (North-Holland) 1989. pp. 211-369.
3. D.P. Bertsekas and P. Tseng, Relaxation methods for minimum cost ordinary and generalized network flow problems, *Operations Research*, 36, pp. 93-114, 1988.
4. R.G. Bland, J. Cheriyan, D.J. Jensen, and L. Ladanyi, An Emprical Study of Recent Flow Algorithm. The First DIMACS International Algorithm Implementation Challenge: Network Flows and Matching. DIMACS, NJ., 1991.
5. DIMACS, The First DIMACS International Algorithm Implementation Challenge: Network Flows and Matching. Call for Participation, General Information, and The Core Experiments. DIMACS, NJ., 1990.
6. J. Edmonds and R. M. Karp, Theoretical Improvements in Algorithmic Efficiency for Network Flow Problems, *Journal of ACM*, 19, pp. 248-264, 1972.
7. L.R. Ford and D.R. Fulkerson, *Flows in Networks*, Princeton University Press, Princeton, NJ, 1962.
8. S. Fujishige, A Capacity-rounding Algorithm for the Minimum Cost Circulation Problem: A Dual Framework of the Tardos Algorithm. *Mathematical Programming*, 35, pp. 298-309, 1986.
9. S. Fujishige, K. Iwano, J. Nakano, and S. Tezuka, A Speculative Contraction Method for the Minimum Cost Flows: Toward a Practical Algorithm. The First DIMACS International Implementation Challenge, Oct., 1991; also available as IBM Research Report RT 0061, Aug., 1991.
10. Z. Galil and É. Tardos, An $O(n^2(m + n \log n) \log n)$ Min-cost Flow Algorithm, *Proc. 27th Annual Symposium on the Foundation of Computer Science*, pp. 136-146, 1986.
11. A.V. Goldberg and M. Kharitonov, Experimental Evaluation of the Push-Relabel Method for the Minimum-Cost Flow Problem. The First DIMACS International Algorithm Implementation Challenge: Network Flows and Matching. DIMACS, NJ., 1991.
12. _____, On Implementing Scaling Push-Relabel Algorithms for the Minimum-Cost Flow Problem. Technical Report STAN-CS-92-1418, Stanford University, 1992.
13. A. V. Goldberg, É. Tardos, and R. E. Tarjan, Network Flow Algorithms. In *Paths, Flows, and VLSI-Layout*, Algorithms and Combinatorics 9, B. Korte et al., Eds., Springer-Verlag, New York, NY, pp. 101-164, 1990.
14. A. V. Goldberg and R. E. Tarjan, Finding Minimum-cost Circulations by Successive Approximation, Technical Report CS-TR-106-87, Dept. Computer Science, Princeton University, July 1987.
15. _____, Finding Minimum-Cost Circulations by Canceling Negative Cycles, *Journal of ACM*, vol. 36, no. 4, pp. 873-886, 1989.
16. M.D. Grigoriadis, An Efficient Implementation of the Network Simplex Method, *Mathematical Programming Study*, 26, pp. 83-111, 1986.
17. R. Hassin, Algorithms for the Minimum Cost Circulation Problem in Cycles, Based on Maximizing the Mean Improvement, February 1990.
18. J.L. Kennington and R.V. Helgason, *Algorithms for Network Programming*. John Wiley and Sons, New York, 1980.
19. D. Klingman, A. Napier, and J. Stutz, NETGEN: A Program for Generating Large Scale Capacitated Assignment, Transportation, and Minimum Cost Flow Network Problems. *Management Science*, vol. 20, no. 4, pp. 814-821, 1974.
20. J. B. Orlin, Genuinely Polynomial Simplex and Non-simplex Algorithms for the Minimum Cost Flow Problem, Technical Report No. 1615-84, Sloan School of Management, M. I. T., Cambridge, MA., 1984.
21. _____, A Faster Strongly Polynomial Minimum Cost Flow Algorithm, *Journal of ACM*, vol. 35, no. 2, pp. 374-386, 1988.
22. S. A. Plotkin and É. Tardos, Improved Dual Network Simplex, *Proceeding of the first annual ACM-SIAM Symposium on Discrete Algorithms*, pp. 367 - 376, 1990.
23. É. Tardos, A Strongly Polynomial Minimum Cost Circulation Algorithm. *Combinatorica*, 5, pp. 247-255, 1985.

Appendix A. Additional experimental results

TABLE 14. The netgen density family for $n = 200, 1600$

(n,m)	Threshold	$5n\Delta$	16Δ	8Δ	4Δ	2Δ	PD
(200, 800)	CPU(total)	1.44	.61	.50	.43	.56	.80
	CPU(PD)%	0%	0%	1.7%	9.9%	45.7%	100%
	Speed up	1.00	.43	.35	.30	.39	.56
	CPU(SP)%	79.0%	71.2%	70.6%	66.3%	68.9%	82.9%
	#(SP)	131	66	55	46	52	62
(200, 1060)	CPU(total)	1.69	.78	.65	.55	.73	.87
	CPU(PD)%	0%	.2%	1.8%	9.7%	45.0%	100%
	Speed up	1.00	.46	.38	.33	.43	.52
	CPU(SP)%	78.2%	71.4%	67.2%	68.7%	68.3%	81.7%
	#(SP)	139	67	57	49	59	62
(200, 2500)	CPU(total)	2.69	1.25	1.04	.94	1.05	1.85
	CPU(PD)%	0%	0%	.3%	8.6%	34.6%	100%
	Speed up	1.00	.46	.38	.35	.39	.69
	CPU(SP)%	72.8%	66.3%	64.9%	63.0%	62.6%	79.9%
	#(SP)	133	64	54	48	52	83
(200, 10000)	CPU(total)	7.93	4.07	3.44	3.03	3.97	5.72
	CPU(PD)%	0%	0%	.6%	8.2%	36.3%	100%
	Speed up	1.00	.51	.43	.38	.50	.72
	CPU(SP)%	68.8%	62.8%	61.0%	58.3%	58.0%	77.4%
	#(SP)	129	67	55	47	61	91

(n,m)	Threshold	$5n\Delta$	16Δ	8Δ	4Δ	2Δ	PD
(1600, 6400)	CPU(total)	88.63	23.82	20.41	25.11	54.94	60.03
	CPU(PD)%	0%	1.7%	9.8%	37.9%	68.3%	100%
	Speed up	1.00	.27	.23	.28	.62	.68
	CPU(SP)%	83.3%	70.7%	67.8%	69.8%	75.8%	85.8%
	#(SP)	716	244	203	231	455	414
(1600, 11804)	CPU(total)	128.65	35.68	32.43	33.99	67.49	106.26
	CPU(PD)%	0%	3.6%	12.6%	30.0%	61.0%	100%
	Speed up	1.00	.28	.25	.26	.52	.83
	CPU(SP)%	82.0%	69.5%	66.6%	66.0%	71.5%	84.3%
	#(SP)	789	254	228	227	416	574

TABLE 15. The *washington* Square families

Square-Sparse		$5n\Delta$	16Δ	8Δ	4Δ	2Δ	PD
$(64, 8, 4)$	CPU(total)	0.20	0.11	0.09	0.07	0.07	0.13
	Speed up	1.00	0.57	0.47	0.36	0.35	0.68
	CPU(SP)%	77.95%	64.29%	60.25%	53.90%	52.79%	77.02%
	#(SP)	64.33	43.13	34.93	29.40	26.20	33.33
$(256, 16, 4)$	CPU(total)	3.39	1.28	1.07	0.96	1.08	1.94
	Speed up	1.00	0.38	0.32	0.28	0.32	0.57
	CPU(SP)%	80.30%	72.12%	69.20%	67.84%	68.53%	83.00%
	#(SP)	239.47	120.80	102.93	87.13	92.20	113.40
$(1024, 32, 4)$	CPU(total)	65.86	17.46	14.79	14.16	16.10	33.38
	Speed up	1.00	0.27	0.22	0.21	0.24	0.51
	CPU(SP)%	83.42%	72.88%	69.71%	67.93%	69.72%	85.19%
	#(SP)	1014.20	384.13	319.33	279.47	285.53	429.07

Square-Moderate		$5n\Delta$	16Δ	8Δ	4Δ	2Δ	PD
$(64, 8, 2)$	CPU(total)	0.17	0.08	0.07	0.06	0.05	0.07
	Speed up	1.00	0.49	0.44	0.36	0.30	0.38
	CPU(SP)%	72.92%	69.74%	51.12%	45.12%	45.37%	69.68%
	#(SP)	70.67	46.53	38.20	31.93	27.73	22.07
$(256, 16, 4)$	CPU(total)	3.39	1.28	1.07	0.96	1.08	1.94
	Speed up	1.00	0.38	0.32	0.28	0.32	0.57
	CPU(SP)%	80.30%	72.12%	69.20%	67.84%	68.53%	83.00%
	#(SP)	239.47	120.80	102.93	87.13	92.20	113.40
$(1024, 32, 8)$	CPU(total)	108.56	31.72	25.46	22.20	26.22	58.91
	Speed up	1.00	0.29	0.23	0.20	0.24	0.54
	CPU(SP)%	81.05%	69.51%	65.46%	62.07%	64.58%	84.06%
	#(SP)	1115.73	440.87	353.13	292.13	313.27	555.53

TABLE 16. The *washington* RLG-Long family

RLG-Long		$5n\Delta$	16Δ	8Δ	4Δ	2Δ	PD
$(64, 4)$	CPU(total)	3.73	1.38	1.08	0.97	1.03	1.85
	Speed up	1.00	0.37	0.29	0.26	0.28	0.50
	CPU(SP)%	82.25%	74.87%	70.46%	67.73%	68.95%	82.44%
	#(SP)	287.73	146.60	114.47	101.27	103.13	122.93
$(64, 8)$	CPU(total)	14.50	4.58	3.69	3.38	4.25	7.36
	Speed up	1.00	0.32	0.25	0.23	0.29	0.51
	CPU(SP)%	83.96%	76.34%	72.09%	69.47%	72.51%	85.27%
	#(SP)	497.53	217.20	182.00	157.80	177.80	213.27
$(64, 16)$	CPU(total)	71.20	17.77	14.32	13.88	18.18	34.18
	Speed up	1.00	0.25	0.20	0.19	0.26	0.48
	CPU(SP)%	85.45%	75.99%	72.39%	71.38%	72.91%	86.04%
	#(SP)	1098.27	403.47	321.93	292.67	344.00	462.13
$(64, 32)$	CPU(total)	301.09	70.90	56.42	54.33	71.16	153.73
	Speed up	1.00	0.24	0.19	0.18	0.24	0.51
	CPU(SP)%	87.86%	76.12%	72.21%	70.57%	72.54%	86.56%
	#(SP)	2164.80	750.53	595.53	538.07	625.13	983.13

TABLE 17. The *washington* RLG-Wide family

RLG-Wide		$5n\Delta$	16Δ	8Δ	4Δ	2Δ	PD
(4, 64)	CPU(total)	2.25	0.79	0.65	0.56	0.57	0.74
	Speed up	1.00	0.35	0.29	0.25	0.25	0.33
	CPU(SP)%	78.27%	67.81%	62.91%	58.62%	56.55%	79.63%
	#(SP)	203.60	96.67	78.80	65.80	64.00	54.07
(8, 64)	CPU(total)	10.66	2.98	2.43	2.27	2.80	3.91
	Speed up	1.00	0.28	0.23	0.21	0.26	0.37
	CPU(SP)%	80.58%	69.77%	66.82%	61.79%	63.82%	82.52%
	#(SP)	417.73	165.93	136.27	118.80	135.33	129.60
(16, 64)	CPU(total)	52.22	13.53	13.01	12.39	14.41	21.23
	Speed up	1.00	0.26	0.25	0.24	0.28	0.41
	CPU(SP)%	83.37%	71.47%	69.73%	67.10%	67.74%	84.16%
	#(SP)	898.00	340.73	314.67	283.53	297.20	320.87
(32, 64)	CPU(total)	226.83	59.72	57.42	54.87	63.99	115.79
	Speed up	1.00	0.26	0.25	0.24	0.28	0.51
	CPU(SP)%	86.98%	73.91%	72.21%	69.08%	70.24%	85.50%
	#(SP)	1746.73	662.40	624.33	566.73	593.87	790.80
(64, 64)	CPU(total)	557.98	278.31	233.38	223.27	280.58	602.38
	Speed up	1.00	0.50	0.42	0.40	0.50	1.08
	CPU(SP)%	78.94%	75.87%	72.39%	69.91%	71.47%	86.97%
	#(SP)	1591.40	1390.67	1155.07	1023.27	1147.07	1877.53

TABLE 18. The *washington* RLG-Lean family

RLG-Lean		$5n\Delta$	16Δ	8Δ	4Δ	2Δ	PD
(4, 64)	CPU(total)	2.25	0.79	0.65	0.56	0.57	0.74
	Speed up	1.00	0.35	0.29	0.25	0.25	0.33
	CPU(SP)%	78.27%	67.81%	62.91%	58.62%	56.55%	79.63%
	#(SP)	203.60	96.67	78.80	65.80	64.00	54.07
(4, 128)	CPU(total)	7.57	2.29	1.93	1.80	1.90	2.22
	Speed up	1.00	0.30	0.26	0.24	0.25	0.29
	CPU(SP)%	78.53%	65.63%	60.83%	56.99%	54.31%	80.77%
	#(SP)	322.40	131.87	113.13	100.67	101.87	77.07
(4, 256)	CPU(total)	33.76	6.74	6.42	6.40	7.71	6.63
	Speed up	1.00	0.20	0.19	0.19	0.23	0.20
	CPU(SP)%	79.44%	61.50%	57.61%	53.41%	51.67%	80.76%
	#(SP)	722.13	190.93	180.33	176.40	200.07	113.93
(4, 512)	CPU(total)	150.06	24.39	26.01	28.20	29.90	20.36
	Speed up	1.00	0.16	0.17	0.19	0.20	0.14
	CPU(SP)%	82.27%	59.79%	56.34%	52.11%	49.30%	80.44%
	#(SP)	1478.93	330.73	352.20	372.00	379.33	171.93
(4, 1024)	CPU(total)	448.85	85.83	107.30	128.94	119.32	59.27
	Speed up	1.00	0.19	0.24	0.29	0.27	0.13
	CPU(SP)%	84.19%	56.59%	55.34%	52.17%	48.00%	80.57%
	#(SP)	2055.27	555.80	684.27	788.73	716.73	248.47

TABLE 19. The *washington* Line families

Line-Sparse		$5n\Delta$	16Δ	8Δ	4Δ	2Δ	PD
(64, 4)	CPU(total)	3.68	1.19	0.91	0.84	1.05	1.75
	Speed up	1.00	0.32	0.25	0.23	0.29	0.48
	CPU(SP)%	80.86%	71.08%	69.31%	64.87%	68.69%	81.78%
	#(SP)	263.07	117.87	89.40	78.20	85.33	106.00
(256, 4)	CPU(total)	58.73	12.08	11.30	11.13	15.32	22.61
	Speed up	1.00	0.21	0.19	0.19	0.26	0.39
	CPU(SP)%	82.21%	67.88%	65.78%	64.59%	66.94%	83.17%
	#(SP)	964.13	280.07	250.20	226.07	280.40	320.20
(1024, 4)	CPU(total)	1013.87	132.89	142.62	158.33	195.21	292.57
	Speed up	1.00	0.13	0.14	0.16	0.19	0.29
	CPU(SP)%	86.08%	63.01%	61.77%	60.88%	60.50%	82.92%
	#(SP)	3573.73	694.80	708.73	727.93	860.07	1014.53

Line-Moderate		$5n\Delta$	16Δ	8Δ	4Δ	2Δ	PD
(64, 2)	CPU(total)	1.85	0.65	0.51	0.45	0.45	0.58
	Speed up	1.00	0.35	0.28	0.24	0.24	0.31
	CPU(SP)%	77.52%	67.63%	63.74%	62.12%	58.20%	79.34%
	#(SP)	192.93	91.60	75.53	63.80	61.93	48.53
(256, 4)	CPU(total)	58.73	12.08	11.30	11.13	15.32	22.61
	Speed up	1.00	0.21	0.19	0.19	0.26	0.39
	CPU(SP)%	82.21%	67.88%	65.78%	64.59%	66.94%	83.17%
	#(SP)	964.13	280.07	250.20	226.07	280.40	320.20
(1024, 8)	CPU(total)	1799.75	354.46	310.44	320.71	392.93	670.40
	Speed up	1.00	0.20	0.17	0.18	0.22	0.37
	CPU(SP)%	84.53%	64.62%	61.65%	61.45%	62.39%	83.59%
	#(SP)	4380.79	1217.47	1017.67	946.80	1072.60	1584.40

Appendix B. Comparison study for capacitated networks

In this appendix, we show results of our comparison study done for capacitated networks in the DIMACS new core set [5]. They are the gte-bad instances, grid families, and goto families. We reported an average value of ten random trials for each point for the grid families. For the goto families, we took five random trials in order to aviod long running times.

The GTE-Instances:

All problems in Table 20 are real problems with 49 nodes from [5]. As observed in the table, NETFLO and our code $PT(8\Delta)$ show steady running times, while RELAXT-III produces large variation. Results of $PT(8\Delta)$ are running times for the transformed uncapacitated networks. On average, NETFLO runs faster than $PT(8\Delta)$ (*resp.* RELAXT-III) by 66.0 (*resp.* 1338.1), and our code $PT(8\Delta)$ runs faster than RELAXT-III by 20.3.

The grid families:

The grid families consist of networks laid out on a grid generated by *ggraph* [5] with the following parameters: X: Grid height, Y: Grid width, capacities are

TABLE 20. CPU time (sec) for the GTE-Instances

	Original				Transformed	
	RELAXT-III	(ratio)	NETFLO	(ratio)	$PT(8\Delta)$	(ratio)
gte-bad.2	0.51	12.75	0.04	1.0	2.71	67.75
gte-bad.20	0.04	0.80	0.05	1.0	3.09	61.80
gte-bad.60	0.10	1.67	0.06	1.0	2.85	47.50
gte-bad.510	1.70	34.00	0.05	1.0	1.63	32.60
gte-bad.1160	2.75	91.67	0.03	1.0	1.71	57.00
gte-bad.6830	15.15	505.00	0.03	1.0	1.96	65.33
gte-bad.15710	0.12	6.00	0.02	1.0	2.51	125.50
gte-bad.35620	67.35	1683.75	0.04	1.0	2.05	51.25
gte-bad.469010	141.78	4726.00	0.03	1.0	2.26	75.33
gte-bad.508829	189.59	6319.67	0.03	1.0	2.29	76.33
ave. ratio		1338.13		1.0		66.04

uniform on $[0, 10000]$, and costs are uniform on $[0, 10000]$. Table 21 illustrates CPU time in seconds and index figure to NETFLO. For transformed networks, we observed that NETFLO runs faster than $PT(8\Delta)$ and RELAXT-III by 4.4 and 3.9, respectively. When we compare our code for transformed networks to RELAXT-III and NETFLO for the original networks, our code runs slower than RELAXT-III (resp. NETFLO) by 5.9 (resp. 40.4).

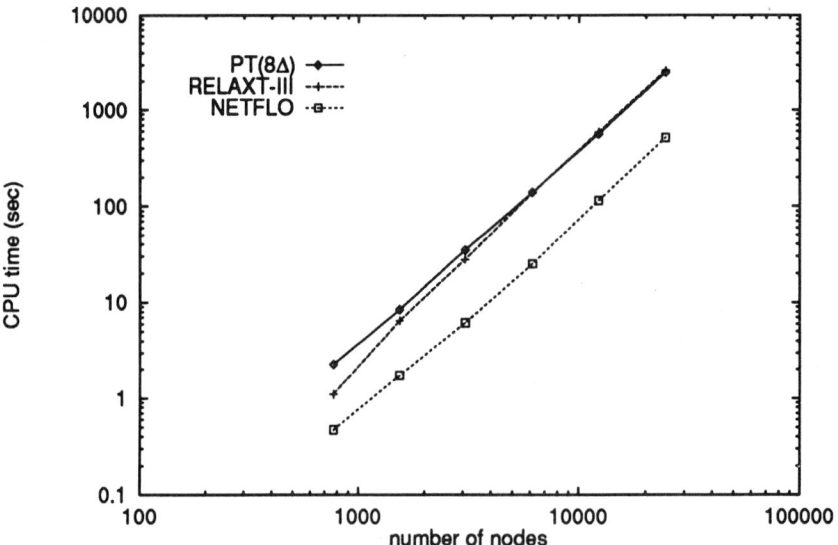

FIGURE 9. The Grid-Wide family (transformed networks)

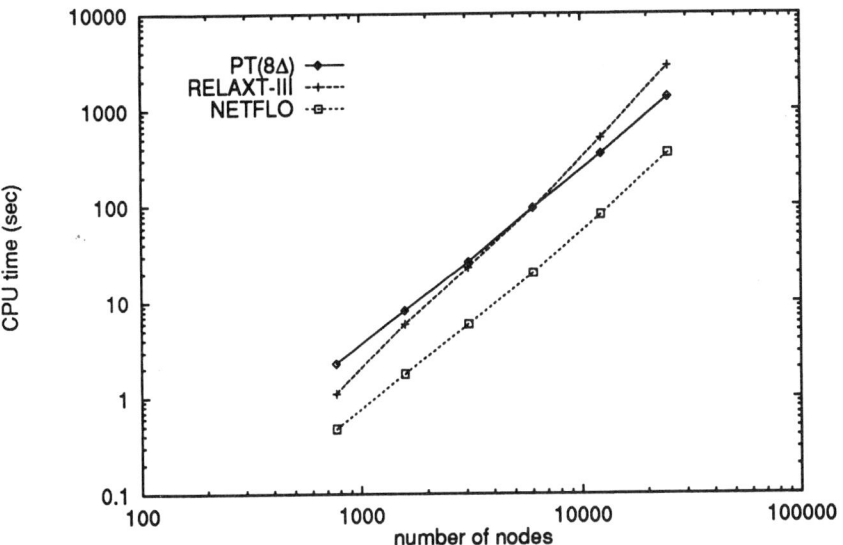

FIGURE 10. The Grid-Square family (transformed networks)

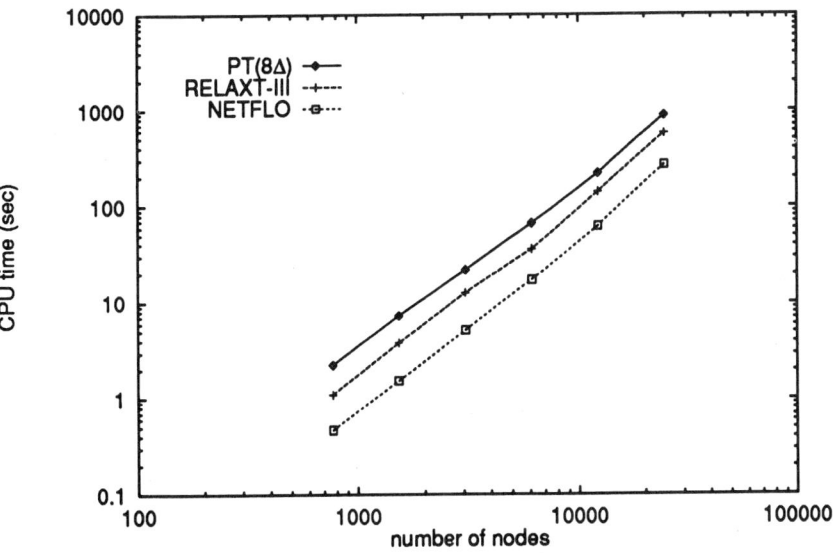

FIGURE 11. The Grid-Long family (transformed networks)

TABLE 21. CPU time (sec) for the Grid families

Grid-Long family

(X,Y)	$PT(8\Delta)$	Transformed		Original	
		RELAXT-III	NETFLO	RELAXT-III	NETFLO
$(16, 32)$	7.35	3.83	1.55	1.01	0.22
ratio	4.74	2.47	1.00	4.59	1.00
$(16, 64)$	22.05	12.75	5.21	3.12	0.57
ratio	4.23	2.45	1.00	5.47	1.00
$(16, 128)$	67.24	36.30	17.36	8.98	1.35
ratio	3.87	2.09	1.00	6.65	1.00
$(16, 256)$	219.16	62.63	141.66	22.32	3.25
ratio	1.55	0.44	1.00	6.87	1.00
$(16, 512)$	877.76	569.24	269.28	67.40	8.05
ratio	3.26	2.11	1.00	8.37	1.00
ave. ratio	3.53	1.91	1.00	6.39	1.00

Grid-Square family

(X,Y)	$PT(8\Delta)$	Transformed		Original	
		RELAXT-III	NETFLO	RELAXT-III	NETFLO
$(16, 16)$	2.27	1.10	0.47	0.39	0.09
ratio	4.83	2.34	1.00	4.33	1.00
$(23, 23)$	8.20	5.90	1.76	1.55	0.31
ratio	4.66	3.35	1.00	5.00	1.00
$(32, 32)$	26.16	23.07	5.85	5.31	0.84
ratio	4.47	3.94	1.00	6.32	1.00
$(45, 45)$	94.90	94.00	20.16	20.00	2.69
ratio	4.71	4.66	1.00	7.43	1.00
$(64, 64)$	347.34	510.93	80.60	92.21	11.63
ratio	4.31	6.34	1.00	7.93	1.00
$(91, 91)$	1353.73	2833.20	350.15	370.23	47.06
ratio	3.87	8.09	1.00	7.87	1.00
ave. ratio	4.47	4.79	1.00	6.48	1.00

Grid-Wide family

(X,Y)	$PT(8\Delta)$	Transformed		Original	
		RELAXT-III	NETFLO	RELAXT-III	NETFLO
$(32, 16)$	8.44	6.50	1.74	1.88	0.33
ratio	4.85	3.74	1.00	5.70	1.00
$(64, 16)$	35.00	28.00	6.20	7.24	1.08
ratio	5.65	4.52	1.00	6.70	1.00
$(128, 16)$	140.35	138.70	25.08	30.72	4.01
ratio	5.60	5.53	1.00	7.66	1.00
$(256, 16)$	563.91	589.74	115.14	136.05	15.14
ratio	4.90	5.12	1.00	8.99	1.00
$(512, 16)$	2507.23	2621.17	514.97	617.67	60.25
ratio	4.87	5.09	1.00	10.25	1.00
ave. ratio	5.17	4.80	1.00	7.86	1.00

The goto families:

The goto families consists of capacitated transportation networks generated by goto.c [5]. This generator takes $(n, m, maxcap, maxcost)$ as an input where n, m, $maxcap$, and $maxcost$ are the number of nodes, the number of arcs, maximum capacity, and maximum arc cost. The Grid-Density-8 (resp. Grid-Density-16) familiy consists of $(n, 8n, 16384, 4096)$ (resp. $(n, 16n, 16384, 4096)$) where $n = 256$, 512, and 1024. As table 22 shows, NETFLO runs faster than $PT(8\Delta)$ (resp. RELAXT-III) by 5.0 (resp. 9.8).

TABLE 22. CPU time (sec) for the goto families

Grid-Density-8 family

n	$PT(8\Delta)$	Transformed		Original	
		RELAXT-III	NETFLO	RELAXT-III	NETFLO
256	48.37	82.37	11.91	25.82	2.56
ratio	4.1	6.9	1.0	10.1	1.0
512	176.16	424.38	56.85	133.74	10.52
ratio	3.1	7.5	1.0	12.7	1.0
1024	886.52	1783.32	222.73	497.09	37.92
ratio	4.0	8.0	1.0	13.1	1.0
ave. ratio	3.7	7.5	1.00	12.0	1.0

Grid-Density-16 family

n	$PT(8\Delta)$	Transformed		Original	
		RELAXT-III	NETFLO	RELAXT-III	NETFLO
256	249.10	393.59	39.34	137.38	6.82
ratio	6.3	10.0	1.0	20.1	1.0
512	1144.75	2423.53	184.82	509.36	24.01
ratio	6.2	13.1	1.0	21.2	1.0
1024	5011.95	10733.77	820.48	1686.87	92.55
ratio	6.1	13.1	1.0	18.2	1.0
ave. ratio	6.2	12.1	1.0	19.9	1.0

Institute of Socio-Economic Planning, University of Tsukuba, Tsukuba, Ibaraki 305, Japan

Current address: Institute of Socio-Economic Planning, University of Tsukuba, Tsukuba, Ibaraki 305, Japan

E-mail address: fujishig@shako.sk.tsukuba.ac.jp

Tokyo Research Laboratory, IBM Japan, 5-19 Sanbancho, Chiyoda-ku, Tokyo 102, Japan

E-mail address: {iwano, nakanoj, tezuka}@trl.vnet.ibm.com

An Experimental Implementation of the Dual Cancel and Tighten Algorithm for Minimum-Cost Network Flow

S. THOMAS MCCORMICK AND LI LIU

February 1992; Revised September 1992

UBC Faculty of Commerce Working Paper 92-MSC-020

ABSTRACT. We present an exerimental implementation of a variant of the Dual Cancel and Tighten (DCT) algorithm for Minimum-Cost Network Flow (MCNF). The algorithm maintains a dual feasible π and a primal x necessarily satisfying only conservation. It then seeks to minimize the largest horizontal violation of complementary slackness on the kilter diagram of any arc. The variant uses a heuristic to find a tight or nearly tight flow for the current dual solution at each iteration. We report on the computational results that led us to abandon the original form of the algorithm, on our experiments to find a fast variant of it, and extensive computational results for the variant that we settled on. We find that this algorithm is an order of magnitude (or more) slower than Kennington and Helgason's NETFLO network simplex code.

1. Introduction

In [6], Ervolina and McCormick introduced the *Dual Cancel and Tighten (DCT)* algorithm for Minimum-Cost Network Flow (MCNF). DCT is a dual version of Goldberg and Tarjan's (primal) Cancel and Tighten algorithm for MCNF [8]. DCT is an appealing MCNF algorithm because it is very simple to describe and implement since each iteration involves only two Dijkstra-type shortest path computations, and some flow pushing on trees that can be done in linear time with no fancy data structures. Also, DCT runs in $O(m \log(nU) SP(m, n))$ time

---

1991 *Mathematics Subject Classification.* Primary 90B10; Secondary 90C35, 90-04.
Key words and phrases. min-cost flow, dual network algorithms, cut cancelling.
This research was partially supported by an NSERC Operating Grant.
This paper is in final form and no version of it will be submitted for publication elsewhere.

(Theorem 3.4), where U is the largest arc bound and $SP(m,n)$ is the time for computing one Dijkstra shortest path. This compares to $O(m \log n SP(m,n))$ for an algorithm of Orlin [14], which is the fastest known shortest path-based MCNF algorithm, and one of the fastest MCNF algorithms over all. Thus, as long as U is not too large (which is common in practice), DCT looks reasonably competitive in theory. These factors led us to conjecture in [6] that "Dual Cancel and Tighten has some potential to be a practical MCNF algorithm".

The convergence proof for DCT is based on showing that a measure of closeness to optimality decreases geometrically towards zero. If DCT is to be practical, it would have to be the case that on "real" instances its convergence is much faster than the convergence proof suggests. Our conjecture was based on the knowledge that the actual performance of many MCNF algorithm far outstrips the theoretical worst-case. A further spur to our efforts was the knowledge that the dual algorithm RELAX and its variants due to Bertsekas and Tseng [2, 3] have performed very well in practice. We briefly compare the strategies used by RELAX and DCT in Section 3.

This paper is a quixotic quest to verify our conjecture by looking for an implementation of some variant of DCT that will be competitive with other MCNF algorithms. We quickly discovered that a straightforward implementation of DCT is far from competitive because its average-case performance is very close to its theoretical worst-case: the convergence to optimality is so slow that it does not matter that each iteration is simple and quick. We tried to fix this by borrowing some ideas from the other MCNF algorithm in [6], namely Max Mean Cut Cancelling, and adding some heuristics to speed up each iteration and decrease the number of iterations. However, we find that our best effort is still an order of magnitude (or more) slower than the (fairly old) network simplex code NETFLO (see Kennington and Helgason [10]) in practice.

We should point out why we are bothering to report on an algorithm that we conclude is terrible in practice. The reason is that many of the techniques that are now being used to great effectiveness in flow codes were originally thought to be worthless in practice, e.g., scaling and dynamic trees. Thus it is valuable to be able to say about this technique that we have tried hard to make it work in practice, and it still doesn't pan out.

The original version of DCT alternates Dual Cancel steps with Dual Tighten steps. Dual Cancel uses a Dijkstra shortest path computation to simultaneously "augment" the dual variables across many cuts at once. Dual Tighten then uses a fast heuristic to change the flows so as to guarantee a small (but bounded away from zero) decrease in the maximum amount by which the flows violate complementary slackness. Our preliminary tests showed that the heuristic decrease in maximum violation in Dual Tighten was usually too small to force a change in the dual variables in Dual Cancel, which led to a huge number of iterations in practice.

To fix this, we turned to using *exact tightens*, which are a byproduct of computing max mean cuts. Exact tightens expend a lot more time in tightening, but produce the "tightest possible" result, which guarantees that Dual Cancel will have to make some change to the dual variables. Our tests showed that trading off more time per iteration (from the exact versus the heuristic tightening) for fewer iterations (from the dual variable change in every iteration instead of almost never) led to a big overall savings in time. We do not know of any way to control the extent of this tradeoff, but we note that every improvement we tried that reduced the number of iterations led to a reduction in overall CPU time, despite causing more work per iteration.

However, our Dual Cancel and Exact Tighten (DCET) implementation was still too slow, partly because an exact tighten involves computing an average of five or so max flows. Thus we tried to find a heuristic way to get the same good effect as an exact tighten at a cost of fewer max flows per iteration. What we ended up with can be explained by noting that the max flows in an exact tighten are effectively searching for the minimum possible maximum violation of complementary slackness by the flows. Our experiments showed that we could predict what this minmax violation would be on the next iteration from its recent past values, and so we could start off the max flows much closer to the optimal value. We predict the new value of the minmax violation as a moving average of past values. If our prediction is too high, we backup by repeatedly cutting it in half until it is low enough, at which point we resume our max flow algorithm. Thus we call this Dual Cancel and Moving Average with Backup (DCMAB).

We also implemented a number of tricks to speed up DCMAB. Based on what we learned at the DIMACS Workshop, we added periodic breadth-first search to our max flow routine, and changed our shortest path routine from Dijkstra with d-heaps to the D'Esopo and Pape variant of the label-correcting method (see [1]). We also took advantage of the fact that the max flows within and between DCMAB iterations are related to each other in such a way that, with care, the previous max flow can be used as a starting point for the next one. All of these changes speeded up our code by a factor of roughly two over what we had at the time of the workshop.

Our experiments showed that DCMAB does indeed reduce the average number of max flows per tighten from roughly five to roughly three, though at the cost of somewhat higher iteration counts. Especially for large instances, this tradeoff favored DCMAB over DCET. However, comparing the running time of our new, improved DCMAB to that of NETFLO (a somewhat out-of-date network simplex code, see [10]) over a common set of instances on the same machine shows that DCMAB is an order of magnitude (or more) slower than NETFLO.

2. Notation

We will use the following notation to describe MCNF problems: Let $\mathcal{D} = (N, A)$ be a directed graph with $|N| = n$ and $|A| = m$. Let $c \in R^A$ be a *cost* function on A and $-\ell, u \in \{R\cup\{+\infty\}\}^A$ be *lower* and *upper bounds* on A, with $\ell \leq u$. Let $d \in R^N$ be external *demands* with $e^T d = 0$ (if $d_i < 0$, then $-d_i$ is a *supply*). The MCNF problem is to find *flow* values x_{ij} between ℓ_{ij} and u_{ij} on each arc $i \to j$ such that the net flow into each node i exactly meets demand d_i, and whose total cost $c^T x$ is minimum.

We shall be considering many x's that violate the bounds. In order to clearly distinguish whether we are requiring bounds to be satisfied, we shall (rigorously) call $x \in R^A$ a *circulation* if only conservation is required, and a *bounded circulation* if both conservation and boundedness are required.

We denote the dual variable on node i by π_i, and the *reduced cost* of arc $i \to j$ by

$$\bar{c}_{ij}^\pi = c_{ij} - \pi_j + \pi_i.$$

We omit the superscript π in $\bar{c}_{ij}$ when there is no confusion.

3. The Basic Algorithm

We need to describe how DCT works in some detail in order to explain our variations on it. DCT always keeps a current π which is dual feasible, and a primal x which is a circulation. In particular, x need not be bounded. DCT's operation is best understood in terms of kilter diagrams.

We recall that a *kilter diagram* like Figure 1 is a graphical representation of the complementary slackness conditions for an arc $i \to j$ w.r.t. a dual feasible π and a circulation x (see, e.g. Lawler [11]). That is, when the current $(x_{ij}, \pi_j - \pi_i)$ for $i \to j$ lies on the heavy *kilter line* in Figure 1, then x_{ij} and $\pi_j - \pi_i$ satisfy complementary slackness, and we say that $i \to j$ is *in kilter*.

Otherwise, we say that $i \to j$ is *out of kilter*, and we can measure how far $i \to j$ is out of kilter in the horizontal direction by defining

(1) $$\delta(\pi, x)_{ij} = \begin{cases} \text{(a)} \ u_{ij} - x_{ij} & \text{if } \bar{c}_{ij} < 0, \\ \text{(b)} \ u_{ij} - x_{ij} & \text{if } \bar{c}_{ij} = 0, \ x_{ij} > u_{ij}, \\ \text{(c)} \ \ell_{ij} - x_{ij} & \text{if } \bar{c}_{ij} > 0, \\ \text{(d)} \ \ell_{ij} - x_{ij} & \text{if } \bar{c}_{ij} = 0, \ x_{ij} < \ell_{ij}, \text{ and} \\ \text{(e)} \ 0 & \text{otherwise,} \end{cases}$$

the *signed horizontal kilter number* (see Figure 1; the letters in (1) are keyed to the labels in Figure 1). For any feasible π and circulation x, define

$$\delta_{max}(\pi, x) = \max_{i \to j} |\delta(\pi, x)_{ij}|,$$

the maximum unsigned horizontal kilter number. Now if $\delta_{max}(\pi, x) = 0$, then π and x are jointly optimal since such an x must be bounded, hence feasible, and all arcs satisfy complementary slackness. However, it is easy to show that π must be optimal when $\delta_{max}(\pi, x)$ gets "close enough" to zero:

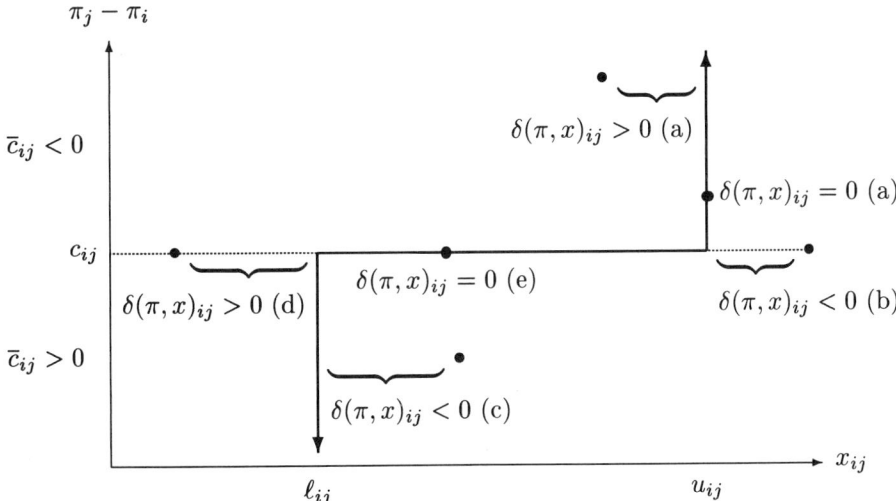

FIGURE 1. Measurement of signed horizontal kilter numbers, $\delta(\pi, x)_{ij}$. The letters in the labels refer to equation (1).

THEOREM 3.1 ([6]). *When ℓ and u are integral, if $\delta_{max}(\pi, x) < 1/m$ then π is optimal.* □

This gives a general outline of DCT: drive $\delta_{max}(\pi, x)$ towards zero until it is less than $1/m$. We decrease $\delta_{max}(\pi, x)$ by alternating the routines Dual Cancel, which modifies π, and Dual Tighten, which modifies x.

In principle, Dual Cancel operates by cancelling (changing π w.r.t.) totally positive cuts (see [6] for the definition) until none remain. In practice, this is accomplished by performing two Dijkstra-type shortest path computations on an *admissible graph* AG, and w.r.t. an arbitrary root node r, as described in [6].

Let $SP(m, n)$ denote the running time of a Dijkstra-type shortest path algorithm. The key properties of Dual Cancel are summarized in

THEOREM 3.2 ([6]). *Dual Cancel runs in $O(SP(m,n))$ time, and the admissible graph AG w.r.t. the π at the end of Dual Cancel is strongly connected.* □

Now Dual Tighten takes the strongly connected AG and finds a pure circulation (i.e., a circulation with $d = 0$) Δx in AG w.r.t. bounds that force x to move in a direction that will decrease $\delta_{max}(\pi, x)$. Dual Tighten then computes the new flow x' via $x' \leftarrow x + \alpha \cdot \Delta x$ for a certain step length α. A fast implementation of Dual Tighten can be obtained by a two-pass procedure called *Tree Circulation* described in [6]. The relevant properties of Dual Tighten using Tree Circulation are collected below:

THEOREM 3.3 ([6]). *Tree Circulation runs in $O(m)$ time, and hence Dual Tighten runs in $O(m)$ time also. The x' output from Dual Tighten satisfies*

$$\delta_{max}(\pi, x') \leq (1 - 1/2m) \cdot \delta_{max}(\pi, x). \quad \square$$

This gives us a geometric rate of decrease in $\delta_{max}(\pi, x)$, which makes DCT converge in polynomial time. We summarize DCT below.

Dual Cancel and Tighten
Input: A feasible dual solution π, and a circulation x.
Step 1: [Dual Cancel]. Cancel all totally positive cuts via two shortest path computations.
Step 2: [Dual Tighten]. Find circulation x' such that $\delta_{max}(\pi, x') \leq (1 - 1/2m)\delta_{max}(\pi, x)$ using Tree Circulation. If $\delta_{max}(\pi, x') \leq 1/m$ then STOP, π is optimal; otherwise go to Step 1.

The geometric decrease leads to these running times:

THEOREM 3.4 ([6]). *One iteration of DCT runs in $O(SP(m,n))$ time. The number of iterations of DCT is at most $O(m \log(nU))$, where U is the largest absolute bound or demand.* $\square$

The paper [6] also shows how to modify DCT so that it has strongly polynomial running time. This involves finding a tight or nearly tight flow from time to time to ensure that significant progress is happening. Our variants in Section 5 use some of these techniques.

It is useful to contrast DCT and its relatives in Section 5 with Bertsekas and Tseng's algorithm RELAX [2, 3], which is a dual MCNF algorithm that is known for having implementations that perform well in practice. Both algorithms are specializations of a generic positive cut cancelling algorithm (see [6]), but they select their cuts to cancel in very different ways.

DCT and its relatives are very fussy about choosing cuts. DCT chooses cuts based on the surrogate measure $\delta_{max}(\pi, x)$ rather than the dual objective value. The algorithms try to choose cuts which lead to optimal or near-optimal reductions in $\delta_{max}(\pi, x)$, and are willing to expend large amounts of time to find these good cuts. The good cuts tend to have a large number of nodes in both S and T, so updating the dual variables is time-consuming. DCT and its relatives are thus using up more time per iteration in the hopes of reducing the number of iterations.

In contrast, the RELAX family of algorithms uses the dual objective value as a guide in choosing cuts instead of the surrogate $\delta_{max}(\pi, x)$. RELAX wants to have very fast iterations at the expense of possibly having a larger number of iterations. The RELAX algorithms attain this goal by growing the S-side of their cut one node at a time starting from a single node, checking at each step if the current cut is positive. As soon as a positive cut is found it is cancelled.

In practice, this means that the great majority of iterations for RELAX consist of cancelling single-node cuts, which can be done extremely cheaply.

RELAX and DCT differ in their use of primal variables (flows) as well. Both use primal-infeasible flows to help in finding positive cuts. However, at each iteration DCT has to compute a new flow essentially from scratch (since the auxiliary graph changes), whereas RELAX maintains the same flow from iteration to iteration. The basic flow operation in the implementations of DCT described in Section 5 is computing a max flow, which is expensive. The basic flow operation in RELAX is pushing flow along a path, which is cheap.

4. A Simple Implementation of DCT

Our base implementation of DCT was written in C, and executed under SunOS Unix on a Sun 386i workstation with 8 Meg of memory (the 386i is about four times slower than a Sparc 1). We keep all the data for each arc together in a structure, and the structures hold pointers that link all arcs with a given tail and with a given head. In addition, we have pointers to the beginning of each list. All data is kept as 32-bit integers except for x, δ_{ij}, α, and $\delta_{max}(\pi, x)$, which are double floats. All runs reported are for a single execution of a single instance. However, our results are so striking that minor variations in reported running time do not change them.

We choose $\pi = 0$ as our initial dual solution. We use a max flow routine to find an initial x that satisfies conservation. We implemented Goldberg and Tarjan's preflow push (see [7]), using the selection rule suggested in [5]. Our adaptation did not explicitly add any extra nodes or arcs to solve this feasibility problem, but rather let the "source" be all the positive excess nodes, and the "sink" be all negative excess nodes.

We use the criterion "is $\delta_{max}(\pi, x) < 1/m$?" from Theorem 3.1 as our stopping rule. Note that DCT produces only an optimal π, not an optimal x. When an optimal flow is desired, we do a max flow feasibility calculation w.r.t. the optimal set of reduced costs to find one.

We implemented the shortest path computations in Dual Cancel via Dikjstra's Algorithm using d-heaps to select the minimum distance label as suggested in [16]. We chose $d = 4$ so that we could multiply and divide by d by shifting. We did not explicitly construct the admissible graph AG, but rather do the shortest path computations directly on the original graph. To facilitate this, we keep the value of δ_{ij} in each arc's structure and update it after each change in π or x. We collected statistics on how often Dual Cancel actually changed π, since we suspected that it would often happen that AG would remain strongly connected after Dual Tighten.

We implemented Dual Tighten using Tree Circulation as described in [6], and again we did not explicitly construct AG.

The results from this simple implementation are in Table 1. We tested this

Name	Nodes	Arcs	CPU Sec	Itns	π Changes	Itns Aft Opt
t1	48	240	82.1	846	91	308
t2	100	500	804.1	3797	198	2380
t3	200	1600	23448.8	41587	564	14923
t4	400	3600	∞	—	—	—

TABLE 1. Results of Simple Dual Cancel and Tighten

Name	Nodes	Arcs	CPU Sec	Itns	MF/itn
t1	48	240	16.4	30	3.90
t2	100	500	78.4	53	4.02
t3	200	1600	855.5	130	4.24
t4	400	3600	4148.2	233	4.56

TABLE 2. Results of Dual Cancel and Exact Tighten

simple DCT on four instances generated with Goldberg's *mesh* generator, using the parameters listed. The "Itns" column gives the number of DCT iterations; the "π Changes" column gives the number of these iterations that actually changed π during Dual Cancel; and the "Itns Aft Opt" gives the number of iterations of Dual Tighten after Dual Cancel made its last nonzero change to π (i.e., the optimal π was already in hand, but we expended some tightens in getting $\delta_{max}(\pi, x)$ below $1/m$). We terminated instance t4 before it finished because it was taking a huge number of iterations.

We see from Table 1 that simple DCT is very slow, and has a bad rate of asymptotic growth. One reason is that it spends a lot of work doing null shortest path calculations. We also experienced some numerical difficulties with this DCT: When $\delta_{max}(\pi, x)$ is close to $1/m$ and α is close to its theoretical minimum of $\delta_{max}(\pi, x)/2m$, we are changing flows by an amount which is only $O(1/m^2)$. This can be indistinguishable from 0 for large m, and this fact made us change our flows variables to double precision floats.

5. Improving DCT, Part I

One part of Table 1 that stands out is the fact that Dual Cancel did not change π on most iterations. In fact, in Table 1 only 1.8% of DCT's iterations on t1, t2 and t3 caused π to change. Intuitively, this is because Dual Tighten did not make x "tight enough" to cause a dual change. We also have no way to recognize when π is optimal (roughly one third of the iterations in Table 1 come after we already have the optimal π). An obvious tack to try is to find an x that is as tight as possible, i.e., to solve

Exact Tighten $\min_{x'} \delta_{max}(\pi, x')$.

We will need to expend more effort to solve Exact Tighten than we did for Dual

Tighten, but $\delta_{max}(\pi, x)$ will get reduced faster, leading to fewer overall iterations. Table 1 shows that reducing the number of iterations is very important in DCT. The counterpart to Theorem 3.3 in this case is

THEOREM 5.1 ([6]). *The x' output from Exact Tighten satisfies*

$$\delta_{max}(\pi, x') \le (1 - 1/m) \cdot \delta_{max}(\pi, x). \quad \square$$

The paper [13] shows that Exact Tighten is equivalent to a max mean cut problem. We need to cover some of the details of this to explain our experiments.

From complementary slackness, π is optimal if and only if there is a feasible flow w.r.t. the *modified bounds* ℓ^π and u^π defined by

$$(2) \quad \text{if } \overline{c}_{ij}^\pi \equiv \pi_i + c_{ij} - \pi_j \text{ is } \begin{cases} > 0 & \text{then } \ell_{ij}^\pi = u_{ij}^\pi = \ell_{ij}, \\ = 0 & \text{then } \ell_{ij}^\pi = \ell_{ij}, u_{ij}^\pi = u_{ij}, \\ < 0 & \text{then } \ell_{ij}^\pi = u_{ij}^\pi = u_{ij}. \end{cases}$$

When π is not optimal, consider the parametric network $\mathcal{N}(\pi, \delta)$ with bounds $\ell^\pi - \delta$ and $u^\pi + \delta$. An equivalent way to express Exact Tighten is that it asks for the minimum δ, call it $\delta(\pi)$, such that $\mathcal{N}(\pi, \delta)$ is feasible.

Thus if we have a δ_0 and we test the feasibility of $\mathcal{N}(\pi, \delta_0)$ with a max flow calculation, we will discover that $\delta_0 \ge \delta(\pi)$ if the network is feasible, and $\delta_0 < \delta(\pi)$ if not. When $\mathcal{N}(\pi, \delta_0)$ is infeasible, the max flow routine gives us a cut (S, T) proving infeasibility. We can then increase δ_0 just enough to ensure that (S, T) no longer blocks flow. This forms the basis of a parametric algorithm for Exact Tighten; the last cut encountered is a max mean cut, and the last flow computed solves Exact Tighten.

The papers [13] and [15] show that variations of this parametric algorithm solve Exact Tighten in $O(\min\{n, \log(nU)\} \cdot MC(m, n))$ time, where $MC(m, n)$ is the time to solve one min cut problem. There is an approximate binary search algorithm for computing max mean cuts which runs in $O(mn \log(nU))$ time (see [9, 12]), which is theoretically faster than the parametric algorithm. However, the average number of iterations of approximate binary search is essentially the same as its worst case bound, whereas (as we shall see) the average number of iterations for the parametric algorithm is much better than its worst case bound. Thus we chose to implement the parametric algorithm.

Note that the time bound on the parametric algorithm is much larger than Dual Tighten's $O(m)$, so in theory we pay a large price for optimality (however, in theory DCT now becomes a strongly polynomial algorithm, and will perform at most $O(m^2 \log n)$ iterations, see [6]). But we will see that the actual number of max flows per iteration is quite small in practice. Using Exact Tighten has the added benefit that it allows us to immediately recognize an optimal π when we get one, since we start with $\delta = 0$ and π is optimal if and only if the network $\mathcal{N}(\pi, 0)$ is feasible.

We ran some tests comparing simple DCT to Dual Cancel and Exact Tighten (DCET), and the results appear in Table 2 (these tests were also done on a Sun 386i). These are the same four instances we had in Table 1. Column "MF/itn" gives the average number of max flows used in computing the exact tighten in each iteration.

These results are better than we expected, particularly in CPU time. Stepping up from a shortest path in each iteration (simple DCT) to several max flows in each iteration (Exact Tighten) has not hurt us that much. One reason is that the average number of max flows per exact tighten is, as expected, much lower than the $O(n)$ bound. We also see that the quality of each DCET iteration is better than simple DCT in that the number of DCET iterations is less than half the number of simple DCT iterations that changed π.

We now thought about trying to minimize the number of max flows incurred by Exact Tighten. In looking at iteration logs from the runs in Table 2, we noticed that although Exact Tighten always starts out with $\delta = 0$, it usually ends up with an optimal value that is very close to the value of $\delta_{max}(\pi, x)$ from the previous iteration. For example, Figure 2 shows such a plot from NETGEN benchmark instance 36, which has 8000 nodes and 15,000 arcs. Recall from Theorem 5.1 that the worst case for $\delta_{max}(\pi, x')$ on the current iteration is $(1 - 1/m)$ times the value of $\delta_{max}(\pi, x)$ from the previous iteration. Figure 2 plots what percent of this theoretical worst case is represented by the actual $\delta_{max}(\pi, x')$ at each iteration, i.e.,

$$100 \frac{\delta_{max}(\pi, x')}{(1 - 1/m)\delta_{max}(\pi, x)}.$$

This suggests that we could short-circuit many max flow iterations by starting out with a value for δ which is much closer to the old $\delta_{max}(\pi, x)$ than to zero.

We call this strategy *Partial Tighten*. It starts the parametric max flow algorithm with an initial guess of $discount \cdot$ (old $\delta_{max}(\pi, x)$), where $discount$ is a parameter we can tune. If the max flow says that the initial network is feasible, we accept that x', which will probably not be a tight flow. If the initial network is infeasible, we continue the parametric algorithm from that point, hoping that it will find a max mean cut in very few additional max flows. In Partial Tighten (and the other variants described below) we also use the heuristic of starting out with two Exact Tightens, and going back to Exact Tightens if $\delta_{max}(\pi, x)$ is less than 5.0 (we want to get a fast start, and we don't want to waste time iterating with a small $\delta_{max}(\pi, x)$ at the end when we are actually optimal).

Results from running Partial Tighten (and the variants described below) for various values of *discount*, averaged over four NETGEN instances appear in Table 3, along with the averages for DCET on the same four instances. (These results were obtained on an HP 730 workstation, so the CPU times are not comparable to other tables.) We see that Partial Tighten was indeed successful in reducing the average number of max flows per iterations. However, the decrease in CPU time over DCET was not as great as we had hoped. This is because there are too many iterations where π does not change, which occurs only when the initial guess at δ is too high. This causes a non-exact tighten to take place, which may leave Dual Cancel with nothing to do.

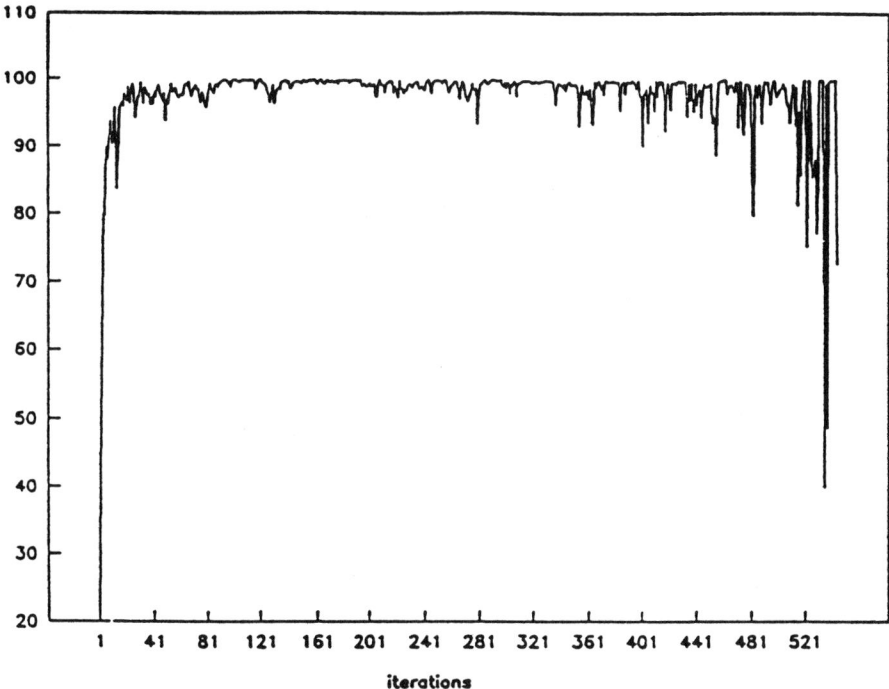

FIGURE 2. Plot of $\delta_{max}(\pi, x)$ as a percent of theoretical worst-case over all iterations of DCET on NETGEN instance 36

We came up with two independent strategies to speed up Partial Tighten. The first is called *Backup Tighten*. Here, when our initial guess at δ is too high, instead of just accepting that x' we cut our guess in half, which we call a backup, and try again. (Thus we are doing binary search as long as our guess is too high, and switching to the parametric algorithm once we find a guess that is too low.) We allowed up to five backups. This ensures that our tighten is almost always exact, because as long as backing up finds a δ below the optimal value, the parametric algorithm will always end with the optimal value.

The second strategy to improve Partial Tighten is called *Moving Average Tighten*. The idea here is that Figure 2 makes it seem plausible that we can get a good guess at the optimal value of δ at the next iteration by looking at the actual values of $\delta_{max}(\pi, x)$ on the last few iterations. That is, towards the

discount	CPU Sec	Iterations	π Changes	Max Flows	MF/Itn
Dual Cancel and Exact Tighten					
	5411.01	420.25	420.25	2384.50	5.69
Dual Cancel and Partial Tighten					
.80	5366.18	428.50	427.75	2005.75	4.68
.90	5106.36	444.50	441.50	1854.00	4.15
.95	4970.97	633.00	591.25	1676.00	2.63
.98	4981.73	633.00	591.25	1676.00	2.63
Dual Cancel and Backup Tighten					
.80	5341.01	426.50	426.50	2009.75	4.71
.90	5070.70	430.25	430.25	1854.25	4.30
.95	5301.89	486.25	486.25	1846.25	3.83
.98	5313.13	486.25	486.25	1846.25	3.83
Dual Cancel and Moving Average Tighten					
.80	5629.87	428.00	424.50	1919.75	4.48
.90	5213.96	441.25	436.75	1776.00	4.01
.95	4832.75	459.25	451.00	1641.75	3.55
.98	5012.39	544.00	521.00	1593.75	2.92
Dual Cancel and Moving Average with Backup					
.80	5818.45	438.75	438.75	1969.25	4.48
.90	5217.45	437.50	437.50	1796.00	4.09
.95	4964.71	445.75	445.75	1672.00	3.74
.98	5076.26	482.00	482.00	1624.75	3.37

TABLE 3. Results of variants of DCT averaged over NETGEN instances N37, N38, N39 and N40

beginning and end of the set of iterations $\delta_{max}(\pi, x)$ appears to be smaller on average than it is in the middle of the set of iterations.

We chose to predict the next iteration's initial δ based on the moving average of the $\delta_{max}(\pi, x)$'s of the previous five iterations. Using a moving average allows the variation that is apparent in Figure 2 to be dampened. The specific formula we used was that the initial δ for the parametric max flow algorithm was *discount·* (moving average of the last five values of $\delta_{max}(\pi, x)$) (because of the initial two Exact Tightens, we wait for seven iterations before starting with this). Once again, *discount* is a tunable parameter.

This gives us five variants of DCT: Exact Tighten (DCET), Partial Tighten (DCPT), Backup Tighten (DCBT), Moving Average Tighten (without backup; DCMAT), and Moving Average with Backup (DCMAB). Each of the last four variants depends on the parameter *discount*. We did some limited tests to determine a good value for *discount* for each of these variants. The data in Table 3 is taken from this testing. A more graphic look at this data appears in Figure 3, where we plot average CPU time over the four instances against *discount*, Figure 4, where we plot the average number of max flows per iteration against the total number of iterations for instance N38 for the four variants, and Figure 5, which is the same as Figure 4 except that it contains all four instances and the single variant DCMAB.

Based on Figure 4, Figure 5 and Table 3 we chose to use Moving Average with Backup (DCMAB) with *discount* = .98 for further testing, despite the fact that in Figure 3 Moving Average Tighten and *discount* = .95 appear to be better. This is because our experience with larger instances suggests that the smaller number of iterations of Moving Average with Backup versus Moving Average Tighten will pay off, and similarly the smaller number of max flows per iteration of *discount* = .98 versus *discount* = .95 will pay off.

Figure 4 and Figure 5 show how a higher value of *discount* trades off a smaller number of max flows per iteration for a slight increase in the number of iterations. Figure 4 also suggests that DCMAB looks the best overall of the four variants.

The results of this testing show that we have indeed achieved what we wanted with Dual Cancel and Moving Average with Backup (DCMAB), namely a decrease of about two max flows per iteration, while changing π on essentially every iteration.

6. Improving DCT, Part II

We came away from the DIMACS Workshop armed with a lot of tips about what techniques work well in practice, and we added them to our code to make our implementation faster. We added a periodic breadth first search (every $n/2$ iterations) to our max flow code, and changed our shortest path code from Dijkstra with d-heaps to label correcting with D'Esopo and Pape's rule (see [1]).

We also modified our code to take advantage of the fact the successive max

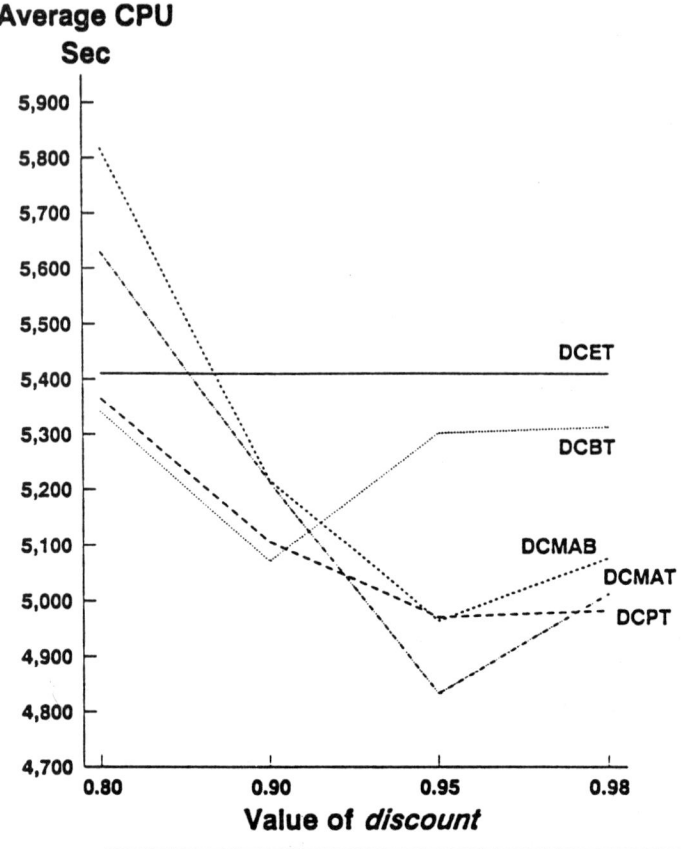

FIGURE 3. Plot of average CPU time over four NETGEN instances as a function of *discount* for the four variants of DCT, with a line at the average CPU time for DCET

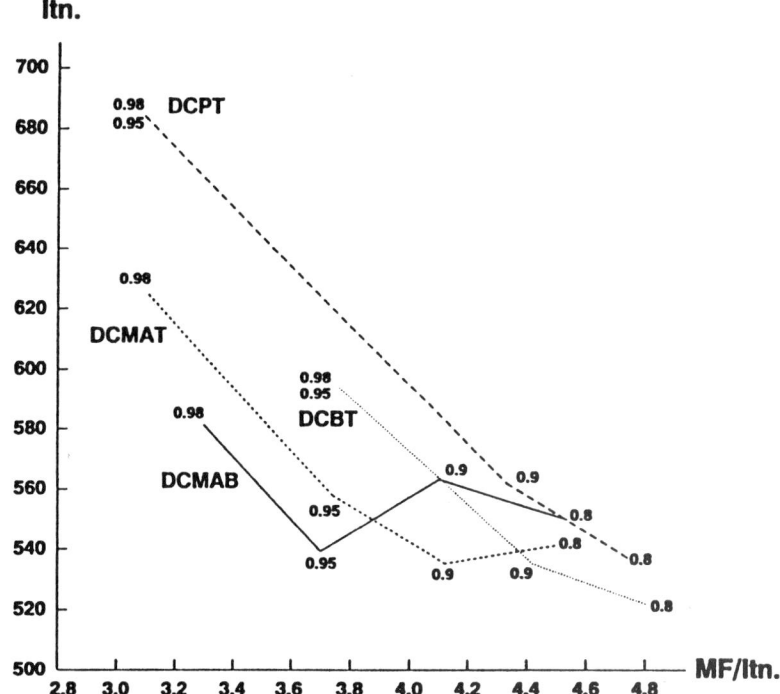

FIGURE 4. Plot of the number of iterations for NETGEN instance N38 against the average number of max flows per iteration for the four variants of DCT, labeled with *discount* values

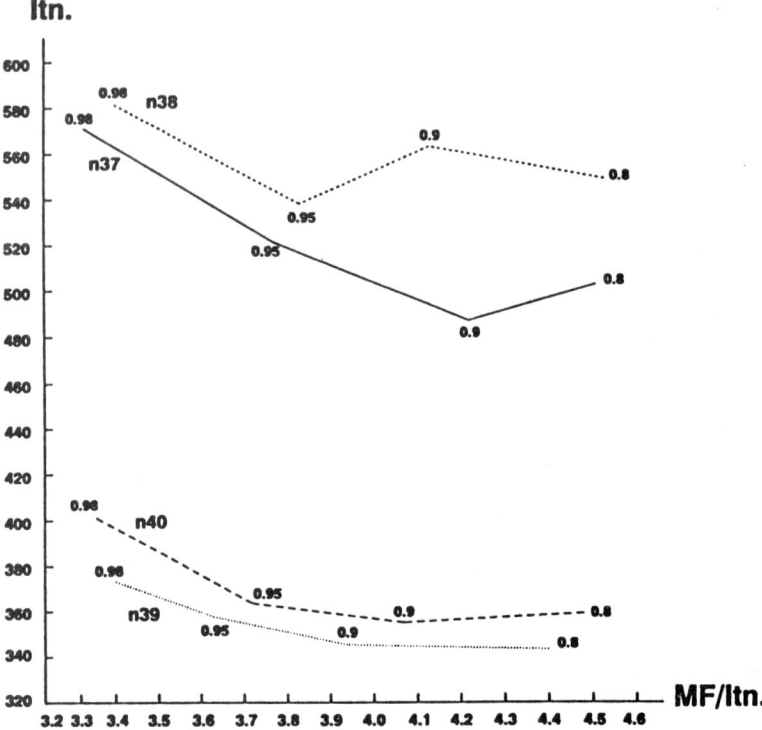

FIGURE 5. Plot of the number of iterations for NETGEN instances N37, N38, N39 and N40 against the average number of max flows per iteration for variant Dual Cancel Moving Average with Backup, labeled with *discount* values

| | Old DCET | | | | New DCET | | | |
Name	Itns	MFs	MF/Itn	CPU Sec	Itns	MFs	MF/Itn	CPU Sec
nge.11	53	221	4.17	16.41	50	198	3.96	8.83
nge.20	89	397	4.46	22.61	84	354	4.21	13.03
nge.27	71	309	4.35	20.39	61	255	4.18	12.87
nge.28	97	483	4.98	103.76	92	438	4.76	43.88
nge.32	158	841	5.32	406.85	137	656	4.79	163.54
nge.36					476	2567	5.39	7296.63
nge.38					532	2989	5.62	13158.78
nge.40					352	2014	5.72	4640.90

TABLE 4. Results of Old and New DCET on NETGEN Instances

flows in the parametric max flow routine are strongly related to each other. In particular, we run max flow only until we identify a min cut, which is what we really want anyway. Thus when we find that a node with positive excess is cut off from the negative excess nodes, we leave its excess in place rather than trying to push it back to one of the original positive excess nodes (recall that our max flow code works directly on the original graph, trying to push excess simultaneously from all positive excess nodes to all negative excess nodes). When we move from one max flow to the next, we keep the old flow, we refresh the distance labels by doing a Breadth First Search from the remaining negative excess nodes (which we keep in a list for efficiency), and we put all the old positive excess nodes (also kept in a list) into the preflow queue.

Unfortunately, we made all three of these changes at the same time, so we are not able to comment on the relative contributions of these three techniques to the speedup we found. The results from running the old versions of both DCET and DCMAB against the new versions of DCET and DCMAB on the DIMACS NETGEN subset appear in Tables 4 and 5; these tables do not include data on all eight instances for the old codes since the old codes took too long to run on some of the larger instances. (Table 5 also includes a column of NETFLO times, discussed below.) We can see that our improvements sped up the code by a factor of roughly two, and that DCMAB is indeed faster than DCET, at least on larger instances. Note that all of our post-Workshop tests were run on an HP 720, which tests out as being roughly four times faster than a Sparc 1 on the DIMACS benchmarks, so the times in these tables may look artificially fast compared to previous tables.

We also ran our new DCMAB code on the suggested DIMACS core experiments (except for the CAPT instances, for which the suggested DIMACS parameters give instances that are misleadingly easy to solve), and ran NETFLO (see [10]) on the same instances. DCMAB did reasonably well on the gte_bad instances, solving them in .21–.71 seconds, 15–30 iterations, and 40–77 max flows, so it doesn't suffer from the same non-robustness as its fellow dual algorithm

Name	Old DCMAB				New DCMAB				NETFLO
	Itns	MFs	MF/Itn	CPU Sec	Itns	MFs	MF/Itn	CPU Sec	
nge.11	53	221	4.17	16.20	50	198	3.96	8.90	0.9
nge.20	98	342	3.49	23.56	97	291	3.00	15.78	0.9
nge.27	79	252	3.19	22.33	82	231	2.82	17.63	1.7
nge.28	112	397	3.54	116.11	102	317	3.11	51.70	2.6
nge.32	174	599	3.44	372.31	160	476	2.98	158.55	8.5
nge.36					492	1734	3.52	5616.16	121.2
nge.38					575	2181	3.79	12539.88	120.5
nge.40					393	1438	3.66	4043.11	122.9

TABLE 5. Results of Old DCMAB, New DCMAB, and NETFLO on NETGEN Instances

RELAX. Otherwise, the results in Table 5 are for the DIMACS NETGEN subset. The results in Table 6 come from the grid-on-torus generator where instance goto8.k has $|A| = 8|N|$ and $|N| = 2^k$ nodes; instance goto16.k is the same but with $|A| = 16|N|$; and instance gotoid.k is the same but with $|A| = \lceil |N|^{3/2} \rceil$. The results in Table 7 come from the gridgraph generator, where instance gl.k is on a $16 \times k$ grid; instance gw.k is on a $k \times 16$ grid; and instance gs.k is on a $k \times k$ grid. In all cases we report the running time on just a single run of a single instance at that size, but we shall see that our results are so striking that we don't need to worry about small variations in running time. We also did not try very large instances, but it is easy to extrapolate from the results we have here that the same behavior will continue. For both programs the time reported is user time, but DCMAB counts it only from after the problem is read in to when the optimal solution is first reached, whereas NETFLO counts total time including all I/O (i.e., this comparison gives DCMAB an unfair advantage).

From this data we can see that our efforts to make a practically useful implementation of DCT have failed. In all cases, DCMAB's running time is an order of magnitude (or more) worse than NETFLO's time. Worse yet, we can see that at fixed densities, the running time of NETFLO is increasing sub-quadratically, whereas the running time of DCMAB is increasing super-quadratically. Thus DCMAB's performance will only get worse as instance sizes get larger.

We have also done some informal comparisons of the number of max flows used by DCMAB versus the number used by SCALE, the Bland, Cheriyan, Jensen and Ladanyi cost-scaling code [4]. These comparisons indicate that DCMAB consistently uses 3–5 times more max flows calls than SCALE on similar instances. This indicates that our poor results are not due to faulty coding, but rather are intrinsic to this method.

7. Conclusion

We have conclusively shown that the fastest variant of DCT we could come up with, Dual Cancel and Moving Average with Backup, runs very slowly compared even to the older algorithm NETFLO. We believe this slowness to be intrinsic to DCT.

Name	Itns	MFs	MF/Itn	CPU Sec	NETFLO
GOTO $A = 8N$ Instances					
goto8.7	145	331	2.28	15.61	1.0
goto8.8	257	604	2.35	69.93	3.0
goto8.9	411	984	2.39	430.64	9.0
goto8.10	909	2401	2.64	2920.83	32.6
goto8.11	1742	4895	2.81	15708.71	121.2
goto8.12	2912	8512	2.92	68514.94	559.0
GOTO $A = 16N$ Instances					
goto16.7	259	584	2.25	52.42	2.4
goto16.8	350	765	2.19	273.09	6.8
goto16.9	560	1259	2.25	1312.70	20.0
goto16.10	1170	3074	2.63	7723.10	78.3
goto16.11	2036	6132	3.01	37502.10	263.7
GOTO $A = N^{3/2}$ Instances					
gotoid.8	350	765	2.19	277.29	6.7
gotoid.9	797	1755	2.20	2813.61	30.7
gotoid.10	1827	4114	2.25	26939.75	222.3

TABLE 6. Results of New DCMAB on the Classes of GOTO Instances

Name	Itns	MFs	MF/Itn	CPU Sec	NETFLO
GGRAPH Long Instances					
gl.32	63	186	2.95	11.87	0.6
gl.64	86	246	2.86	44.31	1.3
gl.128	138	372	2.70	230.74	2.6
gl.256	190	604	3.18	1185.81	5.8
gl.512	267	977	3.66	5299.60	12.0
GGRAPH Wide Instances					
gw.32	84	225	2.68	13.86	0.6
gw.64	224	577	2.58	84.58	1.7
gw.128	417	1082	2.59	485.86	4.7
gw.256	836	2222	2.66	2534.91	12.6
gw.512	1703	4831	2.84	14693.85	47.5
GGRAPH Square Instances					
gs.16	48	133	2.77	3.64	0.3
gs.32	130	351	2.70	54.07	1.5
gs.64	321	852	2.65	1107.63	10.7

TABLE 7. Results of New DCMAB on the Classes of GGRAPH Instances

The success of Bertsekas and Tseng's code RELAX [2, 3] shows that it is possible for a dual algorithm to be competitive. It would be worthwhile to have an alternate practically useful dual MCNF algorithm. We have shown that DCT and its variants do not fit the bill.

Acknowledgement

We thank Noel Paul for his help in generating Table 3 and Figures 3, 4, and 5.

References

1. R.K. Ahuja, T.L. Magnanti, and J.B. Orlin (1989). Network Flows. Chapter IV of *Handbooks in Operations Research and Management Science, Volume 1: Optimization*, eds. G.L. Nemhauser, A.H.G. Rinnooy Kan and M.J. Todd, North Holland, pp. 211-369.
2. D. Bertsekas (1991). *Linear Network Optimization*. MIT Press, Cambridge, MA.
3. D. Bertsekas and P. Tseng (1988). Relaxation Methods for Ordinary and Generalized Network Flow Problems. *Operations Research*, **36**, pp. 93–114.
4. R.G. Bland, J. Cheriyan, D.L. Jensen and L. Ladanyi (1991). An Empirical Study of Recent Min Cost Flow Algorithms. Extended Abstract accepted by the Dimacs Netflow Challenge Workshop; full paper in revision for the proceedings of the workshop.
5. U. Derigs and W. Meier (1989). Implementing Goldberg's Max-Flow Algorithm — A Computational Investigation. *Methods and Models of Operations Research* 33, pp. 383–403.
6. T.R. Ervolina and S.T. McCormick (1992). Two Strongly Polynomial Cut Cancelling Algorithms for Minimum Cost Network Flow. UBC Faculty of Commerce Working Paper 92-MSC-019, Vancouver, BC; to appear in *Discrete Applied Math*.
7. A.V. Goldberg, and R.E. Tarjan (1988). A New Approach to the Maximum Flow Problem. *JACM* 35, no. 4, pp. 921–940.
8. A.V. Goldberg, and R.E. Tarjan (1989). Finding Minimum-Cost Circulations by Canceling Negative Cycles. *JACM* 33, no. 4, pp. 873–886.
9. K. Iwano, S. Misono, S. Tezuka, and S. Fujishige (1990). A New Scaling Algorithm for the Maximum Mean Cut Problem. IBM Research Report RT 0049, Tokyo, Japan; to appear in *Algorithmica*.
10. J.L. Kennington and R.V. Helgason (1980). *Algorithms for Network Programming*, John Wiley, New York.
11. E.L. Lawler (1976). *Combinatorial Optimization: Networks and Matroids*. Holt, Rinehart and Winston, New York.
12. S.T. McCormick (1992). A Note on Approximate Binary Search Algorithms for Mean Cuts and Cycles. UBC Faculty of Commerce Working Paper 92-MSC-021, Vancouver, BC.
13. S.T. McCormick and T. R. Ervolina (1990). Computing Maximum Mean Cuts. UBC Faculty of Commerce Working Paper 90-MSC-011, Vancouver, BC; to appear in *Discrete Applied Math*.
14. J.B. Orlin (1988). A Faster Strongly Polynomial Minimum Cost Flow Algorithm. *Proc. 20th ACM Symp. on the Theory of Comp.*, pp. 377–387.
15. T. Radzik (1992). Minimizing Capacity Violations in a Transshipment Network. *Proceedings of the Third Annual ACM-SIAM Symposium on Discrete Algorithms*, Orlando 1992, pp. 185–194.
16. R.E. Tarjan (1983). *Data Structures and Network Algorithms*. SIAM, Philadelphia, PA.

(S. T. McCormick and L. Liu) FACULTY OF COMMERCE AND BUSINESS ADMINISTRATION, UNIVERSITY OF BRITISH COLUMBIA, VANCOUVER, BC V6T 1Z2 CANADA
E-mail address, S. T. McCormick: stmv@adk.commerce.ubc.ca

A Fast Implementation of a Path-Following Algorithm for Maximizing a Linear Function Over A Network Polytope

ANIL JOSHI, ARTHUR S. GOLDSTEIN, AND PRAVIN M. VAIDYA

December 31, 1992

1. Introduction

Many network optimization problems - Assignment, Maximum Flow, and Minimum Cost Flow - can be cast as Linear Programming (LP) problems. Recently several algorithms have been proposed for LP which belong to the general class of algorithms known as Interior Point methods (See [**Vai90, Vai89, Ren88, Kar84**]). The objective of our implementation is to show that path-following interior point algorithms can work well if the problem peculiarities are taken into consideration and appropriate data-structures are chosen. Computational experiments comparing the performance of Network Simplex and Interior Point methods have been carried out by several researchers. The results can be found in [**AM91, KY91, RV90, Raj89, A**[+]**85**].

We use the notation e for the number of edges and v for the number of vertices in the given network. A network flow problem is defined as the following LP [**FF62**].

$$\max c^T x$$

subject to the lower and upper bound (*i.e.* capacity) constraints

$$l_i \leq x_i \leq u_i$$

and the flow conservation constraints

$$Ax = b$$

1991 *Mathematics Subject Classification.* Primary 90B10, 90C05, 90-04; Secondary 49M35, 65K05.

where $x \in \Re^e$ is the flow vector whose ith element is the flow in edge i. l_i and u_i are the lower and upper bounds respectively on the flow in edge i. In addition, these capacities are assumed to be integral. Matrix $A \in \Re^{(v-1) \times e}$ describes the network under consideration. A is obtained from the node-arc incidence matrix of the network by dropping the row corresponding to a pre-specified ground node. Each column of A corresponds to an edge of the network. If the edge k is directed from node i to node j then there is a -1 in the ith row and a $+1$ in the jth row in column k of A (If edge k is incident on the ground node there is a $+1$ or a -1 in column k of A). All other entries in that column are zero. Vector $b \in \Re^{v-1}$ corresponds to the node flow values (with appropriate sign for a source or a sink). Vector $c \in \Re^e$ is the cost vector.

Any algorithm for LP can be employed to solve this problem. But since we know that this is a network flow problem, it is possible to speed up the LP algorithm by taking advantage of the special structure of matrix A and the form of the upper and lower bound constraints. This formulation is suitable for Assignment, Maximum Flow and Minimum-Cost Flow problems after some minor transformations are applied to the network, the cost vector c and the node-flow vector b. In Section 2 the algorithm used to solve this problem is described (See [**Vai90**] for a detailed description and rate-of-convergence). One of the time consuming computations in the algorithm is to solve a sparse linear system at each step. We use a Pre-conditioned Conjugate Gradient (PCCG) method to solve the linear system (See [**GL89**]). In Section 3, the construction of the pre-conditioner is described. Section 4 describes additional heuristics such as reduction of problem size by fixing flow values in certain edges, approximate centering, etc. that may be employed to speed up the algorithm in practice. The experimental results are presented in Section 5 and analyzed.

2. An Overview of the Path-Following Algorithm

The network polytope is defined by the upper and lower bound constraints which are the bounding hyperplanes of the polytope. It is denoted as

$$P = \{x : Ax = b,\ l_i \leq x_i \leq u_i\}.$$

The path-following algorithm employed in this implementation follows the trajectory of analytic centers of a family of polytopes defined by the intersection of the network polytope and the objective function hyperplane, *i.e.* the centers of the family of polytopes $P(\beta)$ defined as:

$$P(\beta) = P \cap \{x : c^T x \geq \beta\}.$$

The analytic center of $P(\beta)$ is the minimizer of the strictly convex logarithmic barrier function $F(x, \beta)$ where

$$F(x,\beta) = -(\sum_{i=1}^{e}(\ln(x_i - l_i) + \ln(u_i - x_i)) + w\ln(c^T x - \beta)).$$

We put a large weight w on the objective function hyperplane so that the analytic center is pushed away from the objective function hyperplane $c^T x = \beta$ towards the optimal facet of the network polytope. Specifically the weight w equals the number of edges e times a large multiplier (which is chosen to be $\min\{\sqrt{e}, \sqrt{100,000}\}$).

Clearly, when the flow in any edge equals either the capacity, the lower bound or is such that $c^T x = \beta$ then this function goes to ∞. At any point which is strictly in the interior of the polytope $P(\beta)$ the above function is finite. The path-following algorithm proceeds in several phases. Initially we start with a strictly interior point $x^{(0)}$ which is a good approximation to the analytic center of $P(-\infty)$ and $\beta^{(0)} = -\infty$. The path-following algorithm generates a sequence of parameters $\beta^{(0)}, \beta^{(1)}, \ldots, \beta^{(k-1)}, \beta^{(k)}, \ldots$ whose limit is the maximum value of $c^T x$ over the network polytope P, and also a sequence of points $x^{(0)}, x^{(1)}, \ldots, x^{(k-1)}, x^{(k)}, \ldots$ such that $x^{(k)}$ is a good approximation to the analytic center of $P(\beta^{(k)})$. The final β is very close to the optimal value of $c^T x$ which enables us to isolate the optimal facet of the network polytope together with an exact optimum. As β is varied from $-\infty$ to its maximum possible value in a continuous manner, a continuous trajectory is generated that would start from the analytic center of the network polytope P and terminate at a point that maximizes the objective function $c^T x$ over P. This trajectory is characterized by

$$\Delta \tau + tc = A^T \lambda$$

where $t \in \Re$, $t \geq 0$. $\Delta \in \Re^{e \times e}$ is a diagonal matrix whose ith diagonal entry equals $1/(l_i - x_i) + 1/(u_i - x_i)$ (See [**Vai90, Vai89, Ren88**] for more details). $\tau \in \Re^e$ is a vector of all 1's and $\lambda \in \Re^{v-1}$ is a vector of node potentials. Note that here $\Delta \tau$ can be interpreted as a vector of "reduced costs". In each phase, we find $x^{(k)}$ from $x^{(k-1)}$ by applying Newton's method to the logarithmic barrier function $F(x, \beta^{(k)})$ starting from $x^{(k-1)}$.

A Step of Newton's Method. A step of Newton's method applied to minimizing $F(x, \beta)$ proceeds as follows. Let $\tilde{x}$ be the current point. Let η be the gradient and Q be the Hessian of $F(x, \beta)$ at the current point. We first compute a direction ξ by solving the local optimization problem

$$\max -\eta^T \xi$$

subject to

$$A\xi = 0$$

and

$$\xi^T Q \xi \leq 1,$$

and then (approximately) minimize $F(x, \beta)$ on the line $\tilde{x} + t\xi$. For details see [**Vai89**]. During the line minimization, we express the logarithmic barrier function as a function of a single parameter t and then do a bisection search.

Isolating an Exact Optimum. Since the capacities are integral, the analytic center has the property that any edge flow that is not fixed to either the upper or the lower bound on the optimal facet must be at least $1/(e+1)$ away from both the bounds (See [**Vai90, Vai89, Ren88**] for details). Using this property we adaptively try to isolate an optimal point in the following manner. Suppose that we have a good approximation to the center of the polytope $P(\beta)$ for some β (for a measure of goodness of the approximation see section 4). Then any edge flow that is less than $1/(2e)$ from either the upper (resp. lower) bound is guaranteed to be fixed to the upper (resp. lower) bound at the optimum (See [**Vai90, Vai89, Ren88**] for details). We fix each such edge flow to the closer of the two bounds and obtain a flow that may not satisfy the conservation constraints $Ax = b$.

Let

$$x = \begin{bmatrix} x_1 \\ x_2 \end{bmatrix} \quad l = \begin{bmatrix} l_1 \\ l_2 \end{bmatrix} \quad u = \begin{bmatrix} u_1 \\ u_2 \end{bmatrix} \quad A = \begin{bmatrix} A_1 & A_2 \end{bmatrix}$$

where subscript 1 indicates those edges fixed to appropriate bounds. Let $\tilde{x} = \begin{bmatrix} \tilde{x}_1 \\ \tilde{x}_2 \end{bmatrix}$ denote the solution before any edge flows are fixed to upper or lower bounds. Let $\hat{x}_1$ be the flow vector obtained from $\tilde{x}_1$ after fixing edge flows to the closer of the two bounds. Since the flow vector $\begin{bmatrix} \hat{x}_1 \\ \tilde{x}_2 \end{bmatrix}$ may not satisfy conservation constraints we need to obtain an $\hat{x}_2$ satisfying the conservation constraints

$$A_2 \hat{x}_2 = b - A_1 \hat{x}_1 = \hat{b}_2.$$

We obtain $\hat{x}_2$ from $\tilde{x}_2$ by applying a correction to edge flows in a spanning forest (Though any spanning forest will do, we find a maximum spanning forest with edge weights as defined in section 3, subsection **Reducing to a Full-Dimensional Form**, since these edges have large slacks). Next we check if $\hat{x} = \begin{bmatrix} \hat{x}_1 \\ \hat{x}_2 \end{bmatrix}$ satisfies the capacity and lower bound constraints. If it does then we have a feasible solution $\hat{x}_2$ to the reduced problem

$$\min c_2^T x_2$$

subject to

$$A_2 x_2 = \hat{b}_2$$

$$l_2 \leq x_2 \leq u_2.$$

The correction to $\tilde{x}_2$ may be computed as follows. Let

$$x_2 = \begin{bmatrix} x_{2_f} \\ x_{2_o} \end{bmatrix}$$

where x_{2_f} corresponds to edges in a spanning forest and x_{2_o} corresponds to edges outside the forest (Here the spanning forest is in the network defined by the edges corresponding to x_2). We want to find a correction $\varepsilon = \begin{bmatrix} \varepsilon_f \\ 0 \end{bmatrix}$ such that

$$\hat{x}_2 = \tilde{x}_2 + \varepsilon$$

satisfies conservation constraints. Let

$$c_2 = \begin{bmatrix} c_{2_f} \\ c_{2_o} \end{bmatrix} \quad l_2 = \begin{bmatrix} l_{2_f} \\ l_{2_o} \end{bmatrix} \quad u_2 = \begin{bmatrix} u_{2_f} \\ u_{2_o} \end{bmatrix} \quad A_2 = \begin{bmatrix} A_{2_f} & A_{2_o} \end{bmatrix}.$$

The correction ε may be obtained by solving the linear system

$$A_{2_f}\varepsilon_f = \hat{b}_2 - A_2\tilde{x}_2.$$

This may be done in linear time since A_{2_f} corresponds to a spanning forest.

By further eliminating the variables x_{2_f} (i.e. those in a spanning forest) we can reduce the problem to full-dimensional form:

$$\min \; (c_{2_o} - A_{2_o}^T (A_{2_f}^{-1})^T c_{2_f})^T x_{2_o}$$

subject to

$$l_{2_f} \leq A_{2_f}^{-1}(\hat{b}_2 - A_{2_o} x_{2_o}) \leq u_{2_f}$$

$$l_{2_o} \leq x_{2_o} \leq u_{2_o}.$$

If the reduced cost $c_{2_o} - A_{2_o}^T(A_{2_f}^{-1})^T c_{2_f}$ is zero (we test for numerical zero) then the flow $\hat{x}$ is guaranteed to be optimal.

To summarize, we carry out the following procedure after every centering step in the final phases of the algorithm. Fix all the flows that are within $1/(2e)$ from the upper (resp. lower) bound to the upper (resp. lower) bound. Find a maximum spanning forest in the network with the edges corresponding to these fixed variables removed. Apply a correction to the variables in this spanning forest (as described above) and check if the resulting flow is feasible. If this flow is feasible, we have a facet, namely, the one where the variables in x_1 are fixed, which is a candidate for being the optimal facet. If the reduced cost function $c_{2_o} - A_{2_o}^T(A_{2_f}^{-1})^T c_{2_f}$ is zero then this facet is declared to be the optimal and the algorithm stops. Otherwise the algorithm continues. Note that the algorithm always halts with an exact optimum. Specifically, this procedure allows us to isolate the entire optimal facet along with a good approximation to the center of the optimal facet.

If an optimal vertex is desired, the cost vector may be modified by adding a "small" random perturbation. An optimal solution to this modified problem is also optimal for the original problem, and with high probability the optimal facet of the modified problem is a single vertex (See [**Cha52**]). As this procedure requires too many bits of precision in the perturbed cost vector, it was not implemented. Instead our algorithm isolates a good approximation to the analytic center of the optimal facet.

Getting an Initial Feasible Point. To get an initial interior feasible point, the flow in each edge is set midway between the upper and lower bound. Since such a flow may violate the conservation constraints, we introduce extra edges to a distinguished ground node to carry the excess or deficiency from those nodes where the conservation constraint is not satisfied. We put a high cost on each of these edges to ensure that at an optimum the flows in these edges are zero provided the original problem is feasible. We use a couple of simple heuristics to estimate the cost on these newly introduced edges; for example, letting the cost of a new edge be greater than the cost of a Maximum Spanning Tree (MST) of (absolute) costs suffices.

3. Solving the Linear System at Each Step

During each step of the Newton's Method (while computing the center), we have to find a direction ξ which is the solution to the local optimization problem

$$\max -\eta^T \xi$$

subject to

$$A\xi = 0$$

and

$$\xi^T Q \xi \leq 1,$$

Matrix Q is the Hessian of the logarithmic barrier function $F(x, \beta)$ and can be expressed as $D + \hat{w}cc^T$ where D is a diagonal matrix whose ith diagonal entry is

$$D_{ii} = \frac{1}{(x_i - l_i)^2} + \frac{1}{(u_i - x_i)^2}$$

and $\hat{w} = w/(c^T x - \beta)^2$ where w is a problem dependent multiplier. The above problem can be reduced to full-dimensional form by eliminating the flow variables in a spanning tree using the conservation constraints $A\xi = 0$.

Reducing to a Full-Dimensional Form. Let $\tilde{x}$ be the current point and $\tilde{D}$ the matrix D evaluated at $\tilde{x}$. Let the weight of edge i with respect to the current point $\tilde{x}$ be defined as $D_{ii}^{-\frac{1}{2}}$. We reduce the above problem to full-dimensional form by eliminating the flow variables corresponding to the edges in a specific spanning tree, namely a Maximum Spanning Tree(MST) with edge weights defined above. Intuitively the MST edges are those edges that have the "largest" slacks for changing the flow. We split the variables of ξ into two blocks - those belonging to the MST, denoted by ξ_z, and those that do not, denoted by ξ_y. Let

$$\xi = \begin{bmatrix} \xi_z \\ \xi_y \end{bmatrix}$$

$$A = \begin{bmatrix} A_z & A_y \end{bmatrix}$$

$$D = \begin{bmatrix} D_z & 0 \\ 0 & D_y \end{bmatrix}$$

$$c = \begin{bmatrix} c_z \\ c_y \end{bmatrix}$$

be the block representation of the vectors and matrices involved in the above local optimization problem, where subscript z (y) represents the columns/rows corresponding to the MST (non-MST) variables. Let $B = -A_z^{-1} A_y$. Eliminating the MST variables reduces the above local optimization problem to

$$\max -\hat{\eta}^T \xi_y$$

subject to

(1) $$\xi_y^T \hat{Q} \xi_y \leq 1 ,$$

where $\hat{\eta} = B^T \eta_z + \eta_y$, $\hat{Q} = B^T D_z B + D_y + \hat{w}\hat{c}\hat{c}^T$, and $\hat{c} = B^T c_z + c_y$.

The direction ξ_y may be obtained by solving the linear system

(2) $$\hat{Q} \xi_y = -\hat{\eta}$$

and then ξ_z may be computed as $\xi_z = B \xi_y$. We solve the linear system (2) by a Pre-conditioned Conjugate Gradient (PCCG) method (See [**GL89**]) with $M = D_y + \hat{w}\hat{c}\hat{c}^T$ as the pre-conditioner. Note that $\hat{Q}$ may be expressed as

$$\hat{Q} = M + B^T D_z B$$

Since we eliminated variables ξ_z corresponding to an MST, it is guaranteed that the absolute value of any entry in $D_z^{\frac{1}{2}} B D_y^{-\frac{1}{2}}$ is bounded by 1 and thereby the trace of $M^{-\frac{1}{2}} B^T D_z B M^{-\frac{1}{2}}$ is bounded by ve. The bound on any entry in $D_z^{\frac{1}{2}} B D_y^{-\frac{1}{2}} = D_z^{\frac{1}{2}} A_z^{-1} A_y D_y^{-\frac{1}{2}}$ may be obtained as follows:

Consider a column a_y of A_y corresponding to a non-tree edge e. Let Π be the unique path in the MST between the end points of e. Note that a numbering of the columns of A_z naturally induces a numbering of the edges in the MST. The ith entry in $A_z^{-1} a_y$ is ± 1 if the ith edge in the MST lies on the path Π; otherwise the entry is zero. So a non-zero entry in $D_z^{\frac{1}{2}} A_z^{-1} a_y D_y^{-\frac{1}{2}}$ is of the form w_e/w where w_e is the weight of edge e and w is the weight of some edge on path Π. It then follows that the absolute value of any entry in $D_z^{\frac{1}{2}} A_z^{-1} A_y D_y^{-\frac{1}{2}}$ is bounded by 1 (See [**Vai91**] for details).

The above argument ensures that M is a good pre-conditioner for $\hat{Q}$. As we approach a vertex solution, entries in $D_z^{\frac{1}{2}}$ remain bounded because they correspond to edges with non-zero slack whereas entries in $D_y^{-\frac{1}{2}}$ approach zero. Hence, the closer we are to a vertex solution, the better is the pre-conditioner.

In our experiments, it has been observed that the PCCG with this pre-conditioner M halts in about 60-70 iterations on the average (with a solution of desired accuracy) for problems with up to 250,000 edges. Note that solving a linear system $Mu = r$ can be trivially accomplished in $O(e)$ operations, and we never need to form the matrix $\hat{Q}$ (or any other matrix) explicitly. Moreover, multiplication of $\hat{Q}$, A or A^T by a vector may be performed in $O(e)$ operations, and $A_z u = r$ may be solved in $O(v)$ operations, using a standard graph adjacency list data structure.

MST Computation. For the sake of efficiency, we round off the edge weights to multiples of $(1+\delta)$ for a suitable δ and use Bucket Sorting coupled with Kruskal's MST algorithm (See [**TCR90**]). Using such an approximate MST - instead of an exact MST - only slightly degrades the quality of the pre-conditioner M.

Reducing the Constants in PCCG. All the intermediate vectors generated in our implementation of PCCG lie in a Krylov Space of dimension at most $v+1$ rather than e. This may be seen as follows:

Note that $\hat{Q}$ may be written as $\hat{Q} = D_y + UU^T$ where $U = [\hat{w}^{\frac{1}{2}}c \quad B^T D_z^{\frac{1}{2}}]$, and $U \in \Re^{e \times v}$. Let $\tilde{Q} = D_y^{-\frac{1}{2}} \hat{Q} D_y^{-\frac{1}{2}} = I + \tilde{U}\tilde{U}^T$ where $\tilde{U} = D_y^{-\frac{1}{2}} U$. Instead of directly solving $\hat{Q}\xi_y = -\hat{\eta}$ we first compute $\tilde{\xi}$ by solving

$$\tilde{Q}\tilde{\xi} = -D_y^{-\frac{1}{2}}\hat{\eta} = \tilde{\eta}$$

and then compute $\hat{\xi}$ as $\hat{\xi} = D_y^{\frac{1}{2}}\tilde{\xi}$. During the solution of $\tilde{Q}\tilde{\xi} = \tilde{\eta}$ by PCCG successive vectors are generated as linear combinations of previous vectors or by multiplying previous vectors by M^{-1} or $\tilde{Q}$ (See [**GL89**]). The initial vector is $\tilde{\eta}$. Note that if $x \in span\{\tilde{\eta}, \tilde{u}_1, \ldots, \tilde{u}_v\}$ where $\tilde{u}_1, \ldots, \tilde{u}_v$ are the columns of $\tilde{U}$ then both $M^{-1}x$ and $\tilde{Q}x$ lie in the span of $\{\tilde{\eta}, \tilde{u}_1, \ldots, \tilde{u}_v\}$. As a result all the vectors generated during PCCG lie in the Krylov space given by $span\{\tilde{\eta}, \tilde{u}_1, \ldots, \tilde{u}_v\}$ and may be implicitly represented by $v+1$ coefficients.

Thus most of the intermediate calculations may be implicitly performed using vectors of size $v+1$; this leads to reduced constants in the running time as well as storage and to an improvement in practical performance.

4. Some Techniques to Improve Speed and Stability

In our implementation we use several techniques to gain additional speed and improve the stability of the algorithm. We describe them in this section.

Approximate Centering. In the initial phases of the algorithm, it suffices to compute a very coarse analytic center. A good approximation to the center is required only when the algorithm is getting ready to isolate the optimal facet. So in the initial phases we find a very coarse approximation to the center which is sufficiently far from the objective function hyperplane, and eventually switch to an almost exact centering when the width of the polytope in the direction of the cost vector falls below a certain threshold. Closeness to the center is measured by the quantity $\hat{\eta}^T \hat{Q}^{-1} \hat{\eta}$. Initially a coarse approximation x to the center such that $\hat{\eta}^T \hat{Q}^{-1} \hat{\eta} \leq \epsilon e$ (ϵ a small constant) suffices. Later on when we are close to isolating the optimal facet, we switch to a finer approximation x satisfying the condition $\hat{\eta}^T \hat{Q}^{-1} \hat{\eta} \leq 0.5$. Note that for a point x close to the center ω, $\hat{\eta}^T \hat{Q}^{-1} \hat{\eta} \approx 2(F(x, \beta) - F(\omega, \beta))$ (See [**Vai89, Ren88, Kar84**] for theoretical results). The empirical data suggests that in practice the complexity of the two centerings is much better than the theoretical bounds.

Variable PCCG Stopping Threshold. The farther we are from the analytic center the larger may be the error in the Newton direction ξ without degrading the rate of convergence to the center. As a result, stopping (error) threshold for the PCCG can be coarser in the beginning when we are far from the center and may be progressively tightened as we approach the center. The PCCG as implemented in our algorithm stops when

$$r^T M^{-1} r \leq threshold$$

where r is the current residue and M is the preconditioner as defined earlier. When we are not too close to the center i.e. $\hat{\eta}^T \hat{Q}^{-1} \hat{\eta} \geq 0.1 w$ where w is the weight on the objective function plane, the *threshold* is set to $\max\{10^{-4} w, 0.1\}$. When we are close to the center i.e. $\hat{\eta}^T \hat{Q}^{-1} \hat{\eta} < 0.1 w$, the *threshold* is set to 0.1.

Reducing the Problem Size. Since the capacities are integral, the center has the property that any edge flow that is not fixed to either the upper or the lower bound on the optimal facet must be at least $1/(e+1)$ away from both the bounds. This may be argued as follows:

At the analytic center ω of the polytope $P(\beta) = P \cap \{x : c^T x \geq \beta\}$ we have that

$$\sum_{i=1}^{e} \left(\frac{u_i - x_i}{u_i - \omega_i} + \frac{x_i - l_i}{\omega_i - l_i} \right) + w \frac{c^T x - \beta}{c^T \omega - \beta} = w + 2e$$

(See [**Vai90, Ren88**] for details). At any optimum point $x^\star$, we have that $c^T x^\star - \beta \geq c^T \omega - \beta$. Thus

$$\sum_{i=1}^{e} \left(\frac{u_i - x_i^\star}{u_i - \omega_i} + \frac{x_i^\star - l_i}{\omega_i - l_i} \right) \leq 2e$$

Also, each term in the summation on the left hand side in the above inequality is at least 1. Thus for each i and each optimal point $x^\star$ we have

(3)
$$\frac{u_i - x_i^\star}{u_i - \omega_i} + \frac{x_i^\star - l_i}{\omega_i - l_i} \leq e + 1$$

Furthermore, for any coordinate x_i either x_i is fixed to l_i or u_i on the optimal facet or the width of the optimal facet in the direction of x_i is at least 1 since vertices have integer coordinates. Suppose x_i is not fixed on the optimal facet.

Then there exist optimal points x' and x'' such that $u_i - x'_i \geq 1$ and $x''_i - l_i \geq 1$. Then from inequality (3) it follows that $u_i - \omega_i \geq 1/(e+1)$ and $\omega_i - l_i \geq 1/(e+1)$.

Using this property, we fix those edge flows which at the current approximate center are closer than $1/(2e)$ to a bound to that bound. For these edges the optimal flow value is the bound it is fixed to. The edges that are fixed may be removed from further consideration in the algorithm.

Dynamic Shift of Origin for Stability. If either the upper or the lower bound of an edge is large in magnitude there is a chance of numerical instability as the edge flow approaches a bound in the final stages of the algorithm. A typical machine has about 15 to 16 decimal digits of precision. However, if we have a capacity of 10^8 we effectively could loose half the digits of precision. For example, suppose the current flow in an edge is 10^{-8} less than the capacity which is 10^8 and the machine has 16 decimal digits of precision. Now if the flow moves further towards the capacity, the flow value would be set to the capacity as a result of numerical error thereby leading to a division by zero in the intermediate calculations since the algorithm assumes a strictly interior point. To remove instabilities of this kind, we apply a dynamic shift-of-origin changing those bounds on edges which have a flow close to them to zero and working with the offset of the flow on such edges. At the end of the algorithm, the origin is shifted back to get the correct flows. As a result, an edge flow can approach a bound to essentially within machine precision. Note that even though the analytic centers generated during the algorithm cannot be too close to a boundary, the intermediate points generated in moving from one center to the next get quite close to some of the boundaries. It is because of this that dynamic shift of origin is required for large problems.

5. Experimental Results and Analysis

MINFLOW, our implementation for minimum cost flow, was tested on several problems using a Sun Sparc-2 workstation running SunOS 4.1.1 with FPU version 3. It was experimentally observed that vector inner product and saxpy operations ran at about 0.7 to 1.0 MEGAFLOPS. In order to estimate the speed of pointer manipulations several examples of standard breadth-first-search using adjacency list representation were run, and the time for breadth-first-search was estimated to be about 3 microseconds per edge. All maximum flow problems were converted to minimum cost flow problems and then solved using MINFLOW. The following tables and graphs show the experimental data along with timing and iteration counts. I/O times are not included in the timing data. The time and iteration counts to find the optimal solution are averaged over the number of instances of each size that was run.

Seeds were selected according to the following rule for all families reported except mesh-standard-one, mesh-standard-two, mesh-standard-four and mesh-standard-eight. Suppose the total number of instances for a given size is I.

Then the seed for instance i, $1 \leq i \leq I$, is given by the formula $5 + (i-1)*2$. For families mesh-standard-one, mesh-standard-two, mesh-standard-four and mesh-standard-eight the formula is $1 + (i-1)*2$.

One of the referees pointed out that NETFLO uses a slow insertion sort while reading in the data and suggested that we run bench marks to see how this affects the comparisons between MINFLOW and NETFLO. From each family we selected two instances of moderate size and timed the initial insertion sort in the NETFLO code. It was found that for families capt-hi-bits, capt-hi-bits-mod, capt-lo-bits and capt-lo-bits-mod the percentage time spent in the initial insertion sort is in the range 60%–64%. For washing-line-moderate this percentage is 47%. For acyclic-dense, genrmf-long and grid-long this initial time is moderately small (in the range of 13%–25%). For the remaining eleven families, this time is negligible (below 8%). This fact should be taken into consideration while examining the timing data for NETFLO.

The data for each input class consists of:
- A table which gives the absolute timing statistics for MINFLOW, NETFLO and RELAXT III along with the standard deviation.
- A table which gives the iteration count statistics for MINFLOW, NETFLO and RELAXT III along with the standard deviation. Each iteration for MINFLOW is a PCCG iteration in which the work done is proportional to the number of edges. A NETFLO iteration consists of moving from one vertex to to another in the network simplex algorithm [**KH80**]. A RELAXT III iteration corresponds to a relaxation step as described in [**Ber91, BT90, BT88a, BT88b**].
- Graphs for comparison with NETFLO and RELAXT III codes. The average of the running time ratios are plotted against problem size represented by the number of nodes on a log-log scale. This is done because for small problems, both NETFLO and RELAXT III are typically much faster than MINFLOW but as the problem size increases the situation reverses for several input classes. For input sizes where the graph is above the abscissa of 1.0, MINFLOW is slower compared to the other code (NETFLO or RELAXT III) and in the region where the graph is less than the abscissa of 1.0, MINFLOW is faster. Note that we were unable to obtain some of the data points for comparison with RELAXT III because the RELAXT III code failed to solve those problems.

Note that we have shortened RELAXT III to RELAXT in the tables and graphs.

One of the referees also suggested modifying the parameters to the CAPT generator so that the number of sinks is roughly $N/2$ where N is the number of vertices minus two. So the CAPT families were run with two types of parameter settings: one, as per DIMACS recommendations, the results of which are given under capt-lo-bits and capt-hi-bits; and two, as per the referee's recommendations, the results of which are summarized under capt-lo-bits-mod and capt-hi-bits-mod. In particular, the parameters *flow* and *intlen* were chosen as *flow*= 2∗*intlen*= 2000 for capt-lo-bits-mod and capt-hi-bits-mod.

We have included the results for the maximum flow problem families for comparison purposes only. So, while examining the statistics for families acyclic-dense, washing-line-moderate, genrmf-long and genrmf-wide, which are maximum flow problems converted to mincost flow problems, the fact that *the simplex algorithm is known to perform poorly on mincost flow problems derived from maximum flow problems* should be borne in mind. Furthermore, inclusion of these results indicate that Interior Point methods work well uniformly irrespective of the origins of the problem.

The smaller problems give trends for the comparative growth rates for the three codes. Since it was too time consuming to run many instances for the large problem sizes, we typically ran 1 or 2 instances only. The main memory in our machine was sufficient to run problems with a little over a 100,000 arcs. For all input classes we ran as big an instance as would fit in virtual memory without thrashing. As a result only a few data points were obtained for input classes acyclic-dense, capt-lo-bits, capt-lo-bits-mod, capt-hi-bits and capt-hi-bits-mod for which the number of arcs grew quadratically with the number of nodes.

A cursory analysis indicates a growth rate of about $O(e^{1.5})$ for MINFLOW on all the problems.

For input families washing-line-moderate, genrmf-long, genrmf-wide, grid-den-sixteen, grid-inc-den, grid-square, and grid-wide, the graphs indicate that MINFLOW has a slower growth rate than both NETFLO and RELAXT III with increasing problem size.

For grid-den-eight, MINFLOW is faster than NETFLO and RELAXT III, even for small problem sizes.

For acyclic-dense and grid-long, MINFLOW has faster growth rate than both NETFLO and RELAXT III.

For capt-lo-bits, capt-lo-bits-mod, capt-hi-bits and capt-hi-bits-mod only four data points for each family have been obtained due to memory restrictions. The trend seems to be that MINFLOW has lower growth rate than NETFLO but higher than RELAXT III for these families.

For mesh-standard-one, mesh-standard-two, mesh-standard-four and mesh-standard-eight MINFLOW has lower growth rate than NETFLO but seems to have the same growth rate as RELAXT III.

6. Conclusions

The implementation is quite robust, numerically stable, and fairly insensitive to variations in costs and capacities. Furthermore, we can handle problems with non-integer (real) costs since the stopping criterion does not depend on integrality of costs.

Size			Avg. Time (seconds)			Std. Dev. Time (seconds)		
Nodes	Arcs	#	MINFLOW	NETFLO	RELAXT	MINFLOW	NETFLO	RELAXT
64	2016	13	19.2	0.8	0.7	1.0	0.0	0.1
128	8128	13	141.0	3.7	2.8	9.5	0.1	0.7
256	32640	13	1049.1	21.0	21.3	28.3	0.7	14.7

Size			Iterations			Iterations (Std.dev)		
Nodes	Arcs	#	MINFLOW	NETFLO	RELAXT	MINFLOW	NETFLO	RELAXT
64	2016	13	349.9	794.7	103.5	15.8	93.0	46.4
128	8128	13	812.8	2761.5	177.8	63.8	353.9	88.0
256	32640	13	2002.8	4861.9	457.5	53.7	669.5	226.6

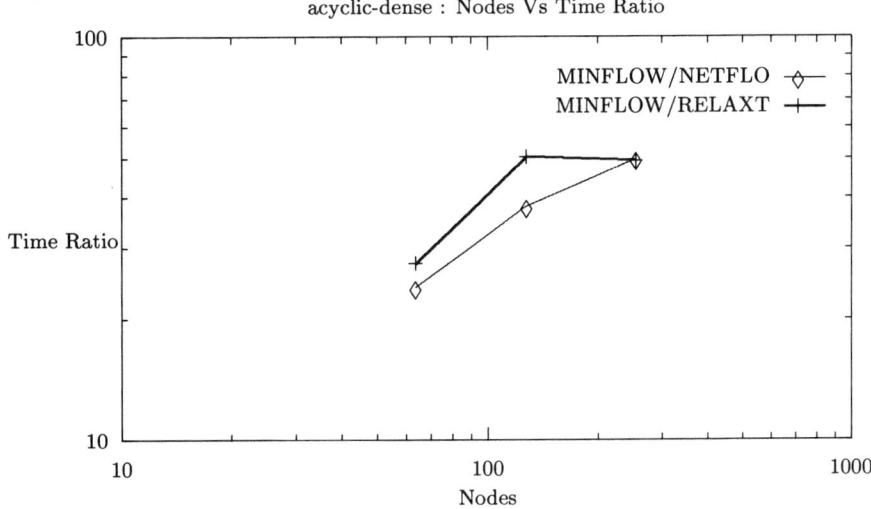

FIGURE 1. Statistics for family **acyclic-dense**

Size			Avg. Time (seconds)			Std. Dev. Time (seconds)		
Nodes	Arcs	#	MINFLOW	NETFLO	RELAXT	MINFLOW	NETFLO	RELAXT
66	127	10	1.7	0.1	0.0	0.2	0.0	0.0
130	373	10	5.3	0.3	0.2	0.4	0.0	0.0
258	1001	10	16.4	1.0	0.9	1.3	0.1	0.2
514	3006	10	67.1	5.4	4.5	15.4	0.3	0.7
1026	8071	10	226.8	27.7	23.6	37.3	1.2	3.3
2050	24276	10	708.8	193.7	136.2	90.5	6.4	20.0
4098	65002	1	3464.6	1088.6	621.7	0.0	0.0	0.0

Size			Iterations			Iterations (Std.dev)		
Nodes	Arcs	#	MINFLOW	NETFLO	RELAXT	MINFLOW	NETFLO	RELAXT
66	127	10	124.6	86.4	25.2	19.2	9.0	7.7
130	373	10	209.4	261.4	76.7	24.4	37.3	24.0
258	1001	10	338.5	580.6	168.9	29.0	108.1	54.6
514	3006	10	627.8	1282.2	357.5	145.4	134.4	69.4
1026	8071	10	1034.3	2696.7	747.1	178.9	165.4	113.3
2050	24276	10	1412.6	5712.0	1538.2	181.9	476.3	232.2
4098	65002	1	3119.0	11063.0	2820.0	0.0	0.0	0.0

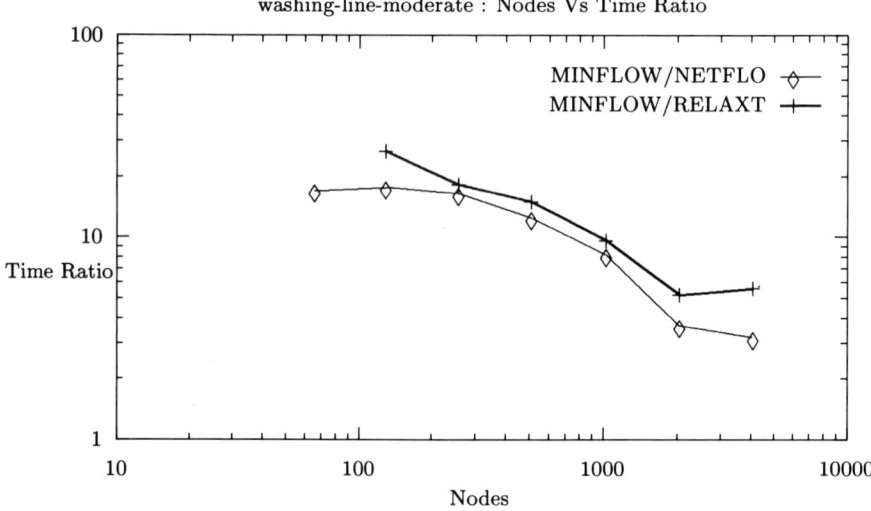

FIGURE 2. Statistics for family **washing-line-moderate**

Size			Avg. Time (seconds)			Std. Dev. Time (seconds)		
Nodes	Arcs	#	MINFLOW	NETFLO	RELAXT	MINFLOW	NETFLO	RELAXT
72	255	10	1.8	0.1	0.1	0.1	0.0	0.0
192	752	10	7.0	0.4	0.5	0.6	0.0	0.0
256	1008	10	10.5	0.6	0.9	0.7	0.0	0.0
575	2390	10	30.7	3.0	4.5	0.7	0.1	0.5
1152	4956	10	80.5	10.8	19.8	3.9	0.3	1.0
2254	9933	10	182.2	41.2	80.7	6.8	1.5	3.6
4096	18368	10	406.7	120.9	276.0	17.0	5.5	24.7
9100	41760	10	1260.0	628.8	1500.2	59.3	33.3	123.6
18432	85872	10	6858.3	2539.0	6958.9	1427.3	65.5	293.3

Size			Iterations			Iterations (Std.dev)		
Nodes	Arcs	#	MINFLOW	NETFLO	RELAXT	MINFLOW	NETFLO	RELAXT
72	255	10	83.6	126.5	41.9	5.5	6.2	7.1
192	752	10	119.0	387.3	123.1	8.2	21.3	13.9
256	1008	10	124.0	496.5	149.8	7.8	21.1	14.3
575	2390	10	165.3	1259.4	394.2	7.8	42.6	41.7
1152	4956	10	225.1	2684.5	857.4	20.9	83.6	54.1
2254	9933	10	250.9	5397.5	1871.4	15.6	159.0	85.8
4096	18368	10	298.5	9641.5	3374.9	20.8	298.7	301.0
9100	41760	10	392.0	22591.8	7763.6	30.8	846.4	590.1
18432	85872	10	730.4	48517.3	16825.2	146.9	862.1	682.3

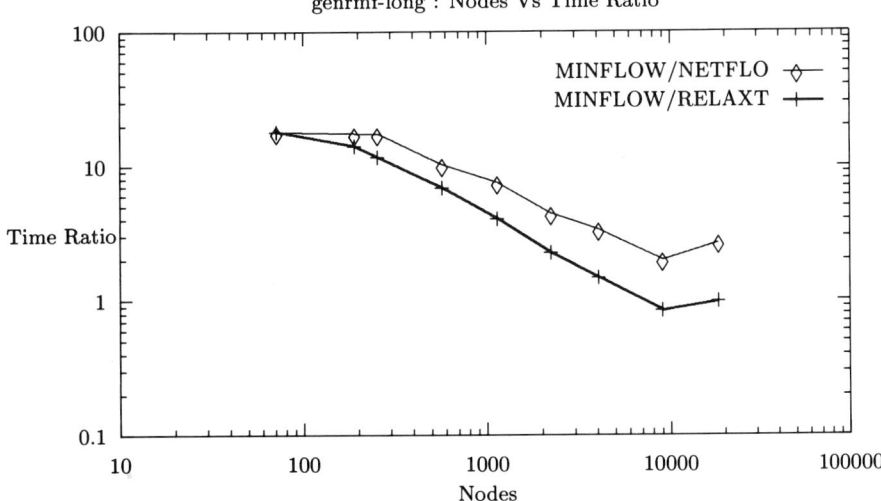

FIGURE 3. Statistics for family **genrmf-long**

Size			Avg. Time (seconds)			Std. Dev. Time (seconds)		
Nodes	Arcs	#	MINFLOW	NETFLO	RELAXT	MINFLOW	NETFLO	RELAXT
72	276	10	2.1	0.1	0.1	0.2	0.0	0.0
147	602	10	5.9	0.3	0.4	0.5	0.0	0.0
300	1280	10	17.0	0.8	1.2	0.9	0.0	0.1
676	3003	10	54.5	3.3	7.0	1.0	0.1	0.5
1024	4608	10	96.5	5.3	16.6	2.1	0.2	1.0
1936	8844	10	243.2	15.4	58.5	12.7	0.5	0.0
3920	18256	10	623.1	59.7		16.2	2.4	
16807	80262	1	10468.2	1308.9		0.0	0.0	

Size			Iterations			Iterations (Std.dev)		
Nodes	Arcs	#	MINFLOW	NETFLO	RELAXT	MINFLOW	NETFLO	RELAXT
72	276	10	124.2	120.4	38.2	9.8	2.2	1.5
147	602	10	179.2	281.0	137.8	12.8	3.7	25.2
300	1280	10	270.6	586.6	330.3	12.9	24.3	22.6
676	3003	10	390.4	1820.8	740.8	8.5	91.9	63.2
1024	4608	10	498.3	2909.8	1142.3	16.5	115.9	56.5
1936	8844	10	708.3	5284.7	2206.0	48.5	208.0	0.0
3920	18256	10	917.2	11785.8		32.9	419.2	
16807	80262	1	3959.0	78634.0		0.0	0.0	

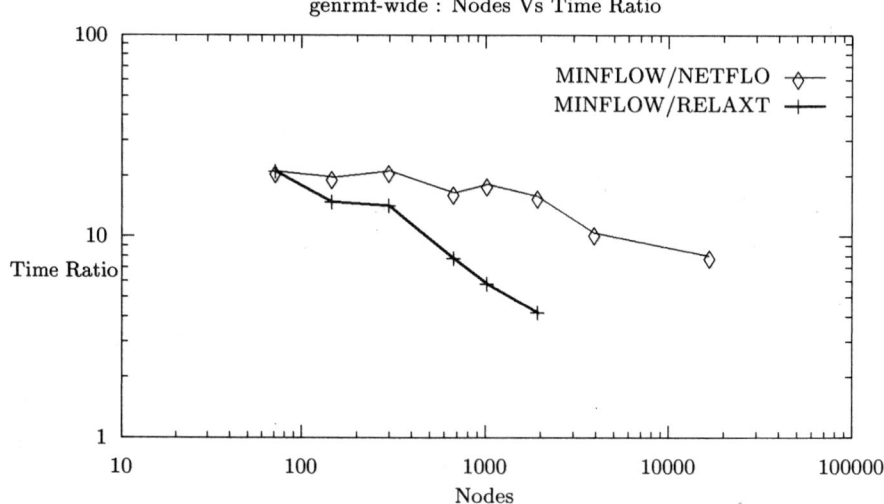

FIGURE 4. Statistics for family **genrmf-wide**

Size			Avg. Time (seconds)			Std. Dev. Time (seconds)		
Nodes	Arcs	#	MINFLOW	NETFLO	RELAXT	MINFLOW	NETFLO	RELAXT
256	2048	10	26.5	2.8	18.2	1.7	0.1	8.8
512	4096	10	56.1	9.9	75.2	1.8	0.4	11.1
1024	8192	10	148.1	34.5	313.4	12.5	0.9	47.1
2048	16384	10	338.5	134.7	1215.2	25.9	7.5	116.2
4096	32768	10	707.2	716.0	4658.0	17.5	27.7	619.1
8192	65536	2	1856.6	2504.3	25876.0	23.7	59.1	1063.1
16384	131072	1	4139.1	10068.9	102480.5	0.0	0.0	0.0
32768	262144	1	16566.1			0.0		

Size			Iterations			Iterations (Std.dev)		
Nodes	Arcs	#	MINFLOW	NETFLO	RELAXT	MINFLOW	NETFLO	RELAXT
256	2048	10	150.2	5878.6	53197.5	20.0	205.7	21464.5
512	4096	10	137.5	15952.7	170049.1	14.2	736.8	26027.1
1024	8192	10	199.8	29290.1	521763.9	18.2	537.4	73833.2
2048	16384	10	235.7	68639.6	1353897.0	40.0	2465.5	131036.8
4096	32768	10	267.4	373951.4	2154553.1	20.5	11898.2	521322.2
8192	65536	2	252.5	331879.0	10645328.0	14.5	2165.0	232605.0
16384	131072	1	354.0	745617.0	29090153.0	0.0	0.0	0.0
32768	262144	1	1576.0			0.0		

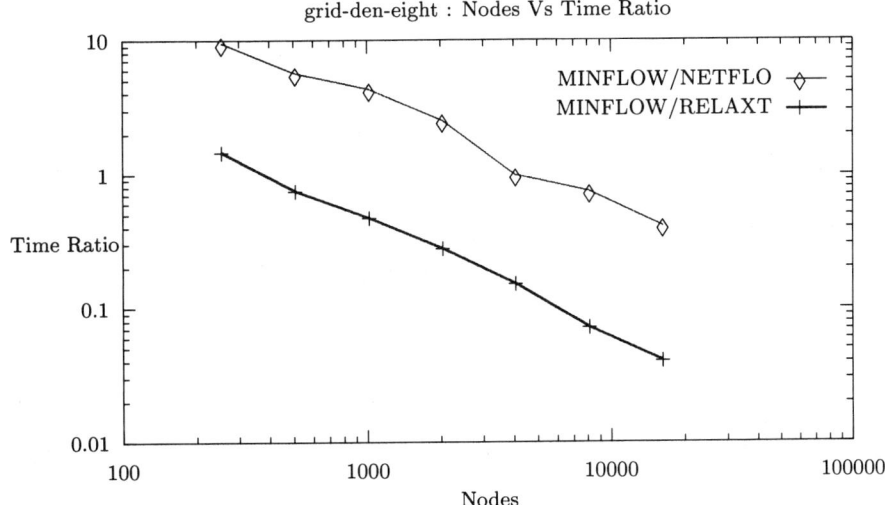

FIGURE 5. Statistics for family **grid-den-eight**

Size			Avg. Time (seconds)			Std. Dev. Time (seconds)		
Nodes	Arcs	#	MINFLOW	NETFLO	RELAXT	MINFLOW	NETFLO	RELAXT
256	4096	8	55.5	5.6	97.7	1.6	0.2	9.5
512	8192	10	132.5	18.7	328.8	7.5	0.3	39.3
1024	16384	10	317.7	74.6	1232.1	16.9	1.6	99.6
2048	32768	10	759.7	287.8	4041.4	20.6	6.0	265.7
4096	65536	10	1666.5	1250.7	15730.9	84.5	38.7	1358.1
8192	131072	2	5517.7	4998.6	58637.1	218.2	63.1	2081.5
16384	262144	1	13211.3	20026.6		0.0	0.0	

Size			Iterations			Iterations (Std.dev)		
Nodes	Arcs	#	MINFLOW	NETFLO	RELAXT	MINFLOW	NETFLO	RELAXT
256	4096	8	202.7	14467.7	165768.2	16.8	701.0	16869.1
512	8192	10	251.9	32430.8	413513.6	17.0	455.5	46136.0
1024	16384	10	347.1	75449.5	1110089.5	16.2	1416.7	129243.0
2048	32768	10	431.7	177102.0	2585272.4	12.8	3244.6	229859.9
4096	65536	10	470.1	459134.6	6166728.4	38.7	13582.9	820072.0
8192	131072	2	697.0	842259.0	14155666.5	4.0	6441.0	577765.5
16384	262144	1	744.0	1949450.0		0.0	0.0	

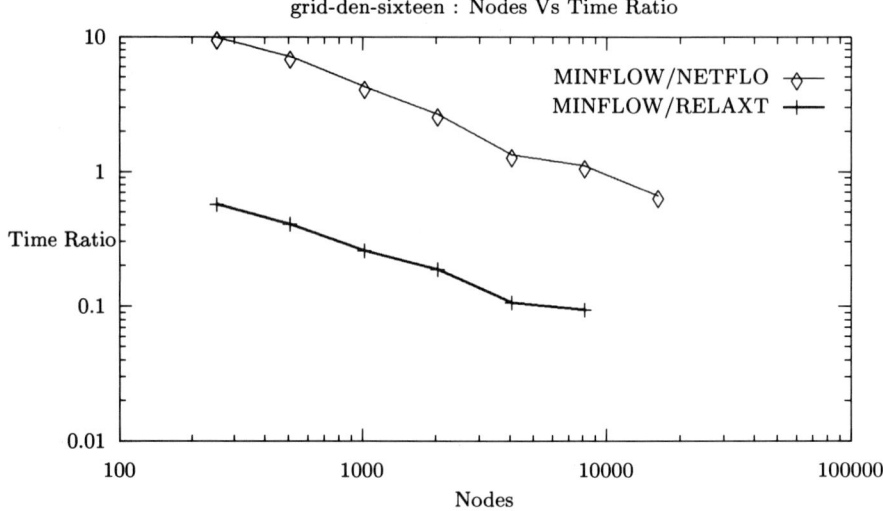

FIGURE 6. Statistics for family **grid-den-sixteen**

Size			Avg. Time (seconds)			Std. Dev. Time (seconds)		
Nodes	Arcs	#	MINFLOW	NETFLO	RELAXT	MINFLOW	NETFLO	RELAXT
256	4096	10	54.9	5.6	97.3	1.8	0.2	9.3
512	11586	10	199.6	29.9	485.1	10.6	0.6	63.6
1024	32768	10	720.0	183.5	2537.8	23.0	3.9	199.1
2048	92682	10	3005.6	1297.4	12802.8	118.2	20.9	457.9
4096	262144	1	12669.0	12088.7	40994.6	0.0	0.0	0.0

Size			Iterations			Iterations (Std.dev)		
Nodes	Arcs	#	MINFLOW	NETFLO	RELAXT	MINFLOW	NETFLO	RELAXT
256	4096	10	201.9	14467.7	165768.2	16.3	701.0	16869.1
512	11586	10	312.4	47781.6	454180.0	20.8	742.0	81517.8
1024	32768	10	527.8	186138.8	1387404.4	21.1	3880.1	165211.7
2048	92682	10	1046.4	637883.4	3666692.0	37.4	34136.0	221686.4
4096	262144	1	1248.0	2514153.0	8293935.0	0.0	0.0	0.0

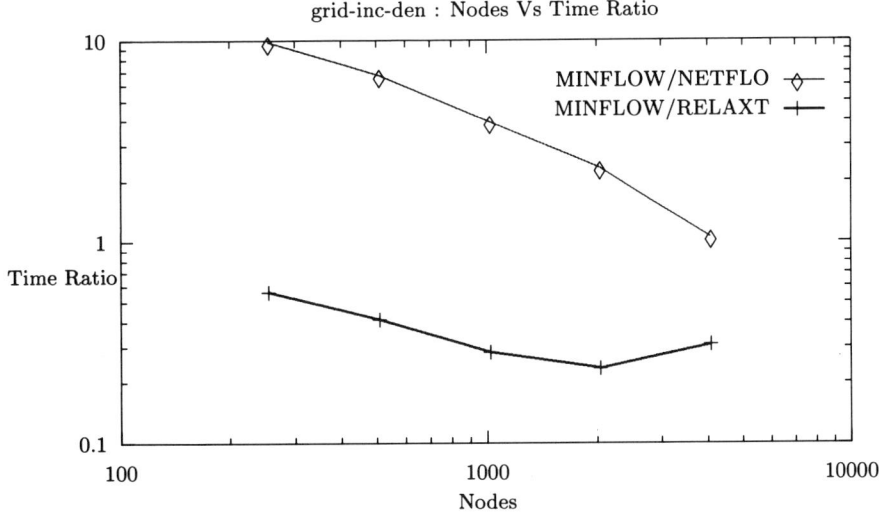

FIGURE 7. Statistics for family **grid-inc-den**

Size			Avg. Time (seconds)			Std. Dev. Time (seconds)		
Nodes	Arcs	#	MINFLOW	NETFLO	RELAXT	MINFLOW	NETFLO	RELAXT
258	512	10	16.5	0.2	0.4	2.4	0.0	0.1
531	1058	10	43.3	0.5	1.3	4.7	0.0	0.1
1026	2048	10	97.5	1.3	4.1	34.3	0.1	0.5
2118	4232	10	280.1	3.4	17.4	20.0	0.3	2.1
4098	8192	10	779.2	9.8	67.8	94.3	0.7	12.6
8283	16562	2	2567.6	36.2	315.5	261.5	4.6	12.0
16386	32768	1	6938.3	118.0	1244.4	0.0	0.0	0.0
33126	66248	1	21711.0	558.9	5608.4	0.0	0.0	0.0

Size			Iterations			Iterations (Std.dev)		
Nodes	Arcs	#	MINFLOW	NETFLO	RELAXT	MINFLOW	NETFLO	RELAXT
258	512	10	309.7	346.5	1357.5	37.0	43.1	334.7
531	1058	10	433.0	826.9	3353.7	47.8	79.2	570.0
1026	2048	10	514.4	1968.3	7620.6	179.0	123.9	1323.2
2118	4232	10	814.9	4916.6	13886.0	46.6	297.1	4367.4
4098	8192	10	1221.3	11509.7	29922.8	134.2	625.8	11600.2
8283	16562	2	2112.5	30226.0	136411.5	208.5	1814.0	3973.5
16386	32768	1	2831.0	70167.0	377766.0	0.0	0.0	0.0
33126	66248	1	4597.0	197148.0	1020525.0	0.0	0.0	0.0

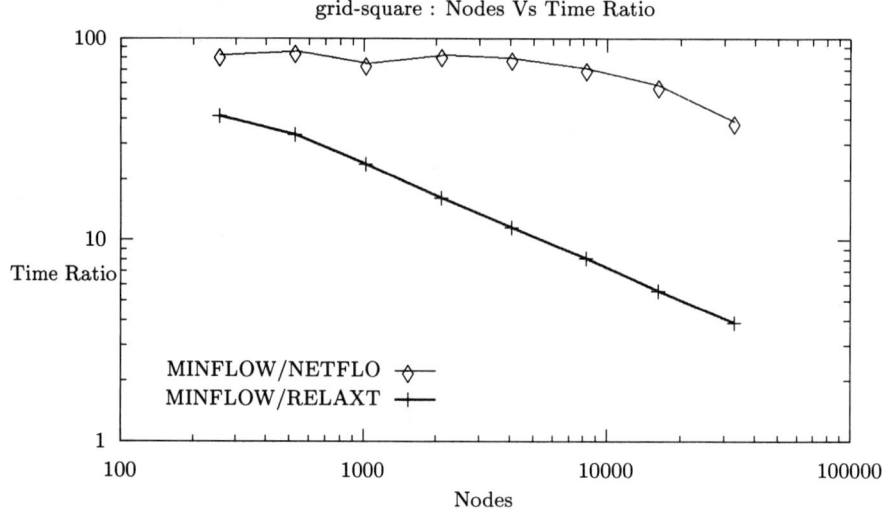

FIGURE 8. Statistics for family **grid-square**

Size			Avg. Time (seconds)			Std. Dev. Time (seconds)		
Nodes	Arcs	#	MINFLOW	NETFLO	RELAXT	MINFLOW	NETFLO	RELAXT
258	512	10	16.6	0.2	0.4	2.3	0.0	0.1
514	1008	10	43.6	0.5	1.0	5.0	0.0	0.2
1026	2000	10	132.3	1.0	2.5	5.8	0.0	0.2
2050	3984	10	364.4	2.2	6.6	43.6	0.1	1.1
4098	7952	10	978.6	5.1	19.7	42.2	0.4	2.9
8194	15888	2	3367.1	10.5	43.4	147.7	0.5	3.3
32770	63504	1	36619.2	57.6	374.4	0.0	0.0	0.0

Size			Iterations			Iterations (Std.dev)		
Nodes	Arcs	#	MINFLOW	NETFLO	RELAXT	MINFLOW	NETFLO	RELAXT
258	512	10	309.7	346.5	1357.5	37.0	43.1	334.7
514	1008	10	463.7	686.4	1726.6	40.0	110.8	936.8
1026	2000	10	776.8	1335.1	790.2	33.6	95.0	238.3
2050	3984	10	1140.2	2353.1	842.3	154.5	137.5	374.0
4098	7952	10	1595.0	5096.2	1500.9	107.1	633.1	878.0
8194	15888	2	2767.5	9056.0	1382.0	147.5	849.0	158.0
32770	63504	1	6500.0	25243.0	6023.0	0.0	0.0	0.0

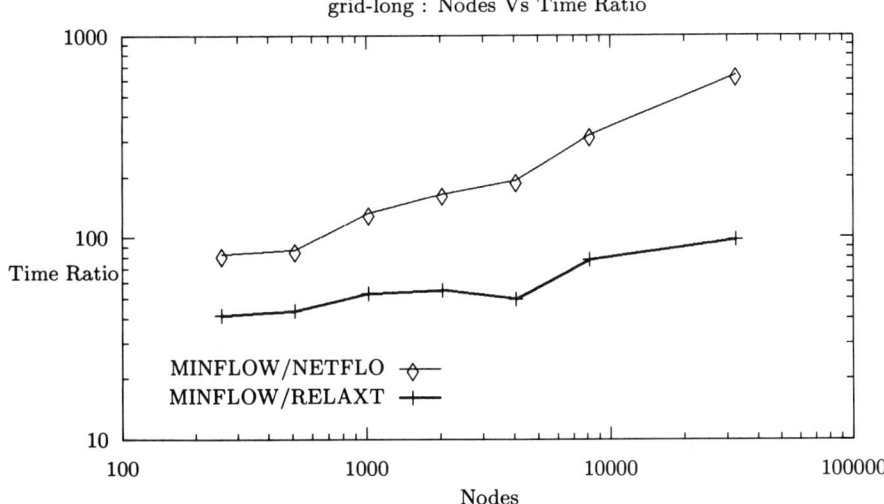

FIGURE 9. Statistics for family **grid-long**

Size			Avg. Time (seconds)			Std. Dev. Time (seconds)		
Nodes	Arcs	#	MINFLOW	NETFLO	RELAXT	MINFLOW	NETFLO	RELAXT
258	512	10	16.4	0.2	0.4	2.3	0.0	0.0
514	1040	10	41.4	0.6	1.5	5.3	0.0	0.1
1026	2096	10	106.1	1.4	6.0	14.6	0.1	0.7
2050	4208	10	222.2	3.9	25.3	25.2	0.3	3.3
4098	8432	10	529.0	12.1	111.7	33.7	0.8	10.8
8194	16880	2	1327.4	41.7	589.3	99.7	0.2	70.2
16386	33776	1	2925.1	166.2	2501.6	0.0	0.0	0.0
32770	67568	1	6617.7	719.1	12477.0	0.0	0.0	0.0
65538	135152	1	20737.5	2578.5	66517.5	0.0	0.0	0.0

Size			Iterations			Iterations (Std.dev)		
Nodes	Arcs	#	MINFLOW	NETFLO	RELAXT	MINFLOW	NETFLO	RELAXT
258	512	10	309.7	346.5	1357.5	37.0	43.1	334.7
514	1040	10	387.1	1090.7	4444.6	49.6	139.7	656.5
1026	2096	10	501.9	2618.6	17748.9	51.3	312.5	3199.2
2050	4208	10	559.9	6132.2	55576.4	52.7	523.0	9823.2
4098	8432	10	688.1	14204.1	156956.9	48.0	770.5	20308.4
8194	16880	2	845.5	33238.5	550904.5	78.5	138.5	85058.5
16386	33776	1	932.0	73542.0	1263055.0	0.0	0.0	0.0
32770	67568	1	1080.0	170984.0	4595085.0	0.0	0.0	0.0
65538	135152	1	1469.0	346385.0	13321493.0	0.0	0.0	0.0

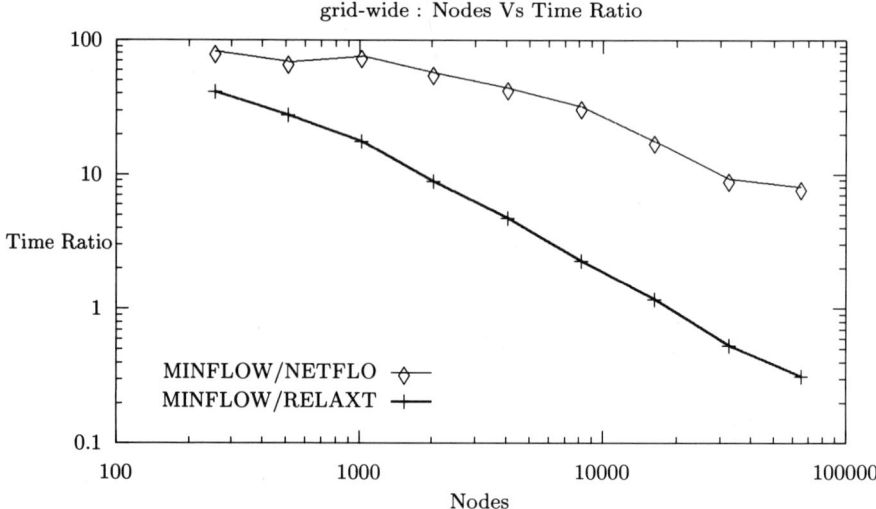

FIGURE 10. Statistics for family **grid-wide**

Size			Avg. Time (seconds)			Std. Dev. Time (seconds)		
Nodes	Arcs	#	MINFLOW	NETFLO	RELAXT	MINFLOW	NETFLO	RELAXT
130	2113	10	16.1	1.5	0.5	2.6	0.0	0.0
258	8321	10	87.4	11.5	2.0	13.3	0.4	0.0
514	33025	10	546.0	93.8	7.9	69.5	2.1	0.1
1026	131585	2	4273.5	812.9	33.8	270.1	43.9	0.4

Size			Iterations			Iterations (Std.dev)		
Nodes	Arcs	#	MINFLOW	NETFLO	RELAXT	MINFLOW	NETFLO	RELAXT
130	2113	10	92.8	1.6	0.3	35.6	0.5	0.5
258	8321	10	175.3	4.2	3.1	44.8	2.6	2.3
514	33025	10	378.5	21.6	20.7	46.7	3.2	3.7
1026	131585	2	1198.0	90.5	83.0	29.0	5.5	7.0

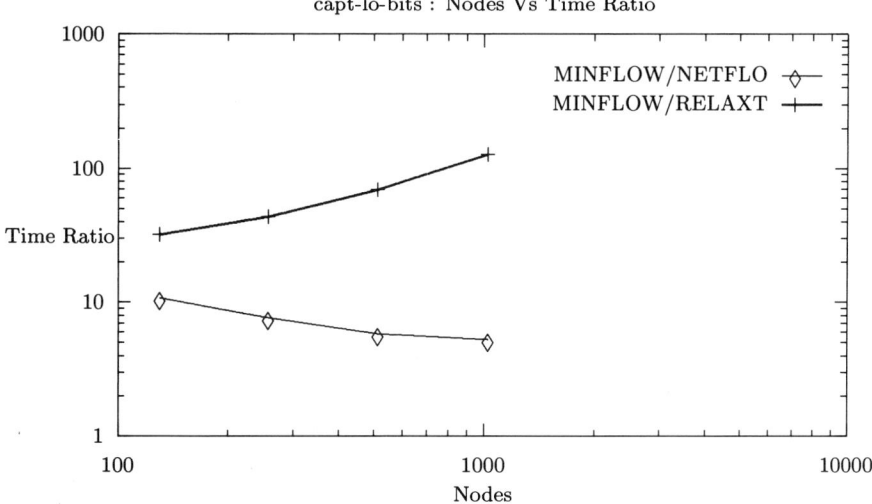

FIGURE 11. Statistics for family **capt-lo-bits**

Size			Avg. Time (seconds)			Std. Dev. Time (seconds)		
Nodes	Arcs	#	MINFLOW	NETFLO	RELAXT	MINFLOW	NETFLO	RELAXT
130	2113	10	24.7	1.6	0.5	2.0	0.1	0.0
258	8321	10	133.0	11.7	2.2	11.3	0.4	0.0
514	33025	10	801.4	94.1	8.7	54.7	2.3	0.1
1026	131585	2	3513.7	820.5	38.5	125.0	5.2	0.1

Size			Iterations			Iterations (Std.dev)		
Nodes	Arcs	#	MINFLOW	NETFLO	RELAXT	MINFLOW	NETFLO	RELAXT
130	2113	10	207.0	335.6	171.3	17.5	34.4	11.9
258	8321	10	312.5	681.7	369.1	22.4	41.0	23.0
514	33025	10	510.2	1316.6	765.1	39.8	45.2	17.1
1026	131585	2	674.5	2702.5	1574.0	29.5	42.5	12.0

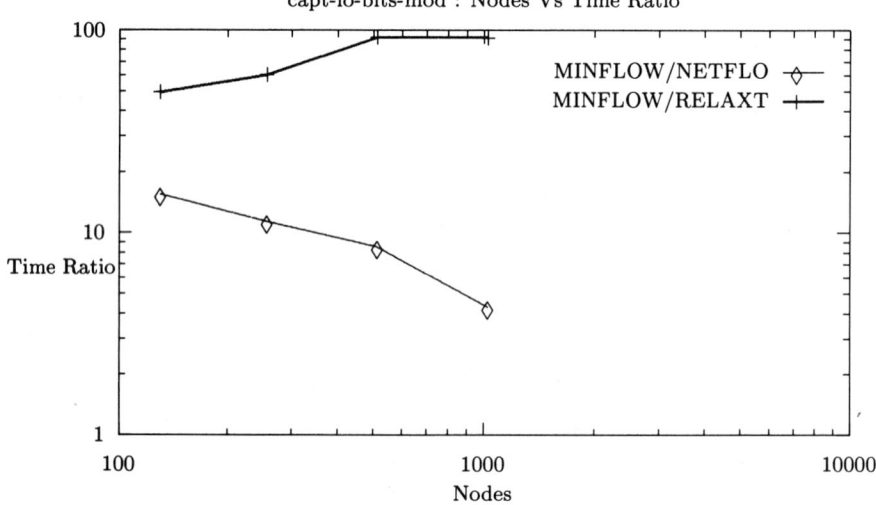

FIGURE 12. Statistics for family **capt-lo-bits-mod**

Size			Avg. Time (seconds)			Std. Dev. Time (seconds)		
Nodes	Arcs	#	MINFLOW	NETFLO	RELAXT	MINFLOW	NETFLO	RELAXT
130	2113	10	16.1	1.5	0.5	2.4	0.0	0.0
258	8321	10	81.5	11.7	2.0	10.3	0.4	0.0
514	33025	10	529.7	93.6	8.0	97.6	2.3	0.1
1026	131585	2	3983.9	771.6	32.2	452.3	1.7	0.0

Size			Iterations			Iterations (Std.dev)		
Nodes	Arcs	#	MINFLOW	NETFLO	RELAXT	MINFLOW	NETFLO	RELAXT
130	2113	10	97.1	1.8	0.2	24.2	0.4	0.4
258	8321	10	154.5	4.2	3.0	40.1	2.1	2.4
514	33025	10	351.9	20.0	17.6	56.1	5.9	5.6
1026	131585	2	1137.0	89.0	73.5	91.0	1.0	1.5

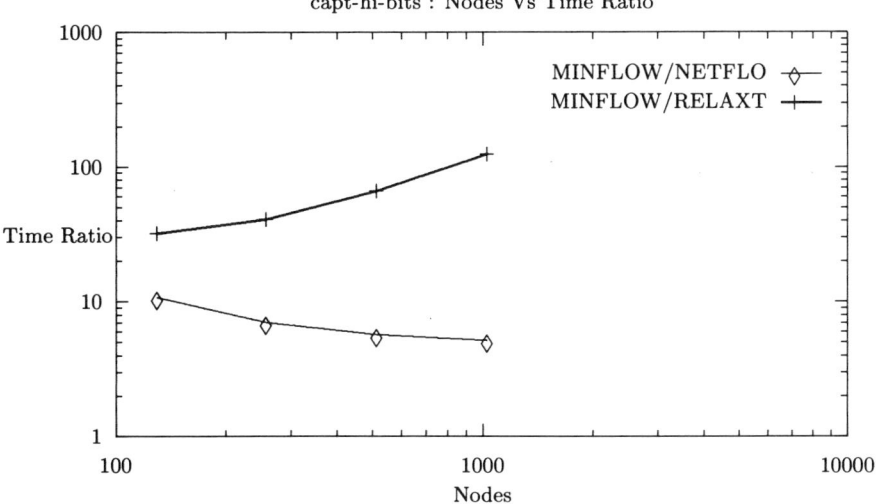

FIGURE 13. Statistics for family **capt-hi-bits**

Size			Avg. Time (seconds)			Std. Dev. Time (seconds)		
Nodes	Arcs	#	MINFLOW	NETFLO	RELAXT	MINFLOW	NETFLO	RELAXT
130	2113	10	26.3	1.6	0.6	3.0	0.0	0.0
258	8321	10	142.8	11.7	2.2	17.2	0.4	0.2
514	33025	10	895.9	94.3	8.7	60.9	2.4	0.0
1026	131585	2	3713.4	852.9	37.6	228.2	1.3	0.2

Size			Iterations			Iterations (Std.dev)		
Nodes	Arcs	#	MINFLOW	NETFLO	RELAXT	MINFLOW	NETFLO	RELAXT
130	2113	10	203.7	361.1	159.1	16.2	39.8	14.7
258	8321	10	278.4	758.9	322.9	21.1	46.9	23.1
514	33025	10	455.7	1525.1	657.2	14.3	67.2	33.7
1026	131585	2	594.5	3197.0	1234.5	5.5	75.0	24.5

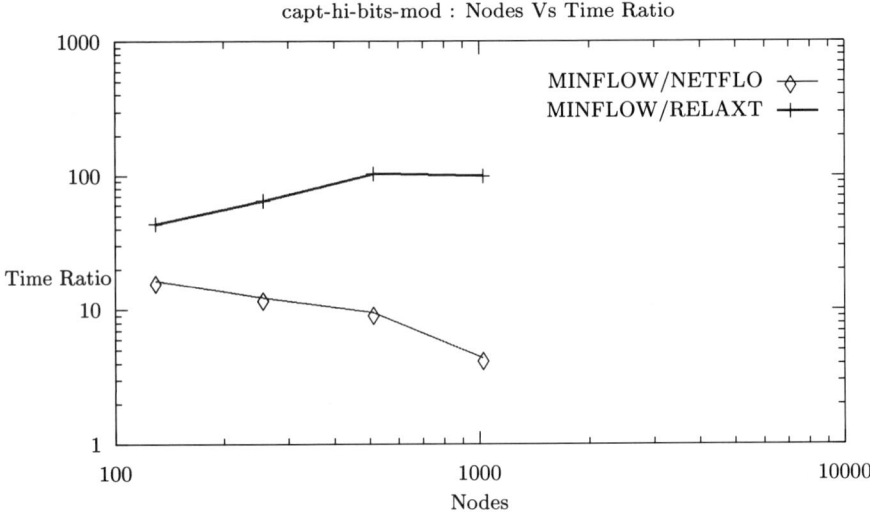

FIGURE 14. Statistics for family **capt-hi-bits-mod**

Size			Avg. Time (seconds)			Std. Dev. Time (seconds)		
Nodes	Arcs	#	MINFLOW	NETFLO	RELAXT	MINFLOW	NETFLO	RELAXT
256	512	3	9.8	0.4	0.2	0.1	0.0	0.0
529	1058	3	26.1	1.3	0.7	0.1	0.0	0.1
1024	2048	3	62.5	4.2	1.4	0.5	0.2	0.1
2116	4232	3	171.5	16.2	4.4	7.8	1.0	0.5
4096	8192	3	436.7	55.8	10.4	25.9	1.8	0.8
8281	16562	3	1199.1	239.8	27.3	54.0	2.0	2.8
16384	32768	3	3060.4	1046.9	78.0	98.4	27.2	8.0
33124	66248	3	8887.6	4667.0	169.1	90.9	138.3	26.2
65536	131072	3	24706.9	19616.1	540.2	182.2	1031.3	49.0

Size			Iterations			Iterations (Std.dev)		
Nodes	Arcs	#	MINFLOW	NETFLO	RELAXT	MINFLOW	NETFLO	RELAXT
256	512	3	219.3	1016.7	672.0	6.6	58.9	30.6
529	1058	3	282.3	2531.7	1760.3	6.8	90.7	205.5
1024	2048	3	355.7	5833.7	3110.3	5.4	138.0	215.3
2116	4232	3	527.3	13784.7	6949.3	2.6	469.3	194.3
4096	8192	3	751.3	28993.3	13129.7	12.0	1005.1	483.9
8281	16562	3	1115.7	67341.3	27472.7	26.4	499.2	282.2
16384	32768	3	1558.3	151026.0	57056.7	15.0	2766.8	640.7
33124	66248	3	2378.7	351469.7	113439.3	35.7	6161.2	1252.2
65536	131072	3	3483.3	779332.3	226449.3	33.2	8654.5	1996.9

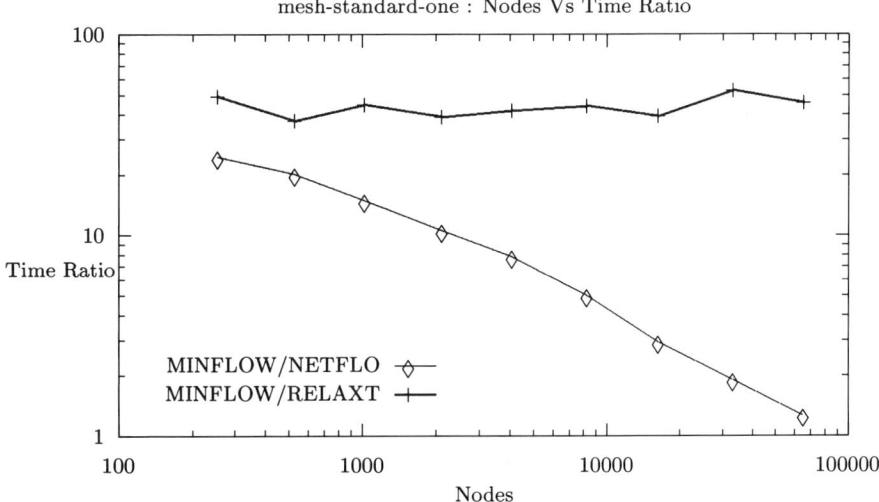

FIGURE 15. Statistics for family **mesh-standard-one**

Size			Avg. Time (seconds)			Std. Dev. Time (seconds)		
Nodes	Arcs	#	MINFLOW	NETFLO	RELAXT	MINFLOW	NETFLO	RELAXT
256	1024	3	16.7	0.9	0.4	0.4	0.0	0.0
529	2116	3	44.9	3.6	1.1	1.3	0.1	0.0
1024	4096	3	110.0	13.0	2.2	1.7	0.3	0.3
2116	8464	3	306.5	54.2	6.2	6.4	0.9	0.3
4096	16384	3	761.2	211.8	15.2	8.7	5.4	1.5
8281	33124	3	2035.9	930.4	46.5	14.9	6.1	4.3
16384	65536	3	5958.4	3892.7	100.3	39.2	20.4	21.4
33124	132496	3	16779.6	18285.5	346.3	553.8	1163.5	14.2
65536	262144	3	33900.3			23984.5		

Size			Iterations			Iterations (Std.dev)		
Nodes	Arcs	#	MINFLOW	NETFLO	RELAXT	MINFLOW	NETFLO	RELAXT
256	1024	3	269.3	3028.3	903.7	1.2	109.4	74.4
529	2116	3	377.0	8143.3	1979.7	5.0	281.3	162.1
1024	4096	3	530.0	18902.3	3614.7	11.3	180.9	414.5
2116	8464	3	764.0	44960.0	8056.0	5.7	795.4	348.5
4096	16384	3	1091.7	100814.0	17023.3	2.5	539.2	1135.2
8281	33124	3	1659.3	236127.3	38345.0	27.5	1227.2	367.5
16384	65536	3	2444.0	526127.0	71911.7	23.3	3452.9	3701.2
33124	132496	3	3751.3	1249134.7	156234.0	146.4	23765.4	2685.1
65536	262144	3	3626.3			2564.2		

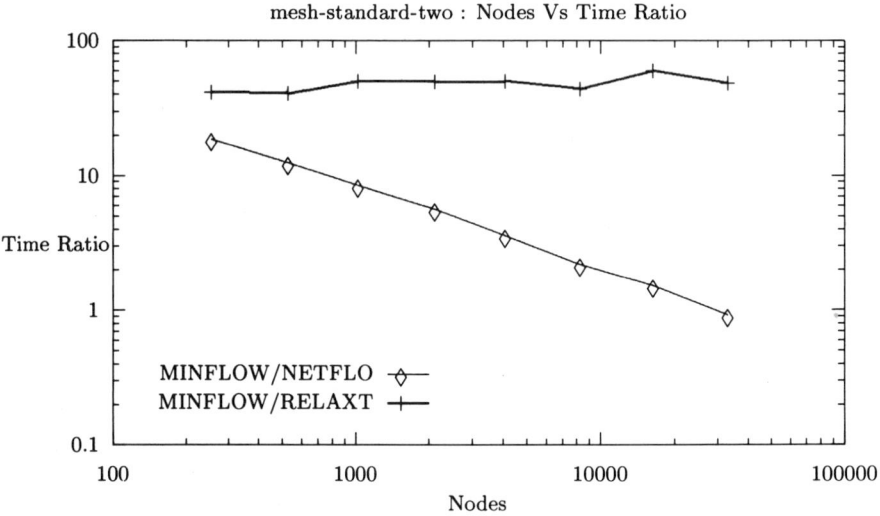

FIGURE 16. Statistics for family **mesh-standard-two**

Size			Avg. Time (seconds)			Std. Dev. Time (seconds)		
Nodes	Arcs	#	MINFLOW	NETFLO	RELAXT	MINFLOW	NETFLO	RELAXT
256	2048	3	27.7	1.9	0.7	0.4	0.0	0.0
529	4232	3	74.3	7.9	1.8	1.2	0.1	0.1
1024	8192	3	190.4	31.7	4.0	14.5	1.1	0.1
2116	16928	3	472.0	147.7	10.1	13.7	2.1	0.9
4096	32768	3	1143.2	605.5	29.1	26.6	11.8	3.1
8281	66248	3	3295.4	2736.6	67.8	102.2	57.1	2.2
16384	131072	3	9191.0	12084.4	182.8	180.4	92.7	4.2
33124	264992	3	18096.9			12796.4		

Size			Iterations			Iterations (Std.dev)		
Nodes	Arcs	#	MINFLOW	NETFLO	RELAXT	MINFLOW	NETFLO	RELAXT
256	2048	3	301.3	6717.7	975.0	17.2	158.1	82.0
529	4232	3	414.3	19854.0	2092.0	4.8	505.1	193.8
1024	8192	3	625.7	50809.7	4354.0	12.7	1087.9	141.4
2116	16928	3	890.0	126310.3	10060.7	21.5	758.8	603.5
4096	32768	3	1265.0	286119.0	20588.3	28.3	3267.3	2547.8
8281	66248	3	1993.0	664587.7	42485.0	39.9	3579.5	423.3
16384	131072	3	2974.0	1500231.7	87171.7	32.5	11059.1	2153.2
33124	264992	3	3084.0			2180.8		

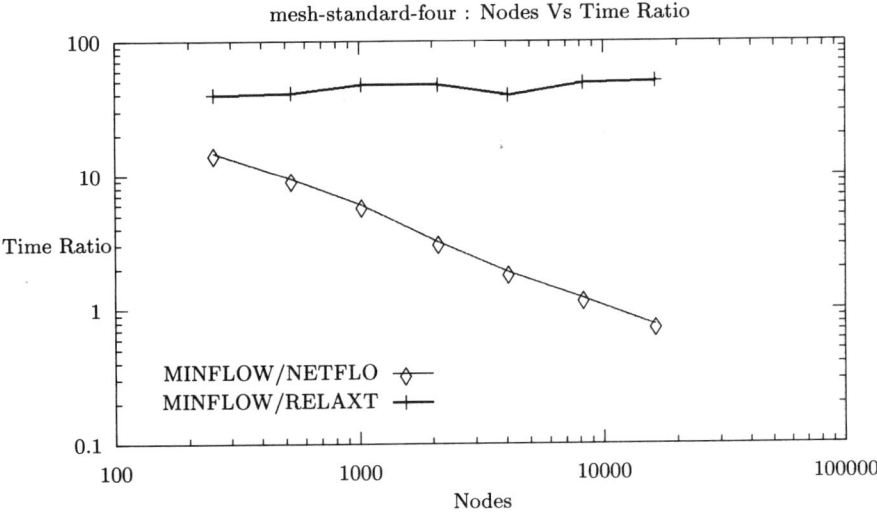

FIGURE 17. Statistics for family **mesh-standard-four**

Size			Avg. Time (seconds)			Std. Dev. Time (seconds)		
Nodes	Arcs	#	MINFLOW	NETFLO	RELAXT	MINFLOW	NETFLO	RELAXT
256	4096	3	51.4	3.6	1.5	2.5	0.1	0.0
529	8464	3	132.1	16.4	3.5	2.4	0.3	0.1
1024	16384	3	327.2	66.6	7.4	16.6	1.0	0.4
2116	33856	3	801.0	347.3	19.5	18.5	5.7	0.2
4096	65536	3	2148.2	1651.9	48.9	56.9	35.3	3.2
8281	132496	3	5160.2	6977.0	116.6	230.6	35.5	6.0
16384	262144	3	10374.1			7363.3		

Size			Iterations			Iterations (Std.dev)		
Nodes	Arcs	#	MINFLOW	NETFLO	RELAXT	MINFLOW	NETFLO	RELAXT
256	4096	3	307.7	12141.3	1152.0	15.3	470.2	84.2
529	8464	3	470.0	41434.0	2794.0	19.6	921.1	211.6
1024	16384	3	649.3	114986.0	5397.3	17.4	1190.6	125.2
2116	33856	3	933.3	309586.0	11669.7	13.3	6483.7	472.8
4096	65536	3	1363.0	711413.3	24505.0	26.2	9406.9	1226.1
8281	132496	3	1983.7	1672519.7	50600.7	42.6	986.2	1407.7
16384	262144	3	2154.3			1526.1		

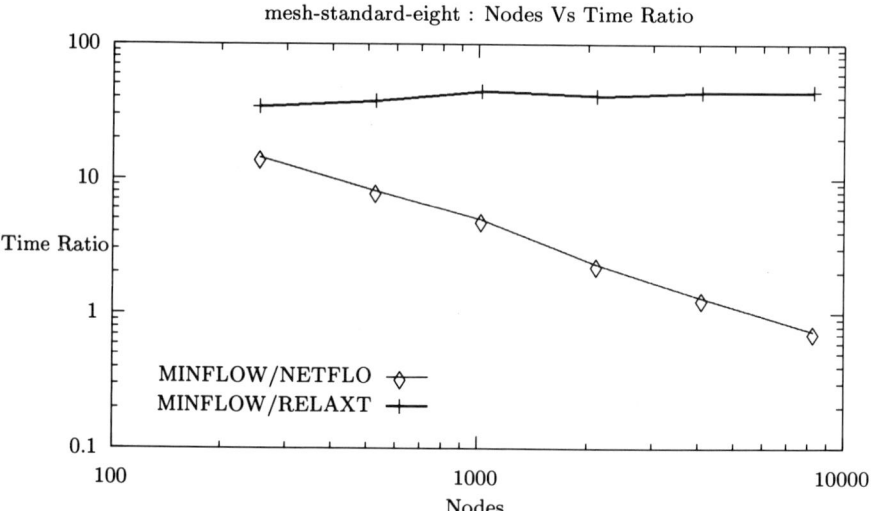

FIGURE 18. Statistics for family **mesh-standard-eight**

References

[A+85] J. Aronson et al. The projective transformation algorithm of Karmarkar: A computational experiment with assignment problems. Technical Report 85-OR-3, Southern Methodist University, Department of Operations Research, Southern Methodist University, Dallas, Texas, 1985.

[AM91] A. Armacost and S. Mehrotra. Computational comparison of the network simplex method with the affine scaling method. *OpSearch*, 28:26–43, 1991.

[Ber91] D.P. Bertsekas. *Linear Network Optimization: Algorithms and Codes*. The MIT Press, 1991.

[BT88a] D.P. Bertsekas and P. Tseng. RELAX: A computer code for minimum cost network flow problems. *Annals of Operations Research*, 13:127–190, 1988.

[BT88b] D.P. Bertsekas and P. Tseng. Relaxation methods for minimum cost ordinary and generalized network flow problems. *Operations Research*, 36:93–114, 1988.

[BT90] D.P. Bertsekas and P. Tseng. RELAXT-III: A new and improved version of the relax code. Technical Report P-1990, Massachusetts Institute of Technology, Laboratory for Information and Decision Systems, MIT, Cambridge, Massachusetts, 1990.

[Cha52] A. Charnes. Optimality and degeneracy in linear programming. *Econometrica*, 20:160–170, 1952.

[FF62] Lestor R. Ford Jr. and D.R. Fulkerson. *Flows in Networks*. Princeton University Press, 1962.

[GL89] Gene H. Golub and Charles F. Van Loan. *Matrix Computations*. The Johns Hopkins University Press, Baltimore, MD, 1989.

[Kar84] Narendra Karmarkar. A new polynomial algorithm for linear programming. *Combinatorica*, 4(4):373–395, 1984.

[KH80] J.L. Kennigton and R.V. Helgason. *Algorithms for Network Programming*. John-Wiley & Sons, 1980.

[KY91] J.A. Kalinski and Y. Ye. A decomposition variant of the potential reduction algorithm for linear programming. Technical Report 91-11, University of Iowa, Department of Management Sciences, University of Iowa, Iowa City, Iowa, 1991.

[Raj89] A. Rajan. An empirical comparison of KORBX against REALXT, a special code for network flow problems. Technical Report (Not available), AT&T Bell Laboratories, Holmdel, NJ, 1989.

[Ren88] J. Renegar. A polynomial-time algorithm, based on Newton's method, for linear programming. *Mathematical Programming*, 40:59–93, 1988.

[RV90] M.G.C. Resende and G. Veiga. An implementation of the dual affine scaling algorithm for minimum cost flow on bipartite uncapacitated networks. Technical Report (Not available), AT&T Bell Laboratories, Murray Hill, NJ, 1990. (To appear in Siam Journal on optimization).

[TCR90] C.E. Leiserson T.H. Cormen and R.L. Rivest. *Introduction to Algorithms*. The MIT Press, 1990.

[Vai89] Pravin M. Vaidya. A simple Newton-type method for finding the analytical center of a polytope. In Nimrod Megiddo, editor, *Progress in Mathematical Programming : Interior Point and Related Methods*, pages 79–90. Springer-Verlag, New York, 1989.

[Vai90] Pravin M. Vaidya. An algorithm for linear programming which requires $O(((m+n)n^2 + (m+n)^{1.5}n)L)$ arithmetic operations. *Mathematical Programming*, 47:175–201, 1990.

[Vai91] Pravin M. Vaidya. Solving linear equations with symmetric diagonally dominant matrices by constructing provably good preconditioners. Technical Report (In preparation), Department Of Computer Science, University of Illinois at Urbana-Champaign, 1991.

Department of Computer Science, University of Illinois, Urbana, Illinois
E-mail address: joshi@cs.uiuc.edu

Department of Computer Science, University of Illinois, Urbana, Illinois
Current address: Department of Mathematics and Computer Science,
Ben-Gurion University, Beer-Sheva 84105, Israel
E-mail address: arthur@cs.bgu.ac.il

Department of Computer Science, University of Illinois, Urbana, Illinois
E-mail address: vaidya@cs.uiuc.edu

An Efficient Implementation of a Network Interior Point Method

MAURICIO G.C. RESENDE AND GERALDO VEIGA

ABSTRACT. We describe DLNET, an implementation of the dual affine scaling algorithm for minimum cost capacitated network flow problems. We compare DLNET with network simplex code NETFLO and relaxation code RELAXT-3 on an extensive range of minimum cost network flow problems, including minimum cost circulation, maximum flow and transshipment problems. The computational results show that DLNET offers more predictable running times than those of NETFLO and RELAXT-3. Its performance, relative to the other codes, improves with problem size for most problem classes. Furthermore, it can outperform NETFLO and RELAXT-3 by a wide margin in some of the classes. The network simplex solver CPLEX 2.0 is tested on a subset of the test problems and its performance compared to NETFLO and the general linear programming solver CPLEX 2.0. While the CPLEX network optimizer is significantly faster than the CPLEX linear programming solver, it is only marginally faster than NETFLO on most classes. The conclusions made with respect to NETFLO also hold with respect to the CPLEX network optimizer.

1. Introduction

Consider a network with an underlying directed graph $G = (V, E)$, where V is a set of m vertices and E a set of n edges. Let (i,j) denote a directed edge from vertex i to vertex j. For each vertex $i \in V$, let b_i denote the net flow out of vertex i. If $b_i > 0$ vertex i is a source, if $b_i < 0$ vertex i is a sink and, otherwise, vertex i is a transshipment vertex. For each edge $(i,j) \in E$, let c_{ij}, l_{ij} and u_{ij} denote, respectively, the unit flow cost, lower bound and upper bound on flow in edge (i,j). All data are assumed to be integer. A feasible solution of a network flow problem (often referred to as *flow*) is given by the n-dimensional vector x, where component x_{ij} is the flow in edge (i,j), satisfying flow conservation

---

1991 *Mathematics Subject Classification.* Primary 90C05, 90C06, 90C35; Secondary 65F10, 65K05, 65Y05.

Key words and phrases. Interior point algorithm, network flows, linear programming, computer implementation, simplex method, network simplex method, conjugate gradient.

constraints for all vertices and flow lower bound and capacity constraints on all edges.

The minimum cost network flow (MCNF) problem consists of finding a flow of minimum cost, as expressed in the following classical linear programming formulation:

$$\min \sum_{ij \in E} c_{ij} x_{ij} \tag{1}$$

subject to:

$$\sum_{jk \in E} x_{jk} - \sum_{kj \in E} x_{kj} = b_j, \ j \in V \tag{2}$$

$$l_{ij} \leq x_{ij} \leq u_{ij}, \ (i,j) \in E. \tag{3}$$

More compactly, the linear program in (1-3) can be expressed as

$$\min \{c^\top x \mid Ax = b, \ l \leq x \leq u\},$$

where A is the incidence matrix of G. We denote the i-th column of A by A_i, the i-th row of A by $A_{\cdot i}$ and a submatrix of A formed by columns with indices in set S by A_S. If graph G has p components, there are exactly p redundant flow conservation constraints, which are sometimes removed from the problem formulation. We rule out a trivially infeasible problem by assuming

$$\sum_{j \in V^k} b_j = 0, \ k = 1, \ldots, p, \tag{4}$$

where V^k is the set of vertices for the k-th component of G.

Often, it is further required that x_{ij} be integer, i.e. we replace (3) with

$$l_{ij} \leq x_{ij} \leq u_{ij}, \ x_{ij} \text{ integer}, \ (i,j) \in E. \tag{5}$$

In the remainder of this paper we assume, without loss of generality, that $l_{ij} = 0$ for all $(i,j) \in E$ and that $c \neq 0$.

Variants of the simplex method [9] can be customized to solve the MCNF problem (e.g. [15, 21]) based on two special properties of the graph incidence matrix. Firstly, since a graph incidence matrix is totally unimodular, every primal feasible basis corresponds to an integer flow. Hence, even if an integer flow is required, one can relax the integrality constraints in (5) and solve the resulting linear program by any simplex variant. Second, in the resulting constraint matrix, there is a one to one correspondence between basic sequences and maximal forests of G, which provides a block triangular ordering for the basic matrix, once redundant constraints are removed. Data structures in implementations of algorithms for solving MCNF problems rely heavily on this property, implying that only integer arithmetic is used and, unlike implementations of the simplex methods for general linear programs, costly factorizations of the basic matrix are unnecessary.

The motivation of this study is that, in practice, the number of iterations taken by interior point algorithms for linear programming appears to grow slowly with problem size. Furthermore, empirical evidence suggests that interior point methods are not affected by degeneracy as much as the simplex method [32]. Most direct comparisons between interior point algorithms and the simplex method (e.g.[2, 24, 26]), conclude that as problem size grows the advantage increasingly tilts toward interior point methods.

To replicate the improved performance observed for the network simplex method over the general simplex method, an interior point implementation dedicated to MCNF problems must use some of the distinguishing properties of the structure of the problem. For example, double precision multiplications are eliminated in operations involving the coefficient matrix and specialized preconditioners based on the network structure can be devised for the conjugate gradient algorithm. Also, the detection of an optimal solution can be based on the integer data.

Several studies compare implementations of interior point algorithms with specialized network codes [3, 4, 27] and conclude that interior point algorithms are not competitive with the specialized network codes. In this paper, we show that a network interior point implementation outperforms specialized network codes on several classes of large MCNF problems. Furthermore, in most problem classes, as the size of the instances grows, so does the difference in solution times.

The dual affine scaling (DAS) algorithm [10] (see also [5, 30, 36]) was among the first interior point methods to be shown to be a competitive alternative to the simplex method [1, 2]. Let A be an $m \times n$ matrix, c, u, and x be n-dimensional vectors and b an m-dimensional vector. The DAS algorithm solves the linear program

$$\min \{c^\top x \mid Ax = b,\ 0 \leq x \leq u\}$$

indirectly by solving its dual

(6) $$\max \{b^\top y - u^\top z \mid A^\top y - z + s = c,\ z \geq 0, s \geq 0\}$$

where z and s are an n-dimensional vectors and y is an m-dimensional vector. The algorithm starts with an initial interior solution $\{y^0, z^0, s^0\}$ such that

$$A^\top y^0 - z^0 + s^0 = c,\ z^0 > 0,\ s^0 > 0,$$

and iterates according to

$$\{y^{k+1}, z^{k+1}, s^{k+1}\} = \{y^k, z^k, s^k\} + \alpha \{\Delta y, \Delta z, \Delta s\},$$

where the search directions $\Delta y, \Delta z$, and Δs satisfy

$$A(Z_k^2 + S_k^2)^{-1}A^\top \Delta y = b - AZ_k^2(Z_k^2 + S_k^2)^{-1}u,$$
$$\Delta z = Z_k^2(Z_k^2 + S_k^2)^{-1}(A^\top \Delta y - S_k^2 u),$$
$$\Delta s = \Delta z - A^\top \Delta y,$$

where

$$Z_k = \text{diag}(z_1^k, \ldots, z_n^k) \text{ and } S_k = \text{diag}(s_1^k, \ldots, s_n^k)$$

and α is such that $z^{k+1} > 0$ and $s^{k+1} > 0$, i.e. $\alpha = \gamma \times \min\{\alpha_z, \alpha_s\}$, where $0 < \gamma < 1$ and

$$\alpha_z = \min\{-z_i^k/(\Delta z)_i \mid (\Delta z)_i < 0, \ i = 1, \ldots, n\}$$
$$\alpha_s = \min\{-s_i^k/(\Delta s)_i \mid (\Delta s)_i < 0, \ i = 1, \ldots, n\}.$$

The dual problem (6) has a readily available initial interior point solution:

$$y_{i^0} = 0, \ i = 1, \ldots, n$$
$$s_i^0 = c_i + \lambda, \ i = 1, \ldots, n$$
$$z_i^0 = \lambda, \ i = 1, \ldots, n,$$

where λ is a scalar such that $\lambda > 0$ and $\lambda > -c_i$, $i = 1, \ldots, n$. In the implementation described in this study (called DLNET), we use $\lambda = 2 \|c\|_2$.

The bulk of the work in the DAS algorithm is related to building and updating the matrix $AD_k A^\top$ and solving the system of linear equations

(7) $$AD_k A^\top \Delta y = b - AZ_k^2 D_k u,$$

where $D_k = (Z_k^2 + S_k^2)^{-1}$. This system determines the ascent direction at each iteration of the algorithm. Whereas for a large class of linear programs system (7) can be handled efficiently by direct factorization methods, this is not the case for MCNF problems. In [28], a direct method and an iterative approach based on the conjugate gradient method are compared on randomly generated assignment problems. That study illustrates, for MCNF problems, the gains observed with an iterative approach over a direct factorization method. In a companion paper [29], the relative performance of the interior point approach using a preconditioned conjugate gradient algorithm to the network simplex code NETFLO [21] and the relaxation algorithm code RELAX [6] was shown to improve with the size of the instance. However, the study left unanswered whether an interior point implementation could outperform a network simplex or relaxation method implementation on MCNF problems.

This paper builds on [29], where we use a conjugate gradient with diagonal and spanning tree preconditioners. Whereas the old implementation was limited to handling uncapacitated bipartite MCNF problems, DLNET can solve capacitated MCNF problems as formulated in (1–5). In addition, we implement two new stopping criteria and a more stable preconditioned conjugate gradient procedure. The paper is organized as follows. In Section 2, we describe a generic

```
procedure pcg(A, D_k, b̄, ε_cg, Δy)
1    Δy_0 := 0;
2    r_0 := b̄;
3    z_0 := M^{-1} r_0;
4    p_0 := z_0;
5    i := 0;
6    do stopping criterion not satisfied →
7        q_i := A D_k A^⊤ p_i;
8        α_i := z_i^⊤ r_i / p_i^⊤ q_i;
9        Δy_{i+1} := Δy_i + α_i p_i;
10       r_{i+1} := r_i - α_i q_i;
11       z_{i+1} := M^{-1} r_{i+1};
12       β_i := z_{i+1}^⊤ r_{i+1} / z_i^⊤ r_i;
13       p_{i+1} := z_{i+1} + β_i p_i;
14       i := i + 1
15   od;
16   Δy := Δy_i
end pcg;
```

FIGURE 1. The preconditioned conjugate gradient algorithm

preconditioned conjugate gradient algorithm used in DLNET. The preconditioners applied to the conjugate gradient algorithm defined in Section 2 are described in Section 3. In Section 4, we describe the stopping strategies implemented in DLNET. Computational results on a wide range of MCNF problems are given in Section 5. Concluding remarks are made in Section 6.

2. Computing the Ascent Direction

The computational efficiency of DLNET relies heavily on a preconditioned conjugate gradient algorithm to solve the direction finding system at each iteration. We differ slightly from the preconditioned conjugate gradient algorithm described in [2]. Here, the preconditioned conjugate gradient algorithm is used to solve

(8) $$M^{-1}(AD_kA^\top)\Delta y = M^{-1}\bar{b}$$

where M is a positive definite matrix and $\bar{b} = b - AZ_k^2 D_k^{-1} u$. The objective is to make the preconditioned matrix

$$M^{-1}(AD_kA^\top)$$

less ill-conditioned than $AD_kA^\top$, improving the convergence of the conjugate gradient algorithm.

The preconditioned conjugate gradient algorithm is presented in the pseudo-code in Figure 1. The computationally intensive steps in the preconditioned conjugate gradient algorithm are lines 3, 7 and 11 of the pseudo-code. Those lines correspond to a matrix-vector multiplication (7) and solving systems of linear equations (3 and 11). Line 3 is computed once and lines 7 and 11 are computed once every conjugate gradient iteration. The matrix-vector multiplications are of the form $AD_k A^\top p_i$, carried out without forming $AD_k A^\top$ explicitly. It is more efficient to compute the above matrix-vector multiplication by decomposing it into three sparse matrix-vector multiplications. Let

$$\zeta' = A^\top p_i \quad \text{and} \quad \zeta'' = D_k \zeta'.$$

Then

$$(A (D_k (A^\top p_i))) = A\zeta''.$$

Note that the matrix-vector multiplications are $\mathcal{O}(n)$, involving n additions, $2n$ subtractions and n floating point multiplications. Note further that while we limit ourselves to serial implementation, this computation can be carried out in parallel. See [29] for numerical results of a parallel implementation of the matrix-vector multiplication in the conjugate gradient algorithm.

The preconditioned residual is computed in lines 3 and 11 and amounts to solving the system of linear equations

(9) $$M z_{i+1} = r_{i+1},$$

where M is a positive definite matrix such that the system can be easily solved. Such preconditioners are the subject of Section 3.

It was pointed out in [2] that the DAS algorithm is particularly well suited to use approximate solutions of the ascent direction linear system. To determine when the direction Δy_i produced by the conjugate gradient algorithm is satisfactory, we use the suggestion made in [18] and compute the angle θ between $(AD_k A^\top)\Delta y_i$ and $\bar{b}$ and stop the conjugate gradient procedure when $|1 - \cos\theta| < \epsilon_{cos}$, where ϵ_{cos} is some small tolerance. The computation of

$$\cos\theta = \frac{|\bar{b}^\top (AD_k A^\top)\Delta y_i|}{\|\bar{b}\| \cdot \|(AD_k A^\top)\Delta y_i\|}$$

has the complexity of one conjugate gradient iteration and is carried out only every l_{cos} iterations.

3. Preconditioners

A useful preconditioner for the conjugate gradient algorithm must be such that the system of linear equations (9) is easy to solve while reducing the number of conjugate gradient iterations. A diagonal matrix constitutes the most straightforward and perhaps the most common preconditioner used in conjunction with the conjugate gradient algorithm [14]. They are simple to compute, taking $\mathcal{O}(n)$ double precision operations and can be very effective [28, 29, 38]. The diagonal preconditioner used in DLNET is $M = \text{diag}(AD_k A^\top)$. The preconditioned residue systems of lines 3 and 11 of the conjugate gradient pseudo-code in Figure 1 can each be solved in $\mathcal{O}(m)$ double precision divisions.

Karmarkar and Ramakrishnan [19] and Vaidya [35] have suggested using a maximum weighted spanning tree preconditioner for network flow problems. Since, in our presentation, the graph G is not necessarily connected, we identify a maximal forest using as weights the diagonal elements of the current scaling matrix,

(10) $$w = D_k e,$$

where e is a unit n-vector. The maximal forest is computed by approximately ordering the edges with a bucket sort and applying Kruskal's algorithm [31].

At the k-th iteration of the DAS algorithm, let $\mathcal{S}_k$ be the submatrix of A with columns corresponding to edges in the maximal forest, $t_1, \ldots, t_q$. The preconditioner can be written as

$$M = \mathcal{S}_k \mathcal{D}_k \mathcal{S}_k^\top,$$

where

$$\mathcal{D}_k = \mathrm{diag}(1/z_{t_1}^2 + 1/s_{t_1}^2, \ldots, 1/z_{t_q}^2 + 1/s_{t_q}^2).$$

For simplicity of notation, we include in $\mathcal{S}_k$ the linear dependent rows corresponding to the redundant flow conservation constraints. At each conjugate gradient iteration, the preconditioned residue system

(11) $$(\mathcal{S}_k \mathcal{D}_k \mathcal{S}_k^\top) z_{i+1} = r_{i+1}$$

is solved with the variables corresponding to redundant constraints set to zero. As with the diagonal preconditioner, (11) can be solved in $\mathcal{O}(m)$ time, as the system coefficient matrix can be ordered into a block triangular form. The spanning tree preconditioner has been previously used in [16, 17, 28, 29].

In practice, the diagonal preconditioner is effective during the initial iterations of the DAS algorithm. As the DAS iterations progress, the spanning tree preconditioner is more effective as it becomes a better approximation of matrix $AD_k A^\top$. In the DLNET implementation, we begin with the diagonal preconditioner and monitor the number of iterations required by the conjugate gradient algorithm. When the conjugate gradient takes more than $\beta\sqrt{m}$ iterations, where $\beta > 0$, DLNET switches to the spanning tree preconditioner. We also set upper and lower limits to the number of DAS iterations using a diagonal preconditioned conjugate gradient.

4. Stopping with an Optimal Flow

The simplex method restricts the sequence of solutions it generates to vertices of the linear programming polytope. Since the constraint matrix is totally unimodular, and assuming integrality of the data, when a simplex variant is applied to a MCNF problem, the optimal solution is integer. On the other hand, the DAS algorithm generates a sequence of dual interior solutions. For general linear programs, a tentative primal solution is computed based on each dual iterate [33]. Under very mild conditions, these sequences converge, respectively, to the relative interiors of the primal and dual optimal faces [11, 34]. Unless the primal optimal solution is unique, the primal solution returned by DAS is not guaranteed to be integer. Furthermore, we wish to use MCNF-specific properties to stop the algorithm earlier than its theoretical convergence. We discuss below the stopping strategies implemented in DLNET.

4.1. Stopping with Basic Solution. As explained in Section 3, computing the spanning tree preconditioner involves identifying a basic sequence for the MCNF problem. Under a dual nondegeneracy assumption, as DAS converges, this basic sequence corresponds to an optimal one. At the end of each iteration of DAS, the maximal forest used to build the preconditioner can be used to compute a tentative primal optimal solution in $\mathcal{O}(m)$ operations. Under dual degeneracy this technique can still be useful if only a small number of degeneracies is present in the optimal dual face. Also, a linear program can be made dual nondegenerate by applying the classical perturbation scheme of [7] to the cost vector. See [29] for details on how this has been implemented in DLNET.

Let $\mathcal{T} = \{t_1, \ldots, t_q\}$ denote the set of edge indices in the maximal forest used to compute the preconditioner. To obtain a tentative primal basic solution, we first set flow of edges not in the forest to either its upper or lower bound. For all $i \in E \setminus \mathcal{T}$:

$$x_i^* = \begin{cases} 0 & \text{if } s_i^k > z_i^k \\ u_i & \text{otherwise,} \end{cases}$$

where s^k and z^k are the current iterates of the dual slack vectors as defined in (6). The remaining basic edges have flows that satisfy the linear system

$$(12) \qquad A_\mathcal{T} x_\mathcal{T}^* = b - \sum_{i \in \Omega^-} u_i A_i,$$

where $\Omega^- = \{i \in E \setminus \mathcal{T} | s_i^k \leq z_i^k\}$. Linear system (12) can be solved in $\mathcal{O}(m)$ time. If $u_\mathcal{T} \geq x_\mathcal{T}^* \geq 0$ then the primal solution is feasible.

Optimality can be verified by producing a dual feasible solution (y^*, s^*, z^*) that is either complementary or implies in a duality gap less than 1. We build the tentative optimal dual solution by first identifying the set of dual constraints defining the supporting affine space of the dual face complementary to x^*, defined by

$$\mathcal{F} = \{i \in \mathcal{T} \mid 0 < x_i^* < u_i\},$$

the set of edges with zero dual slacks. To ensure a complementary primal dual pair, we project orthogonally the current dual interior vector y^k onto this affine space,

$$(13) \qquad \min_{y^* \in \Re^m} \{\|y^* - y^k\| \mid A_\mathcal{F}^\top y^* = c_\mathcal{F}\}.$$

A similar scheme that uses orthogonal projection to attempt to identify the optimal face has been independently investigated in Kalinski and Ye [17] and Mehrotra and Ye [25]. Ye [37] has analyzed that procedure to prove finite convergence of interior point algorithms for linear programming.

Let $G_\mathcal{F} = (V, \mathcal{F})$ be the subgraph of G with $\mathcal{F}$ as its set of edges. Since this subgraph is a forest, its incidence matrix, $A_\mathcal{F}$, can be reordered into a block triangular form, with each block corresponding to a tree in the forest. Assume $G_\mathcal{F}$ has p components, with $T_1, \ldots, T_p$ as the sets of edges in each component tree. After reordering, the incidence matrix can be expressed as

$$A_\mathcal{F} = \begin{bmatrix} A_{T_1} & & \\ & \ddots & \\ & & A_{T_p} \end{bmatrix}.$$

The supporting affine space of the dual face can be expressed as the sum of orthogonal one-dimensional subspaces. The operation in (13) can be performed by computing the orthogonal projections onto each individual subspace independently, and therefore can be completed in $\mathcal{O}(m)$ time. For $i = 1, \ldots, p$, m_i is the number of edges in T_i, V_i is the set of vertices spanned by those edges, A_{T_i} is an $(m_i + 1) \times m_i$ matrix and each subspace

$$\Psi_i = \{y_{V_i} \in \Re^{m_i+1} \mid A_{T_i}^\top y_{V_i} = c_{T_i}\}$$

has dimension one. Then, for all $y_{V_i} \in \Psi_i$, we have

(14) $$y_{V_i} = y_{V_i}^0 + \alpha_i y_{V_i}^h,$$

where $y_{V_i}^0$ is a given solution in Ψ_i and $y_{V_i}^h$ is a solution of the homogeneous system $A_{T_i}^\top y_{V_i} = 0$. Since A_{T_i} is the incidence matrix of a tree, the unit vector is a homogeneous solution. The given solution $y_{V_i}^0$ can be computed by selecting $v \in V_i$, setting $y_v^0 = 0$, removing the row corresponding to vertex v from matrix A_{T_i} and solving the resulting triangular system

$$\tilde{A}_{T_i}^\top y_{V_i \setminus \{v\}} = c_{T_i}.$$

With the representation in (14), the orthogonal projection of y_{V_i} onto subspace Ψ_i is

$$y_{V_i}^* = y_{V_i}^0 + \frac{e_{V_i}^\top (y_{V_i} - y_{V_i}^0)}{m_i} e_{V_i}$$

where e is the unit vector.

The orthogonal projection, as indicated in (13), is obtained by combining the projections onto each subspace,

$$y^* = (y_{V_1}^*, \ldots, y_{V_q}^*).$$

We build a feasible dual solution by computing the slacks as

$$z_i^* = \begin{cases} -\delta_i & \text{if } \delta_i < 0 \\ 0 & \text{otherwise} \end{cases} \qquad s_i^* = \begin{cases} 0 & \text{if } \delta_i < 0 \\ \delta_i & \text{otherwise,} \end{cases}$$

where $\delta_i = c_i - A^\top y^*$.

The primal and dual solutions, x^* and (y^*, s^*, z^*), are optimal if complementary slackness is satisfied, i.e. if for all $i \in E \setminus \mathcal{T}$ either $s_i^* > 0$ and $x_i^* = 0$ or $z_i^* > 0$ and $x_i^* = u_i$. Otherwise, the primal solution, x^*, is still optimal when the duality gap is less than 1, i.e. if $c^\top x^* - b^\top y^* + u^\top z^* < 1$.

4.2. Stopping with Maximum Flow Solution. Determining the magnitude in the classical perturbation technique poses a major obstacle. The theoretical acceptable perturbation can be too small to resolve all dual degeneracies in a reasonable number of DAS iterations. A larger perturbation, on the other hand, can change the combinatorial structure of the optimal dual face. An alternative stopping procedure consists of identifying the optimal dual face with the dual interior solution and computing an optimal primal solution by solving a maximum flow problem on a restricted network. Compared to solving spanning tree based linear systems, the maximum flow problem displays a high theoretical complexity. However, the low practical complexity of new maximum flow algorithms makes this procedure an attractive option. Furthermore, the stopping test does not need to be performed at every iteration of the DAS algorithm.

As described in the previous section, the dual iterates generated by the DAS algorithm converge to the relative interior of the optimal dual face. In practice, the algorithm tries to identify a set of active edges defining the supporting affine space of the optimal dual face. As implemented in DLNET, an edge is marked as active if its corresponding dual slack variables, s_i and z_i are either both very small or of the same order of magnitude, i.e.,

$$\mathcal{F} = \{i \in E \mid |s_i^k - z_i^k| < \epsilon \text{ or } \gamma \leq s_i^k/z_i^k \leq 1/\gamma\},$$

where $\epsilon > 0$ and $1 > \gamma > 0$ are small tolerances. Unless the DAS algorithm is close to convergence, there is no guarantee that $\mathcal{F}$ actually defines a dual face, as the set $\{y \in \Re^m \mid A_\mathcal{F}^\top y = c_\mathcal{F}\}$ can be empty. Instead, the supporting affine space is defined by a maximal forest $\mathcal{T}$ of graph $G_\mathcal{F} = (V, \mathcal{F})$. We select this maximal forest by computing maximum weighted spanning trees for each component of $G_\mathcal{F}$, using as weights the diagonal elements of the current scaling matrix. Whenever $\mathcal{F}$ defines a dual face, edges in $\mathcal{F} \setminus \mathcal{T}$ correspond to redundant hyperplanes, and the dual face is unique. A tentative dual optimal solution is computed by projecting the current dual interior vector y^k onto the supporting affine space of the dual face defined by $\mathcal{T}$,

$$\min_{y^* \in \Re^m} \{\|y^* - y^k\| \mid A_\mathcal{T}^\top y^* = c_\mathcal{T}\}.$$

This operation is identical to the orthogonal projection described in Subsection 4.1. The dual slacks for the tentative dual optimal solution are computed as

$$z_i^* = \begin{cases} -\delta_i & \text{if } \delta_i < 0 \\ 0 & \text{otherwise} \end{cases} \qquad s_i^* = \begin{cases} 0 & \text{if } \delta_i < 0 \\ \delta_i & \text{otherwise,} \end{cases}$$

where $\delta_i = c_i - A^\top y^*$.

Based on the projected dual solution y^*, we select a refined tentative optimal face by redefining the set of active edges as

$$\tilde{\mathcal{F}} = \{i \in E \mid |c_i - A_{.i}^\top y^*| < \epsilon\}.$$

Next, we attempt to build a primal feasible solution, x^*, complementary to the tentative dual optimal solution by setting the inactive edges to lower or upper bounds, i.e., for $i \in E \setminus \tilde{\mathcal{F}}$,

$$x_i^* = \begin{cases} 0 & \text{if } i \in \Omega^+ = \{i \in E \setminus \tilde{\mathcal{F}} \mid c_i - A_{.i}^\top y^* > 0\} \\ u_i & \text{if } i \in \Omega^- = \{i \in E \setminus \tilde{\mathcal{F}} \mid c_i - A_{.i}^\top y^* < 0\}. \end{cases}$$

By considering only the active edges, we build a *restricted network*, represented by the constraint set

$$(15) \qquad A_{\tilde{\mathcal{F}}} x_{\tilde{\mathcal{F}}} = \tilde{b} = b - \sum_{i \in \Omega^-} u_i A_i,$$

$$(16) \qquad 0 \leq x_i \leq u_i, \quad i \in \tilde{\mathcal{F}}.$$

Clearly, from the flow balance constraints (15), if a feasible flow $x_{\tilde{\mathcal{F}}}^*$ for the restricted network exists, it defines, along with $x_{\Omega^+}^*$ and $x_{\Omega^-}^*$, a primal feasible solution complementary to y^*. A feasible flow for the restricted network can be determined by solving a maximum flow problem on the *augmented network* defined by underlying graph $\tilde{G} = (\tilde{V}, \tilde{E})$, where

$$\tilde{V} = \{\sigma\} \cup \{\theta\} \cup V$$

and

$$\tilde{E} = \Sigma \cup \Theta \cup \tilde{\mathcal{F}}.$$

In addition, for each edge $(i,j) \in \tilde{\mathcal{F}}$ there is an associated capacity u_{ij}. The additional edges are such that

$$\Sigma = \{(\sigma, i) \mid i \in V^+\},$$

with associated capacity $\tilde{b}_i$ for each edge (σ, i), and

$$\Theta = \{(i, \theta) \mid i \in V^-\},$$

with associated capacity $-\tilde{b}_i$ for each edge (i, θ), where $V^+ = \{i \in V \mid \tilde{b}_i > 0\}$ and $V^- = \{i \in V \mid \tilde{b}_i < 0\}$.

PROPOSITION 1. Let $\mathcal{M}_{\sigma,\theta}$ be the maximum flow value from σ to θ, and $\tilde{x}$ a maximal flow on the augmented network. Then, $\mathcal{M}_{\sigma,\theta} = \sum_{i \in V^+} \tilde{b}_i$ iff $\tilde{x}_{\tilde{\mathcal{F}}}$ is a feasible flow for the restricted network.

PROOF. Firstly, we observe that if the original network problem is feasible, we have $\sum_{i \in V^+} \tilde{b}_= - \sum_{i \in V^-} \tilde{b}_i$ in the restricted network. By construction of $\tilde{G}$, the sets of edges Σ and Θ are σ-θ cutsets with capacity $\sum_{i \in V^+} \tilde{b}_i$. Hence, if $\mathcal{M}_{\sigma,\theta} = \sum_{i \in V^+} \tilde{b}_i$, for a maximal flow $\tilde{x}$, edges in Σ and Θ are at capacity, and $\tilde{x}_{\tilde{\mathcal{F}}}$ satisfies the flow conservation constraints for the restricted network. Since the maximal flow always satisfies the capacity constraints for edges in $\tilde{\mathcal{F}}$, this proves the first direction of this proposition. The reverse direction is immediate. If $\tilde{x}_{\tilde{\mathcal{F}}}$ is a feasible flow for the restricted network, setting flow for edges in σ-θ cutsets Σ and Θ at capacity gives a maximal flow. □

From Proposition (1), we conclude that finding a feasible flow for the restricted network involves the solution of a maximum flow problem. Furthermore, this feasible flow is integer, as we can select a maximum flow algorithm that provides an integer solution.

5. Computational Investigation

The aim of this investigation is to compare network implementations of interior point, simplex, and relaxation algorithms on a wide range of MCNF problems, including minimum cost circulation, maximum flow, and transshipment. In the main experiment, we compare our interior point code with the network simplex code NETFLO [21], and RELAXT-3 [6], an implementation of the relaxation algorithm. We also compare NETFLO with modern network simplex and general simplex codes from the commercial mathematical programming system CPLEX^{TM} 2.0 [8].

The problem classes that make up the computational experiment were suggested by several participants in the The First DIMACS International Algorithm Implementation Challenge [12]. All problem instances are generated with problem generators distributed for the challenge. The problems are grouped into five categories: minimum cost circulation, transshipment on a mesh, minimum cost-maximum flow on a grid, maximum flow on a random layered graph, and problems generated with the standard NETGEN [22] generator. All instances and/or generators were obtained via anonymous ftp from dimacs.rutgers.edu.

5.1. Computing Environment. Most of the computational experiments are conducted on a Silicon Graphics IRIS computer, model 4D/240, with four 25 MHz IP7 processors, a MIPS R2010A/R3010 FPU, a MIPS R2000A/R3000 CPU, 64 Kbytes instruction cache, 64 Kbytes data cache, 256Kbytes of secondary data cache, and 256 Mbytes of main memory. The swapping device is configured to allow processes over 1 Gbytes in size to run. The operating system is IRIX System V Release 4.0.1.

TABLE 1. Summary of C-language DIMACS challenge machine benchmarks

Compiler Optimization		MIPS R3000 33MHz			MIPS R3000 25MHz	
	time	Test C-1	Test C-2	time	Test C-1	Test C-2
no	real	2.44s	24.49s	real	2.43s	25.93s
	user	2.08s	22.99s	user	2.29s	25.52s
	sys	0.25s	0.66s	sys	0.12s	0.39s
yes	real	2.44s	24.92s	real	2.08s	22.74s
	user	1.90s	21.10s	user	1.92s	22.32s
	sys	0.24s	0.78s	sys	0.16s	0.41s

TABLE 2. Summary of Fortran-language DIMACS challenge machine benchmarks

Compiler Optimization		MIPS R3000 33MHz			MIPS R3000 25MHz	
	time	Test F-1	Test F-2	time	Test F-1	Test F-2
no	real	3.35s	4.84s	real	2.40s	4.24s
	user	2.07s	3.63s	user	2.19s	3.92s
	sys	0.39s	0.48s	sys	0.19s	0.30s
yes	real	2.83s	4.63s	real	1.65s	2.69s
	user	1.36s	2.56s	user	1.34s	2.39s
	sys	0.34s	0.52s	sys	0.27s	0.30s

The experiments comparing NETFLO with CPLEX 2.0 were done on a slightly faster Silicon Graphics IRIS computer, model 4D/240, with four 33 MHz IP7 processors, a MIPS R2010A/R3010 FPU, a MIPS R2000A/R3000 CPU, 64 Kbytes instruction cache, 64 Kbytes data cache, 256 Kbytes of secondary data cache, and 128 Mbytes of main memory. The operating system is IRIX System V Release 4.0.1.

The DIMACS machine benchmark suite was run on both machines, first with no compiler optimization and then with the same optimization used in the experiments. Tables 1 – 2 summarize those benchmarking runs for the C-language and Fortran benchmarks.

DLNET is written mostly in Fortran, with only memory management and input/output routines written in C, yacc and lex. The experiments were done with version 1.4b (30 Jan 92) of DLNET. Version 1.4b of the code contains 5337 lines of Fortran (3434 of which are comments), 5623 lines of C, 515 lines of yacc and 116 lines of lex. This includes the implementation Dinic's algorithm of Goldfarb and Grigoriadis [13], used to solve the maximum flow problems on the restricted network described in Section 4. NETFLO has 1072 lines of Fortran code (290 of which are comments) while RELAXT-3 has 2559 lines (653 of which are comment lines). We modified NETFLO and RELAXT-3 to avoid integer overflow when computing the optimal objective function value. These computations were originally done in integer arithmetic and are carried out here in double precision floating point arithmetic. They are computed once, upon termination of the algorithm. The compilers f77 and cc are used with optimization level -O2 -Olimit 800. Running times are measured with the UNIX routine times() and exclude the time required to input the problem description.

```
begin
        mode                         minimize
        seed                         999 333
        maximum iterations           300
        maximum perturbation         1.e-3
        warm iterations              2
        maximum switch iteration     15
        active tolerance             1.e-3 1.e-4
        zero tolerance               1.e-20
        primal basic optimality      yes 10 1 1.e-8
        absolute gap optimality      yes 10 1 1.
        dual interior optimality     yes 10 1 1.e-15
        max flow optimality          yes 10 10 1.e-8
        sort buckets                 1
        cg tolerance                 1.e-3
        cg maximum iterations        1
        cg maximum diagonal          0.5
        cg residual check            5
end
```

FIGURE 2. DLNET specification file

5.2. DLNET Parameter Settings. DLNET obtains from a specification file run-time parameters that control the execution of the algorithm. Fine tuning these parameters for each individual problem could lead to faster execution times. However, for the experiments reported here, we run DLNET on all instances with identical parameter settings listed in Figure 2. The DAS step back factor γ is hard-wired in the code and cannot be set through the specification file, with $\gamma = 0.99$ for the first 10 iterations and $\gamma = 0.95$ thereafter.

5.3. Experimental Results. In this subsection, we present experimental results on the five problem classes mentioned in the beginning of this section. Several of the classes are further broken down into subclasses, for which we present an experiment summary composed of two tables and a figure. The first table displays running times and iteration counts for each of the three network codes. In most cases, in an attempt to reduce variance, values listed are averaged over three instances of each problem size, generated with different random seeds. For some larger problems, we were forced to consider data from only one instance, as multiple runs would be too time consuming. The second table summarizes the output for DLNET runs, listing averages for conjugate gradient iterations and running times, running times for spanning tree approximate sorting and Kruskal's algorithm, number of calls to the maximum flow optimality testing procedure, the average running time of each call, the number of edges in the restricted graph ($|\tilde{G}|$) and a measure of the density of this graph, ($|\tilde{E}|/(|V|-1)$). The figure shows running time ratios, in log-log scale, for NETFLO to DLNET, and RELAXT-3 to DLNET.

In all runs carried out in this experiment, the optimal objective function values found by DLNET (primal value), NETFLO and RELAXT-3 were identical.

TABLE 3. Summary of DLNET optimal solutions

Problem Class	Instances	PB-Stopping #	PB-Stopping %	MF-Stopping #	MF-Stopping %	PD-Opt Sol'n #	PD-Opt Sol'n %
Grid-Density-8	23	19	83	4	17	23	100
Grid-Density-16	18	5	28	13	72	18	100
Grid-Increasing-Density	18	3	17	15	83	18	100
RLG-Wide	7	0	0	7	100	7	100
Grid-Square	14	12	86	2	14	14	100
Grid-Wide	28	28	100	0	0	26	93
Grid-Long	28	27	96	1	4	28	100
Mesh_1	16	14	88	2	12	16	100
Mesh_2	16	11	69	5	31	16	100
Mesh_4	16	10	63	6	37	16	100
Mesh_8	12	8	67	4	33	12	100
Netgen-Lo	27	19	70	8	30	22	81
Netgen-Hi	27	23	85	4	15	25	93
All instances	250	179	72	71	28	241	96

In most cases, the dual objective function value obtained by DLNET was equal to the primal value. However, because we allowed DLNET to stop with a primal dual gap of less than one, but not necessarily zero, in a few cases, the dual objective value did not equal the primal. By changing the specification file to disallow stopping with absolute duality gap, all DLNET solutions could be made primal-dual complementary. Alternatively, a Bellman-Ford shortest-path computation can be added to the stopping routine, determining a dual optimal solution after primal optimality has been detected. Table 3 summarizes how DLNET stopped for all problem instances. For each problem class, the table lists the number of instances solved, the number of instances in which DLNET stopped with the primal estimate stopping criterion (PB-Stopping), the corresponding percentage of all instances for which this occurred, the number of instances in which DLNET stopped with the maximum flow stopping criterion (MF-Stopping), the corresponding percentage, the number of instances that DLNET produced a primal-dual complementary solution (PD-Opt Sol'n) and the percentage of instances for which this occurred.

5.3.1. *Transshipment Problems on a Grid.* The networks in this class of problems are obtained by removing the minimum number of edges from a grid graph embedded on a torus such that all vertical paths wrap around and no horizontal path wraps around. Next, we add source and sink vertices with edges going from the source to all vertices on one side of the resulting tube network and from all vertices on the other side to the sink. Network density is controlled by adding extra edges in both the horizontal and vertical directions. All vertices other than the source and sink vertices are transshipment vertices. Costs and capacities are uniformly distributed in the intervals $[0, 4096]$ and $[0, 16384]$, respectively.

These instances are generated with the MCNF problem generator goto.c by Goldberg [12]. Three subclasses of problems are generated: Grid-Density-8,

TABLE 4. CPU times for problem class Grid-Density-8

SIZE		DLNET		NETFLO		RELAXT-3	
$|V|$	$|E|$	time	itr	time	itr	time	itr
256	2048	5.6	22.7	1.6	5613.7	18.8	68620.0
512	4096	12.5	23.7	7.0	15472.0	106.1	214289.7
1024	8192	39.7	29.0	26.8	29750.0	432.5	600647.3
2048	16384	102.4	32.7	112.8	68479.7	1625.5	1248580.7
4096	32768	296.8	36.0	721.3	360894.0	6695.2	1890436.7
8192	65536	842.7	35.3	2740.7	330211.0	43694.8	9761998.3
16384	131072	2545.6	40.0	12116.9	755820.3	217432.2	29034906.7
32768	262144	18882.0	90.0	293822.2	18718769.0	did not run	
65536	524288	21587.1	60.0	381780.4	3715680.0	did not run	

TABLE 5. DLNET statistics for problem class Grid-Density-8

SIZE		Conj. Grad.		Span. Tree		Max Flow			
$|V|$	$|E|$	itr	time	sort	Kruskal	calls	time	$|\tilde{E}|$	$\frac{|\tilde{E}|}{|V|-1}$
256	2048	6.6	0.07	0.03	0.01	2.0	0.06	257.3	1.01
512	4096	6.2	0.15	0.06	0.01	2.0	0.15	514.3	1.01
1024	8192	7.0	0.38	0.13	0.03	2.3	0.42	1029.9	1.01
2048	16384	7.4	0.99	0.28	0.08	3.0	1.37	2063.3	1.01
4096	32768	8.5	2.95	0.59	0.06	3.3	5.09	4147.7	1.01
8192	65536	10.8	9.85	1.24	0.42	3.0	43.92	8210.2	1.00
16384	131072	12.2	27.55	2.64	0.97	3.3	155.31	16424.3	1.00
32768	262144	19.6	133.87	5.83	1.14	9.0	266.45	33131.6	1.01
65536	524288	14.4	156.03	11.12	5.54	6.0	1052.66	65842.7	1.00

Grid-Density-16 and Grid-Increasing-Density. Let m and n denote the number of vertices and edges of the network, respectively. For Grid-Density-8, $n = 8m$; for Grid-Density-16, $n = 16m$; and for Grid-Increasing-Density, $n = m^{1.5}$. For Grid-Density-8, instances having $256, 512, \ldots, 65536$ vertices were generated. For Grid-Density-16, instances with $256, 512, \ldots, 32768$ vertices are generated. For Grid-Increasing-Density the networks have $256, 512, \ldots, 8192$ vertices.

The random number generator seeds 270001, 270002, and 270003 were used to generate different instances for each problem size. For Grid-Density-8, 3 instances of $256, 512, \ldots, 16384$ vertices and single instances of 32768 and 65536 vertices were generated. RELAXT-3 was not run on the 2 largest instances. For Grid-Density-16, 3 instances of $256, 512, \ldots, 8192$ vertices and single instances 16384 and 32768 vertices were generated. For Grid-Increasing-Density, 3 instances were generated for each problem size. RELAXT-3 was run on only a single 8192 vertex instance.

Tables 4-5 and Figure 3 summarize runs for problem class Grid-Density-8, Tables 6-7 and Figure 4 for problem class Grid-Density-16 and Tables 8-9 and Figure 5 for problem class Grid-Increasing-Density.

We make the following observations regarding this family of problems.
- DLNET was faster than RELAXT-3 on all instances in subclasses Grid-Density-8, Grid-Density-16 and Grid-Increasing-Density. In one instance of Grid-Density-8, a speedup ratio of over 85 was observed, where RELAXT-3 took 60.4 hours

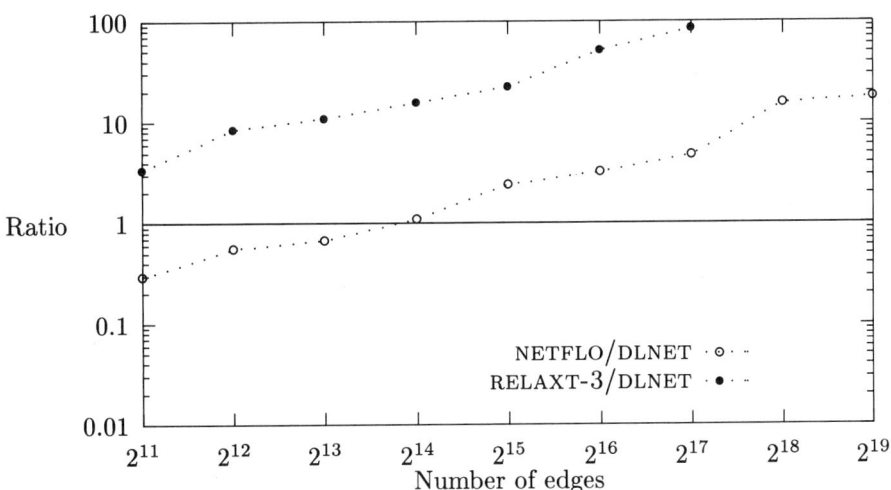

FIGURE 3. CPU time ratios for problem class Grid-Density-8

TABLE 6. CPU times for problem class Grid-Density-16

SIZE		DLNET		NETFLO		RELAXT-3	
$|V|$	$|E|$	time	itr	time	itr	time	itr
256	4096	16.4	32.3	4.3	14224.7	85.3	137897.7
512	8192	48.8	39.7	15.7	32477.7	355.5	335681.7
1024	16384	137.3	48.3	64.9	74794.3	1655.0	1092594.3
2048	32768	426.0	60.0	275.0	172897.7	5905.8	2614008.3
4096	65536	1007.0	66.7	1307.2	453328.0	27352.2	6324750.3
8192	131072	3069.9	80.0	5082.8	839747.0	110404.4	13928929.0
16384	262144	6866.8	70.0	26550.8	1948729.0	536569.2	46306590.0
32768	524288	26572.8	90.0	281157.8	15453366.0	2262740.0	107449245.0

TABLE 7. DLNET statistics for problem class Grid-Density-16

SIZE		Conj. Grad.		Span. Tree		Max Flow			
$|V|$	$|E|$	itr	time	sort	Kruskal	calls	time	$|\tilde{E}|$	$\frac{|\tilde{E}|}{|V|-1}$
256	4096	7.1	0.16	0.06	0.01	3.0	0.07	271.4	1.06
512	8192	7.3	0.34	0.13	0.01	3.7	0.16	543.6	1.06
1024	16384	7.9	0.92	0.28	0.06	4.7	0.47	1085.1	1.06
2048	32768	9.0	2.27	0.59	0.14	6.0	1.56	2234.1	1.09
4096	65536	8.7	5.34	1.27	0.08	6.7	4.39	4446.3	1.08
8192	131072	9.9	15.74	2.70	0.85	8.0	23.93	8788.8	1.07
16384	262144	11.9	43.18	5.76	1.70	7.0	97.86	17051.8	1.04
32768	524288	17.2	171.12	12.30	1.01	9.0	302.67	34520.9	1.05

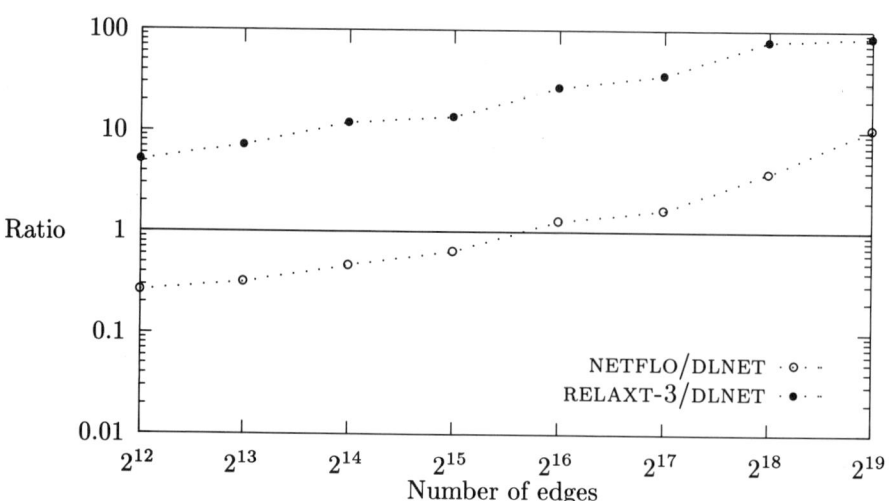

FIGURE 4. CPU time ratios for problem class Grid-Density-16

TABLE 8. CPU times for problem class Grid-Increasing-Density

SIZE		DLNET		NETFLO		RELAXT-3	
$|V|$	$|E|$	time	itr	time	itr	time	itr
256	4096	17.0	31.7	6.3	14224.7	99.4	137897.7
512	11585	79.4	50.0	38.5	47668.0	716.2	444212.0
1024	32768	513.9	70.0	297.2	188971.3	4411.3	1480909.3
2048	92682	2214.6	103.3	2191.1	679631.7	23886.3	3826721.3
4096	262144	9277.8	150.0	18383.2	2516757.3	132558.0	14827709.0
8192	741455	44623.1	200.0	121526.8	7467246.7	1138682.0	30883918.0

TABLE 9. DLNET statistics for problem class Grid-Density-Increasing

SIZE		Conj. Grad.		Span. Tree		Max Flow			
$|V|$	$|E|$	itr	time	sort	Kruskal	calls	time	$|\tilde{E}|$	$\frac{|\tilde{E}|}{|V|-1}$
256	4096	9.0	0.20	0.06	0.01	3.0	0.06	271.6	1.06
512	11585	10.1	0.70	0.19	0.01	5.0	0.17	569.3	1.11
1024	32768	15.3	3.30	0.60	0.07	7.0	0.75	1203.3	1.18
2048	92682	20.2	13.35	1.87	0.33	10.3	2.46	2643.2	1.29
4096	262144	12.9	27.88	5.58	0.13	15.0	6.74	4735.1	1.16
8192	741455	22.2	150.68	18.36	3.97	20.0	20.65	12561.4	1.53

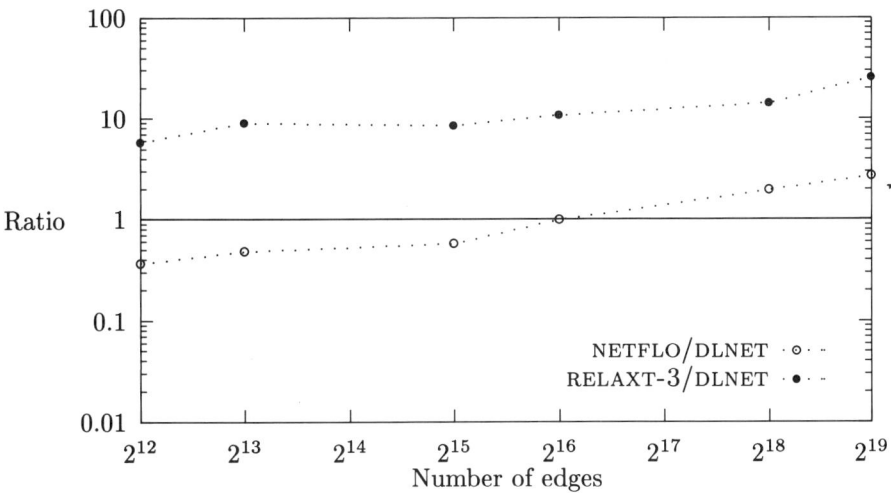

FIGURE 5. CPU time ratios for problem class Grid-Density-Increasing

(2.5 days) while DLNET took 42.4 minutes. In an instance of Grid-Density-16, a speedup ratio of over 85 was also observed, where RELAXT-3 took 628.5 hours (26.2 days) while DLNET took 7.4 hours. On instances from Grid-Increasing-Density, the greatest speedup was over 25, with DLNET solving the 8192 vertex instances on average in 12.4 hours while RELAXT-3 took 316.3 hours (13.2 days).

- DLNET was faster than NETFLO on the larger instances in each subclass. For Grid-Density-8, a speedup ratio of over 17 was observed, where NETFLO took 106.1 hours (4.4 days) while DLNET took 6.0 hours. For Grid-Density-16, a speedup ratio of over 10 was observed, with NETFLO taking 78.1 hours (3.2 days) while DLNET took 7.4 hours. In the subclass Grid-Increasing-Density, a speedup ratio of 2.7 was observed, with NETFLO taking 33.8 hours to solve a 8192 vertex instance while DLNET took 12.4 hours.
- DLNET-to-RELAXT-3 and DLNET-to-NETFLO solution time ratios decrease with network density.
- The conjugate gradient algorithm took on average 14.4, 17.2 and 22.2 iterations for the largest instances of classes Grid-Density-8, Grid-Density-16 and Grid-Increasing-Density, respectively. Those instances have, respectively, 65536, 32768, and 8192 vertices.
- For the largest instances of classes Grid-Density-8, Grid-Density-16 and Grid-Increasing-Density, the computation of the maximum weight spanning tree (sorting and Kruskal's algorithm) took respectively 0.6%, 0.7% and 13% of the total time taken by the conjugate gradient algorithm (computing the spanning

tree plus carrying out the conjugate gradient iterations). Sorting accounted for 67%, 92% and 84% of the total spanning tree computation time for classes Grid-Density-8, Grid-Density-16 and Grid-Increasing-Density, respectively.
- On the largest instances of Grid-Density-8, Grid-Density-16 and Grid-Increasing-Density, DLNET spent on solving maximum flow problems in the maximum flow stopping test the equivalent of respectively 61%, 16.4% and 1.2% of the time spent on the conjugate gradient algorithm.
- The density of the restricted network of the maximum flow stopping criterion increased with the density of the original network. The largest restricted network densities for Grid-Density-8, Grid-Density-16 and Grid-Increasing-Density had $|\tilde{E}|/(|V|-1)$ values of 1.00, 1.10 and 1.53, respectively.
- The instances had optimal objective function values with 10 to 12 digits.

5.3.2. *Maximum Flow Problems.* The class RLG-Wide consists of finding the maximum flow across a wide random leveled network. In these networks, vertices are arranged on a grid in 2^{x-6} rows by 64 columns ($x = 11, 12, \ldots$), with an additional source vertex and sink vertex. Edges go from the source vertex to each vertex in the first column of the grid and from each vertex in the last column to the sink. Furthermore, each vertex in the first 63 columns has an edge going to exactly three randomly selected vertices in the next column. Edge capacities in the grid are uniformly distributed in $[1, 10000]$. Edges out of the source and into the sink edges are uncapacitated.

We transform the maximum flow problem into a MCNF problem by adding an uncapacitated edge from the sink to the source with cost -1, and cost zero for all other edges. Often, these instances display a high degree of dual degeneracy, providing an interesting testbed for linear programming codes. None of the MCNF codes used in the experiments reported here take advantage of special properties specific to maximum flow problems. In particular, the maximum flow routine included in DLNET is not invoked in the initial stages of the algorithm, providing no shortcut to identify a maximum flow solution of the original network. Of course, in practice, specialized maximum flow codes are likely to offer much better performance in the solution of these problems.

The instances are generated with the generator washington.c by Anderson [12]. Seven instances on $2^{x-6} \times 64$ grids ($x = 11, 12, \ldots, 17$) are generated. Unfortunately, after performing the computational experiments described in this subsection, we noticed that the version of washington.c distributed for the Dimacs challenge uses a function of the time of day as the random number generator seed. Consequently, different instances are generated for separate invocations of the generator with the identical input parameters. The numbers of vertices and edges, however, do not change for a given input file. Tables 10-11 and Figure 6 summarize runs for problem class RLG-Wide.

We make the following observations regarding this family of problems.
- Except for the smallest problem category (with $x = 11$), DLNET was faster

TABLE 10. CPU times for problem class RLG-Wide

SIZE		DLNET		NETFLO		RELAXT-3	
$|V|$	$|E|$	time	itr	time	itr	time	itr
2050	6113	40.9	20.0	37.5	6848.0	67.3	2283.0
4098	12225	125.9	20.0	181.7	17640.0	287.9	4460.0
8194	24449	493.0	30.0	917.0	45469.0	1582.8	9824.0
16386	48897	1328.7	30.0	4806.0	120058.0	7891.2	19472.0
32770	97793	3854.5	30.0	25584.1	325919.0	52689.2	46386.0
65538	195585	14496.8	50.0	141157.0	849836.0	252447.2	95147.0
131074	391169	51293.8	70.0	962987.5	2408010.0	1460680.0	189577.0

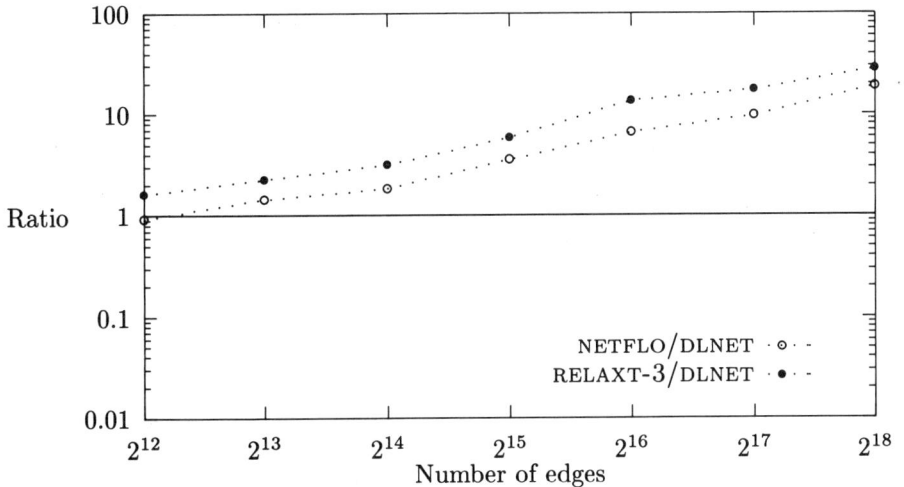

FIGURE 6. CPU time ratios for problem class RLG-Wide

TABLE 11. DLNET statistics for problem class RLG-Wide

SIZE		Conj. Grad.		Span. Tree		Max Flow			
$|V|$	$|E|$	itr	time	sort	Kruskal	calls	time	$|\tilde{E}|$	$\frac{|E|}{|V|-1}$
2050	6113	20.0	1.26	0.10	0.04	2.0	1.81	5869.0	2.86
4098	12225	25.8	4.56	0.20	0.09	2.0	5.10	11684.0	2.85
8194	24449	27.7	11.45	0.41	0.22	3.0	24.09	23076.0	2.82
16386	48897	29.3	31.07	0.87	0.54	3.0	75.74	46680.3	2.85
32770	97793	39.2	95.16	1.86	1.29	3.0	213.00	93035.0	2.84
65538	195585	27.6	151.17	4.11	3.31	5.0	1125.56	184545.2	2.82
131074	391169	31.6	385.24	8.64	7.46	7.0	2917.88	366764.4	2.80

than the other codes on all problem sizes. On the largest size category (131074 vertices and 391169 edges), DLNET was over 18 times faster than NETFLO and 28 times faster than RELAXT-3. On that size category, DLNET took 14.2 hours to solve the problem while NETFLO and RELAXT-3 took 267.5 hours (11.1 days) and 405.7 hours (16.9 days), respectively.

- On the largest instance, the conjugate gradient algorithm took on average 31.6 iterations.
- On the largest instance, the computation of the spanning tree (sorting plus Kruskal's algorithm) accounted for 0.4% of the total time taken by the conjugate gradient (computing the preconditioner plus conjugate gradient iterates). Sorting accounted for 54% of the total spanning tree computation time.
- On the largest instance of RLG-Wide, DLNET spent on solving maximum flow problems in the maximum flow stopping test the equivalent of 72.7% of the time spent on the conjugate gradient algorithm.
- Since these instances are highly dual degenerate, it is expected that the restricted networks of the maximum flow stopping criterion will be dense. In fact, the least dense restricted network had on average 2.8 ($|V| - 1$) edges.
- In all instances, DLNET stopped using the maximum flow stopping criterion. All solutions produced by DLNET were primal-dual complementary.
- The instances had optimal objective function values with 6 to 8 digits.

5.3.3. Minimum Cost-Maximum Flow Problems on a Grid.

The networks in this class are formed on a grid of vertices of height h and width w. Two additional vertices complete the vertex set: a source vertex S and a sink vertex T. Edges go from the source to each vertex in the first column of the grid and from each vertex in the last column of the grid to the sink vertex. On the grid, edges go from vertex to nearest neighbor vertex oriented left to right and top to bottom. Grid edge costs and capacities are generated uniformly in the interval $[1, 10000]$. Edges from the source and into the sink have cost zero and are uncapacitated. All vertices, except for source and sink, are transshipment vertices. The source has a supply of M_{ST}, the maximum flow from S to T, and the sink has a demand of M_{ST}.

These instances are generated with the MCNF problem generator ggraph1.f of Resende [12]. Three subclasses of problems are generated: Grid-Square, Grid-Wide, and Grid-Long. Grid-Square uses parameters $h = w = 16, 32, \ldots$, Grid-Wide $w = 16$ and $h = 32, 64, \ldots$ and Grid-Long $h = 16$ and $w = 32, 64, \ldots$. Karmarkar and Ramakrishnan [20] solved instances similar to those of Grid-Square.

The random number generator seeds 270001, 270002, and 270003 were used to generate different instances for each problem size. For Grid-Square, 3 instances of $258, 1026, \ldots, 16386$ vertices and single instances of 65538 and 262146 vertices were generated. For Grid-Wide, 3 instances of $514, 1026, \ldots, 131074$ vertices and a single instance of 262146 vertices were generated. RELAXT-3 was run on a

TABLE 12. CPU times for problem class Grid-Square

SIZE		DLNET		NETFLO		RELAXT-3	
$|V|$	$|E|$	time	itr	time	itr	time	itr
258	512	1.6	15.0	0.1	355.3	0.3	1354.3
1026	2048	12.7	25.0	0.7	1940.3	4.2	7417.3
4098	8192	149.2	38.3	8.6	10393.7	103.8	18453.0
16386	32768	2192.8	61.3	159.2	72847.3	2295.4	392862.0
65538	131072	17618.4	90.0	2625.9	486330.0	36947.8	2715011.0
262146	524288	255332.0	180.0	67189.7	3482255.0	920103.1	23519410.0

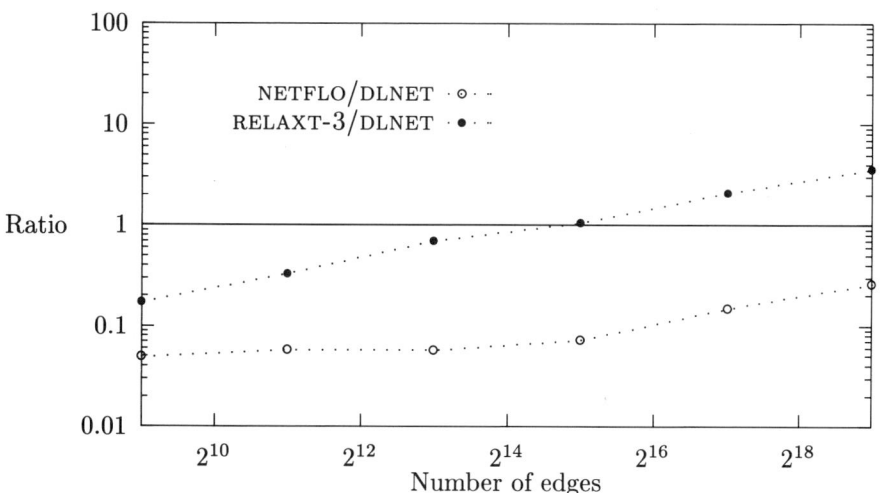

FIGURE 7. CPU time ratios for problem class Grid-Square

single 131074 vertex instance and was not run on the 262146 vertex instance. For Grid-Long, 3 instances of $258, 1026, \ldots, 131074$ vertices and a single instance of 242146 vertices were generated.

Tables 12-13 and Figure 7 summarize runs for problem class Grid-Square, Tables 16-17 and Figure 9 for problem class Grid-Long and Tables 14-15 and Figure 8 for problem class Grid-Wide.

We make the following observations regarding this family of problems.

- On both Grid-Square and Grid-Wide, the interior point code's performance relative to the other codes improved with problem size. On Grid-Long, it initially decreased slowly with size, but later leveled off. On Grid-Square, a speedup ratio of 3.6 relative to RELAXT-3 was observed on the 262146 vertex instance. RELAXT-3 took 255.6 hours (10.6 days) to solve the instance, while DLNET took 70.9 hours (2.9 days). On all instances, NETFLO was faster than DLNET, with a speedup ratio of 3.8 on the largest instance tested. On that instance, DLNET took 70.9 hours, while NETFLO solved the problem in 18.7

TABLE 13. DLNET statistics for problem class Grid-Square

SIZE		Conj. Grad.		Span. Tree		Max Flow			
$|V|$	$|E|$	itr	time	sort	Kruskal	calls	time	$|\tilde{E}|$	$\frac{|\tilde{E}|}{|V|-1}$
258	512	9.9	0.04	0.01	0.00	1.0	0.03	257.0	1.00
1026	2048	12.7	0.28	0.04	0.01	2.0	0.19	1025.0	1.00
4098	8192	17.4	2.61	0.14	0.07	3.3	1.83	4052.2	0.99
16386	32768	32.6	27.70	0.67	0.43	5.7	20.22	16159.8	0.99
65538	131072	47.0	158.46	2.62	1.97	9.0	146.94	64214.6	0.98
262146	524288	70.0	1193.90	12.46	10.19	18.0	1185.79	254344.3	0.97

TABLE 14. CPU times for problem class Grid-Wide

SIZE		DLNET		NETFLO		RELAXT-3	
$|V|$	$|E|$	time	itr	time	itr	time	itr
514	1040	3.7	19.7	0.2	956.3	1.1	3758.3
1026	2096	10.8	26.0	1.0	2471.3	5.6	13963.7
2050	4208	32.1	34.0	3.6	6143.7	29.4	48901.7
4098	8432	88.3	35.7	14.0	14625.0	167.3	140421.0
8194	16880	241.5	44.7	59.7	33214.0	1013.0	467138.7
16386	33776	720.6	49.7	232.0	71172.7	6341.8	1607827.3
32770	67568	1511.8	57.0	1003.3	156852.7	35670.5	5464625.0
65538	135152	3756.6	65.3	4207.0	351092.3	178462.5	12719770.7
131074	270320	11012.5	85.3	24619.6	786468.0	1677121.0	64887141.0
262146	540656	26810.8	113.0	107067.8	1736823.0	did not run	

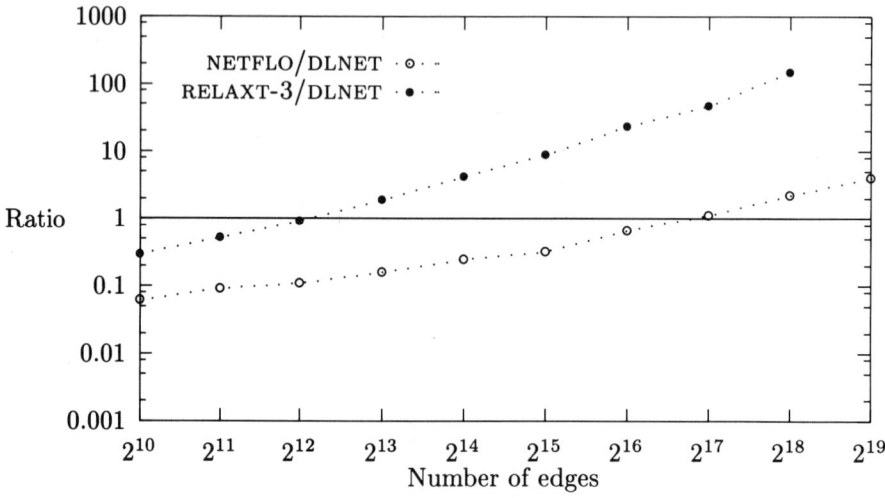

FIGURE 8. CPU time ratios for problem class Grid-Wide

TABLE 15. DLNET statistics for problem class Grid-Wide

SIZE		Conj. Grad.		Span. Tree		Max Flow			
$|V|$	$|E|$	itr	time	sort	Kruskal	calls	time	$|\tilde{E}|$	$\frac{|E|}{|V|-1}$
514	1040	9.9	0.08	0.02	0.00	1.3	0.06	512.7	1.00
1026	2096	10.8	0.20	0.03	0.01	2.0	0.16	1025.0	1.00
2050	4208	11.2	0.48	0.06	0.03	3.0	0.50	2047.3	1.00
4098	8432	10.3	1.47	0.14	0.07	3.0	1.27	4097.3	1.00
8194	16880	8.7	3.02	0.29	0.16	4.0	5.31	8189.5	1.00
16386	33776	9.3	8.61	0.63	0.40	4.3	14.73	16363.7	1.00
32770	67568	7.6	14.79	1.35	0.91	5.0	33.14	32765.0	1.00
65538	135152	6.6	28.97	3.01	2.33	6.0	89.13	65424.5	1.00
131074	270320	5.5	62.24	6.86	5.73	7.7	210.68	131020.2	1.00
262146	540656	4.8	103.29	13.44	10.76	11.0	469.69	261395.9	1.00

TABLE 16. CPU times for problem class Grid-Long

SIZE		DLNET		NETFLO		RELAXT-3	
$|V|$	$|E|$	time	itr	time	itr	time	itr
514	1008	4.7	20.3	0.2	656.7	0.7	1952.0
1026	2000	14.3	27.0	0.4	1207.3	2.2	696.7
2050	3984	50.2	35.3	1.2	2771.3	6.4	902.0
4098	7952	191.9	49.7	3.0	5259.7	19.0	545.3
8194	15888	599.0	59.3	6.9	8808.3	56.9	883.7
16386	31760	2318.5	88.3	18.1	17732.7	159.5	1878.0
32770	63504	7328.8	111.0	61.2	32102.0	411.2	2650.3
65538	126992	18340.7	142.3	199.3	65849.3	1294.9	4258.0
131074	253968	57773.6	150.3	888.6	169126.7	3626.9	5096.7
262146	507920	177017.6	164.0	2656.6	327205.0	11492.1	15892.0

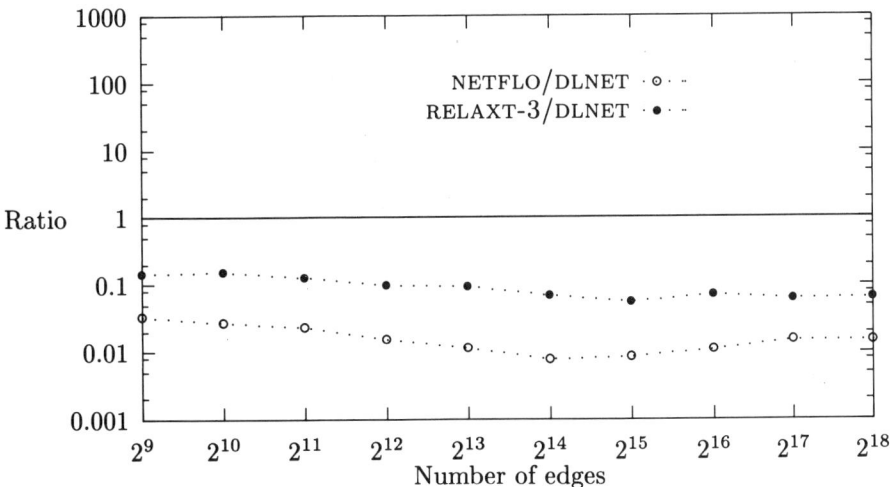

FIGURE 9. CPU time ratios for problem class Grid-Long

TABLE 17. DLNET statistics for problem class Grid-Long

SIZE		Conj. Grad.		Span. Tree		Max Flow			
$|V|$	$|E|$	itr	time	sort	Kruskal	calls	time	$|\tilde{E}|$	$\frac{|\tilde{E}|}{|V|-1}$
514	1008	12.4	0.11	0.02	0.01	1.3	0.06	511.8	1.00
1026	2000	13.7	0.31	0.03	0.01	2.0	0.19	1025.2	1.00
2050	3984	18.4	0.94	0.07	0.03	3.0	0.52	2041.7	1.00
4098	7952	19.5	2.84	0.14	0.06	4.7	1.62	3858.1	0.94
8194	15888	23.1	7.90	0.29	0.13	5.3	4.64	7796.4	0.95
16386	31760	25.8	21.53	0.61	0.35	8.7	11.78	15116.0	0.92
32770	63504	29.4	55.26	1.23	0.73	10.7	29.19	29620.6	0.90
65538	126992	33.5	109.16	2.39	1.69	13.3	56.24	58405.7	0.89
131074	253968	44.5	337.44	5.10	3.87	14.7	129.22	113328.8	0.86
262146	507920	61.5	976.76	10.50	8.29	16.0	324.62	231394.9	0.88

hours. On Grid-Wide, DLNET was over 152 times faster than RELAXT-3, solving a 131074 vertex instance in 3.1 hours while RELAXT-3 took 465.9 hours (19.4 days). On the largest instance, (262146 vertices) DLNET was 4 times faster then NETFLO, solving the problem in 7.5 hours while NETFLO took 29.7 hours. On Grid-Long, NETFLO was the fastest code, solving the 262146 vertex instance in only 44.3 minutes while RELAXT-3 took 3.2 hours and DLNET took 49.2 hours (2.1 days). For that instance, NETFLO was 66.6 times faster than DLNET and 4.3 times faster than RELAXT-3. RELAXT-3 was 15.4 times faster than DLNET.

- The conjugate gradient algorithm took on average 70, 4.8, and 61.5 iterations on the largest instances of classes Grid-Square, Grid-Wide and Grid-Long, respectively. Those instances have 262146 vertices.
- For the largest instances of classes Grid-Square, Grid-Wide and Grid-Long, the computation of the maximum weight spanning tree (sorting and Kruskal's algorithm) took respectively 1.9%, 19% and 1.9% of the total time taken by the conjugate gradient algorithm (computing the spanning tree plus carrying out the conjugate gradient iterations). Sorting accounted for 55%, 56% and 56% of the total spanning tree computation time for classes Grid-Square, Grid-Wide and Grid-Long, respectively.
- On the largest instances of Grid-Square, Grid-Wide and Grid-Long, DLNET spent on solving maximum flow problems in the maximum flow stopping test the equivalent of respectively 9.9%, 44.3% and 3.2% of the time spent on the conjugate gradient algorithm.
- All restricted networks of the maximum flow stopping criterion were very sparse. The largest restricted network densities for Grid-Square, Grid-Wide and Grid-Long had $|\tilde{E}|/(|V|-1)$ values of 1.00, 1.00 and 1.00, respectively.
- The instances had optimal objective function values with 10 to 13 digits.

5.3.4. *Minimum Cost Circulation Problems.* Networks in this class are formed on a grid of vertices embedded on a torus. Edges connect vertices in the same row or column of the grid. All horizontal edges have the same orientation.

TABLE 18. CPU times for problem class Mesh-1

SIZE		DLNET		NETFLO		RELAXT-3	
$|V|$	$|E|$	time	itr	time	itr	time	itr
256	512	1.6	15.7	0.2	1022.3	0.1	671.7
1024	2048	8.6	17.7	3.7	5833.7	0.9	3077.3
4096	8192	110.8	25.0	64.9	28962.3	10.3	13083.7
16384	32768	1291.6	32.0	1426.1	151025.7	98.6	56668.7
65536	131072	11462.6	49.3	29567.3	780700.7	857.2	226535.3
262144	524288	95445.9	80.0	506716.0	4174524.0	13122.6	958396.0

Similarly, all vertical edges are oriented in the same direction. Networks in the four subclasses Mesh-1, Mesh-2, Mesh-4 and Mesh-8 differ only with respect to vertex degree. All vertices in a network from a specific subclass have the same degree. In Mesh-1, each vertex has an edge going to each nearest neighbor (one in the horizontal direction, the other in the vertical direction). In Mesh-2, Mesh-4 and Mesh-8, edges go from a vertex to, respectively, its 4,8 and 16 nearest neighbors. Costs are generated uniformly in the interval $[-1000, 1000]$. Capacities are generated in the interval $[-1000, 1000]$ with a bias that makes longer edges have smaller capacities.

The instances are generated with the MCNF generator mesh.c by Goldberg [12]. All vertex sets are $h \times w$ grids. We generate all instances with $h = w$. In Mesh-1, Mesh-2 and Mesh-4, instances are generated with parameters $h = w = 16, 32, \ldots, 512$, while in Mesh-8, $h = w = 16, 32, \ldots, 128$.

The random number generator seeds 270001, 270002, and 270003 were used to generate different instances for each problem size. For Mesh-1, Mesh-2 and Mesh-4, 3 instances with $256, 1024, \ldots, 65536$ vertices and a single instance of 262144 vertices were generated. NETFLO was not run on two of the 65536 vertex instances and on the 262144 vertex instance of Mesh-4. For Mesh-8, 3 instances with $256, 1024, \ldots, 16384$ vertices were generated.

Since NETFLO requires lower bounds $l = 0$ we applied the change of variables $x' = x - l$ to
$$\min \{c^\top x \mid Ax = b,\ l \leq x \leq u\},$$
resulting in
$$\min \{c^\top x' + c^\top l \mid Ax' = b - Al,\ 0 \leq x' \leq u - l\}.$$

Tables 18-19 and Figure 10 summarize runs for problem class Mesh-1, Tables 20-21 and Figure 11 for problem class Mesh-2, Tables 22-23 and Figure 12 for problem class Mesh-4 and Tables 24-25 and Figure 13 for problem class Mesh-8.

We make the following observations regarding this family of problems.
- RELAXT-3 was the fastest code on all instances. However, on all subclasses, the interior point code's performance relative to the other codes improved with problem size. The asymptotic advantage over RELAXT-3 has little significance,

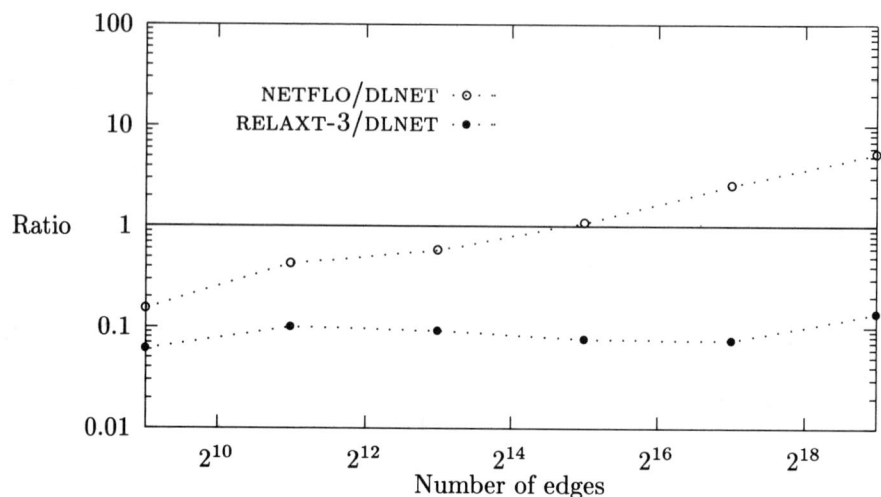

FIGURE 10. CPU time ratios for problem class Mesh-1

TABLE 19. DLNET statistics for problem class Mesh-1

SIZE		Conj. Grad.		Span. Tree		Max Flow			
$|V|$	$|E|$	itr	time	sort	Kruskal	calls	time	$|\tilde{E}|$	$\frac{|E|}{|V|-1}$
256	512	8.5	0.04	0.01	0.00	1.0	0.05	255.0	1.00
1024	2048	11.2	0.25	0.03	0.01	1.0	0.23	1023.3	1.00
4096	8192	13.4	2.99	0.14	0.06	2.0	3.85	4096.2	1.00
16384	32768	16.5	28.79	0.65	0.40	3.0	48.12	16387.4	1.00
65536	131072	20.1	169.16	3.15	2.33	4.3	313.55	65553.5	1.00
262144	524288	19.7	772.43	13.70	11.41	8.0	2630.95	262212.4	1.00

TABLE 20. CPU times for problem class Mesh-2

SIZE		DLNET		NETFLO		RELAXT-3	
$|V|$	$|E|$	time	itr	time	itr	time	itr
256	1024	2.7	17.0	0.7	3054.7	0.2	904.3
1024	4096	16.3	21.7	13.9	18847.0	1.3	3645.0
4096	16384	144.9	28.0	280.4	100593.7	16.4	16878.7
16384	65536	1615.8	42.7	5201.7	527599.0	155.0	71540.7
65536	262144	12297.3	53.3	106817.3	2828038.3	1408.6	295908.3
262144	1048576	102695.2	80.0	1794491.0	14929957.0	16627.2	1271727.0

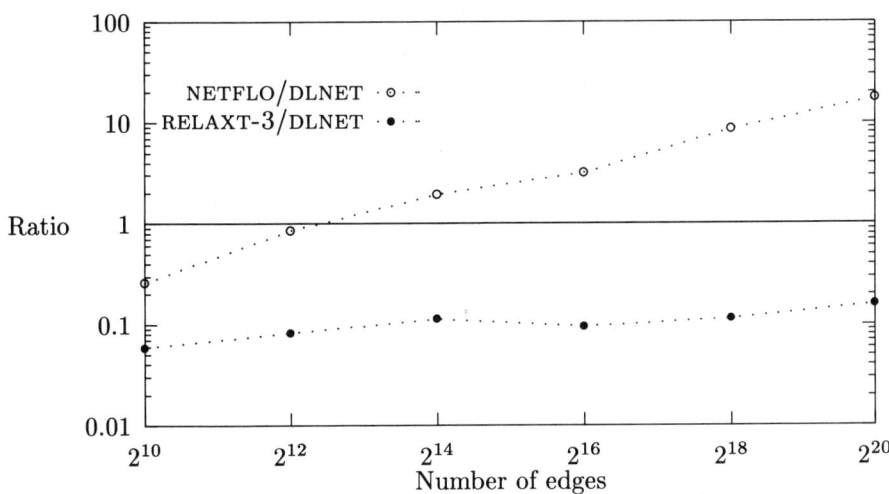

FIGURE 11. CPU time ratios for problem class Mesh-2

TABLE 21. DLNET statistics for problem class Mesh-2

SIZE		Conj. Grad.		Span. Tree		Max Flow			
$|V|$	$|E|$	itr	time	sort	Kruskal	calls	time	$|\tilde{E}|$	$\frac{|\tilde{E}|}{|V|-1}$
256	1024	9.2	0.06	0.02	0.00	1.0	0.04	255.0	1.00
1024	4096	9.4	0.32	0.07	0.01	1.7	0.26	1024.8	1.00
4096	16384	12.2	2.86	0.30	0.08	2.0	2.61	4101.5	1.00
16384	65536	14.8	22.77	1.40	0.49	3.7	33.92	16404.0	1.00
65536	262144	17.1	139.79	6.46	3.00	5.3	310.95	65614.0	1.00
262144	1048576	25.7	846.37	26.41	13.43	8.0	2073.83	262505.1	1.00

TABLE 22. CPU times for problem class Mesh-4

SIZE		DLNET		NETFLO		RELAXT-3	
$|V|$	$|E|$	time	itr	time	itr	time	itr
256	2048	5.4	19.7	1.6	6761.0	0.3	994.3
1024	8192	37.8	25.3	33.8	50395.0	2.7	4354.3
4096	32768	309.1	34.3	798.4	285664.3	35.0	20556.3
16384	131072	2554.1	45.0	18435.3	1512214.7	311.1	87593.0
65536	524288	19409.6	63.3	302213.5	7659533.0	2511.8	358163.3
262144	2097152	168938.0	130.0	did not run		20091.4	1417442.0

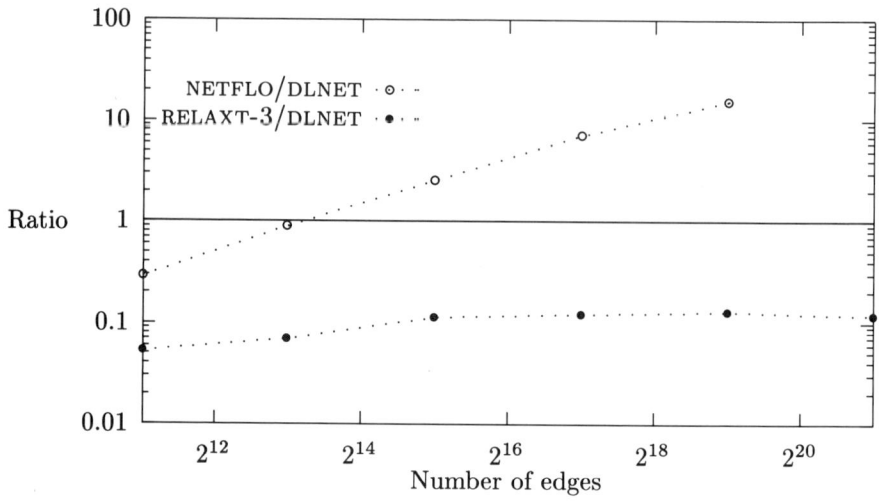

FIGURE 12. CPU time ratios for problem class Mesh-4

TABLE 23. DLNET statistics for problem class Mesh-4

SIZE		Conj. Grad.		Span. Tree		Max Flow			
$|V|$	$|E|$	itr	time	sort	Kruskal	calls	time	$|\tilde{E}|$	$\frac{|\tilde{E}|}{|V|-1}$
256	2048	9.7	0.10	0.03	0.00	1.7	0.05	255.3	1.00
1024	8192	9.8	0.54	0.13	0.02	2.0	0.24	1025.8	1.00
4096	32768	11.7	4.01	0.61	0.08	3.0	2.33	4108.0	1.00
16384	131072	14.4	31.35	2.90	0.52	4.3	27.81	16430.7	1.00
65536	524288	16.1	172.38	12.90	3.34	6.3	273.03	65737.2	1.00
262144	2097152	23.8	777.67	53.15	17.70	13.0	1411.27	262999.8	1.00

TABLE 24. CPU times for problem class Mesh-8

SIZE		DLNET		NETFLO		RELAXT-3	
$|V|$	$|E|$	time	itr	time	itr	time	itr
256	4096	12.3	21.0	2.9	12368.0	0.6	1133.7
1024	16384	93.1	27.7	74.7	115804.7	5.3	5365.3
4096	65536	730.2	37.0	1806.6	712742.0	51.9	24169.3
16384	262144	5822.2	56.7	41255.4	3688382.0	530.2	102671.0

TABLE 25. DLNET statistics for problem class Mesh-8

SIZE		Conj. Grad.		Span. Tree		Max Flow			
$|V|$	$|E|$	itr	time	sort	Kruskal	calls	time	$|\tilde{E}|$	$\frac{|\tilde{E}|}{|V|-1}$
256	4096	10.9	0.24	0.06	0.00	2.0	0.06	256.3	1.00
1024	16384	12.7	1.50	0.28	0.02	2.0	0.31	1028.7	1.00
4096	65536	16.4	9.96	1.32	0.11	3.3	2.96	4123.0	1.01
16384	262144	19.5	56.67	5.88	0.58	5.7	30.26	16497.7	1.01

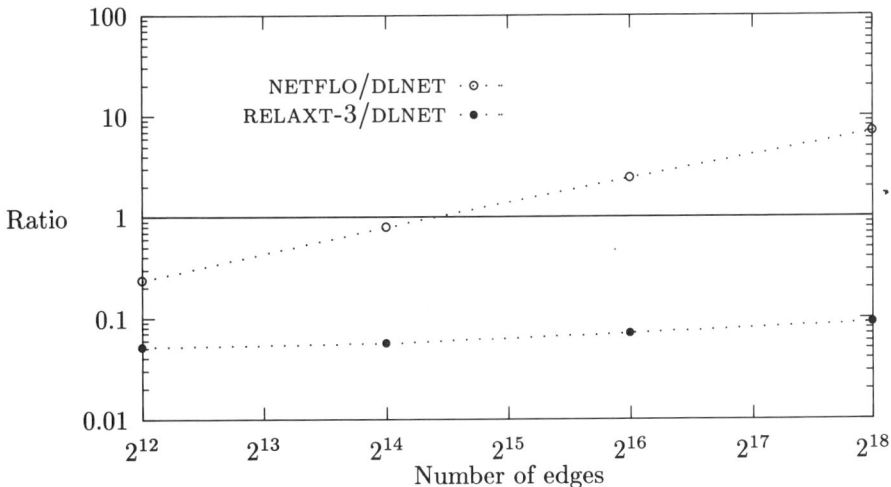

FIGURE 13. CPU time ratios for problem class Mesh-8

since the crossover should occur only for very large networks. On Mesh-1, DL-NET was 5.3 times faster than NETFLO on the largest instances. NETFLO took, on average, 140.8 hours (5.9 days), while DLNET took 26.5 hours. RELAXT-3 was 7.2 times faster than DLNET on those instances taking 3.6 hours. On Mesh-2, DLNET was over 17 times faster than NETFLO, taking, on average, 28.5 hours to solve the largest instances, compared to 498.5 hours (20.8 days) for NETFLO and 4.6 hours for RELAXT-3. On average, RELAXT-3 was 6.2 times faster than DLNET on those largest instances. On the 65536 vertex instances of class Mesh-4, DLNET was, on average, 15.6 times faster than NETFLO, taking 5.6 hours, compared to 83.9 hours (3.5 days) for NETFLO and 41.9 minutes for RELAXT-3. On the 262144 vertex instance RELAXT-3 was 8.4 times faster than DLNET, taking 5.6 hours to solve the instance while DLNET took 46.9 hours (1.9 days). On the largest instances of Mesh-8, DLNET was, on average, 7.1 times faster than NETFLO, solving the problems in 1.6 hours, while NETFLO took 11.5 hours and RELAXT-3 took 8.8 minutes. RELAXT-3 was 11 times faster than DLNET in those instances.

- The conjugate gradient algorithm took on average 19.7, 25.7, 23.8 and 19.5 iterations on the largest instances of classes Mesh-1, Mesh-2, Mesh-4, and Mesh-8, respectively. Those instances have, respectively 262144, 262144, 262144, and 16384 vertices.
- For the largest instances of classes Mesh-1, Mesh-2, Mesh-4 and Mesh-8, the computation of the maximum weight spanning tree (sorting and Kruskal's algorithm) took respectively 3.1%, 4.7%, 8.3% and 10.2% of the total time

seed	Random number seed:	270001, 270002, 270003
problem	Problem number (for output):	1
nodes	Number of nodes:	$m = 2^x$
sources	Number of sources:	2^{x-2}
sinks	Number of sinks:	2^{x-2}
density	Number of (requested) arcs:	2^{x+3}
mincost	Minimum arc cost:	0
maxcost	Maximum arc cost:	4096
supply	Total supply:	$2^{2(x-2)}$
tsources	Transshipment sources:	0
tsinks	Transshipment sinks:	0
hicost	Skeleton arcs with max cost:	100%
capacitated	Capacitated arcs:	100%
mincap	Minimum arc capacity:	1
maxcap	Maximum arc capacity:	16

FIGURE 14. NETGEN specification file for Netgen-Lo

taken by the conjugate gradient algorithm (computing the spanning tree plus carrying out the conjugate gradient iterations). Sorting accounted for 54.6%, 66.3%, 75.0% and 91.0% of the total spanning tree computation time for classes Mesh-1, Mesh-2, Mesh-4, and Mesh-8, respectively.
- On the largest instances of Mesh-1, Mesh-2, Mesh-4, and Mesh-8, DLNET spent on solving maximum flow problems in the maximum flow stopping test the equivalent of respectively 33.0%, 24.5%, 18.1% and 4.8% of the time spent on the conjugate gradient algorithm.
- All restricted networks of the maximum flow stopping criterion were very sparse. The largest restricted network densities for Mesh-1, Mesh-2, Mesh-4, and Mesh-8, had $|\tilde{E}|/(|V|-1)$ values of 1.00, 1.00, 1.00 and 1.01, respectively.
- The instances had optimal objective function values with 8 to 11 digits.

5.3.5. *NETGEN Problems.* Most computational studies of network optimization codes in the past have used the MCNF generator NETGEN [22] of Klingman, Napier and Stutz. In this subsection, we test the codes on two classes of networks generated with NETGEN: Netgen-Lo and Netgen-Hi. Figure 14 lists the NETGEN parameters used to generate the class Netgen-Lo. Networks in class Netgen-Hi are generated with the same parameters except for maxcap, which is set to 16384. For both subclasses 3 instances of each size are generated. Sizes correspond to the settings $x = 8, 9, \ldots, 16$.

Tables 26-27 and Figure 15 summarize runs for problem class Netgen-Lo and Tables 28-29 and Figure 16 for problem class Netgen-Hi.

We make the following observations regarding this family of problems.
- With respect to NETFLO, the relative speedup of DLNET increases with size for both subclasses. On Netgen-Lo, an average speedup of 10.5 was observed for the 65526 vertex instances with DLNET taking an average of 8.2 hours while NETFLO took 86.0 hours (3.6 days). On Netgen-Hi, an average speedup of 6.6 was observed for the 65526 vertex instances with DLNET taking an

TABLE 26. CPU times for problem class Netgen-Lo

SIZE		DLNET		NETFLO		RELAXT-3	
$|V|$	$|E|$	time	itr	time	itr	time	itr
256	2048	6.5	21.3	0.7	3517.0	0.3	958.3
512	4096	16.7	23.7	2.5	9258.7	1.4	2528.0
1024	8204	41.0	29.3	9.4	21541.7	5.0	5707.3
2048	16415	116.3	33.7	40.2	49142.7	28.2	13035.3
4096	32877	308.4	37.3	202.5	113142.7	83.2	31832.3
8192	65750	1025.8	47.0	1214.1	264955.0	306.1	69652.0
16384	131442	3414.9	55.7	8447.5	601164.3	1445.2	148248.0
32768	262909	11833.7	76.7	53656.2	1413134.3	3680.2	316751.3
65536	525792	29366.2	83.3	309572.9	3273724.0	19888.8	656165.0

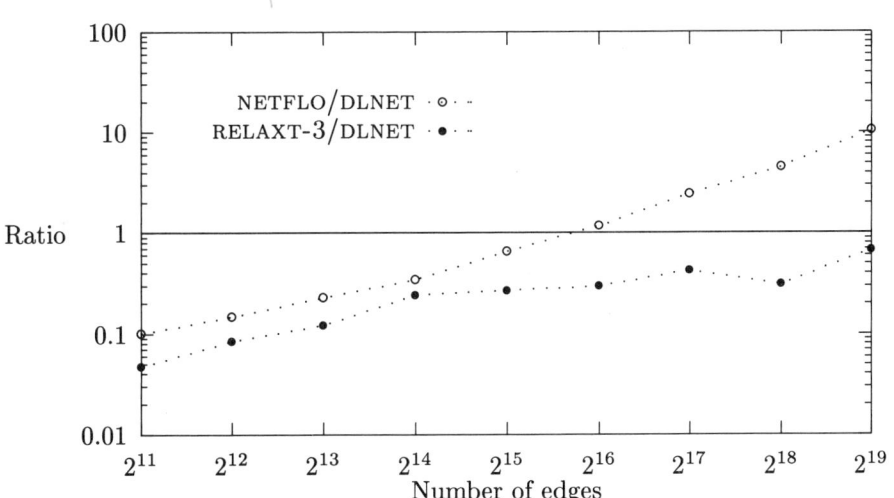

FIGURE 15. CPU time ratios for problem class Netgen-Lo

TABLE 27. DLNET statistics for problem class Netgen-Lo

SIZE		Conj. Grad.		Span. Tree		Max Flow			
$|V|$	$|E|$	itr	time	sort	Kruskal	calls	time	$|\tilde{E}|$	$\frac{|\tilde{E}|}{|V|-1}$
256	2048	9.8	0.11	0.04	0.01	1.7	0.03	255.3	1.00
512	4096	10.1	0.32	0.07	0.02	2.0	0.08	512.3	1.00
1024	8204	9.6	0.62	0.14	0.03	2.3	0.19	1033.0	1.01
2048	16415	10.2	1.78	0.31	0.08	3.0	0.69	2060.7	1.01
4096	32877	11.1	4.86	0.60	0.17	3.0	2.21	4132.6	1.01
8192	65750	12.6	13.60	1.31	0.46	4.3	12.13	8233.7	1.00
16384	131442	14.2	40.32	2.95	1.28	5.3	46.18	16438.5	1.00
32768	262909	15.8	101.37	6.41	3.03	7.7	158.25	33029.0	1.01
65536	525792	17.5	237.19	12.83	6.58	8.3	409.82	66054.9	1.01

TABLE 28. CPU times for problem class Netgen-Hi

SIZE		DLNET		NETFLO		RELAXT-3	
$|V|$	$|E|$	time	itr	time	itr	time	itr
256	2048	8.1	29.7	0.2	1102.0	0.2	431.3
512	4096	22.3	34.3	0.8	2980.3	0.6	806.3
1024	8204	58.1	38.0	3.3	7768.0	2.5	1769.7
2048	16415	172.5	42.7	18.0	20193.3	6.8	3955.0
4096	32877	632.8	61.3	101.4	52000.7	16.6	8107.0
8192	65750	1430.0	58.3	705.9	140144.0	50.3	17596.7
16384	131442	5889.2	58.7	5010.3	397343.0	139.2	39109.0
32768	262909	19176.1	124.7	57912.7	1151210.0	380.5	90430.0
65536	525792	67475.8	200.0	404331.0	3456137.7	1136.5	211132.3

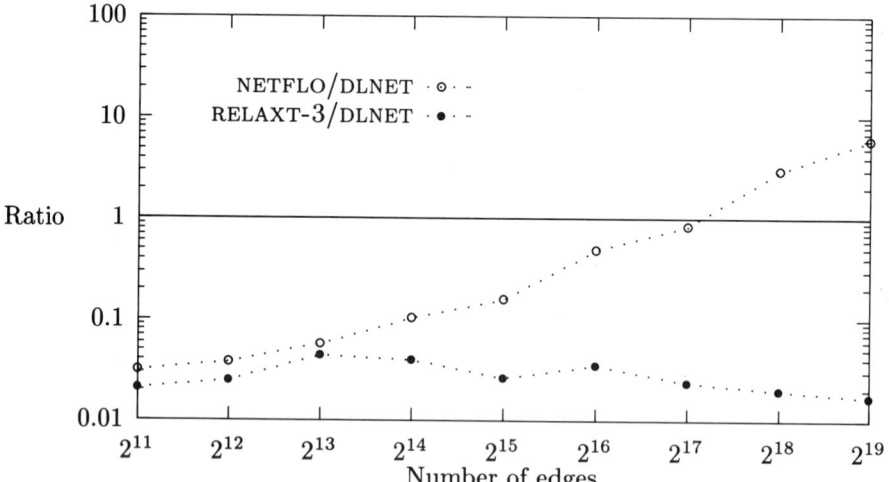

FIGURE 16. CPU time ratios for problem class Netgen-Hi

TABLE 29. DLNET statistics for problem class Netgen-Hi

SIZE		Conj. Grad.		Span. Tree		Max Flow			
$|V|$	$|E|$	itr	time	sort	Kruskal	calls	time	$|\tilde{E}|$	$\frac{|\tilde{E}|}{|V|-1}$
256	2048	10.5	0.11	0.03	0.01	2.3	0.01	247.6	0.97
512	4096	10.9	0.30	0.06	0.02	3.0	0.03	499.8	0.98
1024	8204	13.4	0.81	0.13	0.04	3.0	0.07	1023.2	1.00
2048	16415	14.8	2.49	0.29	0.09	4.0	0.26	1944.3	0.95
4096	32877	15.7	7.10	0.60	0.20	5.3	0.55	3980.9	0.97
8192	65750	15.6	17.13	1.31	0.50	5.3	3.19	7768.2	0.95
16384	131442	31.7	84.12	2.86	1.26	5.7	8.88	15598.0	0.95
32768	262909	29.0	167.00	6.30	2.70	12.3	42.75	30670.5	0.94
65536	525792	21.1	268.14	12.86	7.04	20.0	124.58	63939.0	0.98

average of 18.7 hours while NETFLO took 123.9 hours (5.2 days). With respect to RELAXT-3, the relative speedup of DLNET increased with size for Netgen-Lo, but decreased for Netgen-Hi. On the 65536 vertex Netgen-Lo instances, RELAXT-3 was on average 1.5 times faster than DLNET, with DLNET taking an average of 8.2 hours while RELAXT-3 took 5.5 hours. On the 65536 vertex Netgen-Hi instances, an average speedup of 59.4 was observed with DLNET taking an average of 18.7 hours while RELAXT-3 took only 18.9 minutes.
- RELAXT-3 was the most sensitive code to changes in the edge capacities. For the 65536 vertex networks RELAXT-3 was 17.5 faster on the Netgen-Hi networks than on the Netgen-Lo. DLNET was 2.3 times faster on the 65536 vertex Netgen-Lo instances than on Netgen-Hi. NETFLO was 1.4 times faster on the 65536 vertex Netgen-Lo instances than on Netgen-Hi.
- The conjugate gradient algorithm took on average 17.5, and 21.1 iterations for the 65536 vertex instances of classes Netgen-Lo, and Netgen-Hi, respectively.
- For the 65536 vertex instances of classes Netgen-Lo and Netgen-Hi the computation of the maximum weight spanning tree (sorting and Kruskal's algorithm) took respectively 7.6%, and 6.9% of the total time taken by the conjugate gradient algorithm (computing the spanning tree plus carrying out the conjugate gradient iterations). Sorting accounted for 66% and 65% of the total spanning tree computation time for classes Netgen-Lo and Netgen-Hi, respectively.
- On the 65536 vertex instances of Netgen-Lo and Netgen-Hi, DLNET spent on solving maximum flow problems in the maximum flow stopping test the equivalent of respectively 15.9%, and 4.3% of the time spent on the conjugate gradient algorithm.
- The restricted networks of the maximum flow stopping criterion were very sparse. The largest restricted network densities for Netgen-Lo and Netgen-Hi had $|\tilde{E}|/(|V|-1)$ values of 1.01 and 1.00, respectively.
- The instances had optimal objective function values with 7 to 13 digits.

5.4. Discussion. Before we describe computational results comparing NETFLO with CPLEX 2.0, we make several comments based on statistics taken from all instances solved in the main experiment. Two tables and two figures illustrate this discussion. Table 3 shows how DLNET stopped and what type of optimal solutions were generated. Table 30 shows the regression models for each problem class and for the combined set of problems. Figure 17 plots running times for all codes on all instances in the experiment. Figure 18 plots conjugate gradient iterations for all instances tested.

We make the following observations regarding the experimental results.
- All codes solved all instances on which the were run to optimality. Instances in the problem set had up to 13-digit optimal objective function values.
- Over the complete set of instances, DLNET was asymptotically the fastest code and had the most predictable running times. A linear regression of each family-

TABLE 30. Regression coefficients for all family-code combinations

Problem Class	Code	β_0	β_1	R^2	n
Grid-Density-16	DLNET	-4.150	1.49	0.9982	18
	NETFLO	-7.331	2.18	0.9944	18
	RELAXT-3	-5.643	2.09	0.9982	18
Grid-Density-8	DLNET	-4.429	1.54	0.9920	23
	NETFLO	-7.368	2.26	0.9910	23
	RELAXT-3	-6.041	2.21	0.9954	21
Grid-Increasing-Density	DLNET	-4.213	1.51	0.9976	18
	NETFLO	-6.179	1.92	0.9996	18
	RELAXT-3	-4.272	1.74	0.9964	16
Grid-Long	DLNET	-4.456	1.72	0.9981	28
	NETFLO	-5.496	1.53	0.9910	28
	RELAXT-3	-4.736	1.54	0.9965	28
Grid-Square	DLNET	-4.591	1.74	0.9982	14
	NETFLO	-6.496	1.93	0.9950	14
	RELAXT-3	-6.521	2.18	0.9990	14
Grid-Wide	DLNET	-3.657	1.42	0.9976	28
	NETFLO	-6.907	2.06	0.9987	28
	RELAXT-3	-7.606	2.51	0.9990	25
Mesh-1	DLNET	-4.344	1.64	0.9969	16
	NETFLO	-6.419	2.12	0.9993	16
	RELAXT-3	-5.570	1.67	0.9982	16
Mesh-2	DLNET	-4.302	1.55	0.9980	16
	NETFLO	-6.590	2.14	0.9998	16
	RELAXT-3	-5.840	1.66	0.9965	16
Mesh-4	DLNET	-4.234	1.49	0.9990	16
	NETFLO	-7.193	2.23	0.9995	13
	RELAXT-3	-5.935	1.64	0.9974	16
Mesh-8	DLNET	-4.268	1.48	0.9994	12
	NETFLO	-7.831	2.30	0.9999	12
	RELAXT-3	-6.089	1.62	0.9982	12
RLG-Wide	DLNET	-4.821	1.70	0.9986	7
	NETFLO	-7.663	2.43	0.9992	7
	RELAXT-3	-7.407	2.42	0.9992	7
Netgen-Lo	DLNET	-4.399	1.55	0.9961	27
	NETFLO	-8.174	2.36	0.9942	25
	RELAXT-3	-6.942	1.96	0.9972	27
Netgen-Hi	DLNET	-4.534	1.62	0.9944	27
	NETFLO	-9.583	2.61	0.9911	27
	RELAXT-3	-5.883	1.57	0.9914	27
All instances	DLNET	-4.105	1.52	0.9556	250
	NETFLO	-7.081	2.12	0.9188	245
	RELAXT-3	-5.362	1.74	0.6614	243

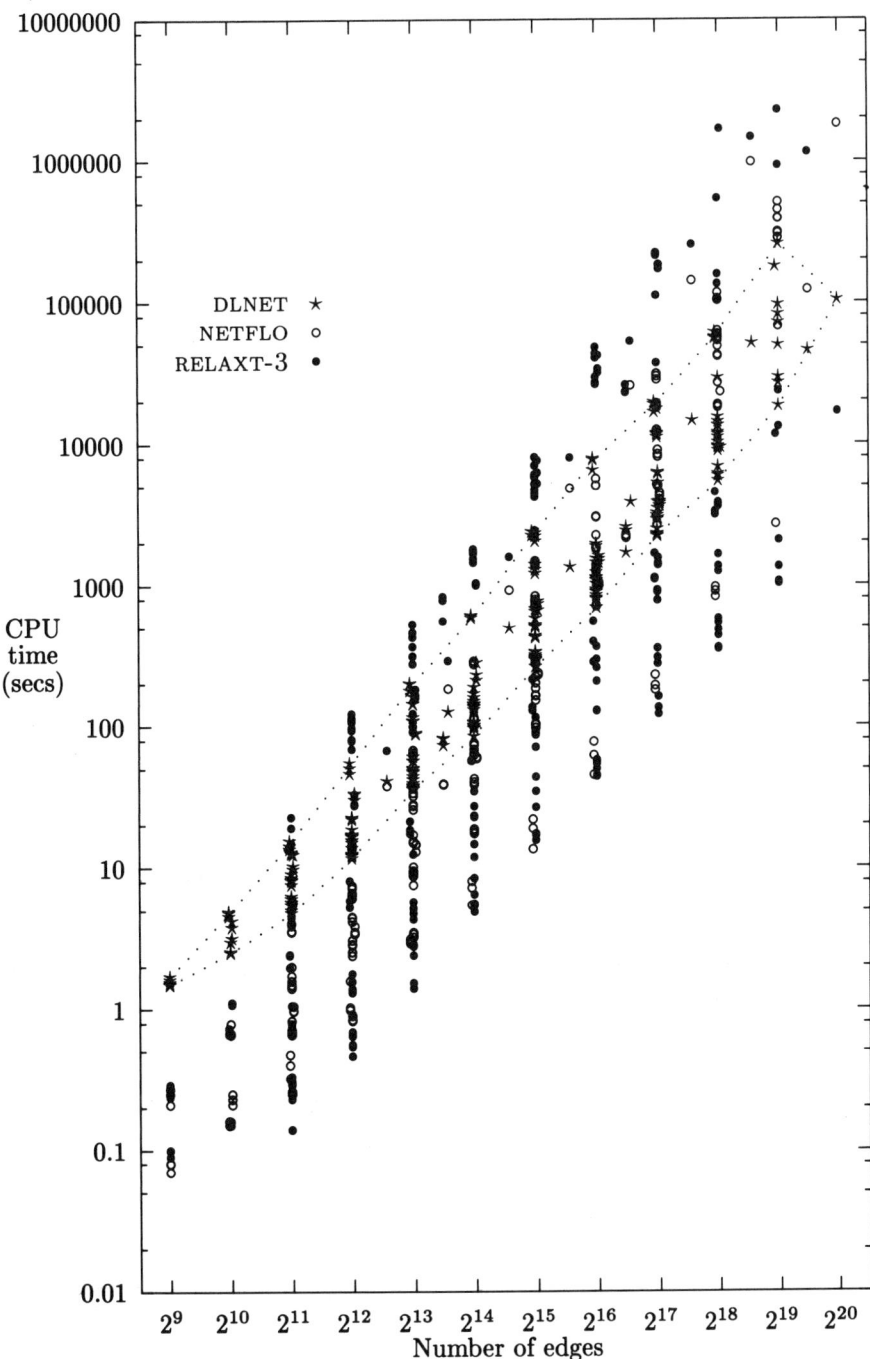

FIGURE 17. CPU times for all problem classes

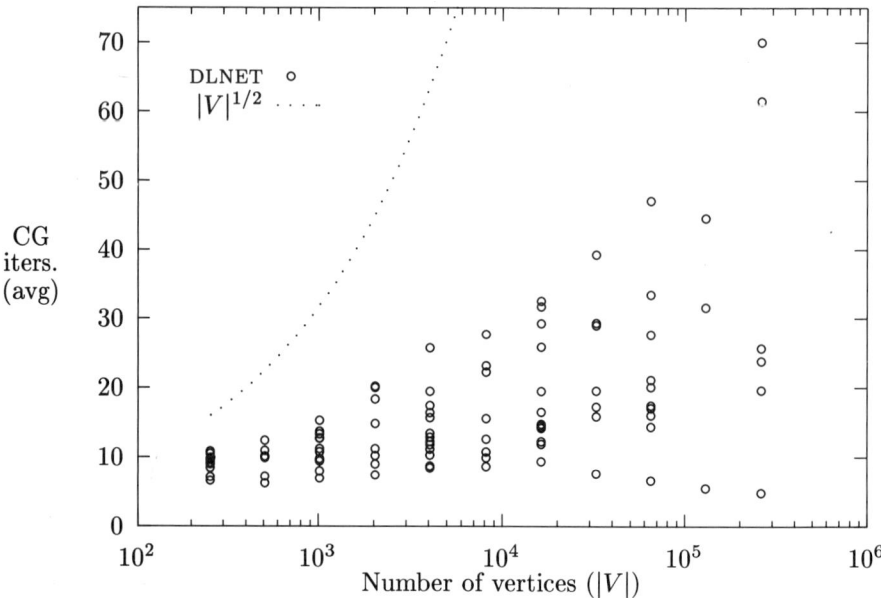

FIGURE 18. Average CG iterations for all problem classes

code data set was computed, using model

$$T = 10^{\beta_0} |E|^{\beta_1},$$

where T is the CPU time in seconds, $|E|$ is the number of edges and β_1, β_0 are the regression parameters to be estimated. The regression used all individual data points (not means). Table 30 summarizes the results of the regression, listing the model coefficients, the coefficient of multiple determination (R^2) and the sample size (n). The fit was good ($R^2 > 0.99$) for all family-code combinations. On the complete set of data, the regression models were

$$T_D = 10^{-4.105}|E|^{1.52} \ (R^2 = .9556)$$
$$T_R = 10^{-5.362}|E|^{1.74} \ (R^2 = .6614)$$
$$T_N = 10^{-7.081}|E|^{2.12} \ (R^2 = .9188)$$

indicating that asymptotically DLNET was the fastest, followed by RELAXT-3 and then by NETFLO. For small networks, the model ranks the codes in the reverse order. The coefficients of multiple determination R^2 show that DLNET had the most predictable running times, while RELAXT-3 had the most variability. Figure 17 illustrates well the variability of the running times. A region indicated by a dotted contour contains all DLNET CPU times.

TABLE 31. Instances with running time over 1 CPU day

CPU days	1	2	3	4	5	10	15	20	25	30
DLNET	5	2	0	0	0	0	0	0	0	0
NETFLO	19	11	11	7	4	2	1	1	0	0
RELAXT-3	17	12	6	6	6	5	3	1	1	0

- Restricted to the largest instance of each subclass, RELAXT-3 was the fastest code in 6 subclasses (4 of which were Mesh subclasses), DLNET was the fastest in 5 subclasses (3 of which Grid-Density), and NETFLO was the fastest in 2 subclasses (Grid-Long and Grid-Square). On the other hand, DLNET was the slowest in a single subclass, while NETFLO and RELAXT-3 were the slowest in 6 subclasses each.
- The longest running times for DLNET, NETFLO and RELAXT-3 were 2.9 days, 20.8 days and 26.2 days, respectively. In 5, 19, and 17 instances DLNET, NETFLO and RELAXT-3, respectively, took over one CPU day to terminate. NETFLO and RELAXT-3 took over 10 CPU days in 2 and 5 instances, respectively. Table 31 displays the number of instances taking more than 1, 2, 3, 4, 5, 10, 15, 20, 25 and 30 CPU days, by code.
- It is well known that interior point algorithms take few iterations in practice. To take advantage of this phenomenon the computation of the ascent direction must be carried out efficiently. Figure 18 illustrates the effectiveness of the preconditioners implemented in DLNET. It shows that all average conjugate gradient iteration counts fell well below the $|V|^{1/2}$ level, considered to be the boundary between good and poor preconditioners.
- The computation of the maximum weight spanning tree took anywhere from 0.4 to 19% of the total conjugate gradient running time. Sorting, in most cases, accounted for over half of the spanning tree computation time, going from a minimum of 1.4% to a maximum of 91%. Kalinski and Ye [17] have implemented a sorting scheme that makes use of the fact that the dual slacks change very little from iteration to iteration. Using such a sorting scheme could reduce the conjugate gradient running time by perhaps 10% in some instances.
- Because the DAS algorithm allows for inexact ascent directions, a high conjugate gradient stopping tolerance of 10^{-3} was used successfully. Even though tightening this tolerance could reduce the DAS iteration count (by using better directions), the increased conjugate gradient running times do not justify this tightening.
- The primal estimate stopping scheme worked well on instances with no or little dual degeneracy. DLNET stopped in 72% of the instances with this scheme. In those, the projected dual solution identified the optimal face 95% of the time. In the remaining 5%, DLNET produced an optimal primal integer solution and a dual interior solution with a duality gap of less than 1. By not allowing DLNET

to stop with the absolute duality gap criterion, primal-dual complementary solutions could have been produced for all of the instances.

- For dual degenerate instances, DLNET stopped with the maximum flow stopping criterion. This occurred in 28% of the instances. By definition, all of those solutions are primal-dual complementary. The restricted networks used in the maximum flow stopping scheme were all very sparse (about the size of a tree) with the exception of the highly dual degenerate RLG-Wide instances. It is possible that the maximum flow scheme running times could be improved by changing maximum flow algorithms. Clearly, starting the maximum flow stopping scheme at iteration 10 of DAS and repeating every 10 DAS iterations was wasteful. DAS iterations were almost always greater than 40 and in some cases up to 200. By starting the maximum flow stopping scheme at iteration 40 and using a larger interval, of say 15 or 20 DAS iterations between tests, DLNET running times could be halved in some instances. A good illustration of this waste is on the largest instance from problem class Grid-Density-8, where DLNET spent on solving maximum flow problems, the equivalent of 61% of the time it spent on solving linear systems with the conjugate gradient algorithm.

5.5. CPLEX 2.0 Network Simplex Solver. Shortly after the main computational experiments described in this paper were completed, CPLEX Optimization, Inc. released its CPLEX 2.0 linear optimizer. This release contains a network simplex optimizer as well as the general linear programming optimizer. We refer to the two CPLEX solvers by the CPLEX commands used to run them, i.e., NETOPT for the network simplex solver and OPTIMIZE for the standard simplex solver. In this subsection, we compare NETFLO with NETOPT on selected instances from the full computational test suite to observe how NETFLO (which is over ten years old) compares with a modern commercial network simplex implementation. To further observe the performance of NETOPT, we compare NETOPT with OPTIMIZE.

Since the CPLEX executable available to us runs only on a 33 MHz MIPS Silicon Graphics IRIS computer, we have rerun NETFLO on the same machine. All problem classes of the main experiment have been included in this comparison. Due to time limitations, we have only run a single instance of each size, using random seed 270001, and have not run some of the largest instances tested in the main experiment.

The CPLEX network solver was run with the default parameters. The CPLEX general linear programming solver was also run with the default parameters, except for maximum iterations, which was set to one million.

For each problem class we present a table. The tables list instances by number of vertices and edges, and for each instance gives running time (in seconds) and number of iterations. Running times exclude input time and the time taken by CPLEX to recover the optimal solution at the end.

TABLE 32. Statistics for problem family Grid-Density-8

SIZE		NETFLO		CPLEX 2.0			
				NETOPT		OPTIMIZE	
$|V|$	$|E|$	time	itr	time	itr	time	itr
256	2048	1.7	5508	0.9	4331	5.0	1670
512	4096	8.1	15209	5.4	13872	16.3	3400
1024	8192	34.5	30076	29.9	33490	54.1	6196
2048	16384	124.8	67443	140.5	69762	228.8	12608
4096	32768	776.2	349594	649.7	183357	1631.8	28653
8192	65536	2622.4	327499	4284.2	395906	4386.6	49656
16384	131072	10718.4	750620	23052.3	852100	24086.7	99729

TABLE 33. Statistics for problem family Grid-Density-16

SIZE		NETFLO		CPLEX 2.0			
				NETOPT		OPTIMIZE	
$|V|$	$|E|$	time	itr	time	itr	time	itr
256	4096	4.5	14802	2.0	9562	10.5	3646
512	8192	14.2	31241	8.9	25637	78.1	7759
1024	16384	56.7	73355	89.4	71839	178.5	14452
2048	32768	232.4	172574	261.7	140250	926.0	30860
4096	65536	1110.2	457549	1568.8	376031	7317.0	60700
8192	131072	4103.5	839747	10165.4	894432	did not run	
16384	262144	26921.1	1948729	40204.2	2023332	did not run	

5.5.1. *Transshipment Problems on a Grid.* Tables 32–34 summarize results for subclasses Grid-Density-8, Grid-Density-16 and Grid-Increasing-Density, respectively. For Grid-Density-8 and Grid-Density-16 we generated instances with 256, 512, ..., 16384 vertices, while for Grid-Increasing-Density instances having 256, 512, ..., 4096 vertices are generated. NETFLO and NETOPT are run on all instances, while OPTIMIZE is not run on the 8192 and 16384 vertex instances of Grid-Density-16.

We make the following remarks regarding these runs.

- On Grid-Density-8, NETFLO and NETOPT have comparable number of iterations, with NETFLO solving the largest instance 2.2 times faster than NETOPT. OPTIMIZE solved the instances in much fewer iterations than NETOPT, resulting in comparable running times for the largest instance. As problem sizes increase OPTIMIZE does better with respect to NETOPT.
- On Grid-Density-16, the number of NETFLO and NETOPT iterations are com-

TABLE 34. Statistics for problem family Grid-Increasing-Density

SIZE		NETFLO		CPLEX 2.0			
				NETOPT		OPTIMIZE	
$|V|$	$|E|$	time	itr	time	itr	time	itr
256	4096	4.5	14802	2.1	9562	10.6	3646
512	11585	26.7	47231	15.5	31284	85.4	9980
1024	32768	195.3	187714	86.2	97426	728.0	30222
2048	92682	1429.7	680063	687.4	364678	5967.6	82446
4096	262144	12714.6	2511345	4671.4	1110467	177800.2	303809

TABLE 35. Statistics for problem family RLG-Wide

SIZE		NETFLO		CPLEX 2.0			
				NETOPT		OPTIMIZE	
$|V|$	$|E|$	time	itr	time	itr	time	itr
2050	6113	30.9	7594	12.2	8163	245.3	6474
4098	12225	122.0	17557	74.4	22657	1207.6	16170
8194	24449	628.5	46484	409.6	55789	6664.7	38581
16386	48897	3096.4	119477	2335.2	136926	did not run	

parable, resulting in a speedup for NETFLO with respect to NETOPT of 1.5. OPTIMIZE took over 6 times fewer iterations than NETOPT on the 4096 vertex instance, but this reduction in number of iterations did not result in a low running time. OPTIMIZE took 4.7 times longer to solve that instance.

- On Grid-Increasing-Density, the pricing heuristics used in NETOPT resulted in a 2.3 fold improvement in number of iterations on the largest instance, compared to NETFLO. This reduction was also accompanied by a 2.7 times reduction in running time. In this subclass OPTIMIZE does not experience a reduction in number of iterations as large as in Grid-Density-8 and Grid-Density-16, resulting in a running time over 38 times longer than NETOPT. The NETOPT to OPTIMIZE speedup grows with problem size.
- The following statistics are for the largest instances in each subclass run on each code. On Grid-Density-8 NETFLO, NETOPT and OPTIMIZE performed 70.0, 37.0 and 4.1 iterations per second, respectively. On Grid-Density-16 NETFLO, NETOPT and OPTIMIZE performed 72.4, 50.3 and 8.3 iterations per second, respectively. On Grid-Increasing-Density NETFLO, NETOPT and OPTIMIZE performed 197.5, 237.7 and 1.7 iterations per second, respectively.

5.5.2. *Maximum Flow Problems.* Table 35 summarizes running times and iterations for problem class RLG-Wide. Four problem categories were generated, having $2050, 2098, 8194$ and 16386 vertices. NETFLO and NETOPT were run on all categories, while OPTIMIZE was run on all but the 16386 vertex category.

We make the following remarks regarding runs on this problem class.

- NETFLO and NETOPT running times and iterations were comparable on these instances, with NETOPT requiring about 15% more iterations than NETFLO to solve the largest instance, while NETFLO took 1.3 times longer. The speedup appears to decrease with problem size.
- OPTIMIZE had a slight advantage in number of iterations, compared to NETOPT, resulting in a speedup of NETOPT with respect to OPTIMIZE of over 16 times. The speedup does not appear to increase with problem size.
- The following statistics are for the largest instances in the subclass run on each code. NETFLO, NETOPT and OPTIMIZE performed 38.6, 58.6 and 5.8 iterations per second, respectively.

5.5.3. *Minimum Cost-Maximum Flow Problems on a Grid.* Tables 36–38 summarize running times and iterations for problem subclasses Grid-Square, Grid-

TABLE 36. Statistics for problem family Grid-Square

SIZE		NETFLO		CPLEX 2.0			
				NETOPT		OPTIMIZE	
$|V|$	$|E|$	time	itr	time	itr	time	itr
258	512	0.1	388	0.1	584	0.9	282
1026	2048	0.5	1869	0.8	2808	11.9	1175
4098	8192	6.2	10697	10.7	12923	261.7	5917
16386	32768	111.8	73179	148.5	58149	5649.9	25998
65538	131072	1988.9	486330	2278.8	264975	did not run	

TABLE 37. Statistics for problem family Grid-Wide

SIZE		NETFLO		CPLEX 2.0			
				NETOPT		OPTIMIZE	
$|V|$	$|E|$	time	itr	time	itr	time	itr
514	1040	0.2	917	0.1	1093	2.8	613
1026	2096	0.8	2549	0.5	2517	12.7	1424
2050	4208	2.7	6260	1.8	5541	47.4	2900
4098	8432	9.0	13210	5.0	10614	218.2	6007
8194	16880	41.5	33306	12.9	18763	846.5	12124
16386	33776	158.6	73829	25.8	37471	4684.8	24528
32770	67568	634.6	156573	59.1	75221	did not run	

Wide and Grid-Long, respectively. Instances of having 258, 1026, ..., 65538 vertices were generated for subclass Grid-Square. For subclasses Grid-Wide and Grid-Long we generated instances with 514, 1026, ..., 32770 vertices. NETOPT and NETFLO were run on all instances. OPTIMIZE was run on all but the 65538 vertex instance of Grid-Square and the 32770 vertex instance of Grid-Wide.

We make the following comments regarding runs on these subclasses.

- On Grid-Square, running times for NETFLO and NETOPT were comparable even though NETOPT had fewer iterations (1.8 times fewer on the largest instance). OPTIMIZE took less than half the number of iterations taken by NETOPT. However, it was much slower than NETOPT, with NETOPT exhibiting a 38-fold speedup with respect to OPTIMIZE on the 16386 vertex instance.
- On Grid-Wide NETOPT was clearly faster than NETFLO, taking less than half the number of iterations on the 32770 vertex instance and solving the instance more than 10 times faster. With respect to OPTIMIZE the speedup was over

TABLE 38. Statistics for problem family Grid-Long

SIZE		NETFLO		CPLEX 2.0			
				NETOPT		OPTIMIZE	
$|V|$	$|E|$	time	itr	time	itr	time	itr
514	1008	0.1	578	0.2	1116	2.8	563
1026	2000	0.4	1366	1.0	2302	11.2	992
2050	3984	0.8	2282	4.4	3869	34.2	1772
4098	7952	2.7	5707	20.5	8120	251.6	4462
8194	15888	6.5	8761	84.5	13602	1092.9	9028
16386	31760	19.0	17129	372.5	30390	6167.8	18202
32770	63504	41.0	30659	1402.9	47419	26790.6	35509

TABLE 39. Statistics for problem family Mesh-1

SIZE		NETFLO		CPLEX 2.0			
				NETOPT		OPTIMIZE	
$|V|$	$|E|$	time	itr	time	itr	time	itr
256	512	0.2	1082	0.1	870	1.5	383
1024	2048	3.4	7002	1.4	4143	24.6	1691
4096	8192	69.5	44241	32.1	21800	558.7	8433
16384	32768	1373.2	264541	713.1	97834	15114.1	41298

TABLE 40. Statistics for problem family Mesh-2

SIZE		NETFLO		CPLEX 2.0			
				NETOPT		OPTIMIZE	
$|V|$	$|E|$	time	itr	time	itr	time	itr
256	1024	0.7	3533	0.3	2034	3.3	1271
1024	4096	13.7	21872	4.2	10626	66.1	6702
4096	16384	276.1	118477	114.1	54183	2073.0	35869
16384	65536	5360.8	686840	2121.3	265458	57071.4	159473

181 for the 16384 vertex instance. OPTIMIZE and NETOPT had comparable iteration counts.

- On Grid-Long NETFLO had a slightly better iteration count than NETOPT, but was much faster, solving the largest instance over 34 times faster than NETOPT. It should be noted that this class was the only one in which NETFLO was asymptotically faster than DLNET in the main experiment. Though iteration counts were similar for the two CPLEX codes, the network code was over 19 times faster than the general linear programming code on the largest instance.
- On Grid-Square NETFLO, NETOPT and OPTIMIZE performed 244.5, 116.3 and 4.6 iterations per second, respectively. On Grid-Wide NETFLO, NETOPT and OPTIMIZE performed 246.7, 1272.8 and 5.2 iterations per second, respectively. On Grid-Long NETFLO, NETOPT and OPTIMIZE performed 747.8, 33.8 and 1.3 iterations per second, respectively.

5.5.4. *Minimum Cost Circulation Problems.* Tables 39–42 summarize running times and iterations for subclasses Mesh-1, Mesh-2, Mesh-4 and Mesh-8, respectively. Instances having 256, 1024, 4096 and 16384 vertices were generated for all subclasses. NETFLO and NETOPT were run on all instances, while OPTIMIZE was not run on the largest instance of Mesh-4 and Mesh-8.

We make the following remarks about these runs.

TABLE 41. Statistics for problem family Mesh-4

SIZE		NETFLO		CPLEX 2.0			
				NETOPT		OPTIMIZE	
$|V|$	$|E|$	time	itr	time	itr	time	itr
256	2048	1.3	7907	0.8	4525	12.3	3061
1024	8192	34.2	58057	12.6	27419	237.9	17682
4096	32768	807.9	340665	322.6	146368	6429.1	105928
16384	131072	15541.0	1832436	9050.7	885264	did not run	

TABLE 42. Statistics for problem family Mesh-8

SIZE		NETFLO		CPLEX 2.0			
				NETOPT		OPTIMIZE	
$|V|$	$|E|$	time	itr	time	itr	time	itr
256	4096	1.8	10817	1.8	10053	24.3	6366
1024	16384	71.5	147577	36.0	59439	635.0	39850
4096	65536	2313.3	999076	761.5	364260	16396.6	204215
16384	262144	55527.4	5911689	22654.5	2268046	did not run	

TABLE 43. Statistics for problem family Netgen-Lo

SIZE		NETFLO		CPLEX 2.0			
				NETOPT		OPTIMIZE	
$|V|$	$|E|$	time	itr	time	itr	time	itr
256	2048	0.4	3435	0.4	2671	7.9	2344
512	4101	1.6	8879	1.6	6778	29.1	5599
1024	8214	6.3	21904	5.1	16629	126.5	11477
2048	16414	27.3	49400	19.4	28154	650.6	28029
4096	32858	146.0	115787	101.5	68405	2919.3	58121
8192	65734	773.5	261682	482.6	149076	15308.0	124560
16384	131409	4121.7	630182	1530.8	259576	did not run	

- NETOPT was consistently about twice as fast as NETFLO on all problem classes, with a slight reduction for the largest instances. NETOPT consistently took less than half the number of iterations taken by NETFLO.
- NETOPT and OPTIMIZE have comparable iteration counts (OPTIMIZE takes 2.4 times fewer iterations than NETOPT) resulting in a NETOPT solution time about 20 times faster than OPTIMIZE on the largest instances of each subclass. This speedup increases with problem size.
- The following statistics are for the largest instances in each subclass run on each code. On Mesh-1 NETFLO, NETOPT and OPTIMIZE performed 192.6, 137.2 and 2.7 iterations per second, respectively. On Mesh-2 NETFLO, NETOPT and OPTIMIZE performed 128.1, 125.1 and 2.8 iterations per second, respectively. On Mesh-4 NETFLO, NETOPT and OPTIMIZE performed 117.9, 97.8 and 16.5 iterations per second, respectively. On Mesh-8 NETFLO, NETOPT and OPTIMIZE performed 106.5, 100.1 and 12.5 iterations per second, respectively.

5.5.5. *NETGEN Problems.* Tables 43–44 summarize running times and iterations for subclasses Netgen-Lo and Netgen-Hi, respectively. Instances having 256, 512, ..., 16384 vertices were generated for Netgen-Lo, while instances having 256, 512, ..., 32768 vertices were generated for Netgen-Hi. NETFLO and NETOPT were run on all instances, while OPTIMIZE was run on all but the 16384 vertex instance of Netgen-Lo and the 16384 and 32768 vertex instances of Netgen-Hi.

We make the following remarks regarding the runs for problem subclasses Netgen-Lo and Netgen-Hi.
- On the largest instance of Netgen-Lo, NETFLO took 2.4 times as many iterations as NETOPT and 2.7 times longer to solve the instance. NETOPT to

TABLE 44. Statistics for problem family Netgen-Hi

SIZE		NETFLO		CPLEX 2.0			
				NETOPT		OPTIMIZE	
$|V|$	$|E|$	time	itr	time	itr	time	itr
256	2048	0.2	1117	0.2	1167	3.2	877
512	4101	0.6	2991	0.7	2791	12.5	2086
1024	8214	2.5	7567	2.9	7044	60.5	4641
2048	16414	12.8	19521	12.0	11957	334.0	11850
4096	32858	76.5	51519	65.6	30798	1595.7	27687
8192	65734	501.8	139777	411.4	77745	8837.7	69872
16384	131409	3236.6	396359	1661.5	143458	did not run	
32768	262903	23366.6	1167935	10453.9	392896	did not run	

NETFLO time ratios slightly increase asymptotically. NETOPT and OPTIMIZE have comparable iteration counts, resulting in a NETOPT solution time over 31 times faster than that of OPTIMIZE for the 8192 vertex instance.

- On Netgen-Hi, NETOPT consistently takes fewer iterations than NETFLO, resulting in a speedup, on the largest instance, of 2.2. NETOPT and OPTIMIZE iterations are comparable, with NETOPT solving the 8192 vertex instance over 21 times faster than OPTIMIZE.

- The following statistics are for the largest instances in each subclass run on each code. On Netgen-Lo NETFLO, NETOPT and OPTIMIZE performed 152.9, 169.6 and 8.1 iterations per second, respectively. On Netgen-Hi NETFLO, NETOPT and OPTIMIZE performed 50.0, 37.6 and 7.9 iterations per second, respectively.

5.5.6. *Comments on the CPLEX Experiment.* NETFLO is a rather old code, and we were somewhat concerned that it might not be a "good representative" of network simplex solvers. Relying only on data obtained from NETFLO could raise doubts about the conclusions made with respect to the network simplex in the main experiment. The runs in this subsection serve to contrast the performance of NETFLO with a more recent implementation of the network simplex method. We selected CPLEX 2.0 network solver, which was readily available to us, based on the reputation of previous releases of its general purpose simplex-based linear programming system.

We conclude this subsection with the following remarks.

- The codes were run on the 13 subclasses of the experiment. Instances of size $16K$ vertices and smaller were run in every class except Grid-Increasing-Density, where the largest had $4K$ vertices and $262K$ edges. The largest instance tested in this experiment had $65K$ vertices.

- In 9 out of the 13 subclasses the CPLEX network solver was faster than NETFLO on the largest instances. However, in only one subclass was it over 3 times faster (it was 10 times faster on the largest instance of Grid-Wide). The other 8 speedup factors varied from 1.3 to 2.8. In the 4 subclasses where NETFLO was the fastest, speedups were 1.1, 1.5, 2.2 and 34. Lustig [23] estimates that a speedup factor of up to 1.6 can be achieved by simply using C language pointers in place of the indirect indexing used in Fortran network simplex implementations on a Silicon Graphics workstation. CPLEX is implemented in C, while NETFLO is in Fortran.

- In 9 out of the 13 subclasses the CPLEX network solver took fewer iterations than NETFLO on the largest instances. This is perhaps due to a more sophisticated pricing scheme used in CPLEX. The reduction in iterations, however, was not substantial. For the largest instances in each subclass, NETFLO never took over 3 times the number of iterations taken by CPLEX and CPLEX never took over 1.6 times the number of iterations taken by NETFLO.

- The pricing scheme implemented in CPLEX's OPTIMIZE appears to be more effective in reducing the number of iterations than the one in NETOPT. In 3 (the 3 Grid-Density subclasses) out of the 13 subclasses, for the largest instances in each subclass, NETOPT took over 3 times the number of iterations (the largest increase in number of iterations was by a factor of 8.5). In the remaining 10 subclasses, OPTIMIZE always took fewer iterations, but the reduction was less dramatic, with NETOPT to OPTIMIZE iteration ratios ranging from 1.1 to 2.2. Since iterations in the general simplex are much more computationally intensive, it may be worthwhile investing in a more sophisticated pricing scheme and, hopefully, reducing the number of iterations. However, in most cases, the reduction in the number of iterations cannot offset the expensive nature of basis updates and periodic refactorizations required in a general purpose simplex implementation. Hence, OPTIMIZE to NETOPT time ratios of up to 180 were observed.

From the above, we conclude that though CPLEX is slightly faster than NETFLO on the majority of problem classes, the conclusions made in the main experiment with respect to NETFLO should still hold with respect to CPLEX's network optimizer.

6. Concluding Remarks

Efficient implementations of variants of the network simplex algorithm and the relaxation method currently make up the set of tools used by analysts to solve large scale MCNF problems. In this study, we have introduced a new tool: the network interior point code DLNET.

In the computational experiments described in this study, DLNET proved to be robust, not failing to find an integer primal optimal solution a single time while using the same parameter settings throughout. Furthermore, it produced integer primal-dual complementary pairs for the majority of problems solved. When it did not produce a complementary pair, DLNET found a primal integer optimal solution. For those instances, forcing a few more DAS iterations would probably produce the primal-dual pair.

We showed that DLNET can be more efficient than NETFLO and RELAXT-3 in several classes of large MCNF problems. In some experiments DLNET was up to 152 times faster than RELAXT-3 and 18 times faster than NETFLO. On the other hand, in other classes, RELAXT-3 and NETFLO outperform DLNET by up to a factor of 59 (RELAXT-3 in Netgen-Hi class) and 66 (NETFLO in Grid-Long class). This difference in relative performance across classes of problem can be explained by the wide variation in the running times of NETFLO and RELAXT-3. There were 11 and 6 instances where NETFLO and RELAXT-3, respectively, had running times of over 3 CPU days. The longest running times for DLNET, NETFLO and RELAXT-3 were, respectively, 2.9, 20.8 and 26.2 CPU days. In classes of problems where DLNET was not the fastest, it proved to be competitive with the third code (with the exception of problem class Grid-Long).

Acknowledgment

One of the authors (M.G.C. Resende) acknowledges several insightful discussions with N. Karmarkar, K.G. Ramakrishnan and P. Vaidya. A discussion with S.T. McCormick led the authors to the idea of solving a maximum flow problem to find an integer primal-dual solution. Comments and suggestions made by an anonymous referee improved the paper greatly. Finally, the authors wish to thank M.D. Grigoriadis, D.S. Johnson, C. McGeogh, C. Monma and R.E. Tarjan for their efforts in organizing the First DIMACS International Algorithm Implementation Challenge.

References

1. I. Adler, N. Karmarkar, M.G.C. Resende, and G. Veiga, *Data structures and programming techniques for the implementation of Karmarkar's algorithm*, ORSA Journal on Computing **1** (1989), 84–106.
2. ———, *An implementation of Karmarkar's algorithm for linear programming*, Mathematical Programming **44** (1989), 297–335.
3. A. Armacost and S. Mehrotra, *Computational comparison of the network simplex method with the affine scaling method*, Opsearch **28** (1991), 26–43.
4. J. Aronson, R. Barr, R. Helgason, J. Kennington, A. Loh, and H. Zaki, *The projective transformation algorithm of Karmarkar: A computational experiment with assignment problems*, Tech. Report 85-OR-3, Department of Operations Research, Southern Methodist University, Dallas, TX, August 1985.
5. E.R. Barnes, *A variation on Karmarkar's algorithm for solving linear programming problems*, Mathematical Programming **36** (1986), 174–182.
6. D.P. Bertsekas and P. Tseng, *Relaxation methods for minimum cost ordinary and generalized network flow problems*, Operations Research **36** (1988), 93–114.
7. A. Charnes, *Optimality and degeneracy in linear programming*, Econometrica **20** (1952), 160–170.
8. CPLEX Optimization, Inc., *Using the $CPLEX^{TM}$ callable library and $CPLEX^{TM}$ mixed integer library including the $CPLEX^{TM}$ linear optimizer and $CPLEX^{TM}$ mixed integer optimizer – Version 2.0*, Incline Village, NV, 1992.
9. G.B. Dantzig, *Maximization of a linear function of variables subject to linear inequalities*, Activity Analysis of Production and Allocation (T.C. Koopsmans, ed.), John Wiley and Sons, 1951, pp. 339–347.
10. I.I. Dikin, *Iterative solution of problems of linear and quadratic programming*, Soviet Mathematics Doklady **8** (1967), 674–675.

11. _____, *Determination of interior point of a system of linear inequalities*, Tech. report, Siberian Energy Institute, Irkutsk, USSR, 1991.
12. DIMACS, *The first DIMACS international algorithm implementation challenge: The benchmark experiments*, Tech. report, DIMACS, New Brunswick, NJ, 1991.
13. D. Goldfarb and M.D. Grigoriadis, *A computational comparison of the Dinic and network simplex methods for maximum flow*, Annals of Operations Research **7** (1988), 83–123.
14. G.H. Golub and C.F. van Loan, *Matrix computations*, The Johns Hopkins University Press, Baltimore, MD, 1983.
15. M.D. Grigoriadis, *An efficient implementation of the network simplex method*, Mathematical Programming Study **26** (1986), 83–111.
16. A. Joshi, A.S. Goldstein, and P.M. Vaidya, *A fast implementation of a path-following algorithm for maximizing a linear function over a network polytope*, Tech. report, Dept of Computer Science, University of Illinois, Urbana, IL, 1991.
17. J.A. Kalinski and Y. Ye, *A decomposition variant of the potential reduction algorithm for linear programming*, Tech. Report 91-11, Dept of Management Sciences, The University of Iowa, Iowa City, Iowa, 1991.
18. N.K. Karmarkar and K.G. Ramakrishnan, *Implementation and computational results of the Karmarkar algorithm for linear programming, using an iterative method for computing projections*, Tech. report, AT&T Bell Laboratories, Murray Hill, NJ, 1988.
19. _____, *Private communication*, 1988.
20. _____, *Computational results of an interior point algorithm for large scale linear programming*, Mathematical Programming **52** (1991), 555–586.
21. J.L. Kennington and R.V. Helgason, *Algorithms for network programming*, John Wiley and Sons, New York, NY, 1980.
22. D. Klingman, A. Napier, and J. Stutz, *Netgen: A program for generating large scale capacitated assignment, transportation, and minimum cost flow network problems*, Management Science **20** (1974), 814–821.
23. I.J. Lustig, *The influence of computer language on computational comparisons: An example from network optimization*, ORSA Journal on Computing **2** (1990), 152–161.
24. K.A. McShane, C.L. Monma, and D.F. Shanno, *An implementation of a primal-dual interior point method for linear programming*, ORSA Journal on Computing **1** (1989), 70–83.
25. S. Mehrotra and Y. Ye, *On finding the optimal facet of linear programs*, Tech. Report 91-10, Department of Industrial Engineering and Management Sciences, Northwestern University, Evanston, IL 60208, 1991.
26. C.L. Monma and A.J. Morton, *Computational experiments with a dual affine variant of Karmarkar's method for linear programming*, Operations Research Letters **6** (1987), 261–267.
27. A. Rajan, *An empirical comparison of KORBX against RELAXT, a special code for network flow problems*, Tech. report, AT&T Bell Laboratories, Holmdel, NJ, 1989.
28. M.G.C. Resende and G. Veiga, *Computational study of two implementations of the dual affine scaling algorithm*, Tech. report, AT&T Bell Laboratories, Murray Hill, NJ, 1990.
29. _____, *An implementation of the dual affine scaling algorithm for minimum cost flow on bipartite uncapaciated networks*, Tech. report, AT&T Bell Laboratories, Murray Hill, NJ, 1990, To appear in SIAM Journal on Optimization.
30. L.P. Sinha, B.A. Freedman, N.K. Karmarkar, A. Putcha, and K.G. Ramakrishnan, *Overseas network planning*, Proceedings of the Third International Network Planning Symposium – NETWORKS'86, June 1986, pp. 8.2.1–8.2.4.
31. R.E. Tarjan, *Data structures and network algorithms*, Society for Industrial and Applied Mathematics, Philadelphia, PA, 1983.
32. M.J. Todd, *The effects of degeneracy and unbounded variables on variants of Karmarkar's linear programming algorithm*, Large-scale Numerical Optimization (T.F. Coleman and Y. L, eds.), SIAM, 1990, pp. 81–91.
33. M.J. Todd and B.P. Burrell, *An extension to Karmarkar's algorithm for linear programming using dual variables*, Algorithmica **1** (1986), 409–424.
34. T. Tsuchiya and M. Muramatsu, *Global convergence of the long-step affine scaling algo-*

rithm for degenerate linear programming problems, Tech. report, The Institute of Statistical Mathematics, Tokyo, January 1992.
35. P.M. Vaidya, *Solving linear equations with symmetric diagonally dominant matrices by constructing good preconditioners*, Tech. report, Department of Computer Science, University of Illinois at Urbana-Champaign, Urbana, IL, 1990.
36. R.J. Vanderbei, M.S. Meketon, and B.A. Freedman, *A modification of Karmarkar's linear programming algorithm*, Algorithmica **1** (1986), 395–407.
37. Y. Ye, *On the finite convergence of interior-point algorithms for linear programming*, Tech. Report 91-5, Dept of Management Sciences, The University of Iowa, Iowa City, Iowa, 1991, To appear in Mathematical Programming B.
38. Quey-Jen Yeh, *A reduced dual affine scaling algorithm for solving assignment and transportation problems*, Ph.D. thesis, Columbia University, New York, NY, 1989.

(M.G.C. Resende) MATHEMATICAL SCIENCES RESEARCH CENTER, AT&T BELL LABORATORIES, MURRAY HILL, NJ 07974 USA

(G. Veiga) DEPARTMENT OF INDUSTRIAL ENGINEERING AND OPERATIONS RESEARCH, UNIVERSITY OF CALIFORNIA, BERKELEY, CA 94720 USA

E-mail address, M.G.C. Resende: mgcr@research.att.com

On the Massively Parallel Solution of Linear Network Flow Problems

SOREN S. NIELSEN, STAVROS A. ZENIOS

February 1992
Revised August 1992

Abstract

We report on a comparative computational study of two algorithms for the massively parallel solution of linear network problems: the ϵ-relaxation algorithm, and the Proximal Minimization algorithm with D-functions. Problems with up to 16 million variables are solved in about 1.5 hours. The massively parallel algorithms are compared with simplex and relaxation codes on vector supercomputers as well as small-scale parallel machines.

1. Introduction

In this paper, we report on a comparative computational study of two algorithms for the massively parallel solution of *linear* network programs: the ϵ-relaxation algorithm of Bertsekas [1986a, 1986b] and the Proximal Minimization algorithm with D-functions (PMD) of Censor and Zenios [1992]. We also compare the massively parallel algorithms with solution techniques on vector supercomputers as well as small-scale parallel machines. Network flow problems with *strictly convex*, separable objective functions have been solved efficiently on massively parallel computers by solving the dual problem. For the strictly convex problem, the dual functional is differentiable, and the dual problem can be solved using row-action methods or simple dual ascent methods. The success of the massively

---

1991 Mathematics Subject Classification. Primary 90B10, 65Y05.
Partially supported by NSF grant CCR-9104042 and AFOSR grant 91-0168.
Final version. No other version will be submitted for publication elsewhere.

parallel implementations is due to the fact that dual ascent steps can be taken concurrently along all dual coordinate directions.

Differentiability of the dual functional does not, however, hold in the case of the linear network problem. Here, situations arise where dual ascent cannot be achieved by steps along any single dual coordinate direction. Steps must be taken along multiple coordinates, and this destroys the parallel decomposition nature of dual ascent algorithms.

Two algorithms to overcome this difficulty have been suggested. The ϵ-relaxation algorithm deals with the non-differentiability of the dual functional by maintaining only approximate complementary slackness. Steps are taken along a dual coordinate direction even if they result in a small deterioration in the dual objective value. The algorithm is, for integral problem data, finitely convergent. The PMD algorithm solves the linear program by introducing a non-linear objective perturbation, namely a certain directed distance measure to a "proximal point". This results in having to solve a sequence of strictly convex subproblems, which can be done using the row-action algorithm of Censor and Lent [1981]. By using a quadratic and entropic distance measure, respectively, we get as special cases of the PMD algorithm the quadratic proximal point algorithm (QPP) of Rockafellar [1976a, 1976b], and the entropic proximal point (EPP) algorithm of Eriksson [1985].

Common for these algorithms is that they decompose in such a way that dual steps, corresponding to updates of the dual prices associated with each network node, can be performed in a massively parallel fashion. By operating on all network nodes concurrently, highly parallelizable Jacobi-type algorithms are obtained (Jacobi-type algorithms, in contrast to Gauss-Seidel algorithms, may operate at each iteration on data which are several steps old. They have poorer convergence properties, but in a parallel implementation they typically have less communication and synchronization overheads).

The dual prices are, in the case of dual relaxation algorithms, updated using a linesearch procedure, Nielsen and Zenios [1992b]. In the case of the row-action algorithm, a simultaneous primal-dual update is performed using a *Bregman projection*, i.e., a primal projection, using a certain distance measure, onto a constraint hyperplane and a corresponding dual update (for details of the row-action algorithms, see, e.g., Nielsen and Zenios [1992b], Linesearch 1).

The objective of our study is to establish the suitability of massively parallel computers for the solution of huge, linear network problems. The algorithms studied in this paper are implemented on a Connection Machine CM-2 with up to 32K processing elements and problems with up to 16 million variables are solved. To further establish the performance of these algorithms on the CM-2, we compare them to a state-of the art network simplex code executing on a supercomputer, the Cray Y-MP, and to a sequential implementation of the Relax code of Bertsekas [1991]. In order to estimate the efficiency of the parallel implementations, we also compare the PMD algorithms on the CM-2 to a small-scale, parallel implementation of the Gauss-Seidel variant of the same algorithm, and in addition show how the massively parallel algorithm scales with different numbers of CM-2 processors.

This paper was written as part of the DIMACS challenge on the practical implementation of algorithms for network flow problems. However, the computational results have been compiled over the period 1989-1992, as part of ongoing projects on parallel network optimization, and the test problems are not those of DIMACS. Hence, our results are not directly comparable with the results of other studies in the DIMACS challenge, but they are still based on publicly available data.

The paper is organized as follows. In Sections 2.1 and 2.2 we describe the ϵ-relaxation algorithm and the PMD algorithms, respectively. Section 3 is a computational study of the algorithms. We test the performance of the ϵ-relaxation and the PMD algorithms, respectively, in Sections 3.2 and 3.3. In Sections 3.4 and 3.5 we apply the PMD algorithms for the solution of problems with up to 16 million variables, and then investigate the question of scalability of the algorithms on the CM-2. In Section 3.6 we compare the CM-2 implementation of the QPP algorithm with a small-scale, parallel implementation. In Section 4 we compare the ϵ-relaxation and PMD algorithms to each other, and to a state-of-the-art network simplex solver running on a Cray Y-MP. Section 5 summarizes the conclusions of our study.

2. The Linear Network Optimization Algorithms

We consider the directed graph $\mathcal{G} = (\mathcal{N}, \mathcal{E})$, where $\mathcal{N}$ is the set of nodes with cardinality m and $\mathcal{E}$ is the set of arcs with cardinality n. Finding a feasible flow at minimum cost on this graph can be formulated as the following linear program:

$$[\text{LNFP}] \quad \underset{x \in \Re^n}{\text{Minimize}} \quad \sum_{(i,j) \in \mathcal{E}} c_{ij} x_{ij} \tag{1}$$

$$\text{Subject to} \quad \sum_{\{j|(i,j) \in \mathcal{E}\}} x_{ij} - \sum_{\{k|(k,i) \in \mathcal{E}\}} x_{ki} = b_i, \quad i \in \mathcal{N} \tag{2}$$

$$l_{ij} \le x_{ij} \le u_{ij} \quad (i,j) \in \mathcal{E} \tag{3}$$

where $c = (c_{ij}) \in \Re^n$ is the vector of cost coefficients, $b = (b_i) \in \Re^m$ is the vector of exogenous supplies or demands, and $l = (u_{ij}) \in \Re^n$ and $u = (u_{ij}) \in \Re^n$ are the vectors of lower and upper bounds on the flows. We associate with the equality constraints (flow conservation constraints) the dual prices $\pi = (\pi_i) \in \Re^m$.

A primal/dual pair (x, π) is optimal if x satisfies the flow conservation constraints (2) and (x, π) satisfy the following complementary slackness (CS) conditions:

If $c_{ij} - \pi_i + \pi_j > 0$ then $x_{ij} = l_{ij}$ and arc (i,j) is termed *inactive*.
If $c_{ij} - \pi_i + \pi_j = 0$ then $l_{ij} \le x_{ij} \le u_{ij}$ and arc (i,j) is termed *balanced*.
If $c_{ij} - \pi_i + \pi_j < 0$ then $x_{ij} = u_{ij}$ and arc (i,j) is termed *active*.

We describe next two algorithms for solving this problem.

2.1 The ϵ-Relaxation Algorithm. The ϵ-relaxation algorithm iteratively adjusts the dual prices and computes flows based on small violations (ϵ) of the

CS conditions. The price adjustment for a particular node is chosen such that the computed flows satisfy conservation of flow for that particular node. Since the algorithm allows small violations (ϵ) of the CS conditions, each step is not guaranteed to improve the dual cost. However, it improves the dual cost of a perturbed problem where some of the arc coefficients are modified by a small constant ϵ. It can be shown, see Bertsekas [1979], Bertsekas and Tsitsiklis [1989, Proposition 3.1, p. 357], that if all problem data are integral then a solution that violates CS by $\epsilon < \frac{1}{m}$ is optimal. Furthermore, the ϵ-relaxation algorithm will obtain an exact solution in a finite number of steps. It has been shown – see Goldberg [1986], Goldberg and Tarjan [1987] – that a polynomial variant of the algorithm is possible when ϵ is successively scaled from a large initial value to less than $\frac{1}{m}$.

In order to formally state the algorithm we introduce first the ϵ-CS conditions. For any price π and a scalar constant $\epsilon > 0$ we characterize an arc (i,j) as:

$$\begin{array}{ll} \epsilon - \text{inactive}, & \text{if } c_{ij} - \pi_i + \pi_j > \epsilon \\ \epsilon^- - \text{balanced}, & \text{if } c_{ij} - p_\pi + \pi_j = \epsilon \\ \epsilon - \text{balanced}, & \text{if } -\epsilon \leq c_{ij} - \pi_i + \pi_j \leq \epsilon \\ \epsilon^+ - \text{balanced}, & \text{if } c_{ij} - \pi_i + \pi_j = -\epsilon \\ \epsilon - \text{active}, & \text{if } c_{ij} - \pi_i + \pi_j < -\epsilon. \end{array}$$

The ϵ-CS conditions (ϵ-CS) for a primal/dual pair (x, π) are defined as:

$$\begin{array}{ll} x_{ij} = l_{ij} & \text{if arc } (i,j) \text{ is } \epsilon - \text{inactive} \\ l_{ij} \leq x_{ij} \leq u_{ij} & \text{if arc } (i,j) \text{ is } \epsilon - \text{balanced} \\ x_{ij} = u_{ij} & \text{if arc } (i,j) \text{ is } \epsilon - \text{active}. \end{array}$$

The ϵ-relaxation algorithm iteratively chooses nodes that violate conservation of flow and adjust the incident flows to maintain ϵ-CS while reducing the violation. If such a flow adjustment is not possible then a price adjustment takes place. Violations of the flow conservation constraints are measured as the surplus of a node defined by

$$g_i = b_i - \sum_{\{j|(i,j)\in\mathcal{E}\}} x_{ij} + \sum_{\{k|(k,i)\in\mathcal{E}\}} x_{ki}. \tag{4}$$

Flow adjustments on an arc are possible only if one of the following conditions are satisfied:

1. The arc is ϵ^+-balanced and $x_{ij} < u_{ij}$. In this case a non-zero increase of flow is possible, without violating ϵ-CS. This arc is ϵ-unblocked.

2. The arc is ϵ^--balanced and $x_{ij} > l_{ij}$. In this case a non-zero decrease of flow is possible, without violating ϵ-CS. This arc is ϵ^--unblocked.

We have now all the components needed to define the basic iterative step of ϵ-relaxation for nodes with positive surplus.

Positive Surplus Up Iteration:

Step 0 Let (x, π) satisfy ϵ-CS, and let i be a node with $g_i > 0$.

Step 1 (Scan arcs incident to node i).
Select a node j such that (i, j) is ϵ^+-unblocked and go to Step 2, or select a node j such that (j, i) is ϵ^--unblocked and go to Step 3. If no unblocked arc can be found go to Step 4.

Step 2 (Decrease surplus of node i by increasing x_{ij} on outgoing arcs).
Let $\delta = \min\{g_i, u_{ij} - x_{ij}\}$, update x_{ij}, g_i and g_j as follows:
$$x_{ij} \leftarrow x_{ij} + \delta, \quad g_i \leftarrow g_i - \delta, \quad g_j \leftarrow g_j + \delta.$$
If $g_i = 0$ and $x_{ij} < u_{ij}$, stop; else go to Step 1.

Step 3 (Decrease surplus of node i by reducing x_{ji} on incoming arcs).
Let $\delta = \min\{g_i, x_{ji} - l_{ji}\}$, update x_{ij}, g_i and g_j as follows:
$$x_{ji} \leftarrow x_{ji} - \delta, \quad g_i \leftarrow g_i - \delta, \quad g_j \leftarrow g_j + \delta.$$
If $g_i = 0$ and $x_{ji} > l_{ji}$, stop; else go to Step 1.

Step 4 (Increase price of node i). Set $\pi_i \leftarrow \min_{\xi \in R_i^+ \cup R_i^-} \xi$, where
$$R_i^+ = \{\pi_j + c_{ij} + \epsilon \mid (i, j) \in \mathcal{E}, x_{ij} < u_{ij}\}$$
$$R_i^- = \{\pi_j - c_{ji} + \epsilon \mid (j, i) \in \mathcal{E}, x_{ji} > l_{ji}\}$$
If $R_i^+ \cup R_i^- = \emptyset$, the problem is infeasible; else go to Step 1.

For nodes with negative surplus, a symmetric down iteration can be defined. Our method for alternating between up and down iterations is described in Section 3.2.

2.1.1 Parallelizing the Flow Calculation Procedure. In order to obtain a highly parallelizable version of the algorithm we need to modify the sequential algorithm given above. For any node i, flows may be pushed out along all ϵ^+-unblocked outgoing arcs or ϵ^--unblocked incoming arcs. We want to push out flows simultaneously on all eligible arcs. However, the upper/lower bounds and the current flows of the arcs are different, leading to different "capacity of update" ($u_{ij} - x_{ij}$ or $x_{ij} - l_{ij}$) for each arc. Therefore we distribute the surplus/deficit g_i over the arcs in proportion to their capacities of update. Let

$$F_i = \{j \mid (i,j) \text{ is } \epsilon^+ - \text{unblocked}\},$$
$$B_i = \{j \mid (j,i) \text{ is } \epsilon^- - \text{unblocked}\},$$
$$\omega = \max\{g_i, \sum_{j \in F_i}(u_{ij} - x_{ij}) + \sum_{j \in B_i}(x_{ji} - l_{ji})\}.$$

Then for each arc $(i, j) \in \mathcal{E}$ with $j \in F_i$, x_{ij} is increased by the amount $g_i \frac{u_{ij} - x_{ij}}{\omega}$. For each arc $(j, i) \in \mathcal{E}$ with $j \in B_i$, x_{ji} is decreased by the amount $g_i \frac{x_{ji} - l_{ji}}{\omega}$. It is easy to verify that the surplus computed with the updated flows is strictly less than the starting surplus g_i.

This approach was proposed by Li and Zenios [1991]. One transportation problem with 100 nodes and 900 arcs was solved using the sequential flow update and the parallel flow update procedure, respectively, on the CM-2. In the former case, the algorithm converges after 3450 iterations with 440.11 sec CM time, while in the latter case, it terminates after 1800 iterations and 13.25 sec CM time.

2.2 Proximal Minimization Algorithm with D-functions (PMD). The PMD algorithm was proposed in Censor and Zenios [1992], where its convergence was established. Let $S \neq \emptyset$ be an open convex set. Let $f : \Lambda \subseteq \Re^n \mapsto \Re$ be an auxiliary function. We assume that $\bar{S} \subseteq \Lambda$, where $\bar{S}$ is the closure of S, and that f is strictly convex and continuous on $\bar{S}$ and continuously differentiable on S. The set S is called the *zone* of f. The *D-function* corresponding to f is defined by

$$D_f(x,y) = f(x) - f(y) - \nabla f(y)^T(x-y). \tag{5}$$

Consider again the linear network flow problem [LNFP], which we restate as follows:

$$\min_{x \in X} c^T x, \tag{6}$$

where $X \subset \Re^n$ is the feasible region, $X = \{x \in \Re^n \mid Ax = b, l \leq x \leq u\}$, assumed to be nonempty, and A is the node-arc incidence matrix implied by (2). For some suitable choice of the auxiliary function f and a positive sequence $\{\gamma^k\}_{k=0}^{\infty}$ with $\liminf_{k \to \infty} \gamma^k = \gamma < \infty$, the **proximal minimization algorithm with D-functions** proceeds from an arbitrary starting point, $x^0 \in S$, with the following iteration.

The PMD Algorithm.

$$x^{k+1} \leftarrow \arg \min_{x \in X \cap S} c^T x + \frac{1}{\gamma^k} D_f(x, x^k). \tag{7}$$

2.2.1 Quadratic Proximal Terms. Consider the PMD algorithm with the auxiliary function:

$$f(x) = \frac{1}{2} \| x \|^2, \tag{8}$$

where $\| \cdot \|$ denotes the Euclidean norm. The corresponding D-function defined by (5) is then $D_f(x,y) = \frac{1}{2} \| x - y \|^2$. In this case $\Lambda = S = \bar{S} = \Re^n$, and we obtain as a special case of (7) the **quadratic proximal point algorithm (QPP)** of Rockafellar [1976a, 1976b]:

The QPP Algorithm.

$$x^{k+1} \leftarrow \arg\min_{x \in X} c^T x + \frac{1}{2\gamma^k} \| x - x^k \|^2 . \qquad (9)$$

For linear programs, QPP is finitely convergent. Also, there exists a $\bar{\gamma} < \infty$ such that for any $\gamma^k \geq \bar{\gamma}$ and any fixed x^k, the solution to the minimization in (9) is also a solution to (6) (see, e.g., Bertsekas and Tsitsiklis [1989, Section 3.4.3]).

2.2.2 Entropy Proximal Terms. To obtain another variant of the PMD algorithm, let ent $: \Re^n \mapsto \Re$ be the negative entropy function

$$\text{ent}(x) \doteq -\sum_{j=1}^{n} x_j \log\left(\frac{x_j}{a_j}\right), \qquad (10)$$

where $a = (a_j) \in \Re^n$ is a given positive vector, log is the natural logarithm, and $0 \log 0$ is defined to be 0. By letting $f(x) = -\text{ent}(x)$ we obtain the D-function

$$D_f(x, y) = \sum_{j=1}^{n} x_j \left[\log\left(\frac{x_j}{y_j}\right) - 1 \right] + \sum_{j=1}^{n} y_j . \qquad (11)$$

f is defined on $\Lambda = \Re_+^n$, with zone $S = \{x = (x_j) \in \Re^n \mid x_j > 0\}$. It was shown to be a Bregman's function in, e.g., Censor et al. [1990]. Applying iterate (7) we obtain the entropic proximal point algorithm (EPP):

The EPP Algorithm.

$$x^{k+1} \leftarrow \arg\min_{x \in X \cap S} c^T x + \frac{1}{\gamma^k} \left(\sum_{j=1}^{n} x_j \left[\log\left(\frac{x_j}{x_j^k}\right) - 1 \right] + \sum_{j=1}^{n} x_j^k \right) \qquad (12)$$

The constant term $\sum_{j=1}^{n} x_j^k$ can be removed from the optimization.

2.2.3 The Subproblem Algorithm. The PMD algorithm as described by the iteration (7) involves the solution of a series of strictly convex, separable subproblems. In the case of (9) we have a quadratic subproblem, and for (12) we have an entropy subproblem. The quadratic and entropic objective functions both belong to the class of *Bregman's functions*, as defined by Censor and Lent [1981], Hence, we can solve the subproblems using the primal-dual row-action algorithm of Censor and Lent [1981].

With the row-action algorithm, dual feasibility and complementary slackness are maintained throughout, but primal feasibility is obtained only in the limit. One dual variable is associated with each network node. Given dual values, the primal (flow) values satisfying CS are uniquely determined. The algorithm iteratively calculates, for each network node, the current surplus or deficit (4). Based on this value, and on the objective costs of the arcs incident to the nodes, the dual price of the node is updated. Simultaneously, the incident arc flows are updated, resulting in a reduced surplus or deficit. This combined update is known as a *Bregman projection*. We implement the algorithm on the CM-2 as a Jacobi-type algorithm, whereby node prices are updated concurrently across the network. The algorithm and its specialization for the quadratic or entropic network problems, and its implementation on the CM-2 are described in detail in Nielsen and Zenios [1992c].

3. A Comparative Computational Study

In this section we compare the ϵ-relaxation and PMD algorithms to each other, and to a network simplex code running on a Cray Y-MP. We then investigate the question of scalability of these algorithms on the massively parallel architecture, and then compare a small-scale, parallel Gauss-Seidel implementation of the QPP algorithm to GENOS, a sequential network simplex. Finally, we compare the QPP and EPP algorithms on the CM-2 with a sequential implementation of Relax, Bertsekas [1991].

The two algorithms were implemented on a massively parallel Connection Machine CM-2 with up to 64K processors. The implementations were based on the data structures of Zenios and Lasken [1988]. Each network arc is assigned two processors, one at the head, the other at the tail of the arc. These processors handle the computation of flows which together with the current dual iterate satisfy CS, or ϵ-CS, depending on the algorithm. In addition, each network node is assigned a processor, which handles node supply or demand. The processors corresponding to a network node, the processors corresponding to the tails of arcs going out of the node, and to the heads of arcs coming into the node, are grouped together as a *segment*, allowing efficient computation (using *segmented scan* operations) of the surplus or deficit of each node. All processors in a node segment calculate the primal/dual update as a function of the node surplus/deficit and the objective coefficients of the incident arcs. The dual price updates are then communicated (using the general *send* operation) from the head to the tail of each arc, and vice versa. For details of the implementations, see Li and Zenios [1991] or Nielsen and Zenios [1992b]. Comparisons of alternative data structures — that concluded that the scheme outlined above is the most efficient — are given in Nielsen and Zenios [1992a].

3.1 Test Problems. The implementations of the algorithms discussed in Section 2 were tested on a series of problems arising from several large scale applications, and on a set of randomly generated transportation problems. TSENG1-TSENG8 are transportation problems generated by NETGEN of Klingman et al. [1974]. (See Zenios and Censor [1991] for a discussion of nonlinear variants of these models). The coefficients $\{c_{ij}\}$ are random numbers uniformly distributed in the interval [1,5], and supply/demand vectors are randomly generated in the interval [1, 1000]. The second class of problems was obtained from a Military Airlift Command application. They are referred to as the Patient Distribution System (PDS) problems and are used to make decisions on the evacuation of patients from Europe. PDSt denotes a problem that models a scenario of t days. They are linear multicommodity network problems with eleven commodities. We

Problem	Nodes	Arcs	Max. arc cost
TSENG1	500+500	5063	5
TSENG2	750+750	7611	5
TSENG3	1000+1000	10126	5
TSENG4	1250+1250	12665	5
TSENG5	500+500	10051	5
TSENG6	750+750	15086	5
TSENG7	1000+1000	20134	5
TSENG8	1250+1250	25153	5
PDS1.1	126	339	99999
PDS5.1	686	2149	99999
PDS10.1	1399	4433	99999
PDS20.1	2857	10116	99999
PDS30.1	4223	15526	99999
NAVY	3498	6841	1
HUGENAVY	30639	64542	0

Table 1: Medium scale test problem characteristics.

extracted the first commodity from each model to form our linear min-cost network problem. The size and complexity of the models increase with t. NAVY and HUGENAVY are two instances of a Naval personnel assignment model which was formulated as a very large linear program with an embedded network structure. We solve only the network component here. The problem NAVY has a nonzero assignment cost vector c whereas the larger problem HUGENAVY has an assignment cost vector which is identically zero. Pinar and Zenios [1992] describe these models in detail. The characteristics of these test problems are summarized in Table 1. Finally, two sets of very large problems were generated: the TPn problems were generated using NETGEN, while the GenNN problems

Problem	Nodes	Arcs	Max. arc cost	Average supply (per supply node)
TP00	5000	50000	100	100
TP01	5000	100000	100	200
TP1	10000	1000000	100	500
TP2	20000	1000000	100	1000
TP5	50000	1000000	100	400
TP10	50000	2000000	100	800
Gen0.25M	32768	245760	10	1500
Gen0.5M	65536	491520	10	1500
Gen16M	2097152	15728640	10	1500

Table 2: Large scale test problem characteristics.

were generated directly on the CM-2 due to their extremely large size, see Table 2. The parameters specified for TP1, TP2 and TP5 are identical to those used by Barr and Hickman [1990].

The termination criterion is small primal error, $\max_{i\in\mathcal{N}} |g_i| \leq \tau$, where g_i is calculated from (4) and τ is chosen as $\tau = 10^{-3}\frac{1}{m}\sum_{i=1}^{m} |b_i|$. The termination tolerance was checked every 25-50 iterations.

3.2 Testing the ϵ-relaxation algorithm. In this section we describe several implementation techniques that improve the convergence rate of the ϵ-relaxation algorithm, and then give results for the test problems described above.

The Initial Dual Point. The algorithm may be initialized by setting $\pi_i = 0$ for all $i \in \mathcal{N}$ and then setting $x_{ij} = l_{ij}$ or $x_{ij} = u_{ij}$ depending on the ϵ-CS conditions. This starting point works well for the bipartite transportation models, like TSENG1-8. However, it is a bad starting point if the network has a large diameter, with supply and demand nodes at the extremes.

A better initial dual point can easily be obtained by using a shortest path algorithm on the network graph. From the CS conditions for arc $(i,j) \in \mathcal{E}$, we have that if $l_{ij} < x_{ij} \leq u_{ij}$, then $\pi_i - \pi_j = c_{ij}$. If j is a destination node with $\pi_j = 0$, then π_i must be equal to c_{ij}. In general, if there is flow from a node i to a node j, then $\pi_i - \pi_j$ should be more than the length of the shortest path from i to j, where the "length" of arc (i, j) is c_{ij}. We hence implemented a variant of the Bellman-Ford shortest path algorithm and used it for providing an initial dual point.

This procedure is executed very efficiently on the CM. It is used only once at the beginning of the algorithm, and usually takes less than 1 sec of CM time. The initialization procedure is especially beneficial for the networks with large diameter, such as PDS20.1 and PDS30.1, where the overall solution time was reduced by a factor of two. The initialization routine is never detrimental, and is used henceforth.

ϵ-Scaling and Dynamic Tolerance Adjustment. The final value of ϵ needed for a precise solution is very small, and if this value is used throughout the algorithm, progress is extremely slow. This suggests the idea of ϵ-*scaling*, which was first analyzed for the linear network flow problem by Goldberg [1986], and was more fully established by Goldberg and Tarjan [1987]. Wein and Zenios [1991] incorporated ϵ-scaling in the auction algorithm for the solution of the assignment problem. ϵ-scaling consists of applying the algorithm in a series of phases, starting with a large value of ϵ and successively reducing ϵ up to an ultimate value which is less than the critical value $\frac{1}{m}$. Each application of the algorithm provides good initial prices for the next phase of application.

The sequence of values used for ϵ is given by:

$$\epsilon(k) = \max\{\frac{1}{m+1}, \frac{C}{m\theta^{k+1}}\}, \quad k = 0, 1, 2, \ldots .$$

where C is the maximum absolute cost coefficient, and $\theta > 1$ is a parameter set by the user; we used $4 \leq \theta \leq 8$. Experiments show that for problems with larger cost range it is beneficial to choose larger θ.

The second strategy is to introduce *adaptive tolerance* (also referred to as *node scaling*) technique. We give a larger error tolerance, τ, at the beginning of the iterations. During the course of the algorithm, we reduce τ by a factor of 2 in each phase of the ϵ-scaling, until we reach the desired tolerance. We tested a problem with 100 nodes and 900 arcs using the original algorithm, the ϵ-scaling variant and the adaptive tolerance variant. The scaled version terminated almost 20 times faster than the original algorithm, and the total number of iterations was reduced by a factor of about 20. When adding the adaptive tolerance procedure, we further reduced the solution time by 15%-20%.

Mixing Up and Down Iterations. We discussed in Section 2.1 the up iteration for positive surplus nodes. A symmetric iteration can be used for nodes with negative surplus (called *down iteration*). The down iteration only applies to nodes i with $g_i < 0$. Down iterations increase g_i and decrease p_i.

However, the algorithm may not terminate if up and down iterations are mixed arbitrarily. It is therefore necessary to impose some assumptions either on the problem structure or on the method by which up and down iterations are interleaved. Our model works as follows: We use two parameters to control the number of up iterations mixed with the number of down iterations. When we switch from up iteration to down iteration (or vice versa), we need to make sure that the dual objective value at this point does not deteriorate. Otherwise we will stay with our current form of iteration. In our experiments, about 10 up iterations were sufficient before switching to down iterations, and vice versa.

Tests showed that combining up and down iterations speeds up the convergence and is especially useful for transshipment problems. For example, the execution time of PDS1.1 is reduced from 10.28 sec to 8.71 sec, PDS5.1 from 23.55 sec to 16.92 sec and PDS10.1 from 40.56 sec to 26.42 sec. PDS20.1 and PDS30.1 did not converge after 10000 iterations without down iterations. After adding the down iterations PDS20.1 converges after 5600 iterations and PDS30.1 converges after 7300 iterations.

3.2.1 Results for the ϵ-Relaxation Algorithm. We now provide a summary of computational results in order to highlight certain aspects of the algorithm. Table 3 reports the results we have obtained from solving all test problems with up to 1 million arcs. During our experiments we found that PDS problems are relatively more difficult than the others. Two reasons explains this difficulty: one is that they have large cost (i.e., 99999) in the network. We can only avoid this difficulty by resorting to ϵ-scaling. Another reason is that large chains are present in the network. For instance, the diameter of the PDS30.1 network is 63. This observation motivated the use of a shortest path routine for finding a good dual starting point. HUGENAVY is a medium-sized problem. It is well scaled because all arcs have 0 costs. It took very few iterations to converge. This demonstrates that the ϵ-algorithm is much faster on problems with small arc costs. As a final exercise, we solved the large-scale problems: TP00, TP01, TP1, TP2 and TP5. These are very sparse transportation problems (0.8%, 1.6%, 4%, 1% and 0.16% sparse, respectively). They have medium arc costs [1,100], which does not cause much difficulty.

Problem	Time	Itns.	VP ratio
TSENG1	32.80	2650	1
TSENG2	30.04	2200	1
TSENG3	43.15	3100	1
TSENG4	47.30	3350	1
TSENG5	37.25	2750	1
TSENG6	39.60	2800	1
TSENG7	55.27	2900	2
TSENG8	44.56	2350	2
PDS1.1	8.71	500	1
PDS5.1	16.92	950	1
PDS10.1	26.42	1900	1
PDS20.1	103.25	5600	1
PDS30.1	135.90	7300	2
NAVY	0.92	50	1
HUGENAVY	2.47	50	4
TP00	88.43	3250	4
TP01	164.84	3050	8
TP1	760.45	2050	64
TP2	956.26	2950	64
TP5	890.76	2700	64

Table 3: Results for the ϵ-relaxation algorithm.

3.3 Testing the Proximal Minimization Algorithms. By *minor iteration* we mean a complete sweep of the row-action algorithm over all nodes of the network. Of course, in the massively parallel implementation this is done concurrently. A *major iteration* is the execution of a number of minor iterations followed by an update of the proximal point.

The *perturbation gap* δ_1 measures the difference between the objective value of the nonlinearly perturbed program, and the linear objective value. The final solution is denoted by x^k. The perturbation gap for QPP is $\delta_1 = \frac{1}{2\gamma^k} \| x^k - x^{k-1} \|^2$, and for EPP,

$$\delta_1 = \frac{1}{\gamma^k} \left(\sum_{j=1}^n x_j \left[\log\left(\frac{x_j^k}{x_j^{k-1}}\right) - 1 \right] + \sum_{j=1}^n x_j^{k-1} \right).$$

The *primal error* δ_2 measures the maximum absolute surplus or deficit at any network node, $\delta_2 = \max_{i \in \mathcal{N}} \{| g_i |\}$. Besides these statistics, we also report the duality gap $b^T \pi - c^T x$ (i.e., the difference between the primal and dual objective values) at termination. Solution times are in seconds, on a dedicated 16K processor CM-2 using a SUN 4/360 front-end. The CM utilization factor for all runs is above 95%.

The algorithm terminates when the following criterion is satisfied: $\delta_1 \leq \eta_1$ and $\delta_2 \leq \eta_2$ where η_1 and η_2 are user specified parameters.

Problem	γ^0	Minor Iter.	Time	Obj. val.	Duality gap
Tseng1	0.2	3050	18.80	$5.792 \cdot 10^7$	-0.001%
Tseng2	0.2	5400	45.94	$8.798 \cdot 10^7$	0.003%
Tseng3	0.2	4025	32.49	$1.240 \cdot 10^8$	-0.001%
Tseng4	0.2	2750	22.76	$1.456 \cdot 10^8$	-0.002%
Tseng5	0.2	3350	28.50	$3.443 \cdot 10^7$	0.0002%
Tseng6	0.2	4025	35.53	$5.364 \cdot 10^7$	0.0004%
Tseng7	0.2	3750	47.23	$7.732 \cdot 10^7$	0.0005%
Tseng8	0.2	3700	48.54	$9.166 \cdot 10^7$	-0.0016%
PDS1.1	1.0	250	1.43	$2.655 \cdot 10^9$	-0.00002%
PDS5.1	1.0	550	3.73	$2.515 \cdot 10^9$	0.00003%
PDS10.1	1.0	950	5.67	$2.341 \cdot 10^9$	0.0001%
PDS20.1	1.0	2000	15.06	$1.978 \cdot 10^9$	0.0003%
PDS30.1	1.0	3850	45.19	$1.632 \cdot 10^9$	0.0007%
Navy	1000	1450	10.80	0	0.0002
Hugenavy	1000	1225	55.66	0	0.0031
TP00	1.0	2025	57.91	$2.901 \cdot 10^6$	0.0002%
TP01	1.0	4525	232.84	$3.169 \cdot 10^6$	0.0001%

Table 4: Solving the test problems using the Quadratic Proximal Point algorithm on a CM–2 with 16K processors.

The theory for QPP (Rockafellar [1976]) assures convergence if the kth subproblem is solved to an accuracy of $\delta_2^k \leq \epsilon^k$, where $\sum_{k=1}^{\infty} \epsilon^k < \infty$. No corresponding result is available for the EPP algorithm. We explored in Nielsen and Zenios [1992c] various strategies regarding subproblem solution accuracy, and found the simple strategy of executing 25-50 minor iterations for each major iteration (proximal point update) to be most efficient. We use the same strategy in this study. The convergence criteria are checked once every major iteration.

For the quadratic and entropic runs of this section, the penalty parameter γ^k and the proximal point were updated as follows:

$$\gamma^{k+1} = \begin{cases} \gamma^k & \text{if } \delta_2 > \eta_2, \\ 2\gamma^k & \text{if } \delta_2 \leq \eta_2, \end{cases} \quad (13)$$

Correspondingly, the proximal point was updated at each major iteration k as follows:

$$x^{k+1} = \begin{cases} \rho x - (1-\rho)x^k & \text{if } \delta_1 > \eta_1 \\ x^k & \text{if } \delta_1 \leq \eta_1, \end{cases} \quad (14)$$

where x is the current iterate of the subproblem, and $\rho \in (0,2)$ is a relaxation parameter. The proximal point is thus not updated if the perturbation error is within the tolerance.

3.3.1 Results for the PMD Algorithms. In Tables 4 and 5 we give detailed results of solving the medium sized test problems with the quadratic and entropic proximal point methods, respectively. A comparison is shown in Figure 1. All

Problem	γ^0	Minor Iter.	Time	Obj. val.	Duality gap
Tseng1	0.002	1900	19.34	$5.792 \cdot 10^7$	-0.0006%
Tseng2	0.002	1725	22.63	$8.797 \cdot 10^7$	-0.0005%
Tseng3	0.002	1775	22.61	$1.240 \cdot 10^8$	0.001%
Tseng4	0.002	2650	34.05	$1.455 \cdot 10^8$	-0.00006
Tseng5	0.002	2450	31.48	$3.443 \cdot 10^7$	0%
Tseng6	0.001	5250	71.87	$5.364 \cdot 10^7$	-0.0004%
Tseng7	0.002	2250	43.14	$7.733 \cdot 10^7$	-0.0002%
Tseng8	0.002	4375	83.79	$9.163 \cdot 10^7$	-0.0008%
PDS1.1	10^{-4}	400	3.77	$2.655 \cdot 10^9$	0.0009%
PDS5.1	10^{-4}	875	7.22	$2.515 \cdot 10^9$	0.003%
PDS10.1	10^{-4}	1400	14.04	$2.341 \cdot 10^9$	-0.002%
PDS20.1	10^{-4}	2625	33.53	$1.979 \cdot 10^9$	0.002%
PDS30.1	10^{-4}	3925	74.10	$1.634 \cdot 10^9$	-0.0027%
Navy	100	100	1.26	0.000023	-0.000020
Hugenavy	100	150	10.08	0	-0.00024
TP00	1/6	1700	67.41	$2.900 \cdot 10^6$	-0.0004%
TP01	1/6	1625	116.48	$3.165 \cdot 10^6$	0.0021%

Table 5: Solving the test problems using the Entropy Proximal Point algorithm on a CM-2 with 16K processors.

problems were solved to a perturbation tolerance of $\eta_1 = 10^{-5}$ and a primal tolerance of $\eta_2 = 0.5$.

Both algorithms solved all of the test problems in a reasonable time. For the QPP algorithm, the solution times were not very sensitive to the choice of initial penalty parameter γ^0, except for the Navy and Hugenavy problems: For a smaller value than the one shown, the convergence of the proximal point would be very slow. The EPP algorithm required more experimentation with the penalty parameter to obtain good convergence. For both algorithms, the larger the penalty parameter, the faster the proximal point sequence $\{x^k\}$ converged, but the subproblems became more difficult for the dual relaxation algorithm.

The duality gap at the solution point was for the Tseng and Navy problems generally much smaller with the EPP algorithm than with the QPP algorithm, whereas the opposite is true for the PDS problems. It is not clear which characteristics of the test problems account for this consistent difference.

The large-scale problems were solved to a primal error tolerance of 0.5, but (in contrast with the medium-size problems) to a perturbation tolerance of 10^{-4} to keep running times reasonable.

Both the QPP and the EPP algorithms solved the problems in a reasonable time. The results are given in Table 6. The EPP algorithm is uniformly faster than the QPP algorithm for these problems (but see Section 3.4). The main difference between the two algorithms was the tail convergence of the subproblem primal error: The EPP algorithm consistently reduced this error faster than the QPP algorithm.

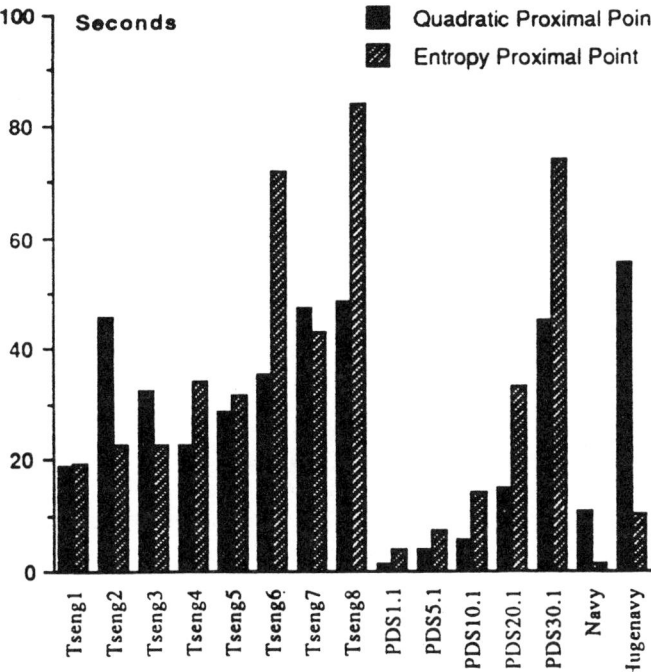

Figure 1: Comparing the Quadratic and Entropic Proximal Point algorithms.

Problem	Algorithm	γ^0	Minor Iter.	Time
TP1	QPP	0.5	9000	2850.18
	EPP	1/12	1550	726.42
TP2	QPP	0.5	5300	1669.73
	EPP	1/12	1275	597.24
TP5	QPP	0.5	3200	1018.32
	EPP	1/24	1300	613.34
TP10	QPP	0.5	4875	3092.86
	EPP	1/24	1725	1616.27

Table 6: Solving large-scale test problems using the proximal point algorithms on a 16K CM–2.

Problem	QPP			EPP		
	ϵ^0	Iter.	Time	ϵ^0	Iter.	Time
Gen0.25M	10	2325	4.6 min.	0.01	9900	25.4 min.
Gen0.5M	10	2400	7.2 min.	0.01	8350	44.9 min.
Gen16M	1	3075	99 min.		N/A	N/A

Table 7: Solving the large-scale, random transportation problems. The smaller runs were executed on a 16K PE CM-2, the Gen16M run on a 32K PE CM-2 at a VP-ratio of 1024.

3.4 Solving extremely large-scale problems. In order to test the PMD algorithms on extremely large, linear network problems, a set of test problems were randomly generated directly on the CM-2. They have an equal number of supply- and demand-nodes, each with a degree of 15. The problem characteristics are given in Table 2. These problems were solved to a perturbation tolerance of 10^{-5} and a primal feasibility tolerance of 0.1.

Results are given in Table 7. It is interesting to note that these problems appear to be easier for the QPP algorithm than the TP problems. The largest problem, consisting of more than 2 million nodes and almost 16 million arcs, was solved using the QPP algorithm in about 1.5 hours on a 32K PE CM-2.

3.5 Scalability of the CM-2 Implementations. In order to see how well the implementations utilize the underlying parallel hardware, we solved a set of test problems of varying size using 8K, 16K and 32K processors. These problems were randomly generated using size as a parameter. Results are shown in Figure 2, where problem size is measured by the number of VPs required for the representation, $2n + m$ (the graph for 64K processors is estimated). We note that, for each machine size, as the problem size is increased by a factor of two, the solution time increases by about 1.8–1.9. Also, as the machine size is doubled, the solution time decreases by about the same factor, 1.8–1.9. That these factors are less than two is evidence that the CM-2 is used more efficiently with higher VP-ratios. However, the factors are close to two, which shows that our implementations scale efficiently with respect to both the machine and problem size.

3.6 Comparison with small-scale parallel algorithm. We now compare a Gauss-Seidel version of the QPP algorithm, running on a 8 processor Alliant FX/8, with the sequential GENOS network simplex. The test problems used are again transportation problems generated using NETGEN, with coefficients generated uniformly in the interval [1,1000]. Total demands and supplies are both $250 \cdot m$, where m is the total number of nodes, randomly allocated to the supply and demand nodes, respectively.

The results of this comparison, shown in Table 8, indicate that even a Gauss-Seidel version of QPP — which is superior to the Jacobi version since it uses at each node iteration data which are up-to-date — is not competitive with a state-

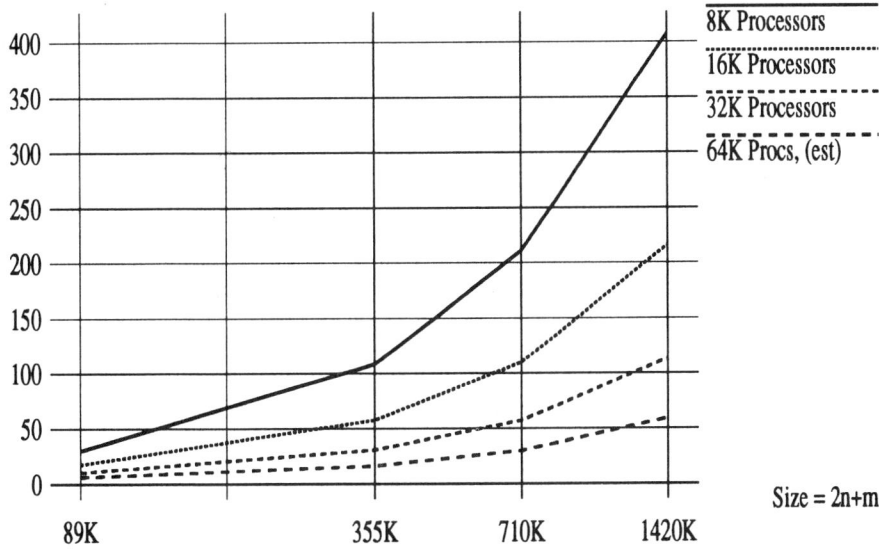

Figure 2: Solving problems of differing size using 8K, 16K and 32K CM–2 processors. Results for the 64K CM–2 are estimated.

Problem	Nodes	Arcs	QPP time	Simplex time
SMTRANS.00	35 + 45	185	18.17	0.343
LTRANS0.00	500 + 500	10046	35.6*	22.47
LTRANS1.00	1250 + 1250	12674	61.3*	68.21
LTRANS2.00	1250 + 1250	25152	112.5*	96.63

Table 8: Comparing a small-scale parallel Gauss-Seidel implementation of QPP to a network simplex. The runs marked * are parallelized to use 8 processors on the Alliant FX/8, the remaining times are using 1 FX/8 processor.

Problem	QPP time	Relax time	Ratio
PDS1.1	1.43	0.012	119
PDS5.1	3.73	0.090	41.4
PDS10.1	5.67	0.164	34.6
PDS20.1	15.06	0.504	29.9
PDS30.1	45.19	1.090	41.5
Navy	10.80	0.125	80.6
Hugenavy	55.66	2.465	22.6

Table 9: Comparing the QPP algorithm to the Relax algorithm. Times are in seconds: QPP on a 16K CM-2 versus the Relax running on a DECStation 5000.

of-the art simplex code when both are executing on serial, or small-scale parallel machines. However, the PMD algorithms are competitive on the massively parallel architectures, and only the PMD algorithms could solve the multi-million variable problems of Table 2. Due to this result, we did not further pursue sequential or small-scale, parallel implementations of the PMD algorithms.

3.7 Comparisons with Relax. We compare the QPP algorithm with the relaxation code Relax of Bertsekas [1991]. Relax was executed on a DECStation 5000 using those of our test problems which are pure networks with integer data. Results, given in Table 9, show that Relax is superior to the PMD algorithms for problems of this size. However, due to memory limitations we could not solve multi-million variable problems with the sequential implementation of Relax.

4. Comparison of the ϵ-relaxation and PMD algorithms

We now compare the ϵ-relaxation algorithm and the two PMD algorithms, based on the results of the preceding sections. The results for the ϵ-relaxation algorithm (Table 3) were obtained on a CM-2 with 32K processors, whereas the results for the PMD algorithms (Tables 4, 5 and 6) are from a 16K CM-2.

The ϵ-relaxation algorithm is very fast on the Navy and Hugenavy problems, and converges much faster than either of the PMD algorithms. These problems are characterized by a very small range of the objective coefficients, which benefits this algorithm. The fastest algorithm overall, however, is the EPP algorithm. On the medium sized problems it is comparable with the QPP algorithm, and they are both roughly twice as fast as the ϵ-relaxation algorithm (when scaling execution times to comparable machine sizes). For the large and extremely large problems, it is not clear whether the QPP or EPP algorithm is superior. This seems to depend on the individual problem characteristics. However, both these algorithms seem to clearly outperform the ϵ-relaxation algorithm, except for problems with very small objective coefficients.

Finally, we provide a comparison with a state-of-the-art network simplex code executing on a CRAY Y-MP, namely GENOS (Mulvey and Zenios [1987]). All CM-2 execution times are estimated for a 64K CM-2, using factors estimated

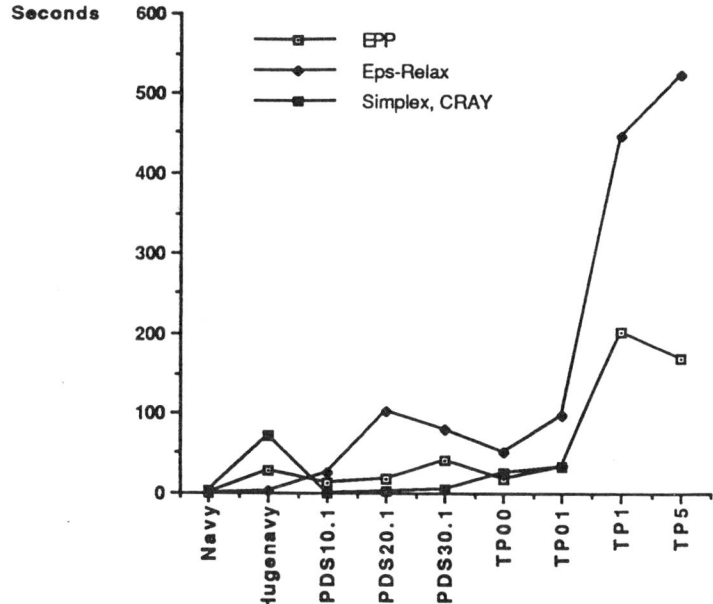

Figure 3: Comparing the EPP, ϵ-relaxation and network simplex algorithms. The figure includes results for the large-scale problems TP1 and TP5. EPP was run on a 16k CM-2, ϵ-relaxation on a 32k CM-2 and network simplex on a CRAY Y-MP.

from Figure 2. Due to the different nature of the hardware, algorithms and solution quality, one must be careful making conclusive claims, based upon solution time alone. With this in mind, we note that the proximal point algorithms, executing on the CM-2, are, with the exception of Hugenavy, not convincingly faster than the simplex algorithm executing on the CRAY Y-MP. However, the proximal point algorithms do appear to become increasingly competitive as the problem size increases. The largest TP or Gen problems could not be solved on the CRAY due to memory limitations.

5. Conclusions

Based on the results of this paper, we can answer affirmatively the question whether massively parallel algorithms can solve large linear network problems. Each of the algorithms investigated solve very large linear network problems efficiently. Generally, the Proximal Minimization Algorithms with D-functions are superior to the ϵ-relaxation algorithm, but much depends on the precise problem characteristics, such as the range of the objective coefficients.

All the algorithms have very similar memory requirements, and memory more than execution time was the limiting factor regarding the size of problems we could solve. It seems reasonable to conclude that even larger problems could be solved given access to a larger CM-2. On a CM-2 with 64K processors, each with 256Kbytes of memory, for instance, we would expect to be able to solve problems with more than 32 million variables.

REFERENCES

1. R.S. Barr and B.L. Hickman. A new parallel network simplex algorithm and implementation for large time-critical problems. Technical report 90-CSE-37, Southern Methodist University, Department of Computer Science and Engineering, 1990.

2. D.P. Bertsekas. A Distributed Algorithm for the Assignment Problem. Laboratory for Information and Decision Systems, Working Paper, Massachusetts Institute of Technology, Cambridge, 1979.

3. D.P. Bertsekas. Distributed asynchronous relaxation methods for linear network flow problems. Report LIDS-P-1606, MIT, Nov. 1986a.

4. D.P. Bertsekas. Distributed relaxation methods for linear network flow problems. *Proc. 25th IEEE Conf. Dec. & Contr.*, pages 2101–2106, 1986b.

5. D.P. Bertsekas. Linear Network Optimization, The MIT Press, Cambridge, Massachusetts, 1991.

6. D.P. Bertsekas and J.N. Tsitsiklis. *Parallel and Distributed Computation: Numerical Methods.* Prentice Hall, Englewood Cliffs, New Jersey, 1989.

7. Y. Censor and A. Lent. An iterative row-action method for interval convex programming. *Journal of Optimization Theory and Applications*, 34:321–353, 1981.

8. Y. Censor, A.R. De Pierro, T. Elfving, G.T. Herman, and A.N. Iusem. On iterative methods for linearly constrained entropy maximization. In A. Wakulicz, editor, *Numerical Analysis and Mathematical Modelling*, volume 24, pages 145–163. Banach Center Publications, PWN - Polish Scientific Publisher, Warsaw, Poland, 1990.

9. Y. Censor and S.A. Zenios. The proximal minimization algorithm with D-functions. *Journal of Optimization Theory and Applications*, 73(3):455–468, 1992.

10. J.R. Eriksson. An iterative primal-dual algorithm for linear programming. Report LiTH-MAT-R-1985-10, Department of Mathematics, Linköping University, S-581 83 Linköping, Sweden, October 1985.

11. A. V. Goldberg. Solving minimum cost flow problems by successive combinatorial algorithms. In *Proc. 18th. ACM STOC.*, pages 136–146, 1986. Extented Abstract.

12. A.V. Goldberg and R.R. Tarjan. Solving minimum cost flow problems by successive approximation. In *Proc. 10th. ACM STOC.*, pages 7–18, 1987.

13. D. Klingman, A. Napier, and J. Stutz. NETGEN - a program for generating large-scale (un)capacitated assignment, transportation, and minimum cost flow network problems. *Management Science*, 20:814–822, 1974.

14. X. Li and S.A. Zenios. Data-level parallel solution of min-cost network flow problems using ϵ-relaxation. Report 91-05-04, Decision Sciences Department, The Wharton School, University of Pennsylvania, Philadelphia, PA 19104, May 1991.

15. M.Ç.Pinar and S.A. Zenios. Naval personnel assignment: An application of linear-quadratic penalty methods. In O. Balci, R. Sharda and S.A. Zenios., eds, *Computer Science and Operations Research: New developments in their Interfaces*, Pergamon Press, pp. 43–58, 1992.

16. J.M. Mulvey and S.A. Zenios. GENOS 1.0: A Generalized Network Optimization System. User's Guide. Report 87-12-03, Decision Sciences Department, the Wharton School, University of Pennsylvania, Philadelphia, PA 19104, 1987.

17. S.S. Nielsen and S.A. Zenios. Data Structures for Network Algorithms on Massively Parallel Architectures, *Parallel Computing* 18:1033–1052, 1992a.

18. S.S. Nielsen and S.A. Zenios. Massively parallel algorithms for singly constrained nonlinear programs. *ORSA Journal on Computing*, 4(2):166–181, 1992b.

19. S.S. Nielsen and S.A. Zenios. Proximal minimization with D-functions and massively parallel solution of linear network programs. *Computational Optimization and Applications*, 1(4), 1992c (to appear).

20. R.T. Rockafellar. Augmented Lagrangians and applications to proximal point algorithms in convex programming. *Math. of Operations Research*, 1:97–116, 1976.

21. R.T. Rockafellar. Monotone operators and the proximal point algorithm. *SIAM Journal on Control and Optimization*, 14:877–898, 1976.

22. J. Wein and S.A. Zenios. On the massively parallel solution of the assignment problem. *Journal of Parallel and Distributed Computing*, 13:228–236, 1991.

23. S.A. Zenios and Y. Censor. Massively parallel row-action algorithms for some nonlinear transportation problems. *SIAM Journal of Optimization*, 1:373–400, 1991.

24. S.A. Zenios and R.A. Lasken. Nonlinear network optimization on a massively parallel Connection Machine. *Annals of Operations Research*, 14:147–165, 1988.

Soren S. Nielsen, Management Science and Information Systems, University of Texas, Austin, TX 78712. *E-mail:* nielsen@bongo.cc.utexas.edu

Stavros A. Zenios, Wharton School of Business, University of Pennsylvania, Philadelphia, PA 19104. *E-mail:* zenios@wharton.upenn.edu

Approximating concurrent flow with unit demands and capacities: an implementation

JAMES M. BORGER, TAEKYUNG S. KANG, AND PHILIP N. KLEIN

ABSTRACT. We have implemented the concurrent flow approximation algorithm described in "Leighton-Rao might be practical: faster approximation algorithms for concurrent flow with uniform capacities," by Klein, Stein, and Tardos, which appeared in STOC 90. The implemented algorithm works for unit demands and unit capacities, and makes use of the idea of randomized path selection, introduced in that paper.

1. Definition of the problem

An instance of multicommodity flow consists of a network G with edge- or node-capacities and a set of *commodity* specifiers, each consisting of a source node s_i, a sink node t_i, and a demand d_i, a positive real. A multicommodity flow consists of a flow of value d_i from s_i to t_i for every commodity i. The multicommodity flow is *feasible* if the sum of the flows through each edge/node does not exceed the edge or node's capacity. An instance is feasible if there exists a feasible multicommodity flow.

The *concurrent flow problem* is, given an instance, to output the minimum positive real λ such that by multiplying all capacities by λ, one obtains a feasible instance. As proof of feasibility, the feasible multicommodity flow in the multiplied network should also be output. The problem was proposed by Matula [16].

The special case handled by our implemented algorithm requires that all demands d_i and all capacities are one. While more general algorithms are available,

---

AMS Categories: 68Q20, 68R10.
Research of first two authors supported by NSF Research Experience for Undergraduates supplements to NSF grants CCR-9012357 and CCR-9157620. Research of third author supported by NSF grant CCR-9012357 and an NSF PYI award (CCR-9157620), together with PYI matching funds from Thinking Machines Corporation and Xerox Corporation. Additional support provided by ONR contract N00014-91-J-4052, ARPA Order 8225.

© 1993 American Mathematical Society
1052-1798/93 $1.00 + $.25 per page

this special case arises in several applications of considerable interest, as we describe in the next section.

2. Applications

Two principal application areas for the algorithm are finding graph separators and VLSI routing.

2.1. Finding graph separators. A crucial ingredient in many divide-and-conquer graph algorithms is a method for choosing a *separator*, a set of edges or nodes whose removal breaks a graph into two pieces. In an important paper that has catalyzed much recent work on algorithms for the concurrent flow problem, Leighton and Rao [12] gave an algorithm to find an approximately *sparsest cut*, a separator that minimizes the ratio of the size of the separator to the size of the smaller piece. The computational bottleneck of their algorithm was solving the linear programming dual of concurrent flow. Their work was foreshadowed by, e.g., that of Matula and Shahrokhi [17].

Leighton and Rao also showed that the algorithm could be used in approximation algorithms for a variety of optimization problems in graph theory. Other researchers [7, 6, 9, 11, 15, 20] have added still more problems approximable by these techniques. Thus the method of Leighton and Rao lends considerable importance to the problem of solving the dual of concurrent flow.

The method involves solving an instance of concurrent flow in which the network is the graph to be separated and the commodities are as follows: there is a commodity with demand 1 between every pair of distinct nodes. For such an instance the number of commodities is too large; in a computationally easier variant proposed by Leighton, each node is the source node s_i for some small number of commodities (e.g. 3) where the sink nodes t_i are randomly chosen nodes. In our computational experiments we therefore generate commodities acccording to this rule.

2.2. VLSI routing. Raghavan and Thompson [18] proposed a method for routing wires in VLSI so as to minimize channel-width. The computational bottleneck is solving the concurrent flow problem. In this case the network represents the VLSI chip being routed through, and the commodities represent wires that must be routed. As in the previous application, all demands are 1.

3. A linear-programming formulation and its dual

The concurrent flow problem can be formulated as a linear program. We are given an undirected network G with capacities $c(e)$ on the edges. By replacing each edge e with two oppositely directed arcs, we obtain a directed graph. For such an arc a, let $\hat{a}$ be its opposite. We are also given a set of commodities specified by a source s_i, a sink t_i, and a demand value d_i. In the linear-programming formulation, there is a variable $f_i(a)$ for every commodity i and arc a. There is also a variable λ representing the maximum total flow on any arc. The goal is to minimize λ subject to the capacity constraints and the conservation constraints.

The capacity constraints are: for each edge e,

$$\sum_i f_i(a) + \sum_i f_i(\hat{a}) \leq \lambda c(e)$$

where a and $\hat{a}$ are the arcs corresponding to e.

The conservation constraints are: for each node v and commodity i,

$$\sum_{a=vw} f_i(a) - \sum_{a=wv} f_i(a) = \begin{cases} d_i & \text{if } v = s_i \\ -d_i & \text{if } v = t_i \\ 0 & \text{otherwise} \end{cases}$$

The dual linear program involves finding an assignment of lengths to edges so as to maximize the weighted sum of source-to-sink shortest path distances, subject to a bound on the weighted sum of lengths. More formally, there is a length-variable $\ell(e)$ for each edge e and a distance-variable $dist_i(v)$ for each commodity i and node v indicating the shortest-path distance of v from t_i. The goal is to maximize $\sum_i d_i dist_i(s_i)$ subject to the usual shortest-path constraints

$$d_i(w) \leq d_i(v) + \ell(vw)$$

for each commodity i and edge vw, and the following constraint on the length-variables:

$$\sum_e c(e)\ell(e) \leq 1.$$

The algorithm we implemented requires that all capacities $c(e)$ have value 1. In this case, the primal LP's capacity constraints state that the sum of flows through each edge must not exceed λ, and the dual LP's constraint on the length-variables states that the sum of lengths must not exceed 1. Our implementation also requires that the demands d_i all have value 1. Hence the dual objective function is to maximize the sum $\sum_i dist_i(s_i)$ of source-to-sink shortest-path distances.

4. The algorithm

4.1. The framework. The framework for the algorithm was originated by Shahrokhi and Matula [24], and further developed by Klein, Stein, and Tardos [10]. It is a primal-dual algorithm; it maintains feasible solutions to both primal and dual linear programs. The primal feasible solution consists principally of the flow assignment; for each commodity i, the flow from source s_i to sink t_i forms a directed graph. The dual feasible solution consists principally of the length assignment. In this algorithmic framework, the lengths are derived directly from the flow assignment using an exponential formula: the length of an edge e is $\exp(\alpha f(e))$, where $f(e)$ is the sum of flows through that edge, and α is a parameter of the algorithm which will be discussed in more detail later.

The algorithm consists of a sequence of iterations; in each iteration, one chooses a commodity i, say from s_i to t_i, finds the longest path in that commodity's flowgraph, computes a shortest path from s_i to t_i, and, depending on the length of the longest flow path compared to that of the shortest path, reroutes some amount of flow from the longest flow path to the shortest path (thus changing the flow graph). Eventually the flow assignment defines a nearly optimal solution to the concurrent flow problem, and the lengths define a nearly optimal solution to the linear programming dual. The details of choosing the flow path and the amount of flow to reroute will be presented later.

The linear-programming dual as formulated in §3 consists in maximizing the sum, over all commodities i, of the shortest-path distance from source s_i to sink t_i, subject to the sum of edge-lengths being at most 1. This is equivalent to maximizing the ratio of the shortest path sum to the sum of lengths, where the length assignment is not subject to the sum-of-lengths constraint. This latter formulation is easier to work with because it obviates the necessity to rescale all lengths when some of them change.

The algorithm uses duality to determine when to terminate. Recall that the objective in the primal is to route the flow so as to minimize the maximum flow on an edge/node. The dual objective is to maximize the sum of shortest path lengths. When the primal objective value is not much more than the dual objective value, both are nearly optimal, and the algorithm may terminate, outputting the flow assignment (the primal solution) and the length assignment (the dual solution).

The approach uses a potential function to show termination. The potential function is the sum of lengths, where each length is computed according to the exponential formula $\exp(\alpha f(e))$. The goal is reroute from those edges/nodes with high flow value $f(e)$ to those with smaller flow value. If done with care, this has the effect of reducing the sum of lengths. In particular, one only reroutes from a flow path to a shortest path if by doing so one significantly reduces the sum of lengths. It can be shown that if no such flow paths exists, then primal and dual are nearly optimal.

4.2. The implemented algorithm. Now we consider the specific algorithm implemented. The measure of quality of solution at any point in the algorithm is the ratio of $|f||\ell|_1$ to the sum of shortest paths, where $|f|$ is the maximum flow on any edge/node, and $|\ell|_1$ is the sum of lengths. This ratio is (barring floating-point roundoff error) always at least one; if it is close to 1, the primal and dual are close to being optimal.

Between $|f||\ell|_1$ and the shortest path sum is another quantity, the sum

$$\sum_x f(x)\ell(x)$$

over all edges/nodes x, of the flow $f(x)$ on x times the length $\ell(x)$ assigned to

x. We can express the above ratio in terms of two ratios

(1) $$\frac{|f|\|\ell\|_1}{\text{shortest path sum}} = \frac{|f|\|\ell\|_1}{\sum_x f(x)\ell(x)} \frac{\sum_x f(x)\ell(x)}{\text{shortest path sum}}$$

Thus when each of the two ratios is not much more than 1, the solutions are nearly optimal.

4.2.1. The relaxed optimality conditions.
In order to show the algorithm progresses towards near-optimality, we use two conditions, derived from the complementary slackness conditions of linear programming. The first condition corresponds to condition 1 in [10]; the second condition is a variant of condition 2 of [10] proposed by Goldberg.

Condition I: $\dfrac{|f|\|\ell\|_1}{\sum_x f(x)\ell(x)} \leq 1+\epsilon$

Condition II: $\sum_P f(P)[\ell(P) - \text{short}(P)] \leq \epsilon|f|\|\ell\|_1$

The first condition states that the first of the ratios in the right-hand side of (1) is nearly 1. The second condition needs some clarification. The sum is over all flow paths P. This includes the paths carrying flow from s_i to t_i for every commodity i. For a flow path P, $f(P)$ denotes the amount of flow carried by P, $\ell(P)$ denotes the length of the path P with respect to the edge/node lengths $\ell(\cdot)$, and short(P) denotes the length of the shortest path with the same endpoints as P. Intuitively, condition II states that the lengths of flow paths are typically not much more than the lengths of corresponding shortest paths. It can be shown that if conditions I and II hold, then the ratio (1) is not much more than one.

4.2.2. Handling condition I.
In order to achieve condition I, we periodically compute the ratio in I. If the ratio is not sufficiently close to 1, we add 0.25 to the parameter α appearing in the length formula $\ell(x) = \exp(\alpha f(x))$. This tends to concentrate more length on those edges/nodes x with particularly high flow $f(x)$, which tends to reduce the ratio. Indeed, it is shown in [10] that if α is sufficiently high, then the ratio is sufficiently close to 1. We do not set α to this sufficiently high value all at once, for reasons that will be discussed later.

4.2.3. Handling condition II.
Recall that the algorithm consists principally of rerouting flow from a flow path P to a shortest path with the same endpoints. The potential function by which we measure progress is the sum of lengths $|\ell|_1$. It is shown in [10] that by rerouting an amount σ of flow from a path P to a shortest path, we reduce $|\ell|_1$ by at least

(2) $$.95\alpha\sigma[(1-\epsilon)\ell(P) - \text{short}(P)]$$

as long as α and σ satisfy the condition

(3) $$1.05\alpha\sigma \leq \epsilon$$

Thus we make significant progress when $[(1-\epsilon)\ell(P) - \text{short}(P)]$ is large for the selected flow path P and α and σ satisfy (3).

To ensure that $[(1-\epsilon)\ell(P)-\text{short}(P)]$ tends to be large, we look to condition II. Condition II is a variant of condition II in [10] which was proposed by Goldberg. We can assume it is not satisfied, for otherwise (assuming I is satisfied) the primal and dual are nearly optimal and we can terminate. When II is not satisfied, for a flow path P randomly selected with probability proportional to $f(P)$, the expected value of $[(1-\epsilon)\ell(P) - \text{short}(P)]$ is at least $\epsilon|f||\ell|_1/D$, where D is the sum of the flows of all paths, i.e. the sum of all demands.

The random path selection technique, first proposed in [10], provides, at least in theory, a major improvement in performance over the deterministic method, which would involve looking at all commodities. One goal of this implementation effort was to evaluate the effectiveness of this method, or, rather, of Goldberg's variant. Initial experimentation suggested that this was an unsatisfactory method in that too much time was spent examining commodities for which a rerouting step would not yield progress. In the current implementation, therefore, we tried an alternative heuristic for selecting commodities to reroute. This heuristic is described in §4.2.5

4.2.4. Granularity of flow.

Recall that rerouting was only guaranteed to achieve a good reduction in the length-sum $|\ell|_1$ when the condition $1.05\alpha\sigma \leq \epsilon$ holds. This condition is problematic in practice. In the algorithm of [10], one assigns α a value such that condition I is guaranteed to hold, and then assigns σ a value so that the condition holds. Taking this seriously would result in too small a σ for practical implementation.

Recall that σ is the amount rerouted in an iteration. There are two disadvantages to σ being small. First, the resulting reduction in $|\ell|_1$ would be too small, so the algorithm would take too long. Second, a small σ would mean that the flow for each commodity would be broken up into too many flow paths. This would especially be a problem if we explicitly decomposed each commodity's flow into distinct flow paths of value σ.

In the implementation, therefore, we made several design decisions intended to alleviate this potential problem. First and foremost, we did not explicitly decompose flows into flow paths of value σ. We maintain each commodity's flow as a directed flow graph.

Second, each commodity has a different σ, and we only decrease these σ and only increase α on an as-needed basis. If rerouting σ_i flow from the longest flow path of commodity i to a shortest path doesn't improve the potential function, then rerouting flow in multiples of σ_i is not fine enough, and, therefore, σ_i is halved. To decide when to update α, every so many iterations (proportional to the number of commodities), we evaluate how well the algorithm is doing. First, we do a check to estimate whether condition II is satisfied (described further in a subsequent section). If the test succeeds, we evaluate the ratio in condition I and increase α by 0.25 if that condition is not satisfied.

Third, whenever possible we reroute more than σ units of flow. Each time a longest path is selected for rerouting, we reroute the amount of flow from that path to the shortest path such that the improvement in the potential function is maximized. This serves to accomplish in one rerouting what would have taken many otherwise.

Fourth, we use a technique from Orlin's min-cost flow algorithm. We inductively maintain that each edge in a flow graph for commodity i has a flow value that is a multiple of σ_i. In particular, this ensures that no flow path has a flow value less than σ_i. To achieve this, first, when we reduce any σ, we reduce it by an integral factor (two), and, second, when we reroute an amount of flow, we make sure that the amount is a multiple of σ_i, rather than rerouting the most flow we could.

4.2.5. Selecting commodities for rerouting.
According to theoretical analysis (Goldberg [8]; see also [10], we should select a flow path with probability proportional to its flow value $f(P)$. In a previous implementation following this approach, however, this rule rarely selects a commodity that benefits from rerouting (nearly always below 5% of the time). To increase our chances of selecting such a commodity, we instead randomly select a commodity based on the predicted improvement of the potential function by rerouting that commodity. More precisely, each commodity is assigned a weight that is equal to $\sigma_i * (long_i - short_i)$, and commodities are chosen with probability proportional to their weight plus a small constant to ensure that every commodity is occasionally selected. ($long_i$ and $short_i$ are estimates of the length of the longest flow path and shortest path save from the last time the commodity was chosen.) The weights are stored in a tree so this can be done in $log(K)$ time. This approach works very well as shown in Figure 2; the percentage of iterations in which rerouting occurs increases greatly as algorithm runs.

4.2.6. Shahrokhi's minimum-spanning-tree heuristic.
A technique Shahrokhi used in his code for concurrent flow [23] provides a method that sometimes obtains from the dual solution an improved dual solution. The idea is to use the current dual solution, i.e. the length assignment, to compute a minimum spanning tree T of the network. Then the maximum-cost edge e of the tree is used to determine a partition of the nodes of the graph, namely that determined by the two components of $T - e$. This partition has the following property: it is the partition maximizing the minimum-length edge crossing from one side to the other. One can obtain a different length assignment by assigning $1/k$ to each of the crossing edges, where k is the number of such edges. Surprisingly, it turns out, as reported by Shahrokhi [22], that the resulting length assignment often yields a better dual value than the length assignment from which it was derived. We used this technique in our implementation to determine closeness to optimality.

5. Experimental results

5.1. Input graphs. In evaluating our program, we chose to test it principally on graphs arising in practice, graphs for which one might want separators.

The graphs beginning with "outc" are ISCAS '85 benchmark circuits [3]. The graph "matrix" was derived from a sparse matrix contained in the Harwell-Boeing collection of sparse matrices [4, 5], matrices that arose in applications in, e.g., structural engineering and finite-element analysis problems.

We also tested our program against some graphs generated by the program GENRMF provided for the DIMACS Implementation Challenge, named "gr1" through "gr3."

For most of the graphs, there are three commodities per source node, each with a random sink. (For one of the graphs, gr1, there are 10 commodities per source node (In applying the separator algorithm to such a small graph, having three commodities per node is too few to guarantee good separation with high probability.)

Target approx is a number used to determine when to terminate the program. At regular intervals the program determines the best primal and dual values, and terminates when the ratio is at most $1 + \epsilon$. Since such checks are only made periodically, the achieved ratio is often less than $1 + \epsilon$.

Times are in seconds on a Sparc 1. Since the algorithm is randomized, we report average statistics averaged over 10 runs, for all the graphs but gr2, outc1355, and outc1908 (the most time-consuming ones). We kept track of running times, number of iterations, number of iterations in which rerouting took place, and the precision actually achieved. The data are in Figures 1 and 2.

5.2. Results of the experiments. In Figures 4 and 5 we have plotted time and iterations versus the quantities they are supposed to be proportional to in theory. The analysis of the algorithm predicts that the time required for a fixed target precision should be $O((k+n)(m+n))$, where k is the number of commodities, m is the number of edges, and n is the number of nodes. The linear trend on the log-log graph (Figure 4) indicates that the observed relationship is nearly linear, and, in fact, a least-squares fit gives the exponent to be 1.05 with $R^2 = 0.88$. (These are averages over the five precisions.) Thus the theoretical linear relationship between time and $(k+n)(m+n)$ is confirmed.

The analysis also predicts that the number of iterations for fixed precision should be $O(k+m)$ (see Figure 5). This relationship is fairly accurate for the least two precise approximation values (0.1 and 0.01); a least-square fit gives an $R^2 = 0.79$ for both approximation values. It is clear from the graph that this breaks down with greater target precision, and that, since the graphs have similar shape, the break-down is due to characteristics of the graphs rather than the randomness in the algorithm. We have not been able to determine what characteristics in some graphs might lead to this departure.

Figure 6 addresses the question of how many iterations must the algorithm complete to produce a desired precision. (Time grows almost perfectly linearly with iterations, as that is all the algorithm does.) Again, the linear nature on the log-log graph indicates and a least-square fit of the data indicate that, on the average, the required number of iterations is proportional to $\epsilon^2.89$ with $R^2 = 0.95$.

5.3. Comparisons, theoretical and actual. There are three programs to which we could make comparisons, that of Leong, Shor, and Stein, that of Kennington, and that of Shahrokhi. Leong, Shor and Stein [13] implemented an algorithm to solve a much more general problem. That algorithm handles arbitrary capacities and arbitrary demands, and is consequently more complicated; each iteration involves finding a minimum-cost flow instead of just a minimum-cost path. One must keep in mind the greater generality of their algorithm when making any comparison in run-times; however, the comparison is still instructive.

We have data from Stein [25] describing his implementation's performance on four of our instances, all tests being run on Sparc 1s. On all the graphs compared, our times are considerably less and achieved better precision. The bar graphs in Figure 3 depict the difference. This is not to suggest that the code of Leong, Shor, and Stein is inferior, only that its greater generality comes at a cost and that in the case of unit capacities it is better to use our code.

Kennington has a simplex-based code for minimum-cost multicommodity flow. His program thus solves an even more general problem; however, we know of no less general code. Leong, Shor, and Stein have made a convincing case that their code performs significantly better than that of Kennington on instances with many commodities, and, as discussed above, our code outperforms that of Leong, Shor, and Stein on uncapacitated instances; thus our code should outperform Kennington's. Our experience bears this out. Although we were able to get solutions from Kennington's code for small graphs, Kennington did not yield a solution in over nine hours when applied to even the simplest instance, matrix, on which we are reporting our code's performance.

The inability of this code to cope with large instances with many commodities is understandable in view of the representation it uses. To formulate multicommodity flow as a linear program, one needs a variable for each pair (edge, commodity) and a constraint for each pair (node, commodity) in addition to the capacity constraints. One can save a little by using a single commodity per source node. For example, consider using linear programming to solving concurrent flow on outc1908, a graph with 938 nodes and 1523 edges. We would need $1523 \cdot 938 = 1,428,574$ variables and more than $938 \cdot 938 = 879844$ constraints.

The code most similar to ours is that of Shahrokhi [23]. Indeed, the algorithm on which our code is based is derived from the algorithm on which his code is based. His results indicate excellent performance on the instances he used in tests. However, his experiments were conducted on very structured graphs, and, more importantly, on instances where there was a commodity with demand 1 between every pair of nodes. No version of his code was available to us that could handle an arbitrary collection of commodities. Moreover, his implementation has a substantial drawback in application to large, sparse graphs, the domain in which we are particularly interested. Namely, he represents the network using a incidence matrix. Thus regardless of the number of edges, his program requires space proportional to the square of the number of nodes.

However, Shahrokhi's implementation contains several heuristics that seem to improve performance, at least in the structured and typically dense instances he studied. We implemented one of these, the minimum-spanning-tree heuristic described in §4.2, and it indeed resulted in improved performance even for our sparse, unstructured instances. This suggests that a future implementation would do well to combine our sparse representation and other techniques with the heuristics of Shahrokhi.

Acknowledgements

We are grateful to Cliff Stein and Farhad Shahrokhi for helpful discussions and suggestions. Technical assistance was provided by Ken Bayse, Jeff Coady, Scott Meyers, and Steve Reiss.

References

1. A. K. Agrawal, *Network Design and Network Cut Dualities: Approximation Algorithms and Applications*, Ph.D. Thesis (1991), available as Brown University Computer Science Department tech report CS-91-60.
2. S. N. Bhatt and F. T. Leighton. A framework for solving VLSI graph layout problems. *Journal of Computer and System Sciences 28* (1984).
3. F. Brglez and H. Fujiwara, "A neutral netlist of 10 combinational benchmark circuits and a target translator in Fortran," *Proc. IEEEE Int. Symposium on Circuits and Systems*, Special Session on ATPG and Fault Simulation (1985).
4. I. Duff, G. Grimes, and J. G. Lewis, "Users' guide for the Harwell-Boeing sparse matrix collection," manuscript (1988).
5. I. Duff, G. Grimes, and J. G. Lewis, "Sparse matrix test problems," *ACM Transactions on Mathematical Software, vol. 15* (1989), pp. 1-14.
6. Mark D. Hansen, "Approximation algorithms for geometric embeddings in the plane with applications to parallel processing problems," *Proceedings, 30th Symposium on Foundations of Computer Science* (1989), pp. 604-609.
7. H. L. Bodlaender, J. R. Gilbert, H. Hafsteinsson, T. Kloks. Approximating treewidth, pathwidth, and minimum elimination tree height. Manuscript (1990).
8. A. V. Goldberg, personal communication (1991).
9. P. Klein, A. Agrawal, R. Ravi, and S. Rao. Approximation through multicommodity flow. In *Proceedings of the 31st Annual Symposium on Foundations of Computer Science* (1990), pp. 726-727.
10. P. Klein, C. Stein, and É. Tardos. Leighton-Rao might be practical: faster approximation algorithms for concurrent flow with uniform capacities. In *Proceedings of the 22nd Annual ACM Symposium on Theory of Computing* (1990), pp. 310-321
11. F. T. Leighton, F. Makedon, and S. Tragoudas. Approximation algorithms for VLSI partition problems. Manuscript (1990).
12. F. T. Leighton and S. Rao, "An approximate max-flow min-cut theorem for uniform multicommodity flow problems with application to approximation algorithms", *Proceedings, 29th Symposium on Foundations of Computer Science* (1988), pp. 422-431.

13. T. Leong, P. Shor, and C. Stein, "Implementation of a combinatorial multicommodity flow algorithm," *Proceedings of DIMACS Implementation Challenge Workshop: Network Flows and Matching* (1992).
14. R. J. Lipton, D. J. Rose, and R. E. Tarjan, "Generalized nested dissection," *SIAM Journal on Numerical Analysis 16* (1979), pp. 346-358.
15. F. Makedon and S. Tragoudas. Approximating the minimum net expansion: near optimal solutions to circuit partitioning problems. In *Proceedings of the 1990 Workshop on Graph Theoretic Concepts in Computer Science* (1990).
16. Matula, D. W., "Concurrent flow and concurrent connectivity in graphs," in *Graph Theory and its Applications to Algorithms and Computer Science*, ed. Alavi, Y., et al., Wiley, New York (1985), pp. 543-559.
17. D. W. Matula. and F. Shahrokhi The maximum concurrent flow problem and sparsest cuts. Technical Report, Southern Methodist University, March 1986.
18. P. Raghavan and C. D. Thompson. Randomized rounding: a technique for probably good algorithms and algorithmic proofs. *Combinatorica 7* (4) (1987), pp. 365-374.
19. S. Rao. personal communication, 1990.
20. R. Ravi, A. Agrawal, and P. Klein. Ordering problems approximated: single-processor scheduling and interval graph completion. In *Proceedings of the 1989 ICALP Conference*, 1991.
21. D. Rose, "Triangulated graphs and the elimination process," *Journal of Math. Anal. Appl. 32* (1970), pp. 597-609.
22. F. Shahrokhi, personal communication (1991)
23. F. Shahrokhi, *Design and Analysis of Efficient Algorithms to Solve the Maximum Concurrent Flow Problem* (1986), Ph.D. Thesis, Western Michigan University, Kalamazoo, Michigan.
24. F. Shahrokhi and D. W. Matula. The maximum concurrent flow problem. *Journal of the ACM 37* (1990) pp. 318-334.
25. C. Stein, personal communication (1991).

JAMES M. BORGER
BROWN UNIVERSITY
PROVIDENCE, RI 02912
jab@cs.brown.edu

TAEKYUNG S. KANG
BROWN UNIVERSITY
PROVIDENCE, RI 02912
tsk@cs.brown.edu

PHILIP N. KLEIN
BROWN UNIVERSITY
PROVIDENCE, RI 02912
pnk@cs.brown.edu

Name	Nodes	Edges	Commodities	Target error	Iterations	Ach. error	Time
gr1	48	352	480	0.1	5,760	0.07465	63
	48	352	480	0.01	23,040	0.009	201
	48	352	480	0.001	91,680	0.000975	655
	48	352	480	0.0001	163,680	0.0000981	1,046
	48	352	480	0.00001	234,720	0.00000978	1,419
gr2	320	2752	960	0.1	15,360	0.08955	1,391
	320	2752	960	0.01	50,880	0.00971	3,567
	320	2752	960	0.001	240,960	0.000981	13,773
	320	2752	960	0.0001	417,600	0.0000979	23,703
	320	2752	960	0.00001	592,320	0.00000976	35,660
gr3	400	3480	1200	0.1	31,200	0.08775	3,257
	400	3480	1200	0.01	56,400	0.00818	5,181
	400	3480	1200	0.001	108,000	0.0009645	7,867
	400	3480	1200	0.0001	200,400	0.0000931	12,298
	400	3480	1200	0.00001	338,400	0.00000961	19,369
matrix	136	354	408	0.1	2,448	0.06475	40
	136	354	408	0.01	6,936	0.008935	104
	136	354	408	0.001	20,400	0.0009365	276
	136	354	408	0.0001	28,968	0.0000939	389
	136	354	408	0.00001	37,944	0.00000888	506
outo432	203	343	609	0.1	6,699	0.06825	125
	203	343	609	0.01	27,405	0.00993	399
	203	343	609	0.001	98,049	0.000957	1,154
	203	343	609	0.0001	149,205	0.0000981	1,718
	203	343	609	0.00001	200,361	0.00000956	2,210
outo499	275	440	825	0.1	14,850	0.09005	305
	275	440	825	0.01	33,000	0.00776	607
	275	440	825	0.001	51,975	0.0009455	861
	275	440	825	0.0001	97,350	0.0000966	1,417
	275	440	825	0.00001	169,125	0.0000095	2,248
outc1355	619	1096	1857	0.1	44,568	0.0992	2,416
	619	1096	1857	0.01	118,848	0.00976	5,758
	619	1096	1857	0.001	> 1 million	0.005	
outc1908	938	1523	2814	0.1	61,908	0.0669	4,453
	938	1523	2814	0.01	> 1 million	0.014	

FIGURE 1. In this table we summarize the performance of the algorithm on a variety of input graphs. Time is measured in seconds. (The algorithm applied to the graphs outc1355 and outc1908 did not reach all the target approximations within one million iterations and was stopped.)

Name	Congestion	Reroutings	% reroutings
gr1	6.665	1,355	23.52%
	6.66	14,717	63.88%
	6.66	69,105	75.38%
	6.66	139,240	85.07%
	6.66	209,022	89.05%
gr2	3.61	5,472	35.63%
	3.59	34,307	67.43%
	3.59	204,019	84.67%
	3.59	378,552	90.65%
	3.59	552,031	93.20%
gr3	3.2	19,015	60.95%
	3.19	39,579	70.18%
	3.19	86,835	80.40%
	3.19	173,832	86.74%
	3.19	306,167	90.47%
matrix	18.3	175	7.15%
	18.3	2,357	33.98%
	18.3	13,245	64.93%
	18.3	19,950	68.87%
	18.3	27,281	71.90%
outo432	14	2,086	31.14%
	14	18,165	66.28%
	14	77,985	79.54%
	14	127,247	85.28%
	14	176,662	88.17%
outo499	14.6	7,442	50.11%
	14.6	21,929	66.45%
	14.6	37,420	72.00%
	14.6	76,716	78.80%
	14.6	141,936	83.92%
outc1355	20.3	27,043	60.68%
	20.2	92,446	77.79%
outc1908	38.6	35,648	57.58%

FIGURE 2. In this table we give some additional information on the tests summarized in Figure 1. The target errors are as in that figure. Note that the value for congestion did not decrease significantly from that achieved in attaining target error .1; subsequent improvement was primarily in the value of the dual. Note also that for target errors smaller than 0.1, the heuristic for selecting commodities to reroute was successful as measured by the percentage of iterations resulting in reroutings.

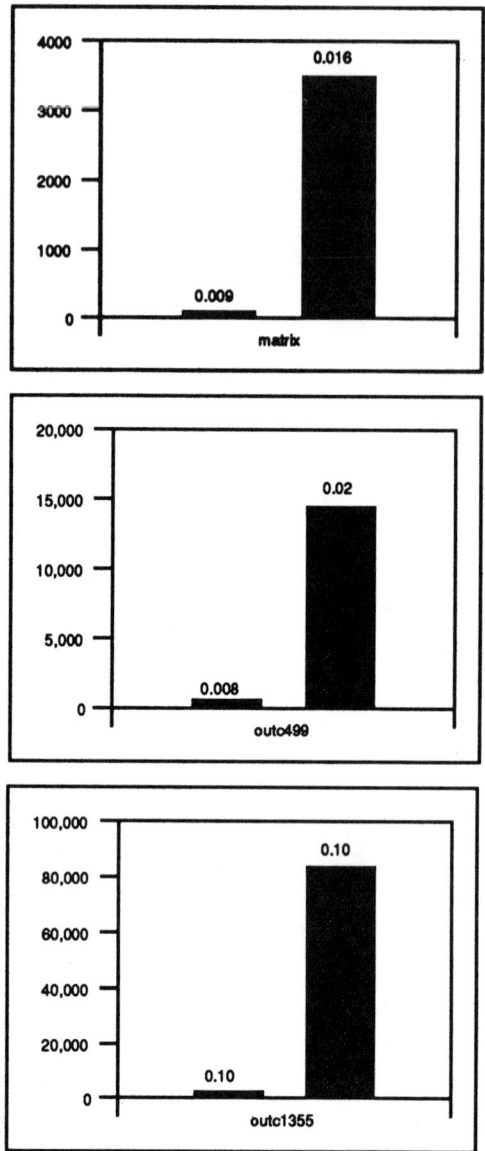

FIGURE 3. Our code (on the left) versus that of Leong, Shor, and Stein on some of our uncapacitated instances. Times are represented graphically (in seconds) and acheived approximations are shown above the corresponding bars.

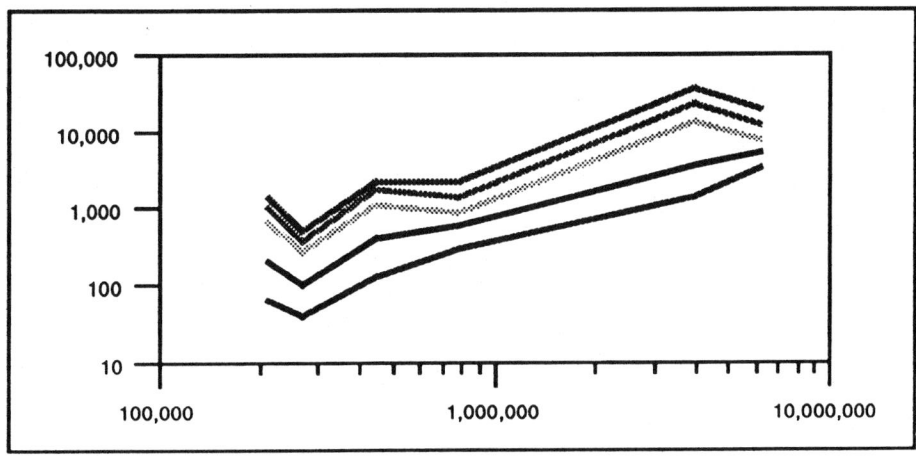

FIGURE 4. Time versus $(k+n) \cdot (m+n)$. Each line represents a different target approximation.

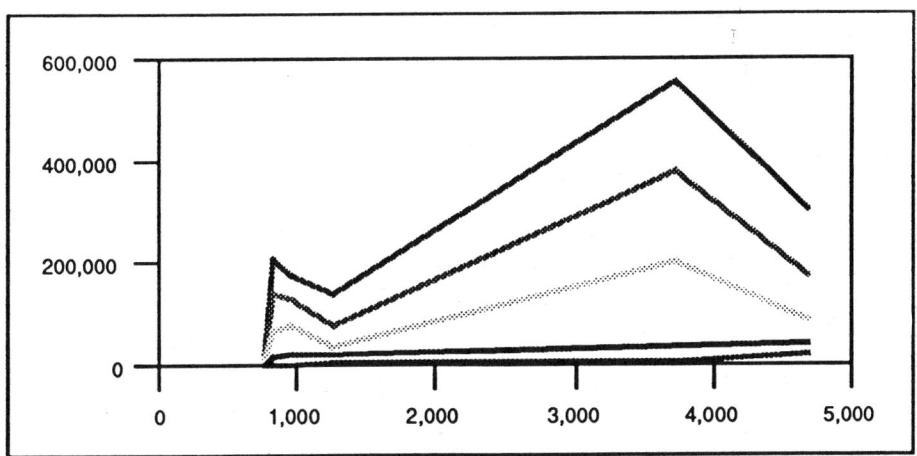

FIGURE 5. Number of iterations versus $(k+m)$. Each line represents a different target approximation.

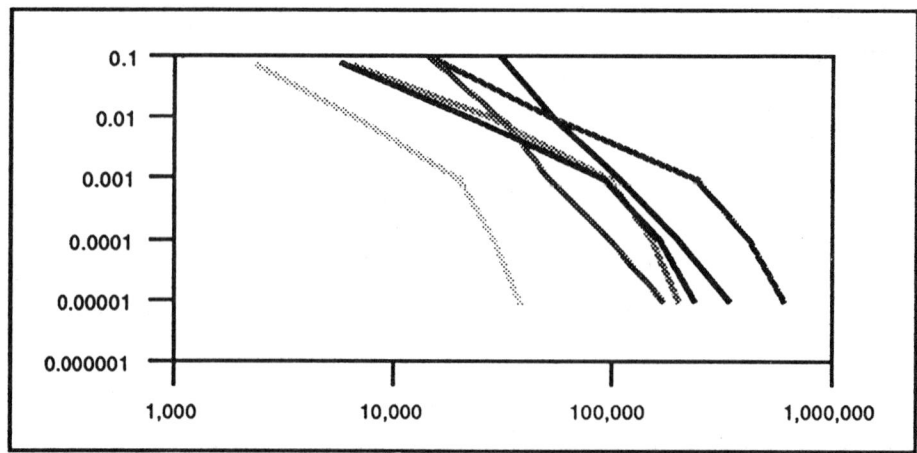

FIGURE 6. Achieved approximation versus number of iterations. Each line represents a different graph. Reading off the lowest points from left to right, the graphs are outc499,gr1,outc432,gr3, matrix, and gr2.

Implementation of a Combinatorial Multicommodity Flow Algorithm

TISHYA LEONG AND PETER SHOR AND CLIFFORD STEIN

January 8, 1993

ABSTRACT. The multicommodity flow problem involves simultaneously shipping multiple commodities through a single network so that the total amount of flow on each edge is no more than the capacity of the edge. This problem can be expressed as a large linear program, and most known algorithms for it, both theoretical and practical, are linear programming algorithms designed to take advantage of the structure of multicommodity flow problems. The size of the linear programs, however, makes it prohibitively difficult to solve large multicommodity flow problems.

In this paper, we describe and examine a multicommodity flow implementation based on the recent combinatorial approximation algorithm of Leighton et al. [13]. The theory predicts that the running time of the algorithm increases linearly with the number of commodities. Our experiments verify this behavior. The theory also predicts that the running time increases as the square of the desired precision. Our experiments show that the running time increases at most this fast, and often slower. We also compare our combinatorial implementation against two different linear programming-based codes. First we compare our code to that of of Kennington [10], which is a network simplex code known to perform well on multicommodity flow problems. For many problems, our combinatorial algorithm outperforms this simplex-based linear programming algorithm. More precisely, as the number of commodities increases, the running time of our algorithm grows much more slowly than that of Kennington's linear programming-based algorithm. Second, we compared our code to an interior point code of Karmarkar and Ramakrishnan. Here too, we achieved similar, but less dramatic results. Our results suggest that our algorithm may be able to solve larger multicommodity flow problems than have been solved in the past.

1991 *Mathematics Subject Classification.* Primary 90B10; Secondary 68Q25, 68A10.
Research of the first author was done at AT&T Bell Laboratories.
Support for the third author provided by NSF PYI Award CCR-89-96272 with matching support from UPS and Sun and by an AT&T Bell Laboratories Graduate Fellowship. Most of this work was done while at AT&T Bell Laboratories.

© 1993 American Mathematical Society
1052-1798/93 $1.00 + $.25 per page

1. Introduction

The multicommodity flow problem involves simultaneously shipping multiple commodities through a single network so that the total amount of flow on each edge is no more than the capacity of the edge. This problem can be expressed as a large linear program, and most known algorithms for it, both theoretical and practical, are linear programming algorithms designed to take advantage of the structure of multicommodity flow problems. The size of the linear programs, however, makes it prohibitively difficult to solve large multicommodity flow problems.

Recently, Leighton, Makedon, Plotkin, Stein, Tardos, and Tragoudas [13] proposed a combinatorial approximation algorithm for the multicommodity flow problem. This algorithm has a faster theoretical running time than the best theoretical linear programming algorithms [7, 18]. Also, the running time of this algorithm is dominated by the computation of minimum-cost flows, which are often efficiently computable in practice. These two facts caused Leighton et al. to conjecture that the algorithm would perform well in practice. In this paper, we describe an implementation based on the algorithm of Leighton et al. and investigate its behavior. We show that the algorithm performs at least as well as the theory predicts. We also compare our combinatorial implementation against two different linear programming-based codes. First we compare our code to that of Kennington [10], which is a network simplex code known to perform well on multicommodity flow problems. For many problems, our combinatorial algorithm outperforms this simplex-based linear programming algorithm. More precisely, as the number of commodities increases, the running time of our algorithm grows much more slowly than that of Kennington's linear programming-based algorithm. Second, we compared our code to a state-of-the-art interior point code of Karmarkar and Ramakrishnan, called ADP. While the difference with ADP was not as dramatic as with Kennington, we still consistently outperformed this code. The number of tests performed, however, was too small to quantify this comparison any further.

1.1. Background. The input to a multicommodity flow problem consists of an n node, m edge graph and k commodities, each with a source, a sink, and a demand. The corresponding linear program has $O(mk)$ variables and $O(nk + m)$ constraints. Even for a graph with average vertex degree Δ, there are $O(\Delta nk + mk) = O(mk)$ non-zero entries in the constraint matrix. The large size of the linear programs makes the general simplex algorithm impractical for all but very small problems. Some algorithms which take advantage of the special structure of multicommodity flow problems have been proposed. These algorithms fall into three main classes: price-directive decomposition, resource-directive decomposition, and partitioning approaches. (See the surveys of Assad [2] and Kennington [8] and the thesis of Schneur [16] for more information on these approaches.) More recent approaches include interior-point methods [1] and a combinatorial scaling algorithm [16]. All of the aforementioned algorithms solve multicommodity flow problems using one of two different objective functions. Some find a minimum-cost multicommodity flow, while others find a flow which maximizes the total amount of flow in the network.

In 1986, Shahrokhi and Matula proposed a different objective function [17]. They defined the *concurrent flow problem*, which involves finding the maximum z such that there exists a flow which satisfies a percentage z of every demand without exceeding the capacity of any edge. This problem can be formulated as a linear program of the same size as that for the multicommodity flow problem, and it is strictly more general than the multicommodity flow problem. The concurrent flow problem is equivalent to the problem of finding the minimum $\lambda = 1/z$ such that there exists a flow which satisfies all demands while using no more than λ times the capacity of each edge.

Shahrokhi and Matula gave a *fully polynomial approximation scheme* for this problem in the special case in which all the capacities and demands are 1. For a given flow f, let λ_f be its *congestion*, i.e., the maximum over all edges of the ratio between the flow on an edge and the capacity of that edge. Their basic approach is first to route flow on an arbitrary path, ignoring capacities, and then gradually to reroute small amounts of flow from highly congested edges onto lightly congested edges. In $O(\epsilon^{-5}nm^7)$ time, the algorithm finds an ϵ-*optimal* flow, so named because its congestion lies within a $(1+\epsilon)$-factor of the minimum possible congestion. Shahrokhi and Matula implemented their algorithm and tested it on small examples. For the same problem, Klein, Plotkin, Stein, and Tardos later proposed a faster algorithm with an expected running time of $O(\epsilon^{-3}\min\{n,k\}(m+n\log n))$ [12].

Building on this framework, Leighton et al. proposed an algorithm for solving the general concurrent flow problem with arbitrary capacities and demands [13]. They also start with an arbitrarily routed flow and gradually improve it by rerouting individual commodities to move flow from highly congested edges to lightly congested edges. To reroute flow, they compute minimum-cost flows in suitably defined auxiliary graphs. They show that the algorithm performs approximately $O(\epsilon^{-2}k)$ minimum-cost flow computations in finding an ϵ-optimal multicommodity flow. The algorithm runs in expected $O(\epsilon^{-3}nmk\log^4 n)$ time. Based on the small number of iterations and the fact that minimum-cost flows can often be solved efficiently in practice, they conjectured that their algorithm might work well in practice. In this paper, we provide support for this conjecture.

2. The Underlying Theory

In this section, we define the concurrent flow problem, relate it to the multicommodity flow feasibility problem, and summarize the algorithm of Leighton et al. from which we derive our algorithm.

Consider an undirected graph $G = (V, E)$ with a positive capacity $u(vw)$ for each edge $vw \in E$. Consider also a set of commodities numbered 1 through k, where each commodity i is specified by a source-sink pair $s_i, t_i \in V$ and a positive demand d_i. For each commodity i, we ship an amount proportional to its demand d_i from its source s_i to its sink t_i. This gives us a *single commodity flow* f_i specified by a set of edge flows $f_i(vw)$ on the edges $vw \in E$, where each edge has an arbitrary direction to keep track of which way the flows travel across it. A positive edge flow $f_i(vw) > 0$ denotes a forward flow of commodity i with respect to the direction of edge vw, while a negative flow $f_i(vw) < 0$ denotes a backwards flow. A *multicommodity flow* f consists of k single commodity flows, one for each commodity. In a multicommodity flow f, the total flow $f(vw)$ on each edge $vw \in E$ equals the sum $\sum_{i=1}^{k} |f_i(vw)|$ of the single commodity flows on that edge.

A multicommodity flow achieves *demand satisfaction* if it ships an amount of each commodity equal to its demand from its source to its sink. It obeys the *capacity constraints* if no flow $f(vw)$ on an edge $vw \in E$ exceeds the capacity $u(vw)$ of the edge. A *feasible* multicommodity flow achieves demand satisfaction while obeying the capacity constraints. The *multicommodity flow feasibility problem* is to determine if a feasible flow exists.

Our algorithm solves a more general problem, the *concurrent flow problem*. Given any multicommodity flow f (which need not obey the capacity constraints), each edge $vw \in E$ has a congestion $\lambda_f(vw)$ equal to the ratio $f(vw)/u(vw)$ of total flow to capacity. The congestion λ_f of the flow is the maximum of these edge congestions. It represents an amount by which we can scale the capacities while achieving demand satisfaction, setting the adjusted capacity $u'(vw)$ of each edge $vw \in E$ equal to $\lambda_f \cdot u(vw)$. The concurrent flow problem is to find the lowest possible congestion, which we call the optimal congestion λ^*.

In solving the concurrent flow problem, we also solve the multicommodity flow feasibility problem. If the optimal congestion is greater than 1, the capacities must be raised to achieve demand satisfaction, and so a feasible flow does not exist. If the optimal congestion is less than or equal to 1, all demands can be met given the original capacities, and a feasible flow does exist. Alternatively, we can solve a concurrent flow problem by solving a logarithmic number of multicommodity flow problems. We perform binary search on the congestions of the multicommodity flows and determine the cutoff between feasible and infeasible flows. The congestion at this cutoff is the solution λ^*.

We now summarize the algorithm of Leighton et al. for approximately solving the concurrent flow problem. Given an error parameter $\epsilon > 0$, the algorithm finds an ϵ-optimal flow, i.e., a flow for which $\lambda_f \leq (1 + \epsilon)\lambda^*$. Because we can make ϵ arbitrarily small, we can find a solution arbitrarily close to optimal.

The algorithm begins with a flow f that achieves demand satisfaction but ignores the capacity constraints. Leighton et al. show that if the flow is not ϵ-optimal, i.e., if $\lambda_f > (1 + \epsilon)\lambda^*$, then there exists at least one "poorly routed" commodity. They then show that by rerouting a fraction of the flow of a poorly routed commodity onto the edges of a minimum-cost flow for that commodity in an appropriately derived auxiliary graph, they cause a decrease in a potential function Φ, which we will define later. Finally, they show that as the potential function decreases, the congestion gradually decreases. The algorithm iteratively reroutes flow, decreasing the potential function and the congestion until the congestion is within a $(1+\epsilon)$-factor of optimal.

The basic idea behind rerouting is to move flow off of highly congested edges. The algorithm achieves this by assigning long lengths to highly congested edges and short lengths to lightly congested edges. Given these lengths, which correspond to the linear programming dual variables, a flow which uses edges of long length marks a poorly routed commodity. Using the lengths as costs, the algorithm finds a minimum-cost flow onto which it reroutes a fraction of the flow of a poorly routed commodity. Leighton et al. assign lengths according to a length function in which $\ell(vw) = e^{\alpha \lambda_f(vw)}/u(vw)$, where α is a constant and $\lambda_f(vw)$ is the congestion on edge vw given the current multicommodity flow f. This length function, being exponential in $\lambda_f(vw)$, clearly penalizes highly congested edges. The minimum-cost flow therefore favors lightly congested edges and utilizes them to the extent allowed by their adjusted capacities. To prevent large increases in the congestion, the capacities have been scaled by the congestion, the capacity of each edge vw in the auxiliary graph being set at $\lambda_f \cdot u(vw)$. After finding a minimum-cost flow, the algorithm reroutes a fraction σ of the current multicommodity flow onto the edges of the minimum-cost flow to create a new multicommodity flow. Choosing the potential function $\Phi = \sum_{vw \in E} u(vw)\ell(vw)$, they can show that the number of iterations of the algorithm is not too large.

In the algorithm as formulated by Leighton et al., the constant α plays a pivotal role. In order to enforce two relaxed optimality conditions which ensure the algorithm's eventual success, Leighton et al. use the values $\alpha = 2(1+\epsilon_0)\lambda_f^{-1}\epsilon_0^{-1}\ln(m\epsilon_0^{-1})$ and $\sigma = \epsilon_0/(8\alpha\lambda_f)$, where $\epsilon_0 \geq \epsilon$ is an error parameter which gradually approaches ϵ and λ_f is the congestion of the current multicommodity flow. Because the fraction σ of flow rerouted depends inversely on α, a smaller α means more flow is rerouted, resulting in a faster decrease of the congestion. However, the value of α also limits how close the algorithm can come to finding the optimal solution. The algorithm is guaranteed to find an ϵ-optimal solution only when α is sufficiently large. As the algorithm progresses, ϵ_0 and λ_f decrease, making α increase. Progress slows as smaller fractions of flow are rerouted, but these choices for α and σ guarantee a solution within a $(1+\epsilon)$-factor of optimal.

We expect the running time of the algorithm to increase as the error parameter ϵ decreases and as the number of commodities k increases. In fact, Leighton et al. prove that given any $\epsilon > 0$, the randomized version of their algorithm finds an ϵ-optimal solution using an expected $O(k(\log k + \epsilon^{-3})\log n)$ minimum-cost flow computations, while the deterministic version uses $O(k^2(\log k + \epsilon^{-2})\log n)$ minimum-cost flow computations. Goldberg [5] and Grigoriadis and Khachiyan

[6] have shown how to reduce the number of computations used by the randomized version to $O(k(\log k + \epsilon^{-2})\log n)$. The running time therefore depends polynomially on ϵ^{-1} and linearly on the number of commodities. In the following sections, we describe an implementation based on the algorithm of Leighton et al., and we compare the running times of our implementation to these theoretical bounds.

3. Our Implementation

We now describe how we have adapted and implemented the algorithm of Leighton et al. Where they have made certain choices in the interest of proving the theoretical bounds, we modify the algorithm for the purpose of improving actual performance. We describe the changes we have made and the motivations behind them. We also point out areas in which our modifications could be fine-tuned with further research.

3.1. Grouping Commodities. First, we group the commodities as suggested by Leighton et al. We place all the commodities with the same source into one commodity group and run the algorithm on the commodity groups instead of on the individual commodities. Under this strategy, the number of commodity groups k' cannot exceed the number of nodes n, and rerouting one commodity group corresponds to rerouting all the commodities in the group, an operation made possible by a minimum-cost flow routine that can handle multiple sinks. The running time, which varies linearly with k, now depends on the number of commodity groups rather than the number of commodities. For problems with large numbers of commodities, this means a significant reduction in running time. Because our algorithm uses $O(km)$ space, commodity grouping also reduces the space requirement, making it possible to run larger problems. The advantages gained by grouping commodities have also been documented by Schneur [16].

3.2. Choosing a Commodity to Reroute. Leighton et al. propose both a deterministic strategy and a randomized strategy for choosing a commodity group (or a commodity) to reroute. Let $f_{i'}$ be the current flow of commodity group i' and $f_{i'}^*$ be the minimum cost flow of commodity group i' in an auxiliary graph in which each edge vw has a capacity $u'(vw) = \lambda_f \cdot u(vw)$ and a cost of $\ell(vw)$. Then, the deterministic method computes the cost $C_{i'} = \sum_{vw \in E} |f_{i'}(vw)|\ell(vw)$ of a commodity group i', its minimum cost $C_{i'}^* = \sum_{vw \in E} |f_{i'}^*(vw)|\ell(vw)$, and the difference $C_{i'} - C_{i'}^*$ between its cost and minimum cost. The commodity group to be rerouted is the first in a predetermined ordering which has a difference $C_{i'} - C_{i'}^*$ greater than $\epsilon_0 C_{i'} + (\epsilon_0 \lambda_f \Phi)/k'$. This method requires k' minimum-cost flow computations per iteration in the worst case. The randomized strategy computes the cost $C_{i'}$ of each commodity group i' and randomly chooses a commodity group with probability proportional to its cost. This method uses an expected ϵ_0^{-1} minimum-cost flow computations per iteration. Once every k' iterations, minimum-cost flows are computed for all the commodity groups, and the congestion λ_f is checked against the lower bound $\sum_{i'=1}^{k'} C_{i'}^*(\lambda_f)/\Phi$ to decide if the algorithm should terminate. This check increases the number of minimum-cost flow computations by at most a factor of 2. Our selection strategy draws from both the deterministic and the randomized methods of Leighton et al. and from the termination check.

To make the most progress per iteration, we attempt to find not only a poorly routed commodity group but the most poorly routed commodity group. We may designate as the most poorly routed commodity group either the group with the highest cost $C_{i'}$ or the group with the largest difference $C_{i'} - C_{i'}^*$ between cost and minimum cost. Using either measure and rerouting larger fractions of flow than the σ of Leighton et al., we have found that an algorithm which deterministically reroutes the most poorly routed commodity group sometimes gets stuck rerouting a single group over and over with no improvement of the congestion. We have also found that when it does not get stuck, such a deterministic algorithm usually progresses faster than a randomized algorithm. We therefore use a partly deterministic, partly randomized selection strategy in which we alternate between $k'/2$ iterations of deterministic selection and $k'/2$ iterations of random selection. By taking advantage of the minimum-cost flow computations performed in the termination check every k' iterations, we can select commodity groups to reroute without computing extra minimum-cost flows. We reroute, in decreasing order, the $k'/2$ groups with the greatest difference between cost and minimum-cost followed by $k'/2$ randomly chosen commodity groups. To prevent domination by a limited number of groups, the random selection weights all commodity groups equally as proposed by Goldberg [5] and Grigoriadis and Khachiyan [6].

3.3. Handling the Minimum-cost Flow. Once the algorithm has chosen a commodity group to reroute, it must find an appropriate minimum-cost flow. For this purpose, we use the RELAXT-III minimum-cost flow code of Bertsekas and Tseng [3]. One drawback of the routine we have chosen is that it requires integer capacities, costs, and demands, making preprocessing and postprocessing necessary each time it is called. Another routine might better suit our algorithm, but we concentrate on the number of iterations of our algorithm and treat the minimum-cost flow routine as a black box.

For the costs used to calculate the minimum-cost flow, we use a length function slightly different from that of Leighton et al. Recall that they set the length $\ell(vw)$ of each edge $vw \in E$ equal to $e^{\alpha \lambda_f(vw)}/u(vw)$. We use a length function in which $\ell(vw) = \lfloor e^{\alpha(\lambda_f(vw) - \lambda_f) + c} \rfloor$, where c is a scaling constant that depends on the largest integer the system can handle. We include the terms $-\lambda_f$ and c because we want to extract real flows from a routine that works only with integers. These terms spread the lengths over the range of viable non-negative integers, giving us the most accurate minimum-cost flow we can procure. We have removed the $u(vw)$ factor so that edges with equally high congestion will have equally high cost in the minimum-cost flow. We have found through limited experimentation that this produces minimum-cost flows which better suit our algorithm.

3.4. Choosing Constants. As noted earlier, the constant α and the fraction σ of flow rerouted greatly affect the running times of the algorithm. Leighton et al. use very large values for alpha and very small values for σ. Their α can easily exceed 1000, and their σ can easily fall below 10^{-5}. For the algorithm to progress at a reasonable rate in practice, we need smaller values for α and larger values for σ. However, when these values begin to block progress, we must raise α and lower σ. We do this by means of a scaling factor s. We set α equal to $c \cdot s/\lambda_f$, where c is the constant $19.1 - \log m$. To decide how much flow to reroute, we sample the values that the potential function Φ would take after the rerouting of various fractions of flow. We reroute the fraction σ_f which gives the lowest Φ. We can find σ_f efficiently because Φ has a positive second derivative with respect to σ, allowing a binary-type search. We sample fractions to the precision $.001/s^2$, and we also use this value as a floor $\sigma_{\min}$ on the fraction of flow that can be rerouted. To avoid wasting time rerouting small amounts of flow, we reroute a commodity only if σ_f is at least as large as $\sigma_{\min}$. We know from [13] that we may have to reroute fractions as small as $O(\epsilon/\alpha\lambda_f)$, and so we must decrease $\sigma_{\min}$ faster than we increase α to lower the minimum value for σ_f. We begin with s equal to .25 and raise it by .25 whenever the maximum fraction rerouted in k' iterations is less than $\sigma_{\min}/(s \cdot k')$ or whenever the ratio of the congestion λ_f to its lower bound $\sum_{i'=1}^{k'} C_{i'}^*(\lambda_f)/\Phi$ increases after k' iterations. We have found that this strategy works well in most instances but scales α too fast in a few instances, slowing the algorithm too much for practical use. In such cases, we rerun the algorithm, scaling α more slowly. Our current heuristic for doing so is to check whether the congestion does not decrease for 2000 consecutive iterations. If this is the case, we consider the algorithm to be "stuck" and decrease the parameter s by a factor of 2. This causes α to grow more slowly. We have not yet discovered the optimal rate at which we should scale α, nor have we discovered exactly when we should scale it. This is the area in which our algorithm would benefit most from further research. Other areas in which it could be further improved include the selection strategy for commodities to reroute and the technique for choosing $\sigma_{\min}$.

4. Experimental Results

We have tested our algorithm on a variety of problems and compared its performance to the theoretical bounds. We used two different random network generators, NETGEN and RMFGEN. When run on random NETGEN and RMFGEN graphs with randomly placed commodities, our algorithm behaved more or less as expected. It took polynomially more time to get closer to the optimal solution and less than linearly more time to handle larger numbers of commodities. Furthermore, for large numbers of commodities, our algorithm outperformed the linear programming-based code of Kennington. It also consistently outperformed the interior point linear programming based codes of Karmarkar and Ramakrishnan. Our algorithm performed poorly on one real problem provided by the GTE Corporation, but we consider this an anomaly arising from a limited number of unusually time-consuming minimum-cost flow computations. This one instance aside, we find our results encouraging and consider it an improvement, in many cases, over the simplex-based algorithms which have preceded it.

4.1. Dependence on the Error Parameter. The theory predicts an inverse polynomial dependence of the running time on the error parameter ϵ. More precisely, as noted in Section 2, it states that the number of minimum-cost flow computations is proportional to ϵ^{-2}. Since our algorithm computes a constant number of minimum-cost flows per iteration, the number of iterations should also depend on ϵ^{-2}. Equivalently, ϵ should depend on $1/\sqrt{\# \text{ of iterations}}$.

We ran our algorithm on various problems and graphed the lowest ϵ achieved against the number of iterations completed. Each run stopped at a final ϵ of .001 or less. To compress the data, we used data points representing ranges of iterations. For each problem, we considered 10 runs and, for each run, the minimum ϵ achieved at each termination check. The aggregate ϵ for a range equaled the average of the minimum ϵ values found at the termination checks falling in the range during each of the 10 runs. We examined a problem with 20 commodities and four problems with 10 commodities using different NETGEN graphs with 50 nodes and 100 edges. We also examined two problems with 10 and 20 commodities, respectively, using an RMFGEN graph with 140 edges and 48 nodes (spread evenly over 12 square planes). While we only ran this particular instance once, due to the large number of iterations, we expect that the variance should be reduced. To test a large problem, we examined a single run on a large RMFGEN problem with 700 commodities, 2075 edges, and 500 nodes (spread over 20 square planes). For all of these problems, we graphed ϵ versus the number of iterations. We also graphed the function $1/\sqrt{\# \text{ of iterations}}$ on which we expected ϵ to depend. As is evident from Figures 3 through 9, our implementation always performed better than the expected bounds. We note that the fact that, in Figure 9, ϵ crosses the function $1/\sqrt{\# \text{ of iterations}}$ is not a contradiction of the theoretical predictions. The predictions are for $O(1/\sqrt{\# \text{ of iterations}})$, and with an appropriately chosen constant the lines would not cross. Some inconclusive attempts at fitting the data to a curve of the form $a * (\# \text{ of iterations})^b + c$ using a regression package lends some additional support to this conclusion as typical values of b were between $-.5$ and -1.

4.2. Dependence on the Number of Commodities. With respect to the number of commodities k, our algorithm also seems to conform to the theoretical bounds. Using 10 runs for each data point and disregarding the shortest and the longest of these runs, we graphed the average number of iterations needed to solve problems with variable numbers of commodities given a fixed graph. In Figure 10, we examined four NETGEN graphs with 50 nodes and 100 edges and values of k between 10 and 70. In Figure 11, we traced the same values of k using an RMFGEN graph with 140 edges and 48 nodes (spread over 12 square planes). In Figure 12, using values of k between 50 and 250, we examined an RMFGEN graph with 752 edges and 192 nodes (spread over 12 square planes). Graphing the number of iterations against the number of commodity groups $k' \leq k$, we observed that the number of iterations either grew linearly or grew linearly to a peak and then dropped. The drops may result from larger numbers of commodities making it possible to route commodities over shorter paths. In trying to find flows which give the edges equal congestions, the algorithm has more commodities at its disposal to congest each edge. In any case, the number of iterations grows no more than linearly with the number of commodity groups and therefore no more than linearly with the number of commodities.

4.3. Comparison to Other Algorithms. Because the running time of our algorithm grows no more than linearly with the number of commodities, it can effectively solve large concurrent flow problems. To the best of our knowledge, our implementation is the first for an algorithm which finds an ϵ-optimal solution to the general concurrent flow problem. Consequently, comparisons to existing algorithms will inherently contain some amount of bias. We have nevertheless compared our algorithm to two others as best we could. The fact that our algorithm runs faster than another on a particular problem instance does not necessarily mean our algorithm is faster in general. However, the comparison reveals sufficiently consistent trends which enable us to draw some general conclusions.

We begin with a brief discussion of the first algorithm to which we have compared our algorithm. The algorithm is MCNF85, a special purpose simplex code for multicommodity flow problems written by Kennington [10], and we chose it for two reasons. First, we had access to the code on our machine. Second, and more importantly, previous tests by Adler, Karmarkar, Resende and Veiga [1] demonstrate its efficiency. Adler et al. compared three different codes for multicommodity flow: MINOS 5.0, an advanced implementation of the simplex method [15], MCNF85, and their own interior point method. Their experiments show that the running time of MINOS grows much faster than that of the other two algorithms and that, for the problems they tested, MCNF85 and the interior point algorithm have comparable running times. Thus we concluded that MCNF85 was one of the best codes available at that time.

We have also compared our algorithm to a state-of-the-art interior point code called ADP[9]. ADP is an approximate dual projective interior point code, written by Karmarkar and Ramakrishnan, and is the latest variant of the interior point algorithm. The algorithm alternates between objective steps and centering steps; the motivation of the algorithm is to stay very close to the "central trajectory" of the polytope. The objective steps are dual affine scaling steps while the centering steps are reciprocal-estimates-improvement step. The implementation of the algorithm took about three years, and uses the iterative technique of preconditioned conjugate gradient at each step to compute the improving direction. The code is especially suited for solving very large linear programs with a particular structure, since in many cases only an approximate solution to the linear system is needed to solve the problem.

We faced two obstacles in comparing our algorithm to MCNF85. First, our algorithm finds an approximate solution while MCNF85 finds an exact solution. Since we could not modify the code for either algorithm to alleviate this problem, we ran our algorithm to both $\epsilon = .01$ and $\epsilon = .001$ before comparing it to MCNF85. The second difficulty in making the comparison is that the algorithms are designed for different problems with objective functions. By using an objective function of 0 for MCNF85 and a cost of 0 on every edge, we can treat it as an algorithm which determines whether a feasible multicommodity flow exists. We could then call this algorithm $O(\log(n\epsilon^{-1}))$ times to find an ϵ-optimal solution to a concurrent flow problem, but this seems too far from the original purpose of the algorithm for fair comparison. Instead, we ran our algorithm to find the maximum z for which there exists a feasible flow satisfying a percentage z of each demand. We then scaled the demands by z to get a problem which we knew to be feasible. This problem corresponds to the problem which MCNF85 would have to solve in the *last* iteration of the binary search procedure defined above. We compared a run of our algorithm to a run of MCNF85 with the input modified as described above. We could better evaluate our algorithm by comparing it to other approximation codes for the same problem. For ADP, we were able to do so, as we just ran the code until the duality gap was less than ϵ.

4.3.1. *The Results.* The results of our experiments appear in Figure 1. The experiments in this table were performed on a Silicon Graphics 4D/340S. They show that as the number of commodities increases, the running time of MCNF85 grows much more rapidly than the running time of our algorithm for graphs of all sizes. The difference does not arise simply because we group the commodities (they could incorporate grouping in their algorithm too). Hardly any grouping occurred in the graphs with 500 nodes and 70 or less commodities, and the running time of our algorithm still grew much more slowly than the time for MCNF85. In fact, as discussed above, the running time of our algorithm grows slower than k while rough analysis of the data shows that the time for MCNF85 grows at least as fast as k^2. Since the size of the linear program grows by k^2, this growth is not particularly surprising.

They also show that our running times are consistently smaller than those of the interior point code ADP. In contrast to the other linear-programming based algorithm, MCNF85, the dependence of the running time on the number

Problem Specification	MCNF85	ADP		Our algorithm	
(n, m, k, GEN)		$\epsilon = .01$	$\epsilon = .001$	$\epsilon = .01$	$\epsilon = .001$
(50, 100, 20, NG)	49	30	39	20	103
(50, 100, 50, NG)	397	118	153	35	43
(50, 100, 70, NG)	857	199	284	29	33
(48, 140, 10, RMF)	8			13	13
(48, 140, 20, RMF)	24	41	56	23	23
(48, 140, 30, RMF)	69			18	35
(48, 140, 40, RMF)	122			25	38
(48, 140, 50, RMF)	216	250	300	21	71
(48, 140, 60, RMF)	316			40	61
(48, 140, 70, RMF)	470	328	347	45	62
(500, 2075, 10, RMF)	87			831	5230
(500, 2075, 20, RMF)	608	2220	2832	1484	2641
(500, 2075, 30, RMF)	1831			2625	3881
(500, 2075, 40, RMF)	6571			3762	6084
(500, 2075, 50, RMF)	15601	7500	9300	4710	7401
(500, 2075, 60, RMF)	18449			3819	6201
(500, 2075, 70, RMF)	34362	19860	13200	4435	8258
(500, 2075, 700, RMF)	B			22411	
(192, 748, 50, RMF)	2702			240	589
(192, 748, 250, RMF)	85754			637	1571
(49, 260, 585, none)	1373			2472 (estimate)	

FIGURE 1. Running time comparison of our algorithm and Kennington's algorithm. Running times are in seconds on a Silicon Graphics machine. NG is NETGEN and generator RMF is RMFGEN. The last problem is the problem defined in Section 4.4 B stands for a breakdown in the computer and a blank signifies that the test was not performed.

of commodities does not appear to grow quadratically. Yet, our algorithm consistently runs faster than ADP. We note that only a small number of tests were performed with ADP and that they did not include the largest problems that we have. We hope to be able to perform more tests in order to allow us to draw more significant conclusions.

Our algorithm will be able to solve large and previously unsolvable multicommodity flow problems. We have already shown that we can solve a 700 commodity problem faster than MCNF85 can solve a 70 commodity problem. For large graphs with small numbers of commodities, our algorithm is slower than MCNF85. However, the rapid growth rate of MCNF85 with respect to the number of commodities makes our algorithm more desirable for problems with more than a few commodities. We note that one of the motivations for this work comes from multicommodity flow problems which arise in approximating a number of NP-hard problems. (See [14],[11],[12], and [13] for details.) These problems have large numbers of commodities, i.e., at least as many commodities as the number of nodes. Our algorithm provides a practical means for solving such problems.

4.4. An Anomaly. In one case, a problem with 49 nodes, 260 edges, and 585 commodities using actual data from GTE, our algorithm performed much more poorly than the linear programming algorithm. Though our algorithm ran for only 3745 iterations, a reasonable number, those iterations took a total of 18.4 hours of CPU time. We attribute this anomaly to inefficiency in the minimum-cost flow routine since minimum-cost flow computations accounted for over 99.8% of the running time. The theory shows that minimum-cost flow computations dominate the running time of the algorithm, but even for the much larger RMFGEN graph with 500 nodes and 1025 edges, minimum-cost flow computations generally took less than 80% of the time. For small graphs, they generally took between 40 and 50 percent of the time. See Figure 2 for a more detailed description of the times. The time spent solving the GTE problem was not equally divided between iterations. Iterations including the termination check aside, most iterations took less than 100 milliseconds. Some iterations, however, took hundreds of seconds, up to 1000 times the normal duration.

With the help of a number of other researchers, we have verified that these are problems on which RELAXT-III takes an inordinately long amount of time. A number of people have run these problems on their codes and observed no anomalous behavior, i.e., the running times for this set of problems are all approximately the same. In order to estimate a more realistic running time for this problem, we will compute an upper bound on the what the running time would have if we were using the RNET code of Grigoriadis. Joseph Cheriyan [4] has reported that on a representative sample of these minimum-cost flow problems, the running time of RNET on a SPARC2 (which is slower than our machine) never exceeds 0.66 seconds. Using the estimate that 50% (see Figure 2) of the

Problem Specification				ϵ	% of time finding
nodes	edges	commodities	generator		Min-cost flows
50	100	20	NG	.001	49.9
50	100	50	NG	.001	50.5
50	100	70	NG	.001	43.6
48	140	30	RMF	.001	44.1
48	140	40	RMF	.001	44.1
48	140	50	RMF	.001	42.4
48	140	60	RMF	.001	44.7
48	140	70	RMF	.001	47.7
500	2075	10	RMF	.01	87.3
500	2075	30	RMF	.01	79.7
500	2075	40	RMF	.01	73.3
500	2075	50	RMF	.01	76.8
500	2075	60	RMF	.01	80.7
500	2075	70	RMF	.01	77.7
192	752	50	RMF	.001	55.8
192	752	250	RMF	.001	59.0
49	260	585	none	.01	**99.8**

FIGURE 2. Percentage of Time that our algorithm spent performing minimum-cost flows. The data is gotten from the UNIX profiling routine **prof**. NG is NETGEN and generator RMF is RMFGEN. The last problem is the problem defined in Section 4.4.

time is spent in the minimum-cost flow computations, we arrive at a figure of 2472 seconds as a "reasonable" upper bound on the running time of this instance. It is intriguing that the one anomaly occurred on the one real-world instance. This gives more evidence to the belief that every effort should be made to find real-world instances on which to test programs.

5. Conclusions

Our algorithm performs as well as, and often better than, the theoretical bounds. The theory predicts the number of iterations of the algorithm to be $O(\epsilon^{-2}k)$. Our experiments show that the number of iterations often grows slower as a function of ϵ. Our experiments also show that for small k, the number of iterations does increase linearly with k. As k approaches the number of nodes, however, the number of iterations grows at most linearly and sometimes actually decreases.

On the problems we tested, the running time of our algorithm grew much slower as a function of k than that of Kennington's algorithm. This implies that our algorithm is preferable to one of the best network simplex based approaches for problems with large numbers of commodities. Our code also performed well against ADP, a state-of-the-art interior point code.

The performance of our algorithm was heavily influenced by our choice of when to scale α. We tested several strategies and found that different strategies performed better for different problems. We therefore believe that more work is needed to find a strategy that works well for all problems.

Our algorithm might be improved by using a different minimum-cost flow algorithm. In fact, we do not require the exact solution to a minimum-cost flow but only an approximate solution. An algorithm which is able to find fast approximations to a minimum-cost flow might significantly improve the running time of our algorithm. Also, the minimum-cost flow problems we solve for the same commodity might have similar solutions. Using the solution to the previous problem as a starting point for the new problem might improve the running time.

We are aware of two other implementations of combinatorial algorithms to which we should compare our algorithm. The first, by Shahrokhi and Matula [17], works only for graphs in which every capacity and demand is 1, but it would still be interesting to see how our algorithm compares to theirs on this class of graphs. The second, by Schneur [16], also works by gradually rerouting flow. She has shown that her algorithm runs well on many problems. We would like to compare the algorithms on the same machine and the same problems.

Acknowledgments

Several people provided us with code and with multicommodity flow problems that were very useful in our work. We thank Dimitri Bertsekas for providing us with his minimum-cost flow code and Mauricio Resende for providing us with the multicommodity flow code MCNF85. We thank K. Ramakrishnan and Mauricio Resende for performing the interior point experiments. We thank David Johnson for an explanation of the code ADP. We thank Rina Schneur and Farhad Shahrokhi for providing us with problem instances and Cathy McGeoch and Mauricio Resende for providing us with problem generators. We thank Debbie Lam and Tom Leighton for work on the early stages of this project. We thank David Johnson, Philip Klein, Cathy McGeoch, and Rina Schneur for helpful discussion and Perry Fizzano for a careful reading of a draft of this paper.

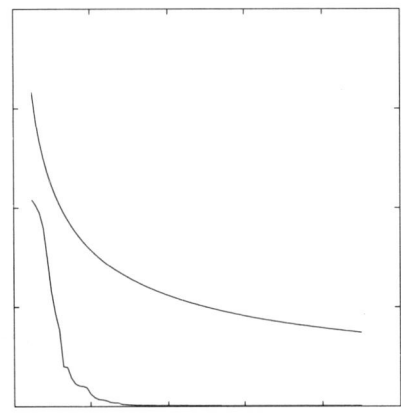

$0 \leq x \leq 5000$, $0 \leq y \leq 0.08$
x-axis is # of iterations.
y-axis is ϵ.
The top curve is $1/\sqrt{\text{# of iterations}}$.
The bottom curve is the minimum ϵ achieved.

FIGURE 3. NETGEN graph with 50 nodes, 100 edges, and 20 commodities.

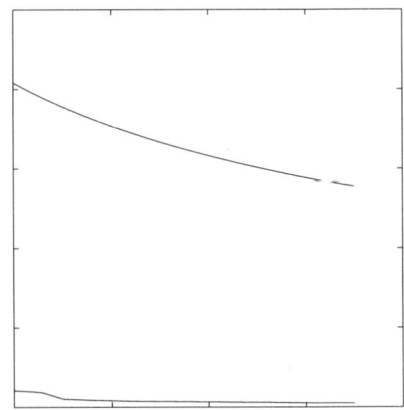

$150 \leq x \leq 350$, $0 \leq y \leq 0.1$
x-axis is # of iterations.
y-axis is ϵ.
The top curve is $1/\sqrt{\text{# of iterations}}$.
The bottom curve is the minimum ϵ achieved.

FIGURE 4. NETGEN graph with 50 nodes, 100 edges, and 10 commodities.

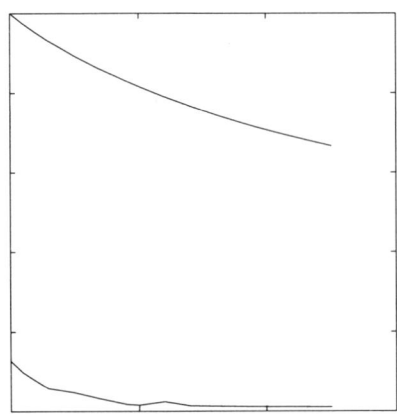

$100 \leq x \leq 250$, $0 \leq y \leq 0.1$
x-axis is # of iterations.
y-axis is ϵ.
The top curve is $1/\sqrt{\text{# of iterations}}$.
The bottom curve is the minimum ϵ achieved.

FIGURE 5. NETGEN graph with 50 nodes, 100 edges, and 10 commodities.

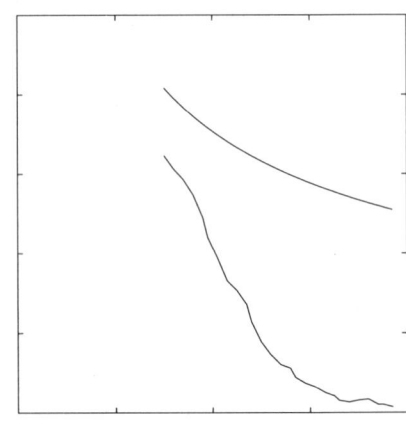

$0 \leq x \leq 400$, $0 \leq y \leq 0.1$
x-axis is # of iterations.
y-axis is ϵ.
The top curve is $1/\sqrt{\text{# of iterations}}$.
The bottom curve is the minimum ϵ achieved.

FIGURE 6. NETGEN graph with 50 nodes, 100 edges, and 10 commodities.

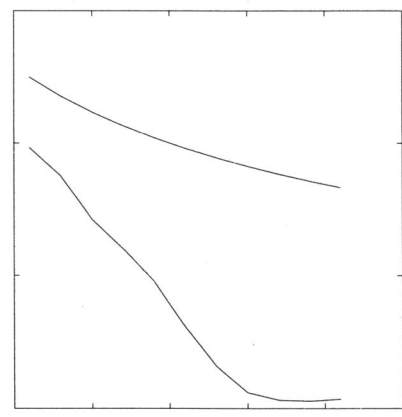

$60 \leq x \leq 160$, $0 \leq y \leq 0.15$
x-axis is # of iterations.
y-axis is ϵ.
The top curve is $1/\sqrt{\text{# of iterations}}$.
The bottom curve is the minimum ϵ achieved.

FIGURE 7. RMFGEN graph with 48 nodes, 140 edges, and 10 commodities.

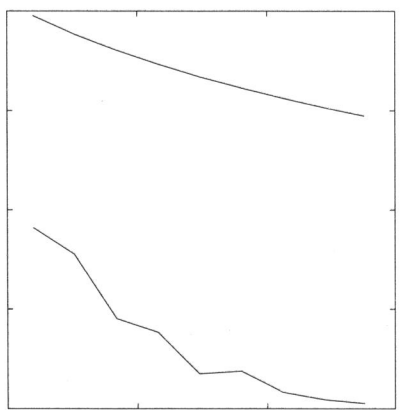

$150 \leq x \leq 300$, $0 \leq y \leq 0.08$
x-axis is # of iterations.
y-axis is ϵ.
The top curve is $1/\sqrt{\text{# of iterations}}$.
The bottom curve is the minimum ϵ achieved.

FIGURE 8. RMFGEN graph with 48 nodes, 140 edges, and 20 commodities.

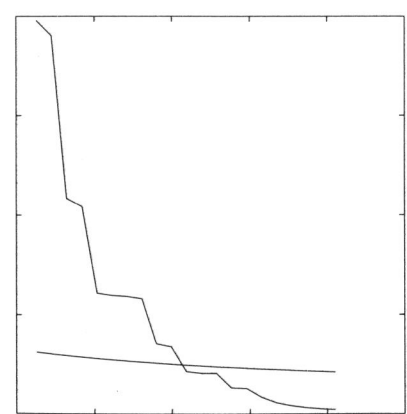

$6000 \leq x \leq 16000$, $0 \leq y \leq 0.08$
x-axis is # of iterations.
y-axis is ϵ.
The top curve is $1/\sqrt{\text{# of iterations}}$.
The bottom curve is the minimum ϵ achieved.

FIGURE 9. RMFGEN graph with 500 nodes, 2075 edges, and 700 commodities.

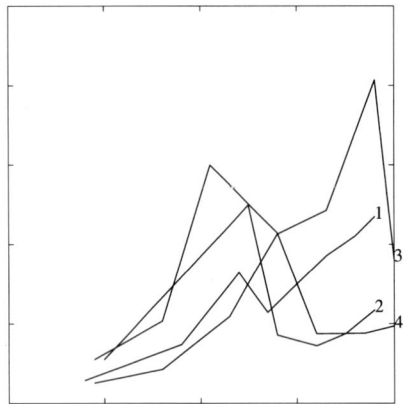

$0 \leq x \leq 40$, $0 \leq y \leq 2500$
x-axis is # of commodity groups.
y-axis is # of iterations.
Each curve represents a set of runs on one of four different underlying graphs.

FIGURE 10. NETGEN graphs with 50 nodes, 100 edges, and from 10 through 70 commodities.

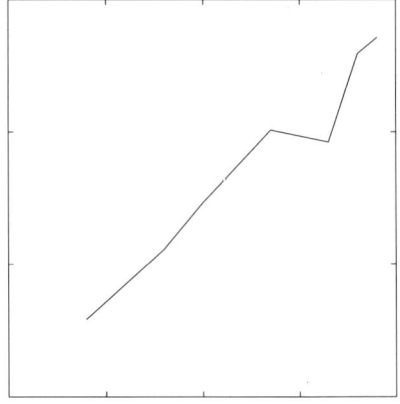

$0 \leq x \leq 40$, $0 \leq y \leq 600$
x-axis is # of commodity groups.
y-axis is # of iterations.
The curve represents a set of runs on one underlying graph.

FIGURE 11. RMFGEN graphs with 48 nodes, 140 edges, and from 10 through 70 commodities.

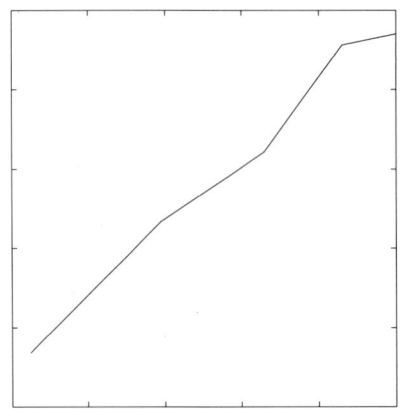

$40 \leq x \leq 140$, $600 \leq y \leq 1600$
x-axis is # of commodity groups.
y-axis is # of iterations.
The curve represents a set of runs on one underlying graph.

FIGURE 12. RMFGEN graphs with 192 nodes, 740 edges, and from 50 through 250 commodities.

References

1. I. Adler, N. Karmarkar, M. Resende, and G. Veiga. An implementation of Karmarkar's algorithm for linear programming. *Mathematical Programming*, 44:297–335, 1989.
2. A. A. Assad. Multicommodity network flows - a survey. *Networks*, 8:37–91, 1978.
3. D. P. Bertsekas and P. Tseng. RELAXT-III: A new and improved version of the RELAX code. Technical Report LIDS–P–1990, MIT, July 1990.
4. J. Cheriyan, October 1991. Private communication.
5. A. V. Goldberg, Personal communication. Jan., 1991.
6. M. D. Grigoriadis and L. G. Khachiyan. Fast approximation schemes for convex programs with many blocks and coupling constraints. Technical Report DCS-TR-273, Department of Computer Science, Rutgers University, New Brunswick, NJ, March 1991.
7. S. Kapoor and P. M. Vaidya. Fast algorithms for convex quadratic programming and multicommodity flows. In *Proceedings of the 18th Annual ACM Symposium on Theory of Computing*, pages 147–159, 1986.
8. J. Kennington. A survey of linear cost multicommodity network flows. *Operations Research*, 26:206–236, 1978.
9. N.K. Karmarkar and K. G. Ramakrishnan. Computational results of an interior point algorithm for large scale linear programming. *Mathematical Programming*, 52:555–586 1991.
10. J. Kennington. A primal partitioning code for solving multicommodity flow problems (version 1). Technical Report Techincal Report 79009, Department of Industrial Engineering and Operations Research, Southern Methodist University, 1979.
11. P. Klein, A. Agrawal, R. Ravi, and S. Rao. Approximation through multicommodity flow. In *Proceedings of the 31st Annual Symposium on Foundations of Computer Science*, pages 726–727, 1990.
12. P. Klein, S. A. Plotkin, C. Stein, and É. Tardos. Faster approximation algorithms for the unit capacity concurrent flow problem with applications to routing and finding sparse cuts. Technical Report 961, School of Operations Research and Industrial Engineering, Cornell University, 1991. A preliminary version of this paper appeared in *Proceedings of the 22nd Annual ACM Symposium on Theory of Computing*, pages 310–321, 1990.
13. T. Leighton, F. Makedon, S. Plotkin, C. Stein, É. Tardos, and S. Tragoudas. Fast approximation algorithms for multicommodity flow problems. In *Proceedings of the 23rd Annual ACM Symposium on Theory of Computing*, pages 101–111, 1991.
14. T. Leighton and S. Rao. An approximate max-flow min-cut theorem for uniform multicommodity flow problems with applications to approximation algorithms. In *Proceedings of the 29th Annual Symposium on Foundations of Computer Science*, pages 422–431, 1988.
15. B.A. Murtaugh and M.A. Saunders. MINOS 5.0 user's guide. Technical Report Technical Report 83-20, Systems Optimization Laboratory, Stanford University, 1983.
16. R. Schneur. *Scaling algorithms for multicommodity flow problems and network flow problems with side constraints*. PhD thesis, MIT, Cambridge, MA, February 1991.
17. F. Shahrokhi and D. W. Matula. The maximum concurrent flow problem. *Journal of the ACM*, 37:318 – 334, 1990.
18. P. M. Vaidya. Speeding up linear programming using fast matrix multiplication. In *Proceedings of the 30th Annual Symposium on Foundations of Computer Science*, pages 332–337, 1989.

Department of Mechanical Engineering, Stanford University, Palo Alto, CA
E-mail address: taleong@leland.stanford.edu

AT&T Bell Laboratories, Murray Hill, NJ
E-mail address: shor@research.att.com

Department of Mathematics and Computer Science, Dartmouth College, Hanover, NH
E-mail address: cliff@cs.dartmouth.edu

Reverse Auction Algorithms for Assignment Problems

DAVID A. CASTAÑON

February 14, 1992

ABSTRACT. The auction algorithm has been shown to be a very effective method for the solution of minimum cost bipartite matching problems from n persons to n objects. Recently, several variations of the auction algorithm have been proposed which use the concept of forward-reverse auction, where the roles of persons and objects are reversed dynamically throughout the course of the algorithm. This paper describes the results of computational experiments conducted as part of the DIMACS algorithm implementation challenge using different variations of auctions algorithms. The results indicate that, for robust performance across problem classes, auction algorithms should use a combination of forward-reverse logic and cost scaling techniques.

1. Introduction

Consider the classical problem of optimal assignment of n persons to n objects. Given a benefit a_{ij} that person i associates with object j, one want to find an assignment of persons to objects, on a one-to-one basis, that maximizes the total benefit. The auction algorithm, a method for solving this problem first proposed in [2], and subsequently developed and extended in [3, 9, 6, 7], has been shown to be very effective in practice. The algorithm operates like an auction. There is a price for each object, and at each iteration, unassigned persons bid simultaneously for their "best" objects (the ones offering maximum benefit minus price), thereby raising the corresponding prices. Objects are then awarded to the highest bidder. The bidding increments must be at least equal to a positive parameter ϵ, and are chosen so as to preserve an ϵ-complementary

---

1991 *Mathematics Subject Classification.* Primary 90C08, 90C27; Secondary 90C05, 90C10.

The author would like to thank Prof. D. Bertsekas of M. I. T. for providing the initial algorithm codes and for his assistance in understanding the performance of the auction algorithms, and DIMACS for their efforts in organizing the workshops and suggesting the experiments.

slackness property. For good theoretical as well as practical performance, it may be important to use ϵ-scaling, which consists of applying the algorithm several times, starting with a large value of ϵ and successively reducing ϵ up to an ultimate value that is less than some threshold ($1/n$ when a_{ij} are integer). Each scaling phase provides good initial prices for the next. For a tutorial presentation of the auction algorithm, refer to [4].

We note that there are several extensions of the auction algorithm; to transportation problems [6], and to minimum cost flow problems (the ϵ-relaxation method of [9] and the network auction algorithm of [7]). Computational studies have shown that the algorithm is very effective, particularly for sparse symmetric assignment problems and special types of transportation problems.

In a recent paper, Bertsekas, Castañon and Tsaknakis [8] proposed new variations of the auction algorithm which were suitable both for inequality-constrained assignment problems as well as the equality-constrained matching problem described above. The central idea is an alternative form of the auction algorithm, called *reverse auction*, where, roughly, the *objects* compete for persons by *lowering* their prices. In particular, objects decrease their prices to a level that is sufficiently low to lure a person away from its currently held object. One can show that forward and reverse auctions are mathematically equivalent, but their combination results in algorithms that can solve problems that forward or reverse auction by themselves either cannot solve at all or can solve but much more slowly.

The work of [8] describes how to combine forward and reverse auction to solve symmetric assignment problems. In particular, mechanisms are designed for switching gracefully between the two types of auction, using a special type of ϵ-complementary slackness condition. The primary motivation for the switching is to avoid a phenomenom called "price wars", which consists of protracted bid sequences involving a small number of persons competing for a smaller number of objects using small bidding increments. In such situations, reversing the auction leads to a smaller number of objects bidding for a larger number of persons, and the resulting auction is easier to resolve. In fact, it is conjectured in [8] that, for many classes of problems, it may not be necessary to resort to ϵ-scaling. However, without ϵ-scaling, the worst-case complexity of the forward-reverse auction algorithm is only pseudo-polynomial.

In addition, the work in [8] extends the reverse auction algorithms for the solution of several types of assignment problems with inequality constraints. Included are asymmetric problems with different numbers of persons and objects. Again, the central new idea in these algorithms is to combine regular auction, where persons bid for objects by raising their prices, with reverse auction, where objects compete for persons by essentially offering discounts. For this class of problems, the use of forward-reverse auction leads to efficient ϵ-scaling approaches which do not require conversion to equality-constrained problems.

To date, there has been limited computational experience with the new variations of the auction algorithm; further experimentation is needed to identify which variations are more efficient, and the dependence of computation performance on different problem structures. The focus of this paper is on evaluating the empirical performance of different forward and forward-reverse auction algorithms on equality-constrained assignment problems. As part of the Center for Discrete Mathematics and Theoretical Computer Science (DIMACS) algorithm implementation challenge, a broad set of computation experiments was conducted using the DIMACS problem generators. The experiments included the following algorithms:

- Unscaled forward-reverse auction algorithm
- Scaled forward-reverse auction algorithm
- Scaled forward auction algorithm [4]
- Jonker and Volgenant's assignment algorithm [12]

The assignment algorithm of Jonker and Volgenant is a sequential shortest path algorithm which has been shown to outperform many other sequential shortest path and Hungarian assignment codes [12]. It was included in the experiments to illustrate the performance differences between auction-based algorithms and alternative algorithm approaches.

This paper documents the results of DIMACS experiments using the above algorithms. A particular focus of the experiments was to determine whether the theoretical advantages of scaling were present in practical implementations of the forward-reverse auction algorithms. The results establish that, for large classes of problems, scaling is an essential part of any efficient implementation of the auction algorithm, including the forward-reverse variation. In particular, certain problem structures lead to prolonged price wars in unscaled variations of auction algorithms. The experimental results also establish that the scaled forward-reverse algorithms are more efficient than the corresponding scaled forward auction algorithms, and that the auction-based algorithms are more efficient than the algorithm of Jonker and Volgenant for most of the problem classes used in the experiments.

The rest of this paper is organized as follows: §2 contains an overview of the theory of the auction algorithm, and describes the combined forward-reverse auction algorithm for equality-constrained assignment problems. §3 contains a discussion of the algorithm implementations, and summarizes the problem classes which were used in testing the algorithms. §4 discusses the computation results obtained. §5 is a summary of the conclusions and indicates areas for further research.

2. Reverse Auction for Symmetric Assignment Problems

In the symmetric assignment problem, there are n persons and n objects. The benefit or value for assigning person i to object j is a_{ij}. The set of objects to which person i can be assigned is a nonempty set denoted $A(i)$. An *assignment*

S is a (possibly empty) set of person-object pairs (i,j) such that $j \in A(i)$ for all $(i,j) \in S$; for each person i there can be at most one pair $(i,j) \in S$; and for every object j there can be at most one pair $(i,j) \in S$. Given an assignment S, person i is *assigned* if there exists a pair $(i,j) \in S$; otherwise i is *unassigned*. An assignment is said to be *feasible* if it contains n pairs, so that every person and every object is assigned; otherwise the assignment is called *partial*. The objective is to find an assignment $\{(1,j_1),\dots,(n,j_n)\}$ with total benefit $\sum_{i=1}^{n} a_{ij_i}$, which is maximal.

2.1. The Auction Algorithm. The auction algorithm for the symmetric assignment problem proceeds iteratively and terminates when a feasible assignment is obtained. At the start of the generic iteration there is a partial assignment S and a price vector $p = (p_1,\dots,p_n)$ satisfying ϵ-*complementary slackness* (or ϵ-CS for short). This is the condition

$$a_{ij} - p_j \geq \max_{k \in A(i)} \{a_{ik} - p_k\} - \epsilon, \qquad \forall\, (i,j) \in S. \tag{2.1}$$

As an initial choice, one can use an arbitrary set of prices together with the empty assignment, which trivially satisfies ϵ-CS. The iteration consists of two phases: the *bidding phase* and the *assignment phase* described in the following.

Bidding Phase:
Let I be a nonempty subset of persons i that are unassigned under the assignment S. For each person $i \in I$:
1. Find a "best" object j_i having maximum value, that is,

$$j_i = \arg\max_{j \in A(i)} \{a_{ij} - p_j\},$$

and the corresponding value

$$v_i = \max_{j \in A(i)} \{a_{ij} - p_j\},$$

and find the best value offered by objects other than j_i

$$w_i = \max_{j \in A(i), j \neq j_i} \{a_{ij} - p_j\}.$$

[If j_i is the only object in $A(i)$, define w_i to be $-\infty$ or, for computational purposes, a number that is much smaller than v_i.]

2. Compute the "bid" of person i given by

$$b_{ij_i} = p_{j_i} + v_i - w_i + \epsilon = a_{ij_i} - w_i + \epsilon.$$

[That is, a person i bid for object j_i, and that object j_i received a bid from person i. The algorithm works if the bid has any value between $p_{j_i} + \epsilon$ and $p_{j_i} + v_i - w_i + \epsilon$, but it tends to work fastest for the maximal choice.]

Assignment Phase:
For each object j:

Let $P(j)$ be the set of persons from which j received a bid in the bidding phase of the iteration.

3. If $P(j)$ is nonempty, increase p_j to the highest bid:

$$p_j := \max_{i \in P(j)} b_{ij},$$

remove from the assignment S any pair (i, j) (if j was assigned to some i under S), and add to S the pair (i_j, j), where i_j is a person in $P(j)$ attaining the maximum above.

Note that there is some freedom in choosing the subset of persons I that bid during an iteration. One possibility is to let I consist of a single unassigned person. This version, known as the *Gauss-Seidel version* in view of its similarity with Gauss-Seidel methods for solving systems of nonlinear equations, usually works best in a serial computing environment. The version where I consists of all unassigned persons is known as the *Jacobi version*, in view of its similarity with Jacobi methods for solving systems of nonlinear equations.

The choice of bidding increment $v_i - w_i + \epsilon$ for a person i is such that ϵ-CS (cf. (2.1)) is preserved. Furthermore, the algorithm is valid in the sense that, if at least one feasible assignment exists, the auction algorithm terminates in a finite number of iterations with a feasible assignment that is within $n\epsilon$ of being optimal, and is optimal if the problem data is integer and $\epsilon < 1/n$ [2, 3].

To obtain polynomial complexity, one uses ϵ-*scaling*, which consists of applying the algorithm several times, starting with a large value of ϵ and successively reducing ϵ up to an ultimate value that is less than $1/n$. For integer data, it can be shown that the worst-case running time of the auction algorithm using scaling and appropriate data structures is $O(nA \log(nC))$, where A is the number of feasible arcs in the assignment graph and C is the cost range; see [9, 10].

2.2. Reverse Auction. Reverse auction uses an alternative form of the auction algorithm where *objects* compete for persons. In particular, objects decrease their prices to a level that is sufficiently low to either attract an unassigned person or lure a person away from its currently held object.

In order to describe reverse auction, a *profit* variable π_i is defined for each person i. Profits play for persons a role analogous to the role prices play for objects. Reverse auction can be described in two equivalent ways; one where unassigned objects lower their prices as much as possible to attract a person without violating ϵ-CS, and another where unassigned objects select a best person and raise his/her profit as much as possible without violating ϵ-CS.

Consider the following ϵ-CS condition for a (partial) assignment S and a profit vector π:

$$a_{ij} - \pi_i \geq \max_{k \in B(j)} \{a_{kj} - \pi_k\} - \epsilon, \quad \forall \, (i, j) \in S, \tag{2.2}$$

where $B(j)$ is the set of persons that can be assigned to object j,

$$B(j) = \{i \mid (i,j) \in \mathcal{A}\}.$$

For feasibility, assume that this set is nonempty for all j. The reverse auction algorithm starts with, and maintains an assignment and a profit vector π satisfying the above ϵ-CS condition. It terminates when the assignment is feasible. At the beginning of each iteration, there is an assignment S and a profit vector π satisfying the ϵ-CS condition.

Typical Iteration of Reverse Auction:

Let J be a nonempty subset of objects j that are unassigned under the assignment S. For each object $j \in J$:

1. Find a "best" person i_j such that

$$i_j = \arg \max_{i \in B(j)} \{a_{ij} - \pi_i\},$$

and the corresponding value

$$\beta_j = \max_{i \in B(j)} \{a_{ij} - \pi_i\},$$

and find

$$\omega_j = \max_{i \in B(j), i \neq i_j} \{a_{ij} - \pi_i\}.$$

[If i_j is the only person in $B(j)$, define ω_j to be $-\infty$ or, for computational purposes, a number that is much smaller than β_j.]

2. Each object $j \in J$ bids for person i_j an amount

$$b_{i_j j} = \pi_{i_j} + \beta_j - \omega_j + \epsilon = a_{i_j j} - \omega_j + \epsilon.$$

3. For each person i that received at least one bid, increase π_i to the highest bid

$$\pi_i := \max_{j \in P(i)} b_{ij},$$

where $P(i)$ is the set of objects from which i received a bid; remove from the assignment S any pair (i,j) (if i was assigned to some j under S), and add to S the pair (i, j_i), where j_i is an object in $P(i)$ attaining the maximum above.

Note that reverse auction is identical to (forward) auction with the roles of persons and objects as well as profits and prices interchanged. Thus, if at least one feasible assignment exists, the reverse auction algorithm terminates in a finite number of iterations. The feasible assignment obtained upon termination is within $n\epsilon$ of being optimal (and is optimal if the problem data are integer and $\epsilon < 1/n$).

2.3. Forward-Reverse Auction Algorithms.

The algorithms of interest in this experimental study combine the ideas from forward and reverse auction, switching back and forth. Such algorithms must simultaneously maintain a price vector p satisfying the ϵ-CS condition (2.1) and a profit vector π satisfying the ϵ-CS condition (2.2). To this end an ϵ-CS condition for the *pair* (π, p) is introduced which implies the other two. Maintaining this condition is essential for switching gracefully between forward and reverse auction.

DEFINITION 2.1. *An assignment S and a pair (π, p) are said to satisfy ϵ-CS if*

$$\pi_i + p_j \geq a_{ij} - \epsilon, \quad \forall\, (i,j) \in \mathcal{A},$$

$$\pi_i + p_j = a_{ij}, \quad \forall\, (i,j) \in S.$$

The above property is maintained in the combined forward/reverse algorithm. The algorithm starts with, and maintains an assignment S and a profit-price pair (π, p) satisfying the above ϵ-CS condition. It terminates when the assignment is feasible.

Combined Forward/Reverse Auction Algorithm

Step 1: (Run forward auction) Execute several iterations of the forward auction algorithm (subject to the termination condition), and at the end of each iteration (after increasing the prices of the objects that received a bid), set

$$\pi_i = a_{ij_i} - p_{j_i}, \tag{2.3}$$

for every person-object pair (i, j_i) that entered the assignment during the iteration. Go to Step 2.

Step 2: (Run reverse auction) Execute several iterations of the reverse auction algorithm (subject to the termination condition), and at the end of each iteration (after increasing the profits of the persons that received a bid), set

$$p_j = a_{i_j j} - \pi_{i_j}, \tag{2.4}$$

for every person-object pair (i_j, j) that entered the assignment during the iteration. Go to Step 1.

Note that the additional overhead of the combined algorithm over the forward or the reverse algorithm is minimal; just one update of the form (2.3) or (2.4) is required per iteration for each object or person that received a bid during the iteration. An important property is that these updates maintain the ϵ-CS condition for the pair (π, p), and therefore also maintain the required ϵ-CS conditions for π and p, respectively [8].

During forward auction, the object prices p_j increase, while the profits π_i decrease, but exactly the opposite happens in reverse auction. For this reason, the termination proof used for forward auction (see e.g. [10], p. 371) does not apply to the combined method. Indeed, it is possible to construct examples of feasible problems where the combined method never terminates if the switch between forward and reverse auctions is done arbitrarily. However, it is easy to guarantee that the combined algorithm terminates in a finite number of steps for a feasible problem; it is sufficient to ensure that some "irreversible progress" is made before switching between forward and reverse auction. The algorithm implementations used in the experiments refrain from switching until at least one more person-object pair has been added to the assignment; it is easily seen that this constitutes "irreversible progress" and guarantees convergence.

3. Algorithms and Computation Experiments

Recently, there have been extensive investigations reporting computational experience with variations of auction algorithms (see e.g. [5, 7, 8]). Based on these results, one can generate the following set of hypotheses for empirical investigation

(i) The performance of auction algorithms degrades with increasing cost range.
(ii) Scaling makes auction algorithm performance robust to problem structure.
(iii) The empirical run-time of auction algorithms scales linearly with the number of arcs in the network.
(iv) Auction algorithms are more efficient than the algorithm of Jonker and Volgenant for sparse assignment problems.
(v) Forward-reverse auction algorithms require less scaling than forward auction algorithms for robust performance, thus leading to more efficient algorithms.

The computation experiments listed below were selected to explore the validity of the above hypotheses.

3.1. Computation Experiments.
In order to test the above hypotheses, the following experiments were conducted:

3.1.1. *Density.* The purpose of these experiments was to determine empirical run-time performance on different classes of random equality-constrained assignment problems. The range of cost values was kept fixed from 1 to 10^5. The random problems were generated using the DIMACS assign.c generator. These experiments included four problem classes:
- **Low-Degree** These problems have between 1000 and 32000 persons, and each person has a constant degree of 8 arcs.

- **Medium-Degree** These problems have between 1000 and 8000 persons; for each person, the degree is given by $\log_2(2s)$, where s is the number of persons.
- **High-Degree** These problems have between 512 and 2048 persons; for each person the degree is given by $s/4$.
- **Dense** These problems have between 256 and 1024 persons, and are fully dense assignment problems.

3.1.2. *Cost Scaling.* The purpose of these experiments is to evaluate the performance of the algorithms for problems of different cost ranges, for moderately sparse assignment problems. These problems were also generated using the DIMACS assign.c generator. The experiments included two problem classes:
- **Low-Cost** These problems have between 128 and 8192 persons; for each person, the degree is given by $2\log_2 2s$, where s is the number of persons. The cost range for the arcs is from 1 - 100.
- **High-Cost** The network structure of these problems is identical to the Low-Cost problems above, but the cost range is from 1 to 10^8.

3.1.3. *Structured Problems.* The purpose of these experiments is to determine empirical performance on classes of difficult assignment problems. Three problem classes were included:
- **Two-Cost** These problems have the same network structure as the Low-Cost and High-Cost problems above; however, the arc costs are such that 50% of the arcs have cost 100, and the other 50% have cost 10^8.
- **Fix-Cost** These problems range from 128 to 2048 persons; for each person, the degree is given by $s/8$. The network structure was generated using the assign.c generator; however, the costs for these problems were fixed (instead of random), so that the cost of arc (i, j) was set to $i*j*100$.
- **Geometric** These problems range from 128 to 1024 persons, and correspond to dense problems for matching two sets of two-dimensional points included in the rectangle $[0, 10^6] \times [0, 10^6]$. The locations of the points were generated using the DIMACS generator dcube.c. The arc costs correspond to the Euclidean distance between the points corresponding to objects and the points corresponding to persons.

3.2. Algorithm Implementations. In order to test the most recent implementations of the auction algorithm, the recently published [5] Fortran 77 implementations of both forward and forward-reverse auction algorithms were used. The experiments also used a Fortran 77 implentation of the algorithm of Jonker and Volgenant (JV) as described in [12]. The JV algorithm consists of three phases:
  (i) An initialization phase, which generates an incomplete assignment S and a set of prices p satisfying strict complementary slackness.

(ii) An augmenting row reduction phase, which is based on using a naive auction algorithm (the forward auction algorithm with $\epsilon = 0$) for a limited number of bids, and increases the cardinality of the assignment S while maintaining prices satisfying complementary slackness.

(iii) A sequential shortest path method phase, which assigns the persons left unassigned by the previous two phases.

In previous computation experiments [12], the JV algorithm has outperformed a variety of alternative sequential shortest path and Hungarian assignment codes.

All of the algorithms were coded using the same basic data structures. The assignment graph was represented in sparse form using a forward-star data (or incidence list) data structure [1]. Since forward-reverse algorithms require a switch in the roles of persons and objects, these algorithms included an extra data structure consisting of linked lists of the arcs going into each object node. Finally, the forward auction algorithms used an additional data structure to keep track of the best three values of the reduced costs evaluated at each bid; these values could be used to reduce the overhead required for computation of subsequent bids by the same person. The detailed implementations are included in [5].

The assignment algorithms in [5] were designed to use integer arithmetic. Unfortunately, the auction algorithm requires the use of integers which are at least as large as the product of the largest cost value times the number of person nodes; for many of the above experiments, this product exceeds the integer range of most computers. In order to avoid integer overflow, the auction and the JV algorithm implementations were modified so that double precision arithmetic would be used for all computations involving dual variables and reduced costs.

The auction algorithms require the selection of a set of input parameters which control the performance of the algorithm. These parameters are:

(i) The initial value of ϵ to be used by scaling.

(ii) The factor by which ϵ is reduced between scaling phases.

(iii) The threshold value of ϵ before it is reduced to a value which is guaranteed to find an optimal solution.

(iv) The initial choice of increment to be used in the bidding.

The last item above is due to the use of an adaptive scaling technique which is used in the implementations in [5], where the increment used for computing bids is slowly increased (up to a maximum of ϵ). The best choice of parameter values depends on the problem class. In the experiments, the parameter settings were chosen based on the algorithm type (scaled vs. unscaled, forward vs. forward-reverse) and on the sparsity structure of the problem class. The parameter choices are described below for each of the implementations, where n denotes the number of persons in the assignment problem, and C denotes the maximum cost value.

- Forward auction: for sparse problems (density less than 15% of the graph), initial ϵ $C/2$, factor 5, threshold $1/(n+1)$, initial increment ϵ. For other problems, initial ϵ $C/5$, factor 5, threshold 0.1, initial increment $1/(n+1)$.
- Forward-reverse auction, unscaled: for all problems, initial ϵ $1/(n+1)$, factor 5, threshold $1/(n+1)$, initial increment $1/(n+1)$.
- Forward-reverse auction, scaled: for sparse problems (density less than 15% of the graph), initial ϵ $C/5$, factor 5 or 10, threshold $1/(n+1)$, initial increment ϵ. For other problems, initial ϵ $C/5$, factor 5 or 10, threshold 0.1, initial increment $1/(n+1)$.

The above settings are similar to the parameter values suggested in the implementations in [5]. Except for the scaling factor in forward-reverse auction, the above settings were used across all of the experiments; in particular, there was no attempt to optimize settings for each problem class.

4. Experiment Results

This section documents the results of the computational experiments using the various assignment algorithms discussed in §3. For each experiment class and problem size, many problem instances were generated, and statistics concerning the computation time required by each algorithm were collected. All of the experiments were conducted on a single-user NeXT 68040 workstation including 20 MBytes of RAM; given the size of the data structures required by the algorithm, the maximum number of arcs which the network problem could contain was 1,200,000 before any virtual memory operations were required. Thus, the experiments were limited to problem sizes limited to under 1,200,000 arcs.

The NeXTstation did not include a Fortran 77 compiler. The algorithm implementations were compiled by first using ATT's Fortran to C translator (f2c) [11], and the using the C compiler on the NeXT with optimized settings. In order to provide some insights into the relative performance of the NeXTstation with these settings, we ran the Fortran benchmark example provided by DIMACS, which consisted of running the NETFLO algorithm of [13] on two minimum cost network flow problems generated by NETGEN [14]. The first problem included 400 nodes, with 4 source nodes and 12 sink nodes and 1382 arcs, whereas the second problem included 500 nodes, with 50 source nodes and 50 sink nodes, and 2000 arcs. Table 1 lists the timing results using the unix *time* command for real, user and system times, compiled from 12 different runs of these problems. The low standard deviations in Table 1 suggest that the algorithm run times for different experiments will reflect accurately the time required to solve the problems.

4.1. Results of Density Experiments. The purpose of the scaling experiments was to determine how computation performance scaled with problem size for the different algorithms. For these experiments, the range of cost values

	Problem 1		Problem 2	
	mean	std. deviation	mean	std. deviation
real time	3.20	0.15	4.80	0.16
user time	2.27	0.05	4.00	0.07
system time	0.53	0.06	0.52	0.07

TABLE 1. Timing in seconds for benchmark problems.

No. Persons	1000		2000		4000		8000	
	mean	s. d.	mean	s. d.	mean	s. d.	mean	s. d.
SF5-old	0.29	0.03	0.66	0.11	1.39	0.05	2.90	0.09
SF5	0.26	0.03	0.59	0.11	1.25	0.04	2.71	0.08
UFR	0.15	0.04	0.34	0.09	0.68	0.16	1.98	0.70
SFR5	0.30	0.02	0.63	0.04	1.79	0.07	3.95	0.25
SFR10	0.23	0.02	0.51	0.03	1.46	0.10	3.19	0.20
JV	0.64	0.11	2.38	0.39	9.42	1.06	34.61	2.79

TABLE 2. Timing in seconds for Low-Degree experiments.

was kept fixed from 1 to 10^5. The first set of experiments was aimed at identifying which variations of the scaled forward auction algorithms and the scaled forward-reverse auction algorithms would be used in the subsequent experiments. Thus, these experiments were conducted using the integer versions of the various algorithms, listed below:

(i) *SF5-old* is the version of the forward auction algorithm without the extra data structure for storing the three best reduced costs.
(ii) *SF5* is the more recent version of forward auction using the extra data structure.
(iii) *UFR* is the unscaled forward-reverse algorithm.
(iv) *SFR5* is the scaled forward-reverse algorithm using a scaling factor of 5.
(v) *SFR10* is the scaled forward-reverse algorith using a scaling factor of 10.
(vi) *JV* is the algorithm of Jonker and Volgenant.

The first set of results was on the sparse problem class *Low-Degree*. Table 2 summarizes the results of these experiments. Each problem was generated using the DIMACS assign.c generator so that each person had degree 8 (8 feasible arcs in the assignment graph). For each problem size, 30 different assignment problems were generated; Table 2 contains the mean and standard deviation of the computation times of each algorithm. Comparison of the times for the two forward auction algorithms suggests that the extra data structure accelerates the performance of the forward auction algorithm even on very sparse problems. Comparing the three forward-reverse algorithms suggests that little or no scaling (scale factor of 10 instead of 5) should be used for these problems.

No. Persons	1000 (11)		2000 (12)		4000 (13)		8000 (14)	
(degree)	mean	s. d.	mean	s. d.	mean	s. d.	mean	s. d.
SF5-old	0.39	0.06	0.91	0.03	2.17	0.08	5.20	0.18
SF5	0.34	0.06	0.77	0.03	1.84	0.08	4.28	0.11
UFR	0.20	0.07	0.45	0.12	1.20	0.32	3.59	1.62
SFR5	0.39	0.04	0.86	0.06	2.93	0.24	7.34	0.79
SFR10	0.32	0.03	0.70	0.07	2.41	0.18	6.00	0.37
JV	0.73	0.17	2.70	0.38	10.05	1.33	39.60	3.83

TABLE 3. Timing in seconds for Medium-Degree experiments.

No. Persons	512 (128)		1024 (256)		2048 (512)	
(degree)	mean	s. d.	mean	s. d.	mean	s. d.
SF5-old	1.73	0.11	8.11	0.50	38.36	1.07
SF5	1.27	0.10	6.41	0.39	30.16	1.00
UFR	0.85	0.26	3.87	1.06	18.6	5.44
SFR5	1.47	0.16	9.27	1.37	50.11	7.12
SFR10	1.30	0.10	6.86	0.81	34.53	4.06
JV	1.10	0.16	5.45	0.65	29.85	4.51

TABLE 4. Timing in seconds for High-Degree experiments.

Tables 3 and 4 contain the results of the *Medium-Degree* and *High-Degree* experiments using the same algorithms. Again, 30 different problems were used for each problem size. In the *Medium-Degree* experiments, the number of arcs for each person is the closest integer to $\log_2(2s)$, where s is the number of persons; the degree is indicated next to the number of persons. In the *High-Degree* experiments, each person has $s/4$ arcs. The results indicate that, across all conditions tested, the new version of the scaled forward auction algorithm is faster than the previous version, and that the speed improvement increases with increasing problem density. Furthermore, a factor of 10 in the scaled forward-reverse algorithm is superior across all conditions to a factor of 5. In subsequent experiments, only the new scaled forward auction algorithm was used, and a scaling factor of 10 was used in the scaled forward-reverse auction algorithm.

It is worthwhile to note the ratio of the standard deviation to the average computation time for each algorithm. In particular, the unscaled forward-reverse auction (UFR) algorithm's performance is much more sensitive to variations in the specific problem structure than other algorithms. Table 5 displays the mean and the maximum computation time across the 30 problems in the Low-Degree experiments with 8000 person nodes. The maximum time of the UFR algorithm is over twice as slow as its mean time, and is worse than the maximum time of the scaled forward-reverse auction (SFR10) algorithm. Similar results were obtained in the Medium-Degree and High-Degree experiments.

Algorithm	SF5-old	SF5	UFR	SFR5	SFR10	JV
mean	2.90	2.71	1.98	3.95	3.19	34.61
max	3.20	2.90	4.41	4.51	3.68	40.85

TABLE 5. Mean and maximum computation times for 8000 person Low-Degree experiments.

No. Persons	1000	2000	4000	8000	16000	32000
UFR	0.19	0.47	1.05	2.78	7.21	16.88
SFR10	0.26	0.59	1.30	3.07	7.30	17.19
SF5	0.67	0.99	2.06	4.76	10.78	24.85

TABLE 6. Average computation time in seconds for Low-Degree experiments.

In order to estimate the growth in computation requirements with problem size, additional experiments were conducted using the double-precision version of the auction algorithms in order to avoid integer overflow. The conjecture was that the empirical run time for the auction algorithms would increase linearly with the number of arcs in the problem. Unfortunately, the assign.c generator produced infeasible sparse problems at large sizes; in order to guarantee feasibility, the Low-Degree problems were modified so that the diagonal assignment was always feasible. Table 6 contains the average computation time in seconds (30 problems for each size) for the auction algorithms in Low-Degree experiments for problems containing up to 32,000 persons. As the numbers indicate, the computation times scale worse than linearly with the number of persons (for a constant degree problem); empirically, the times scale roughly as $s \log_2 s$ in these experiments. This scaling compares favorably with the worst-case complexity scaling of s^2 for problems with constant degree. Table 6 also illustrates the differences in computation time when using the double-precision version of the algorithms on the NeXT workstation; the mean times for comparable problems of the UFR algorithm are roughly 50% slower in the double precision mode.

The final set of experiments was to examine the performance of the algorithms on dense assignment problems (the class *Dense*). Table 7 contains the results of these experiments, averaged across 30 runs at each problem size; the maximum number of persons was chosen to restrict the number of arcs. The results for the *Dense* problems shows that the performance of the JV algorithm improves relative to that of the auction algorithms as problem density increases. Nevertheless, as in all of the previous experiments with random problems, the unscaled forward-reverse auction algorithm had the best performance.

As conjectured, the advantage of the auction algorithms over the JV algorithm increases for sparse problems. There are two principal causes for this: first, the number of assignments made in the first two phases of the JV algorithm decreases with decreasing problem density, leading to a higher number of required

No. Persons	256		512		1024	
	mean	s. d.	mean	s. d.	mean	s. d.
UFR	0.62	0.17	2.54	0.33	12.41	1.68
SFR10	1.24	0.12	5.29	0.53	28.03	3.54
SF5	1.09	0.16	5.20	0.65	28.11	1.09
JV	0.80	0.16	3.34	0.51	17.81	2.32

TABLE 7. Timing in seconds for Dense experiments.

No. Persons	512 (20)		2048 (24)		8192 (28)	
(degree)	mean	s. d.	mean	s. d.	mean	s. d.
UFR	0.14	0.02	0.86	0.26	10.55	13.81
SFR10	0.33	0.03	1.93	0.20	15.27	2.85
SF5	0.36	0.03	2.10	0.19	15.74	3.05
JV	0.30	0.16	3.10	0.27	32.83	1.95

TABLE 8. Timing in seconds for Low-Cost experiments.

shortest path augmentations. Second, the average length of the shortest path augmentations also increases with decreasing density, so that each augmentation is harder to compute. Although the number of bids per person in the auction algorithms also increases with problem sparsity, the additional computations due to this increase appear to be more efficient than the corresponding computations in the JV algorithm.

4.2. Results of Cost Scaling Experiments. The purpose of the cost scaling experiments was to investigate the sensitivity of the auction algorithms to the range of arc costs. Larger cost ranges require the use of more scaling phases in order to reach an optimal value of ϵ. Furthermore, for unscaled algorithms, larger cost ranges may lead to longer 'price wars'. The problems selected for these experiments were sparse assignment problems of different sizes, where the degree of each person node was given by $2 \log_2 2s$ (s is the number of persons). All of the problems were generated using the assign.c generator. Two different sets of experiments were conducted: *Low-Cost* with cost range 1-100 and *High-Cost* with cost range $1\text{-}10^8$.

Table 8 summarizes the results of the *Low-Cost* experiments; for each problem size, the mean and standard deviation were computed based on 30 randomly generated problems. Table 9 summarizes the results of the *High-Cost* experiments. Surprisingly, the auction algorithms perform as well in the *High-Cost* experiments as in the comparable *Low-Cost* experiments, although the *High-Cost* problems require nearly twice as many scaling phases. In particular, the unscaled forward-reverse auction algorithm (UFR) performs efficiently for the large cost range. In contrast, the JV algorithm's performance is slower for the higher cost ranges.

The reason for the auction algorithms' good performance for the high cost

No. Persons	512 (20)		2048 (24)		8192 (28)	
(degree)	mean	s. d.	mean	s. d.	mean	s. d.
UFR	0.16	0.03	0.96	0.12	5.71	0.45
SFR10	0.62	0.05	3.32	0.18	22.65	5.10
SF5	0.47	0.02	2.58	0.08	13.59	0.33
JV	0.89	0.44	9.77	3.54	100.11	14.70

TABLE 9. Timing in seconds for High-Cost experiments.

ranges is that these problems did not generate prolonged 'price wars'. By choosing the arc costs uniformly over the interval $[1, 10^8]$, the differences among any pair of arc costs were of significant size compared to the maximum cost. Thus, the price rises in the auction process were large, leading to quick resolution of any conflicts. In contrast, for the *Low-Cost* experiments, the average price difference was much smaller, so which resulted in small price rises, leading to more difficult auction problems. For these problems, the forward-reverse switching logic of the UFR algorithm is sufficient to shorten price wars, leading to increased efficiency over the scaled auction algorithms which must go through several cost-scaling cycles.

The difficulty of the resulting problems is determined by the number of bids required by the auction algorithms to terminate with an optimal solution. Table 10 contains the average and maximum number of bids across the 30 problems in each experiment for the three auction algorithms. As the results in Table 10 indicate, the large *Low-Cost* problems were much more difficult for the unscaled algorithm than the corresponding *High-Cost* problems; the maximum number of bids was nearly 4 times the average. When scaling is used, prices rise faster; still, for problems with 8192 persons, the ratio of maximum to average for the forward auction algorithm (SF5) was nearly 3:1. The most robust algorithm was SFR10, with a ratio of maximum to average of under 3:2.

For the auction algorithms, computation of a bid depends primarily on the average degree of each person or object node. Thus, the total computation time of the auction algorithms can be predicted accurately from the number of bids generated. However, the average *computation time per bid* is different for the SF5 algorithm than for the other auction algorithms because the extra data structure can allow for computation of each bid without recomputing all the reduced costs of the arcs connected to a person. Thus, for the 2048 person problems, the average computation time per bid for the UFR and SFR10 algorithms is roughly 67.5 microseconds and only 43.3 microseconds for the SF5 algorithm. Thus, although the SFR10 algorithm required less bids to be generated, the SF5 algorithm was faster on average because its bids required 35% less computation.

The difference in the JV algorithm's performance is that, for the *High-Cost* experiments, the number of assignments made in the first two phases (before the sequential shortest path phase) decreased by 20% when compared to similar

No. Persons		2048 (24)		8192 (28)	
(degree)		Low-Cost	High-Cost	Low-Cost	High-Cost
UFR	ave.	10,056	11,1147	75,934	47,918
	max.	19,594	14,456	280,233	55,861
SFR10	ave.	24,504	43,193	144,893	221,901
	max.	29,492	49,203	194,339	287,766
SF5	ave.	37,069	56,391	281,499	246,170
	max.	59,625	59,180	749,786	258,314

TABLE 10. Average and maximum number of bids generated by the auction algorithms.

Low-Cost experiments. This is most likely due to the use of a naive (unscaled) forward auction algorithm in the second phase; as the results above indicate, the use of either the forward-reverse auction logic or a scaling logic makes the auction algorithm performance insensitive to the cost range for these problems.

The results in Table 8 and Table 9 also illustrate that the empirical computation time of the auction algorithms grows worse than linearly with the number of arcs. In these problems, the number of arcs increases by a factor of 4.8 between the 512 and 2048 person experiments, and 4.67 between the 2048 and 8192 person experiments. However, the mean computation times of the auction algorithms grows by larger factors.

As the above discussion indicates, cost range is not a good indicator of problem difficulty for auction algorithms. A better indicator is the ratio of the maximum cost divided by the average cost difference among the three lowest cost arcs for each person; if the average cost difference is small, price rises will also be small, which can lead to lengthy price wars in the auction algorithms (particularly for sparse problems). For each of the structured problem experiments discussed in the next subsection, this ratio is large, leading to difficult problems for the auction algorithms.

4.3. Results of Structured Problem Experiments. The purpose of these experiments was to determine the performance of the different algorithms on classes of difficult assignment problems. In each of these problem classes, the ratio of the maximum cost to the average cost difference of the three lowest cost arcs connected to each person is very large. Thus, the structure of these problems is very different from the random assignment problems used in the previous subsections. Three sets of experiments were conducted:

(i) *Two-Cost* Sparse assignment problems with s persons, degree $2\log_2 2s$ (same as the *High-Cost* and *Low Cost* problems above), but with 50% of the arcs having cost 100, and the other 50% cost 10^8.

(ii) *Fix-Cost* Moderately sparse assignment problems with s persons, degree $s/8$; the cost of arc (i,j) was set to $i*j*100$.

No. Persons	512 (20)		2048 (24)		8192 (28)	
(degree)	mean	s. d.	mean	s. d.	mean	s. d.
UFR	0.25	0.05	2.24	0.33	20.29	3.77
SFR10	2.59	0.19	19.02	1.03	158.44	11.56
SF5	5.23	1.24	75.49	24.34	1195.53	362.99
JV	0.13	0.01	1.18	0.04	13.68	0.22

TABLE 11. Timing in seconds for Two-Cost experiments.

(iii) *Geometric* Dense assignment problems matching two-dimensional points in the rectangle $[0, 10^6] \times [0, 10^6]$; the arc costs correspond to the Euclidean distance between the points.

Table 11 contains the mean and standard deviation across 30 different problems at each size of the *Two Cost* experiments. The results are somewhat surprising in that the JV algorithm is the fastest algorithm for this class of problems. Furthermore, the forward auction algorithm SF5 performance is very slow for these problems; in contrast, the forward-reverse auction algorithms are less sensitive to this problem structure.

One of the principal reasons for the good performance of the JV algorithm in this class of problems is that, for all practical purposes, these problems reduce to a feasible matching problem using only the low cost arcs. Indeed, for each problem instance generated, the optimal assignment did not involve any of the high cost arcs. Since the auction algorithms require the use of a non-zero ϵ which depends on the cost range to achieve optimality, the optimal value of ϵ was set as $10^{-8}/s$ where s was the number of persons. However, for these problems, a value of $\epsilon = 1/s$ would have also produced optimality. Thus, the auction algorithms' performance was degraded by having to use a much smaller value of ϵ than needed. In contrast, the JV algorithm uses no scaling, and was able to exploit the degeneracy in these problems to find optimal solutions efficiently.

A more surprising result is the difference between the SF5 performance and the SFR10 performance, in spite of the more efficient computation of each bid used in the SF5 algorithm. These results illustrate one of the principal advantages of the forward-reverse auction algorithm: the forward-reverse logic deals effectively with the presence of large-cost arcs by finding a direction (either forward or reverse) where the duration of 'price wars' is shortened. The difference in performance between the unscaled and scaled forward reverse algorithms is due to the extra scaling steps required because of the large cost range.

It is useful to contrast the results of Table 11 and Table 9 (the *High-Cost* experiments). Although the cost range in these problems are identical, the ratio of cost range to average cost difference for the three lowest-cost arcs per person is much higher in the *Two-Cost* experiments. Thus, the problems are much harder for auction algorithms, as evidenced by the increase in computation times. Note the decrease in computation time for the JV algorithm due to the special

No. Persons	512 (64)		1024 (128)		2048 (256)	
(degree)	mean	s. d.	mean	s. d.	mean	s. d.
UFR	70.82	5.09	1607.94	158	NA	
SFR10	4.22	0.31	27.92	2.77	190.42	30.10
SF5	3.43	0.44	14.23	0.96	68.34	14.78
JV	41.21	0.46	350.62	2.10	NA	

TABLE 12. Timing in seconds for Fix-Cost experiments.

structure of these problems.

The next set of experiments were the *Fix-Cost* experiments. Table 12 summarizes the mean and standard deviation of the computation times obtained for 30 different problems at each size. The special structure of these problems is that every person's preferred assignments are objects in increasing order, and every object's preferred assignments are persons in increasing order. Thus, these problems can lead to extensive price wars, where all persons bid for the same object until its price rises, then all persons bid for the next object, etc. As the results in Table 12 indicate, this class of problems is extremely hard for both the UFR and the JV algorithms (the 2048 person experiments were not run because of large expected run times!). The structure of these problems is such that the first two phases of the JV algorithm result in very few assignments, thus requiring extensive computations in the sequential shortest path phase.

The *Fix-Cost* class of problems is the first class which demonstrates explicitly the need for scaling in forward-reverse auction algorithms. The symmetric structure of the problems produce price wars which are just as lengthy from either the forward or the reverse side; thus, the UFR algorithm's switching does not facilitate resolution of the price wars. However, the two scaled auction algorithms perform efficiently for this class of problems. In spite of the large cost range, the use of scaling leads to large price rises which accelerate the resolution of price wars. For this class of problems, the SF5 algorithm is more efficient, since the symmetry of the problems eliminates any advantages to using the forward-reverse logic.

Table 13 summarizes the results of the *Geometric* experiments. For each problem size, random positions of person nodes and object nodes in the plane were generated using the DIMACS dcube.c generator, and the Euclidean distances were used to compute the arc costs. Note the excessive times required for the unscaled forward-reverse auction algorithm (UFR). Comparing the results to Table 12, one observes that the two problem classes have comparable complexity; there is an added factor of 8 in Table 13 because the problems are solved as fully dense problems, so that each bid requires scanning 8 times more arcs. Again, the problem symmetry produces price wars both in the forward and the reverse directions.

No. Persons	256		512		1024	
	mean	s. d.	mean	s. d.	mean	s. d.
UFR	68.00	55.12	818.70	557.27	NA	
SFR10	9.19	2.07	73.28	22.55	224.61	55.35
SF5	9.47	3.56	58.23	17.3	157.18	39.28
JV	21.20	6.47	82.69	16.96	362.79	57.35

TABLE 13. Timing in seconds for Geometric experiments.

In contrast, the two scaled auction algorithms perform robustly across all problem sizes. Although the SF5 algorithm is faster in these experiments, the SFR10 algorithm required 15% fewer bids than the SF5 algorithm; since the problems are dense problems, the use of the extra data structure in the SF5 algorithm makes its average bid computation much more efficient.

It is interesting to compare the performance of the JV algorithm in Table 12 and Table 13. Although the problems are of similar difficulty for unscaled auction algorithms (due to the presence of at least one long price war), the *Geometric* problems are much easier for the JV algorithm. This is because there are many more assignments made in the first two phases. In particular, the naive auction algorithm tends to assign each person to its closest object. In geometric problems, this provides a significant number of initial assignments, which reduces the number of augmenting paths which must be computed subsequently.

5. Discussion

This paper presents the results of a detailed investigation of the computation performance of different classes of forward and forward-reverse auction algorithms across a broad set of random and structured problems. The choice of experiments was guided by a set of hypotheses based on previous computational experience with variations of the auction algorithm as reported in [5, 7, 8]. These hypotheses were:

(i) The performance of auction algorithms degrades with increasing cost range.
(ii) Scaling makes auction algorithm performance robust to problem structure.
(iii) The empirical run-time of auction algorithms scales linearly with the number of arcs in the network.
(iv) Auction algorithms are more efficient than the algorithm of Jonker and Volgenant for sparse assignment problems.
(v) Forward-reverse auction algorithms require less scaling than Forward auction algorithms for robust performance, thus leading to more efficient algorithms.

The results in §4 can be summarized in terms of the above hypotheses:

(i) For random problems without structure, the performance of scaled forward and forward-reverse and unscaled forward-reverse auction algorithms is relatively insensitive to cost range. However, unscaled forward-reverse auction algorithms can have slow performance on occasional large random problems or on structured problems such as geometric assignment problems. In essence, the problem difficulty for auction depends on the distribution of the cost values; distributions which are more likely to lead to long "price wars" produce harder problems for all variations of auction algorithms.

(ii) The use of scaling, combined with the use of forward-reverse auction logic, leads to robust performance across all of the problem classes tested.

(iii) In the experiments with sparse problems with constant degree, the empirical performance of auction algorithms scaled worse than linearly with the number of arcs. The key statistic which determines auction algorithm performance is the number of bids. In these experiments, the number of bids increased roughly as $s \log_2 s$, where s is the number of persons in the assignment problem.

(iv) For almost all classes of sparse assignment problems, auction algorithms were more efficient than the algorithm of Jonker and Volgenant. The exception was in the degenerate *Two-Cost* experiments. The advantage of auction algorithms decreases with increasing problem density.

(v) Forward-reverse auction algorithms permit the use of more aggressive scaling factors. In the experiments, the forward-reverse algorithms using a factor of 10 performed comparably or better than forward algorithms using a factor of 5. Previous experiments established that the performance of forward algorithms using a factor of 10 was much worse. Indeed, the combination of forward-reverse logic and scaling produced the most robust performance across all problem classes.

One must note that the performance of auction algorithms is sensitive to the choice of the scaling parameters discussed in §3. In the experiments, two sets of parameters were candidates for use in each algorithm; the choice of parameters was guided by a simple rule concerning the problem density. These choices were based on the experiences reported with randomly-generated problems, where denser problems require less scaling. In contrast, the structured problem classes used in §4 require extensive scaling even for dense problems. Thus, the performance of the auction algorithms reported in this paper may be improved by recognizing the nature of the problem class and selecting algorithm parameters which are better tuned to each class. Since the evaluations focused on robustness of performance, no attempt was made to optimize parameter settings for each problem class.

The results in §4 also suggest three directions for the development of more efficient assignment algorithms. First, the results using the forward auction algorithm establish that significant savings (20-50%) can be achieved in the overhead for computing each bid, by storing the best three reduced cost levels after each bid computation. This feature was not incorporated into the forward-reverse auction algorithms, in part because a straightforward implementation requires monotonicity of object prices (which is not a property of forward-reverse auction). Developing extensions to incorporate this feature would further enhance the performance of scaled forward-reverse auction algorithms.

Second, for most classes of randomly generated problems, the unscaled forward-reverse auction algorithms were most efficient. However, scaling is needed for performance robustness in difficult classes of problem. This suggests that an ideal version of forward-reverse auction should perform with little or no scaling on easier problems, but increase its scaling for hard problems. Adaptive scaling logic would have to be developed which recognized whether to increase the level of scaling based on the algorithm's progress. Although a minimal form of adaptive scaling is included in the current algorithms, it is not sufficient for recognizing when scaling should be minimized.

Finally, new classes of hybrid assignment algorithms (such as the JV algorithm) can be developed combining concepts from forward-reverse auction with other methods such as sequential shortest path methods. Indeed, such hybrid algorithms have been shown to have optimal complexity estimates [15] for assignment problems. By using the more robust forward-reverse auction variations in combination with scaling, the hybrid algorithms would limit the amount of work to be performed in the computation-intensive sequential shortest path phase.

References

1. R. K. Ahuja, T. L. Magnanti and J. B. Orlin, *Network Flows*, Handbook of Operations Research and Management Science, vol. 1: Optimization, G. L. Nemhauser, A. H. G. Rinnooy Kan and M. J. Todd, eds. (1989).
2. Bertsekas, D. P., *A Distributed Algorithm for the Assignment Problem*, Lab. for Information and Decision Systems Working Paper, M.I.T., Cambridge, MA, 1979.
3. Bertsekas, D. P., *The Auction Algorithm: A Distributed Relaxation Method for the Assignment Problem*, Annals of Operations Research **14** (1988)105-123.
4. D. P. Bertsekas, *The Auction Algorithm for Assignment and Other Network Flow Problems: A Tutorial*, Interfaces **20** (1990) 133-149.
5. D. P. Bertsekas, *Linear Network Optimization: Algorithms and Codes*, MIT Press, Cambridge, MA, 1991.
6. D. P. Bertsekas and D. A. Castañon, *The Auction Algorithm for Transportation Problems*, Annals of Operations Research **20** (1989) 67-96.
7. D. P. Bertsekas and D. A. Castañon, *The Auction Algorithm for the Minimum Cost Network Flow Problem*, Laboratory for Information and Decision Systems Report LIDS-P-1925, M.I.T., Cambridge, MA, 1989.
8. D. P. Bertsekas, D. A. Castañon and H. Tsaknakis *Reverse Auction and the Solution of Inequality Constrained Assignment Problems*, March, 1991, submitted to SIAM J. Optimization.
9. D. P. Bertsekas and J. Eckstein, *Dual Coordinate Step Methods for Linear Network Flow Problems*, Math. Progr., Series B **42** (1988) 203-243.
10. D. P. Bertsekas and J. N. Tsitsiklis, *Parallel and Distributed Computation: Numerical Methods*, Prentice-Hall, Englewood Cliffs, N. J., 1989.

11. S. I. Feldman, David M. Gay, M. W. Maimone and N. L. Schryer, *A Fortran-to C Converter*, Computer Science Tech. Report 149, ATT Bell Laboratories, NJ, 1991.
12. R. Jonker and A. Volgenant, *A Shortest Augmenting Path Algorithm for Dense and Sparse Linear Assignment Problems*, Computing, **38** (1987) 325-340.
13. J. L. Kennington and R. V. Helgason, *Algorithms for Network Programming*, Wiley-Interscience, New York, NY, 1980.
14. D. Klingman, A. Napier, and J. Stutz, *NETGEN - A Program for Generating Large Scale (Un) Capacitated Assignment, Transportation, and Minimum Cost Flow Network Problems*, Management Science **20** (1974) 814-822.
15. J. B. Orlin and R. K. Ahuja, *New Scaling Algorithms for Assignment and Minimum Cycle Mean Problems*, Sloan School of Management Report 2019-88, MIT, Cambridge, MA, 1988.

DEPARTMENT OF ELECTRICAL, COMPUTER AND SYSTEMS ENGINEERING, BOSTON UNIVERSITY, BOSTON, MASSACHUSETTS 02115
E-mail address: dac@tawny.bu.edu

An Approximate Dual Projective Algorithm for Solving Assignment Problems

K.G. RAMAKRISHNAN, N.K. KARMARKAR, AND A.P. KAMATH

May 7, 1993

ABSTRACT. This paper discusses a new algorithm for solving the assignment problem. This algorithm, called the Approximate Dual Projective algorithm for assignment (ADP/A), is a variant of the Karmarkar interior point algorithm for linear programming, specialized to solve the assignment problem. Computational results are reported on various classes of assignment problems. These results indicate that this method, holds promise for solving large assignment problems.

1. Introduction

In 1984, Karmarkar proposed a new polynomial time algorithm for solving linear programming (LP) problems [11]. This algorithm, the Karmarkar projective algorithm, and many of its variants are known as interior point methods. These methods have been studied and found to be effective solution techniques for solving large general LP problems [1, 13, 16]. The purpose of the study presented here is to evaluate the performance of one variant of the interior point method for solving a simple class of LP problems, namely, the assignment problem. The variant we study is called the approximate dual projective (ADP) method. The ADP method computes good discrete approximations to the continuous trajectory of the Karmarkar projective method. Asymptotically, ADP trajectory approaches the Karmarkar projective trajectory. We have implemented the ADP method using the preconditioned Conjugate Gradient(CG) technique to solve the linear system that arises in computing the improving direction. Our research software that implements the ADP method has been found to be effective in solving large general LP problems arising in many diverse areas of engineering [13].

1991 *Mathematics Subject Classification.* 90C05, 90C06, 90C35; Secondary 65F10, 65K05.

We have adapted the ADP method to solve large assignment problems. This adapted algorithm, the Approximate Dual Projective algorithm for assignment problems (to be referred to as ADP/A hereafter) is identical to our general purpose ADP method except for the stopping criterion; we simply terminate the dual iterates by identifying the optimal assignment. This identification of optimal assignment is guided by the current dual slacks and the complementary slackness. Fig. 1 shows the overall structure of the algorithm. The algorithm in-

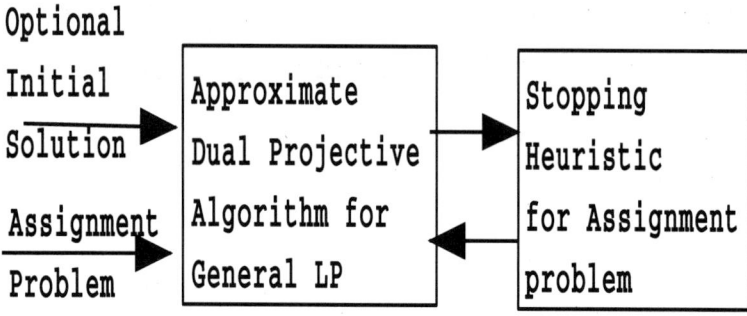

FIGURE 1. Structure of ADP/A Algorithm

side the first box is our ADP method. We give as input to the ADP method, the assignment problem formulated as an LP problem, and optionally an initial dual solution. After the ADP method has made sufficient progress towards optimality, we begin checking, at a set frequency, whether we can identify the optimal assignment from the current solution. If this heuristic succeeds, we terminate the ADP method; otherwise the ADP method continues with the solution procedure. It should be noted that the ADP method has no knowledge that it is solving an assignment problem; in particular, no attempt is made to perform integer arithmetic. There are only two modules that exploit the special structure of the assignment problem, the initial solution evaluation module and the stopping heuristic. These two modules are outside the main ADP algorithm. This is an example of how interior point methods can exploit the special nature of the problem without sacrificing the ability to solve general LP problems. Applications of assignment problems which involve side constraints can be directly solved by our algorithm, unlike specialized combinatorial algorithms for assignment problems.

We report computational results for randomly generated dense assignment problems having as many as 30 thousand nodes and 900 million arcs. We also report on the 4 classes of DIMACS problem generators. We compare our results with a version of the auction algorithm. Section 2 gives an overview of the ADP method. Section 3 discusses the stopping heuristic. Section 4 contains the computational results, followed by a discussion of the results in Section 5. Finally, Section 6 discusses the conclusions of this study.

2. An Overview of the Approximate Dual Projective Algorithm

In this section, we give a brief overview of the Approximate Dual Projective (ADP) algorithm for linear programming. For a complete description of the algorithm please see [13]. As the name implies, ADP method is an approximation to the projective algorithm of Karmarkar [11] applied to the dual. The projective algorithm has very nice stability properties, since the iterates of the projective algorithm always stay on the central trajectory [3]. The ADP method makes discrete approximations to the continuous central trajectory. Asymptotically, after a large number of iterations, the discrete approximations converge to the central trajectory. The discrete approximation consists of two steps; the objective step, and the reciprocal-estimates-improvement (REI) step. Figure 2 depicts the ADP method. In the figure, the optimal vertex is shown as a solid

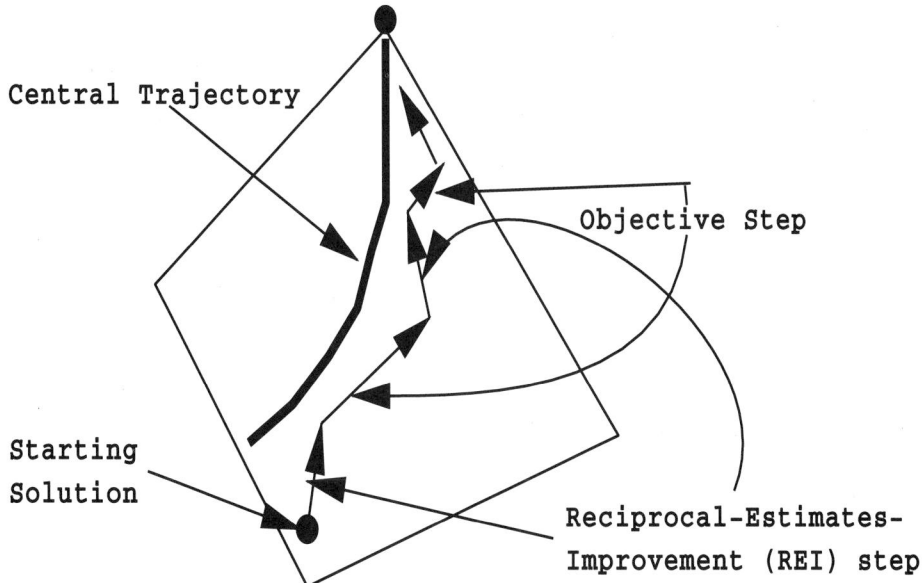

FIGURE 2. Pictorial View of the Approximate Dual Projective(ADP) algorithm

bullet at the top, and the central trajectory is shown as a double solid line. The objective step is the dual affine scaling step [1, 13] that improves the objective function. The REI step has the effect of bringing the iterate closer to the central trajectory. The name reciprocal estimates improvement comes from the fact that on the central trajectory of the dual polytope, the reciprocals of slack variables are exact primal feasible solutions upto a multiple [13]. Thus if we take a step in the dual polytope that improves the primal feasibility of the reciprocals of slacks, then this necessarily implies that the step brings the iterate closer to the central trajectory. The approach of improving reciprocal estimates is an interesting interpretation of the traditional way of centering, i.e., that of minimizing the potential function $-\sum \ln s_i$, where s_i is the dual slack variable. In fact, the linear systems solved by the newton-method-based minimization of the potential function and the REI step are identical. The ADP/A algorithm is similar to the BLCA algorithm of Todd [18], the affine scaling algorithms of Gonzaga [9],Barnes, Chopra, and Jensen [2], and many others. All these algorithms use affine scaling with centering on the primal problem. Recently, dual affine scaling algorithm with centering was studied by Hipolito [10].

Generally the ADP method consists of two phases. In Phase 1, the dual phase, the dual of the LP problem is solved; and in phase 2, a primal solution is obtained, starting with the estimates from the converged dual solution. At the end of execution of both phases, a primal-dual pair with a small relative duality gap is obtained. For the ADP/A algorithm for solving the assignment problems, since we terminate the dual iterates by identifying the optimal assignment, only Phase 1 of the ADP method is executed.

We now give a brief overview of the ADP method. Consider the LP problem, with upper bounds in standard primal form:

$$(2.1) \qquad \min \ c^T x$$

subject to

$$(2.2) \qquad Ax = b$$

$$(2.3) \qquad 0 \leq x \leq u.$$

Here $A \in \mathcal{R}^{m \times n}, c, x, u \in \mathcal{R}^n$, and $b \in \mathcal{R}^m$. Introducing dual variables y and z, corresponding to (2.2) and (2.3) respectively, the dual of (2.1)–(2.3) becomes

$$(2.4) \qquad \max \ b^T y + u^T z$$

subject to

$$(2.5) \qquad A^T y + z \leq c$$

$$(2.6) \qquad z \leq 0.$$

Defining

$$(2.7) \qquad s_2 \triangleq -z$$

and introducing slack variables s_1 in (2.5), the dual (2.4)–(2.6) becomes

$$(2.8) \qquad \max \ b^T y - u^T s_2$$

subject to

$$(2.9) \qquad A^T y - s_2 + s_1 = c$$

(2.10) $$s_1, s_2 \geq 0.$$

We can now develop a variant of the dual projective algorithm for solving (2.8)–(2.10). The scaling matrix D is defined in partitioned form as

(2.11) $$D \triangleq \begin{bmatrix} D_1 & \\ & D_2 \end{bmatrix};$$

where

(2.12) $$D_1^{-1} \triangleq \operatorname{diag}(s_1)$$

and

(2.13) $$D_2^{-1} \triangleq \operatorname{diag}(s_2).$$

Let Δy, Δs_1, and Δs_2 be the ascent directions used in the algorithm. Given below are the formulas for Δy, Δs_1, and Δs_2. The derivation of these formulas is identical to the derivation in [1] except for the treatment of upper bounds on the primal variables. The derivation of these formulas in the presence of upper bounds can be found in [12].

(2.14) $$\Delta y = (A\tilde{D}^2 A^T)^{-1}(b - AD_1^2 D_e^2 u)$$
(2.15) $$\Delta s_1 = -D_e^2 D_2^2 A^T (A\tilde{D}^2 A^T)^{-1}(b - AD_1^2 D_e^2 u) - D_e^2 u$$
(2.16) $$\Delta s_2 = D_e^2 D_1^2 A^T (A\tilde{D}^2 A^T)^{-1}(b - AD_1^2 D_e^2 u) - D_e^2 u$$

where

(2.17) $$D_e^{-2} = D_1^2 + D_2^2$$
(2.18) $$\tilde{D}^2 = D_1^2 - D_1^2 D_e^2 D_1^2.$$

The ADP algorithm is given below.

ALGORITHM A: APPROXIMATE DUAL PROJECTIVE ALGORITHM: DUAL PHASE.

(i) Given an initial interior starting point y^0, s_1^0, s_2^0 and parameters α, and ϵ, set the iteration counter $i = 0$; switch_criterion = false.
(ii) **WHILE** switch_criterion is false **DO**
    2.1 // Objective function improving step.
        a) Compute ascent directions Δy, Δs_1, and Δs_2 using equations (2.11) to (2.18).
        b) // compute maximum step length permissible to keep s_1 and
        // s_2 nonnegative.
$$\beta = \min\left(\min_{\Delta s_{1k} < 0} -\frac{s_{1k}^i}{\Delta s_{1k}}, \min_{\Delta s_{2m} < 0} -\frac{s_{2m}^i}{\Delta s_{2m}}\right)$$
        c) // Compute a fraction of the maximum step length.
          $\beta = \alpha\beta$

d) // Update the solution.
$$y^{i+1} = y^i + \beta \Delta y$$
$$s_1^{i+1} = s_1^i + \beta \Delta s_1$$
$$s_2^{i+1} = s_2^i + \beta \Delta s_2$$
$$i = i + 1$$

2.2 // Check if relative improvement in objective function is less than ϵ

IF $\left((b^T y^i - u^T s_2^i) - (b^T y^{i-1} - u^T s_2^{i-1}) < \epsilon |(b^T y^{i-1} - u^T s_2^{i-1})|\right)$
THEN
   switch_criterion = true
END_IF

2.3 // Reciprocal estimates step.
   a) Apply the reciprocal estimates to the current iterate and obtain direction vectors Δy, Δs_1, and Δs_2.
   b) reexecute steps 2.1.b to 2.1.d to update the solution.

END_WHILE

Several observations about ALGORITHM A are worth making.

- The reciprocal estimates applied in step 2.3.a is based on an elegant mathematical relationship between the dual slacks and the primal variables of the LP problem. The relationship is that, on the central trajectory [3], the reciprocals of dual slacks are primal feasible, up to a scalar multiple. This property is exploited in ADP to perform approximate centering using the reciprocal estimates improvement (REI) algorithm. The details of this algorithm can be found in [13].

- The termination criterion in step 2.2 is the normal termination criterion. This step checks whether the relative improvement in the objective function value is less than ϵ. For the ADP/A method, we have to replace this step by the following step.

   (a) call ALGORITHM ASSIGN, after sufficient progress is made on the dual iterates.
   (b) if ALGORITHM ASSIGN succeeds, stop; else check relative improvement in objective function value (identical to step 2.2 in ALGORTIHM A).

  In all our computational experiments ALGORITHM ASSIGN always succeeded in finding the optimal assignment. Of course, the number of dual iterations that have to be performed before ALGORITHM ASSIGN succeeds depends on the nature of the assignment problem; sometimes we had to perform many iterations of the ADP method before ALGORITHM ASSIGN succeeded while at other instances the algorithm succeeded on its first try (after 10 dual iterations).

- The computationally intensive step is the solution of the positive definite linear system of equations encountered in evaluating the ascent direction. For this step, we have used the preconditioned conjugate gradient method, with a dynamically adaptive preconditioner.

3. Adapting ADP Algorithm for solving the Assignment Problem

We now discuss the ADP/A method, the adapted ADP method to solve the assignment problems. Basically, two items of adaptation are incorporated in ADP/A method; the stopping heuristic and the initial solution. We discuss both of these topics below.

It is well-known that the assignment problem can be formulated as a linear programming problem [17]. In a symmetric assignment problem, we are given n source nodes and n destination nodes. A source node i maybe connected to a destination node j by an edge with cost c_{ij}. Let E be the set of all edges in the problem.

The primal problem can be easily formulated as follows. We associate a 0-1 variable x_{ij} with each edge. Since at least one edge emanating from a source has to be selected, we get the constraint

$$\sum_{j,(i,j)\in E} x_{ij} = 1$$

We get similar constraints for the destination node.

$$\sum_{i,(i,j)\in E} x_{ij} = 1$$

In addition x_{ij} is constrained to be between 0 and 1.

$$0 \leq x_{ij} \leq 1$$

If we need to assign nodes such that the selected edges have minimum cost then the objective of the linear program would be to minimize $\sum_{i,j} c_{ij} x_{ij}$ and the primal form of the linear program can be written in the form

$$\min c^T x$$

$$\text{s.t. } Ax = b$$

$$0 \leq x \leq 1$$

If we ignore the upper bound constraints, which are in any case redundant then the dual obtained is of the form,

$$\min b^T y$$

$$\text{s.t. } A^T y \leq c$$

Here, the dual variables correspond to source and destination nodes and the constraints correspond to the edges in the problem. So if z_i is a variable corresponding to source node i and w_j is a variable corresponding to destination node j then if there exists an edge between these nodes of cost c_{ij} then the dual constraint is of the form

(3.1) $$z_i + w_j \leq c_{ij}$$

3.1. Initial Solution. For general LP problems, ADP method starts with an initial feasible solution by introducing artificial variables, if necessary. In ADP/A method, since the dual constraints are simple, they obviate the need for artificial variables. In ADP/A method, we start from an interior feasible point which is obtained as follows. Since the dual constraints are of the form

(3.2) $$z_i + w_j + s_{ij} = c_{ij}$$

First, we determine
$$\alpha = \min c_{ij}$$
and then set for all i, j and some $\beta > 0$
$$z_i = w_j = (\alpha - \beta)/2$$
$$s_{ij} = c_{ij} - \alpha + \beta$$

It may be easily verified that this is a feasible starting point for the ADP algorithm. For our computational experiments, β was set at .1 times the maximum of the absolute value of costs. This usually resulted in the ratio of maximum to minimum s_{ij} to be about 10.

3.2. Identification of Optimal Assignment from Dual Solution (Stopping Heuristic). As mentioned earlier, we can identify the optimal assignment and prove its optimality from the state of the dual variables by using the complementary slackness principle. Hence there is no need to switch to the primal phase for finding the optimal assignment. The method for identifying the optimal assignment from the dual variables can be described as follows. Assign each source node i to a sink node j, i.e., select edge (i,j), such that
- j is not yet assigned to any source node
- s_{ij} is minimum slack among all sink nodes satisfying above condition.

If the above scheme fails to assign every source node then we quit the stopping heuristic. On the other hand, if we get a feasible assignment with the above scheme then we check the assignment for optimality. For optimality checking, we define temporary dual variables $(\tilde{z}, \tilde{w})$ as follows. For each edge (i,j) selected above, set $\tilde{z}_i = z_i + s_{ij}/2$, and $\tilde{w}_j = w_j + s_{ij}/2$ (note that if we have a feasible assignment then all the temporary dual variables $(\tilde{z}, \tilde{w})$ will be set). We then check whether $\tilde{z}$ and $\tilde{w}$ are feasible solutions of the dual. If they are feasible then we have found the optimal solution and we have also proved optimality (by enforcement of complementary slackness) The optimal dual solution is $(\tilde{z}, \tilde{w})$ and the optimal primal solution is $\tilde{x}$ where $\tilde{x}_{ij}$ is 1 if the edge (i,j) was selected above and 0 otherwise. If $(\tilde{z}, \tilde{w})$ is not feasible, the assignment is declared not optimal and we continue with our ADP algorithm using (z, w) as the current iterate. It can be shown that eventually, the dual phase will converge to a solution from which we can identify the optimal assignment using the scheme described above.

The stopping heuristic discussed above assumes a symmetric *feasible* assignment problem. The heuristic is guaranteed to find the optimal assignment, if it is unique. This is because the slack variables corresponding to the edges in the optimal assignment converge to zero as the dual iterates converge to the optimal solution. When there are multiple optimal assignments, i.e., a facet of the primal polytope is optimal, the ADP/A algorithm goes to the center of the facet. In this case, the heuristic may fail to find an optimal assignment. In all our experiments, we have perturbed the objective vector randomly in the last few bits, to guarantee a unique optimal solution (since ADP/A works with double precision cost coefficients, it is easy to perturb the mantissa of a double precision variable). Of course, the final optimality check is done on the original objective vector.

ALGORITHM ASSIGN shows the pseudocode for the stopping heuristic.

ALGORITHM ASSIGN:.
Input:
(i) m, the number of nodes in each partition of the bipartite graph.
(ii) E, the list of edges in the bipartite graph.
(iii) Current values of dual variables z_i, w_j, and s_{ij}.
Output: return a value of 1 if optimal assignment was found along with the dest array containing the optimal assignment; 0 otherwise

(i) Initialize: $source(k) \leftarrow 0; 1 \leq k \leq m$
(ii) **FOR** $1 \leq i \leq m$; **DO**
    2.1 $\alpha \leftarrow \infty; dest(i) \leftarrow 0$
    2.2 **FOR** j such that $(i,j) \in E$ and $source(j) = 0$ **DO**
        IF $s_{ij} < \alpha$ **THEN** $\alpha \leftarrow s_{ij}; dest(i) \leftarrow j$
        END_IF
    END_FOR
    2.3 **IF** $(dest(i) = 0)$**THEN** return(0);
    ELSE
        $j \leftarrow dest(i)$
        $source(j) \leftarrow i$
        $\tilde{z}_i \leftarrow z_i + s_{ij}/2$
        $\tilde{w}_j \leftarrow w_j + s_{ij}/2$
    END_IF
    END_FOR
(iii) **IF** $(\tilde{z}, \tilde{w})$ is feasible for the dual **THEN** return(1,dest);
    ELSE return(0)
    END_IF

4. Computational Results

We now describe the computational results of the ADP/A method. ADP/A algorithm always takes as input an LP posed in standard primal form; i.e., min $c^T x$ subject to $Ax = b$, $x \geq 0$. By changing the sign on the objective vector coefficients, we converted the maximum assignment problem returned by the generator to an equivalent minimum assignment problem. Ofcourse the optimal solution of the minimum assignment problem is identical to the optimal solution of the maximum assignment problem. Also, the optimal objective value of the maximum assignment problem is equal to the negative of the optimal objective value of the minimum assignment problem.

Our implementation of ADP/A uses a preconditioned conjugate gradient algorithm to solve the linear system at each iteration. The preconditioned conjugate gradient method has a set of parameters which determine its behaviour. Parameters for CG and the range over which the values were varied for the different problem classes are given below in table 1.

parameter type	range of values
step size parameter (α in ALGORITHM A)	.9 to .99
$\cos\theta$ stopping criterion for CG	10^{-6} to 10^{-8}
Residue stopping criterion for CG	10^{-5} to 10^{-7}
Iteration limit for CG	10 to 100

TABLE 1. Parameters of the ADP/A Algorithm that were Used for Different Problem Classes

To evaluate the performance of ADP/A we compare the results with the latest version of the Forward/Reverse Auction code [4, 6, 5] obtained from Prof. D. Castañon. Originally, we ran a version of auction code obtained from Prof. Bertsekas, but found that this code had integer overflows because of initial scaling of costs and a large cost range. When we communicated this problem to Prof. Castañon during the DIMACS conference, he supplied us with a modified auction code that uses double precision locations to store dual variables. Another feature of this auction code is the *absence of initial cost scaling*. This new auction code is what we have used in our comparisons. This code is written in fortran and was compiled using the f77 compiler, with the optimization flag set to -O2.

The parameters that are hardcoded in the program are as follows (all the parameters are defined as real and N is the number of nodes in the bipartite graph):

- BEGEPS = 1.0 / (N + 1)
- ENDEPS = 1.0 / (N + 1)
- STARTINCR = 1.0 / (N + 1)
- FACTOR = 10.0

All runs were carried out on a *Silicon Graphics Machine*, having a Risc-based architecture with 25MHZ clock. The SILICON GRAPHICS(R) uses a variant of the UNIX[1] OPERATING SYSTEM, called the IRIX(R) Operating System. The ADP/A algorithm is implemented in FORTRAN and C. In our experiments, the code was compiled using the f77 and cc compilers with optimization flag -O2. All times reported are user times given by the system call times(). First, a call to times() is made just before the call to the actual solver. This gives us the user cpu time consumed by the program so far. After the call to the solver returns, another call to times() is made to obtain the user cpu time. The difference between the two times is reported as the cpu time consumed by the solver. Thus, all problem generation and problem read times are excluded from the reported times. This remark applies to both the ADP/A and auction experiments.

To benchmark the cpu speed of the SGI machine, the benchmark experiments from pub/netflow/benchmarks were run on the machine. Since the ADP/A implementation uses both C and F77, both sets of benchmarks were run. Tables 2 and 3 show the results of the benchmark experiments. The mean and standard

	Test 1		Test 2	
	Mean	Std. Dev.	Mean	Std. Dev.
real time	2.0s	.03s	22.9s	2.2s
user time	1.9s	.03s	22.4s	2.0s
sys time	.05s	.00s	0.3s	.01s

TABLE 2. Cpu times for Benchmarks run on the SGI machine– C benchmarks

	Test 1		Test 2	
	Mean	Std. Dev.	Mean	Std. Dev.
real time	1.8s	.06s	3.2s	.35s
user time	1.6s	.13s	2.9s	.42s
sys time	.16s	.00s	.23s	.00s

TABLE 3. Cpu times for Benchmarks run on the SGI machine– f77 benchmarks

deviation of times are the results of 4 experiments varying the optimization level of the compilers. (-O1 to -O4). The fortran benchmarks consist of running the NETFLO algorithm [14] on two minimum cost network flow problems generated by NETGEN [15].

There are five assignment problem classes we have studied. The problem classes differ in the cost range and sparsity of the bipartite graphs. The five problem classes are given below.

---

[1] UNIX is a registered trademarks of AT&T.

(i) Assign-Geometric(AGM):In this class, points are distributed on a plane and divided into two groups; red and blue. Edges connect red and blue points with edge costs equal to the Euclidean distance between the points.

(ii) Assign-Fixed-cost(AFX): Problems in this class have a fixed degree of $N/16$ and the edge cost between nodes i and j is set to $100 * i * j$.

(iii) Assign-High-Cost(AHC): Problems in this class have a fixed degree of $2\ log_2 N$ and the edge costs are uniformly distributed between $[0,10^8]$.

(iv) Assign-Low-Cost(ALC): Problems in this class are identical to the problems in AHC, except for the costs being uniformly distributed between $[0,100]$.

(v) Completely dense graphs(CDG): These problems are completely dense with the costs uniformly distributed between $[0,1000]$ or $[0\text{-}10000]$.

The first four problem classes are from the DIMACS problem set. For the problems from these four classes, Appendix A gives the seeds we used to generate the problems along with the optimal objective function value for the problem instances. The fifth problem class (CDG) is generated by our generator. This problem class is similar to the Assign-Xhi-degree problem set in DIMACS. Thus we did not run the Assign-Xhi-degree problem class. Also, Assign-two-cost problem class from DIMACS proved to be too trivial for both ADP/A and auction algorithms. Therefore we do not report on this problem class in this paper.

Now we will describe the computational results. Table 4 summarizes the results for the AGM problem class. The first column indicates the problem size

Nodes	ADP/A		AUCTION		SPEEDUP AUCTION/ADP/A
	Mean	Std. Dev.	Mean	Std. Dev.	
128	12s	.5s	7m24s	4m01s	37
256	47s	1s	1h11m05s	7m40s	78
512	3m34s	42s	39h40m51s	-	667
1024	21m56s	4m48s	**	**	**

** Auction was not run because of projected cpu time requirements

TABLE 4. Performance comparison with Auction on Geometric Cost Assignment Problems(AGM)

in terms of the number of nodes in *each partition*. The mean and standard deviation of cpu times are shown in the table, along with the speedup. The mean and standard deviation are computed using a sample size of three runs with different seeds for the random numbers (See Appendix A). The speedup is the ratio of the means of auction running time to ADP/A running time. For this class of problems ADP/A is seen to be one to two orders of magnitude faster than the version of the auction code obtained from Prof. Castañon.

Table 5 summarizes the results for the AFX class of problems. Problems in

Nodes	ADP/A		AUCTION		SPEEDUP AUCTION ADP/A
	Mean	Std. Dev.	Mean	Std. Dev.	
128	3s	.3s	.2s	.1s	.1
256	11s	.1s	4s	1s	.4
512	46s	1s	1m49s	1m03s	2
1024	4m36s	9s	1h31m50s	1h38m54s	20

TABLE 5. Performance comparison with Auction on Fixed Cost Assignment Problems(AFX)

this problem class have a degree of $N/16$, where N is the number of nodes in the graph. Thus these problems are moderately dense. ADP/A is again seen to perform one order of magnitude better than auction. Interestingly, ADP/A starts performing better than auction after 512 nodes, whereas for AGM class the ADP/A performs better even on smaller problems. One explanation is that problems in AGM class are completely dense. The interior point code is able to quickly isolate interesting edges from this dense problem. Since AFX class is only moderately dense, this nice property of the interior point method takes effect on bigger problems. This phenomenon also happens in CDG class, as discussed below.

Now we describe the CDG class of problems. As mentioned before, this is our own generator for dense assignment problems. To be able to solve very large dense problems, we need to perform some *sifting* during the generation phase. We give as input to the random instance generator the number of source and destination nodes m, the range of the costs (L, U) and an additional parameter called degree d which we describe below. Though we generate dense assignment problems, we drop many unviable edges, in the generation phase. The degree parameter is used to specify how many edges per node have to be retained on an average. Good estimates for degree, such that the resulting reduced graph has a high probability of containing the optimal assignment of the original dense graph, can be derived from the theory of random graphs [7]. The sifting gives us a reduced linear programming problem which can fit into the virtual memory limits of our machine. However, on obtaining the optimal solution, we regenerate all the edges and verify the optimality of the solution for the entire dense assignment problem. A similar approach of dropping low cost edges and later verifying optimality of solution for the entire dense problem was adopted for the auction algorithm. For both algorithms, the generation time is excluded from the running times reported.

Table 6 summarizes the results obtained in these experiments.

Nodes	Cost Distributions					
	0 - 1000			0-10000		
	ADP/A	AUCTION	SPEEDUP AUCTION ADP/A	ADP/A	AUCTION	SPEEDUP AUCTION ADP/A
1000	19s	7s	0.36	21s	8s	0.38
2000	1m14s	26s	0.35	1m30s	29s	0.32
3000	2m26s	2m16s	0.93	3m07s	2m42s	0.86
5000	5m17s	8m17s	1.56	6m05s	7m54s	1.30
7500	12m06s	20m15s	1.67	12m13s	1h39m20s	8.13
10000	18m36s	49m08s	2.64	19m13s	5h07m57s	16.02
20000	1h27m37s	1h58m03s	1.34	1h25m34s	8h38m23s	6.05
30000	3h17m02s	12h04m08s	3.67	3h19m10s	140h50m49s	42.43

TABLE 6. Performance comparison with Auction on Dense Assignment Problems

We report here the timings for two different cost distributions - one in which the costs vary from 0 to 1000 and the second in which the cost distribution is from 0-10000. It should be noted that for both the algorithms, the times include the time spent in verifying the optimality of the solution for the entire dense assignment problem.

The results for the CDG problem class seem to indicate the following.
- ADP/A is slower than auction on smaller instances. This is again the same phenomenon we observed in AFX problem class. For moderately dense problems, auction is faster on small size problems. However for larger instances ADP/A seems to be faster than auction with a speedup of more than 40 times in one instance of the thirty thousand node assignment problem. Observe that even though we are generating completely dense problems, we sift out many edges resulting in a very sparse problem. In fact the average degree for all the problems upto 5000 nodes was less than 30. The key word here is the *average* degree. There were some minor variations between degrees of various nodes. This is different from the *fixed degree* sparse graphs of the AHC and ALC problem classes discussed below. In these latter problem classes, every source node has the same *fixed* degree. Thus the graphs are somewhat regular. As discussed below, auction performs much better than ADP/A on such graphs.
- The change in ADP/A timings with increase in cost range is almost negligible.

Tables 7 and 8 summarize the computational results for the AHC and ALC problem classes.. These two problem classes consist of problems with a fixed degree of $2\log_2 N$. Thus these graphs are somewhat regular and sparse. For

Nodes	ADP/A		AUCTION		SPEEDUP ADP/A / AUCTION
	Mean	Std. Dev.	Mean	Std. Dev.	
1024	17s	.6s	.4s	.1s	43
2048	36s	2s	2s	.4s	18
4096	2m12s	4s	8s	4s	17
8192	3m22s	19s	10s	4s	20
16384	9m05s	29s	32s	10s	17
32768	24m23s	2m19s	51s	4s	29

TABLE 7. Performance comparison with Auction on Assign-High-Cost(AHC) Assignment Problems

Nodes	ADP/A		AUCTION		SPEEDUP ADP/A / AUCTION
	Mean	Std. Dev.	Mean	Std. Dev.	
1024	15s	2s	.3s	.1s	50
2048	39s	1s	1s	.2s	39
4096	1m58s	2s	3s	.2s	39
8192	5m01s	9s	8s	.5s	38
16384	13m23s	38s	40s	17s	20
32768	41m04s	2m19s	2m25s	1m26s	17

TABLE 8. Performance comparison with Auction on Assign-Low-Cost(ALC) Assignment Problems

these classes of problems auction does extremely well. It performs an order of magnitude faster than ADP/A. It appears that the overhead involved in isolating important dual slacks in ADP/A is too excessive and cannot be justified for sparse problems since the number of edges being eliminated is too small.

It is somewhat puzzling why the version of auction code we used takes excessive cpu time for AGM, AFX, and CDG problem classes. There are several possible explanations. It was pointed out to us by A. Goldberg [8] that the auction code we ran implements the unscaled forward reverse auction algorithm. A. Goldberg reports that scaled auction algorithm gives much better results. However, there is still some inconsistency between the results we obtain and the results reported by Castañon for the unscaled forward reverse auction algorithm. This inconsistency can be partially attributed to the fact that Castañon's results are for minimization problem, as opposed to the maximization problem. Further study is required to compare ADP/A with the scaled forward/reverse auction algorithm on assignment problems involving maximization of the objective function.

5. Discussion

The computational results discussed above demonstrate some interesting behavior of the ADP/A algorithm. Firstly, we notice that the coefficient of variation of cpu times (standard deviation / mean) for ADP/A is small. Among all the results presented above, the largest coefficient of variation is .2. This indicates that ADP/A has little sensitivity to the problem instances.

Secondly, the cpu time of ADP/A grows almost linearly with the number of edges in the graph for each of the five problem classes we have studied. For instance, for the Assign-Fixed-Cost problem class (Table 5) the mean cpu time increases by a factor of 4 as you double the number of nodes; i.e., as you increase the number of edges by a factor of 4. Moreover, cpu time taken by ADP/A for a problem instance is roughly proportional to the number of edges regardless of the problem structure. For example, consider the 1024 node problems in different problem classes. The cpu time for Assign-Geometric is 21m56s (Table 4). The number of edges in this problem is about 1 million. In Assign-Fixed-Cost the cpu time is 4m36s (Table 5) and the number of edges is about 1/8 of a million. Also, the cpu time for Assign-High-Cost is 17s (Table 7) and the number of edges is about 20,000. The ratios of cpu times correspond roughly to the ratios of number of edges. This demonstrates uniformity in the behaviour of the ADP/A algorithm over the problem classes studied here. Not only does it show little dependence on the range of costs but also little dependence on the problem structure.

Thirdly, another point we notice is that ADP/A is able to solve very large problems without degradation in performance . Thus we expect ADP/A to asymptotically outperform other algorithms; e.g., for the Assign-Geometric problem class, the speedup of ADP/A increases with problem size (Table 4) and for the Assign-Low-Cost problem class the speedup of auction over ADP/A steadily decreases with increase in problem size (Table 8).

6. Conclusions

We have applied a general purpose approximate dual projective variant of the interior point method to solve large assignment problems. We have made minor adaptations in the code that help early identification of the optimal assignment from the dual iterate.

This general purpose approach for solving assignment problems has some benefits:
- ADP/A method can be directly applied to an assignment problem with side constraints. We simply add these additional constraints to the LP problem and invoke ADP/A. No special algorithmic modification of ADP/A is required.
- ADP/A method can be directly applied to assignment problems with real (nonintegral) costs since ADP/A makes no assumptions about costs being integers.
- The ADP/A method can be easily extended to assignment problems with convex costs. This is because of the general purpose nature of interior point method.

This study has illustrated how one can exploit the special nature of an LP problem without sacrificing the generality of the algorithm. We are currently conducting research on applying the Approximate Dual projective method to solve network flow and multicommodity network flow problems.

7. Appendix A: Problem seeds and Optimal Objective Values

This appendix lists the problem instances we have generated for our experiments. The four problem classes discussed below are from the DIMACS problem set and are in the public domain. For each problem, we have given the random seed used and the optimal objective function value obtained.

7.1. Assign-Geometric Class of Problems.
These problems are generated using the dcube.c generator in the DIMACS library. The parameters used to generate the problems are given below.

Nodes (N)
dimension 2
maxloc 1000000
seed (X)

The various sizes of problems generated, the corresponding seeds and the optimal objective function values are listed in table 9.

Nodes (N/2)	seed (X)	Optimal Objective Value
128	270001	98590148
128	270002	95517946
128	270003	93248933
256	270001	198237217
256	270002	192965978
256	270003	194844097
512	270001	391309422
512	270002	391237799
512	270003	391148927
1024	270001	782311551
1024	270002	786202387
1024	270003	779219446

TABLE 9. Seeds and Optimal Objective Values for Assign-Geometric Problems

Nodes (N/2)	seed (X)	Optimal Objective Value
128	270001	175997100
128	270002	175913300
128	270003	175884000
256	270001	1403859100
256	270002	1403897200
256	270003	1403886100
512	270001	11209411800
512	270002	11209513900
512	270003	11209025600
1024	270001	89580255700
1024	270002	89580094400
1024	270003	89579848400
2048	270001	716240798500
2048	270002	716240324300
2048	270003	716240995800

TABLE 10. Seeds and Optimal Objective Values for Assign-Fixed-Cost Problems

7.2. Assign-Fixed-Cost Class of Problems. This class of problems is generated by assign.c with the parameters given below.
Nodes (N)
sources (N/2)
degree (N/16)
maxcost 100
multiple
seed (X)
Table 10 lists the seeds and the optimal objective values.

7.3. Assign-High-Cost Class of Problems. These problems are generated using assign.c generator, with the following parameters.
nodes (N)
sources (N/2)
degree ($2\log_2(N)$)
maxcost 100000000
Table 11 depicts the seeds and optimal values.

7.4. Assign-Low-Cost Class of Problems. These problems are generated using assign.c generator, with the following parameters.
nodes (N)
sources (N/2)
degree ($2\log_2(N)$)
maxcost 100
Table 12 depicts the seeds and optimal values.

Nodes (N/2)	seed (X)	Optimal Objective Value
1024	270001	94804185690
1024	270002	95326326096
1024	270003	94809125860
2048	270001	190763064590
2048	270002	191079950281
2048	270003	190967830450
4096	270001	384168581871
4096	270002	384756241966
4096	270003	384331263783
8192	270001	772414482375
8192	270002	773314580225
8192	270003	771886833795
16384	270001	1550433745445
16384	270002	1551128099757
16384	270003	1550586814821
32768	270001	3113261955763
32768	270002	3112386719653
32768	270003	3112811065694

TABLE 11. Seeds and Optimal Objective Values for Assign-High-Cost Problems

Nodes (N/2)	seed (X)	Optimal Objective Value
1024	270001	95307
1024	270002	95847
1024	270003	95302
2048	270001	191793
2048	270002	192104
2048	270003	191955
4096	270001	386201
4096	270002	386766
4096	270003	386380
8192	270001	776507
8192	270002	777368
8192	270003	775955
16384	270001	1558512
16384	270002	1559224
16384	270003	1558666
32768	270001	3129458
32768	270002	3128530
32768	270003	3128957

TABLE 12. Seeds and Optimal Objective Values for Assign-Low-Cost Problems

References

1. Ilan Adler, Narendra K. Karmarkar, Mauricio G.C. Resende, and Geraldo Veiga, *An Implementation of Karmarkar Algorithm for Linear Programming*, Mathematical Programming **44** (1989), 297–335.
2. E.R. Barnes, S. Chopra, and D.L. Jensen, *A Polynomial Time Version of the Affine Scaling Algorithm*, Tech. Report 88-101, Graduate School of Business Administration, New York University, Washington Square, N.Y. 1988., 1988.
3. D.A. Bayer and J.C. Lagarias, *The Nonlinear Geometry of Linear Programming I, Affine and Projective Scaling Trajectories*, Transactions of American Mathematical Society **314** (1989), 499–526.
4. D. P. Bertsekas, *The Auction Algorithm for Assignment and Other Network Flow Problems*, Interfaces **20** (1990), no. 4, 133–149.
5. D.P. Bertsekas, *Linear Network Optimization: Algorithms and Codes*, MIT Press, Cambridge, MA, 1991.
6. D. P. Bertsekas and D.A. Castañon, *The Auction Algorithm for Transportation Problems*, Annals of Operations Research **20** (1989), 67–96.
7. Béla Bollobás, *Random Graphs*, Academic Press, London, 1985.
8. A. Godberg and J.R. Kennedy, *Private Communication*, 1993.
9. C.C. Gonzaga, *Polynomial Affine Algorithms for Linear Programming*, Mathematical Programming **49** (1990), 7–21.
10. A. Hipolito, *A Long-step Linear Programming Algorithm Based on Quasi-Newton Inverse Barrier Centering*, Tech. Report 91-12, Dept. of Industrial and Systems Engineering, University of Florida, Gainesville, FL 32611, October 1991.
11. N. Karmarkar, *A new polynomial-time algorithm for linear programming*, Combinatorica **4** (1984), 373–395.
12. N. K. Karmarkar and K.G. Ramakrishnan, *Implementation and Computational Results of the Karmarkar Algorithm for Linear Programming, Using an Iterative Method for Computing Projections*, Tech. Report 11211-891011-10TM, AT&T Bell Laboratories, Murray Hill, NJ, October 1989.
13. _____, *Computational Results of an Interior Point Algorithm for Large Scale Linear Programming*, Mathematical Programming **52** (1991), 555–586.
14. J.L. Kennington and R.V. Helgason, *Algorithms for Network Programming*, Wiley-Interscience Series, New York, NY,, 1980.
15. D. Klingman, A. Napier, and J. Stutz, *NETGEN: A Program for Generating Large Scale (Un) Capacitated Assignment, Transportation, and Minimum Cost Flow Network Problems*, Management Science **20** (1974), 814–822.
16. I.J. Lustig, R.E. Marsten, and D.F. Shanno, *Computational experience with a primal-dual interior point method for linear programming*, Tech. report, Department of Civil Engineering and Operations Research, Princeton University, Princeton, NJ, 1989, Tech report no SOR89-17.
17. C.H. Papadimitriou and K. Steiglitz, *Combinatorial Optimization*, Prentice Hall, Inc., Englewood Cliffs, New Jersey, 07632, 1982.
18. M.J. Todd, *A Low Complexity Interior Point Algorithm for Linear Programming*, SIAM Journal of Optimization **2** (1992), 198–209.

K.G. RAMAKRISHNAN, MATHEMATICAL SCIENCES RESEARCH CENTER, AT&T BELL LABORATORIES, MURRAY HILL, NJ 07974 USA, E-MAIL : KGR@RESEARCH.ATT.COM

N.K. KARMARKAR, MATHEMATICAL SCIENCES RESEARCH CENTER, AT&T BELL LABORATORIES, MURRAY HILL, NJ 07974 USA, E-MAIL: KARMAR@RESEARCH.ATT.COM

A.P. KAMATH, MATHEMATICAL SCIENCES RESEARCH CENTER, AT&T BELL LABORATORIES, MURRAY HILL, NJ 07974 USA AND, COMPUTER SCIENCE DEPT., STANFORD UNIVERSITY, STANFORD, CA 94305, E-MAIL: KAMATH@CS.STANFORD.EDU

An Implementation of a Shortest Augmenting Path Algorithm for the Assignment Problem

JIANXIU HAO and GEORGE KOCUR

August 1991
Revised September 1992

Abstract

We consider the successive shortest augmenting path approach for solving the assignment problem. We propose an algorithm, in which we develop a new initialization procedure and a technique to compute the shortest augmenting paths by using a small subset of the arcs. Computational results show that our algorithm solves dense assignment problems efficiently.

1 Introduction

The *assignment problem* (AP) is defined on a bipartite graph $G(N_1, N_2, A)$. N_1 and N_2 are sets of *right* and *left* nodes, also known as the *row* and *column* nodes. A is set of arcs which connect left and right nodes. Associated with each arc $(u, v) \in A$ there is an integer-valued cost $c_{u,v}$. The objective is to assign each left node to a right node such that the total assignment cost is minimized.

Applications for the AP arise not only in real-world assignment problems, but also as subproblems in algorithms for other combinatorial optimization problems

---

[0]1991 Mathematics Subject Classification. Primary 05C38, 90B10
[0]The authors wish to thank the referees for several helpful comments.

such as the quadratic assignment problem, the traveling salesman problem, crew scheduling and vehicle routing problems (See Ahuja, Magnanti and Orlin [1]).

The literature devoted to the assignment problem is very rich. Kuhn [17] developed the first algorithm for this problem, which is widely known as the *Hungarian algorithm*. The subsequent algorithms for the AP can be roughly classified into the following three basic approaches: the *primal-dual approach*, which includes Kuhn [17], Balinski and Gomory [4], Engquist [10], Bertsekas [8] and Glover et al. [12]; the *dual simplex approach*, which includes Weintraub and Barahona [20], Hung and Rom [15], Balinski [5], Goldfarb [13] and Akgul [3]; and the *primal simplex approach*, which includes Barr et al. [6], Roohy-Laleh [19], Hung [14], Akgul [2] and Orlin [18].

Ahuja, Magnanti and Orlin [1] have noted that, in general, the algorithms based on the primal-dual and dual simplex approaches have better running times in the worst case, while implementations based on the primal-dual, successive shortest augmenting path (SSAP) algorithms tend to be the most efficient approach in average case. The SSAP implementations of Glover et al. [12] and Jonker and Volgenant [16] seem to be the fastest to date. Glover et al. use a threshold shortest path algorithm while Jonker and Volgenant use a label-setting algorithm.

The SSAP solves the assignment problem in stages. Each stage is a *shortest path augmentation* which consists of two fundamental steps. The first step is to find *shortest augmenting paths* from a subset of left nodes to a subset of right nodes. Here the subset of nodes are *unmatched* or *free* nodes. The second step, often called the *flow augmentation* step, consists of matching and unmatching arcs on the shortest augmenting path(s) found in the first step. Here *an augmenting path* from a node u to another node v in the network is a simple path such that forward arcs on this path are *free* and backward arcs are *matched*. An augmenting cycle is an augmenting path such that the starting node is the same as the end node. The cost of an augmenting path (cycle) is the sum of costs of forward arcs minus the sum of costs of backwards arcs on the path (cycle). If the cost of an augmenting cycle is negative, then we call it *a negative augmenting cycle*. If the cost of an augmenting path from a node u to another node v is the minimum among all the augmenting paths from u to v, then we call it the shortest augmenting path from u to v. Note that if the direction of all backward arcs on an augmenting path is reversed, then it is a simple directed path. Hence, finding the shortest augmenting paths is equivalent to finding the shortest paths.

One of the main reasons that the SSAP algorithm for the AP is so successful in practice is effective initialization. Jonker and Volgenant [16] showed that their initialization heuristics match about 90% of nodes in a full dense network. Shortest path augmentations are only used to match the remaining 10% of the nodes.

In this study, we implement a successive shortest augmenting path algorithm. We propose a new initialization procedure. We also propose a technique to compute the shortest augmenting paths using a very small subset of the arcs, in which the shortest augmenting path of the whole network is guaranteed to be contained. When our algorithm is applied to the dense $n \times n$ assignment problem, our initialization procedure, which is simpler than others, also matches a total of about 90% of the nodes in the network. Our shortest augmenting path procedure finds the shortest augmenting paths by examining only $n \log n$ out of a total of n^2 arcs on the average. The algorithm solves the $n \times n$ dense assignment problems in a empirical running time proportional to something between n^2 and $n^2 \log n$. However, for sparse networks, our implementation is not as efficient as Jonker and Volgenant's. One of the main reasons is that our initialization is, though still efficient, not as effective as theirs. We note that an efficient algorithm that solves the dense assignment problem is very important for solving large real TSP problems using the assignment relaxation method.

2   The Initialization Phase

We present a different initialization method which consists of two steps. In step 1 *row* and *column reductions* are performed. The costs of all arcs connected to a left node or a right node are decreased (increased) by the same amount so all costs are greater than or equal to zero, and so there is at least one arc with cost exactly zero connecting each left node and each right node. Then a bipartite matching is found in the zero cost arcs. We use a bipartite matching algorithm developed by Chang and McCormick [16], which is very efficient. The overall time spent on solving the bipartite matching problems is negligible. Step 1 consists essentially of shortest path augmentations, however, all paths consist of one arc and have zero cost. Experimental results show that this matches more than 80% of the nodes in a dense network. Step 1 of the initialization phase is as follows:

Procedure Step1

1. For each left node $u \in N_1$: let $\delta_u = \min\{c_{u,v}|(u,v) \in A\}$; set $c_{u,v} \leftarrow c_{u,v} - \delta_u$, $\forall (u,v) \in A$.

2. For each right node $v \in N_2$: let $\delta_v = \min\{c_{u,v}|(u,v) \in A\}$; set $c_{u,v} \leftarrow c_{u,v} - \delta_v$, $\forall (u,v) \in A$.

3. Find a maximum matching in $(N_1, N_2, \bar{A})$, where $\bar{A} = \{(u,v)|c_{u,v} = 0$ and $(u,v) \in A\}$.

Step 2 of the initialization phase is to make more zero cost paths from free left nodes to free right nodes by adjusting the costs of arcs. Then the matching algorithm is used to make more assignments.

The left and right nodes are partitioned into four groups: *Free* nodes (F), *Zero* nodes (Z), *Matched-Zero* nodes (MZ) and *Remaining* nodes (R). A node is in F if it is not assigned to any node. A node is in Z if it is connected to at least one node in F by a zero cost arc. A node is in MZ if it is matched to one of the nodes in Z. R contains all the remaining nodes. Note that these four sets of nodes are mutually exclusive.

This node partition has several interesting properties:

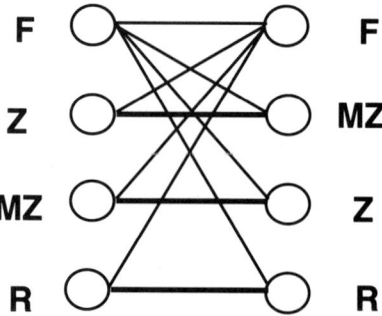

Figure 1: Node partitions

1. The number of nodes in left F equals that in the right F and the number of nodes in left R equals that in the right R

2. The number of nodes in left Z equals that in the right MZ and the number of nodes in right Z equals that in the left MZ

3. All zero cost arcs out of left F connect to right Z and all zero cost arcs out of right F connect to left Z

4. No zero cost arcs connect left MZ to right MZ

In step 2 of the initialization, operations on MZ and F are used to further reduce costs. For each node i in left MZ, the maximum amount is subtracted from the costs of arcs connecting to i such that the new costs are greater than or equal to zero, except for the matched arc. Then the cost of all arcs connecting to the node's matched node in Z is increased by the same amount such that the new cost of the matched arc is still zero.

A similar operation is then applied to each node in right MZ.

Finally, for each node i in right F and left F, the maximum amount is subtracted from the cost of arcs connecting to i so the new costs of arcs are greater than or equal to zero. This ensures that each node in F will still have at least one zero cost arc connecting to it.

The maximum bipartite matching algorithm is then used to find more matches in the new subnetwork of zero cost arcs.

The heuristic in step 2 will probably produce some zero cost augmenting paths with length either one or three. Augmenting paths with length one will consist of new zero cost arcs connecting nodes in left F and nodes in right F.

Augmenting paths with length three will consist of new zero cost paths which indirectly connect nodes in left F with nodes in right F. These includes paths from: F-MZ-Z-F, F-Z-MZ-F and F-R-R-F. This step may also produce some zero cost augmenting paths with length greater than 3 and it is also possible that this step may not produce any new zero cost augmenting paths.

Procedure Step2

1. For each node u in the left MZ: let v be the right node to which u is matched and $\delta_u = \min\{c_{u,j}|(u,j) \in A, j \neq v\}$, set $c_{u,j} \leftarrow c_{u,j} - \delta_u$, $\forall (u,j) \in A$ and $c_{i,v} \leftarrow c_{i,v} + \delta_u$, $\forall (i,v) \in A$.

2. For each node v in the right MZ: let u be the left node to which v is matched and $\delta_v = \min\{c_{i,v}|(i,v) \in A, i \neq u\}$, set $c_{i,v} \leftarrow c_{i,v} - \delta_v$, $\forall (i,v) \in A$ and $c_{u,j} \leftarrow c_{u,j} + \delta_v$, $\forall (u,j) \in A$.

3. For each node u in the left F: let $\delta_u = \min\{c_{u,v}|(u,v) \in A\}$, set $c_{u,v} \leftarrow c_{u,v} - \delta_u$, $\forall (u,v) \in A$.

4. For each node v in the right F: let $\delta_v = \min\{c_{u,v}|(u,v) \in A\}$, set $c_{u,v} \leftarrow c_{u,v} - \delta_v$, $\forall (u,v) \in A$.

5. Find a maximum matching in $(N_1, N_2, \bar{A})$, where $\bar{A} = \{(u,v)|c_{u,v} = 0$ and $(u,v) \in A\}$.

Our computational experiments indicate the heuristic used in step 2 will match about another 10% of the nodes in a random dense bipartite graph, resulting in a total of 90% of the nodes being matched during the initialization stage. The following table shows the computational results on random dense assignment problems with large cost range(1-100,000).

Problem	DHK		DJV	
	After-Step1	After-Step2	After-Step1	After-Step2
128x128	25	15	47	1
256x256	49	29	93	3
512x512	98	58	190	7
1024x1024	197	117	374	18
2048x2048	389	233	755	47

Table 1. Free Node-pairs for Assign-Xhi-Degree Network

Here DHK (SHK) is our shortest augmenting path algorithm for dense (sparse) network and DJV (SJV) is the Jonker and Volgenant's algorithm for dense (sparse) network. For more detailed definitions on the algorithms and problems, see section 4.

On random dense assignment problems with small cost range, our initialization procedures perform even better; the following table is for the random dense problem instances with cost range (1-100).

Problem	DHK		DJV	
	After-Step1	After-Step2	After-Step1	After-Step2
128x128	16	9	51	10
256x256	11	7	115	27
512x512	1	0	295	96
1024x1024	0	0	731	382
2048x2048	0	0	1691	1215

Table 2. Free Node-pairs for Assign-Xhi-Degree-Lo-Cost Network

We note that the initialization procedures of Jonker and Volgenant are performing better on the random instances with large cost range than those with small cost range. This is consistent with what is reported in [16]. We also note that their procedures achieve more matches than ours. But, for dense problems with small cost range, ours are better than theirs.

On random sparse assignment problems, for both small cost range and large cost range, our initialization procedures also achieve about a total of 90% of the matches.

Problem Nodes(Arcs)	SHK After-Step1	SHK After-Step2	SJV After-Step1	SJV After-Step2
512(4608)	45	27	90	10
1024(10240)	91	54	185	25
2048(22528)	180	106	371	48
4096(49152)	361	213	737	102
8192(106496)	720	421	1,485	208
16384(229376)	1,424	844	2,970	429

Table 3. Free Node-pairs for Assign-Lo-Cost Network

Problem Nodes(Arcs)	SHK After-Step1	SHK After-Step2	SJV After-Step1	SJV After-Step2
512(4608)	50	30	90	0
1024(10240)	96	58	184	0
2048(22528)	195	116	367	0
4096(49152)	391	232	733	0
8192(106496)	783	463	1481	1
16384(229376)	1567	922	2957	2

Table 4. Free Node-pairs for Assign-High-Cost Network

However, the performance of our initialization procedures are not as good on the geometric-matching problems. The geometric-matching problem instances are converted from the 2-dimensional random points matching problem instances. Half the points are red and half blue, arcs connect red points to blue, and arc costs correspond to Euclidean distances. About 80% of the nodes in those problem instances get matched by the procedures.

Problem	DHK After-Step1	DHK After-Step2	DJV After-Step1	DJV After-Step2
128x128	36	26	53	11
256x256	71	51	107	22
512x512	141	101	214	47
1024x1024	282	194	424	96

Table 5. Free Node-pairs for Assign-Geometric Network

Two-cost random assignment problem is an sparse assignment problem with two costs either 100 or 100,000,000. See Section 4, for more details on how the problem instances are generated. For the two-cost random assignment problem instances, our initialization procedures solve the problem completely. In fact, procedure Step 1 of the initialization is enough.

Problem	SHK		SJV	
Nodes(Arcs)	After-Step1	After-Step2	After-Step1	After-Step2
512(4608)	0	0	179	44
1024(10240)	0	0	360	93
2048(22528)	0	0	740	186
4096(49152)	0	0	1514	375
8192(106496)	0	0	3089	758
16384(229376)	0	0	6274	1523

Table 6. Free Node-pairs for Assign-Two-Cost Network

The following proposition shows that Procedures Step1 and Step2 are valid.

Proposition 1. If there is a feasible solution to the AP, then the optimal assignment will not change if the costs of all arcs connecting to a left(right) node are changed by the same amount. ∎

3 Shortest Path Augmentation Phase

To solve the assignment problem by the SSAP algorithm, we introduce an artificial node s. Artificial arcs of cost zero are added connecting s to those nodes in left F which are unmatched after the initialization phase.

The SSAP algorithm for the assignment problem is a primal-dual approach in which dual feasibility is always maintained so there is no negative augmenting cycle in the solution. This means that the shortest augmenting paths could be found in a polynomial time. We use a label-correcting shortest path algorithm since it is more efficient than the label-setting algorithm for sparse networks [1]. As we have shown from our computational results on random dense networks in the last section, when the input network is dense, a sparse subnetwork could be identified for the shortest path calculations. The algorithm used in this implementation is essentially a Bellman-Ford's algorithm [7] [11] using a list array to store the candidate nodes whose out-arcs are eligible to be scanned in the next stage. This would reduce the unnecessary scan of nodes and their out-arcs if their labels are not reduced since the last scan.

Since the costs of arcs can be negative in the label-correcting shortest path algorithm, the reduced costs of arcs need not be updated after each shortest path augmentation. Instead we use the final costs from our initialization phase in the shortest path calculations. This is another advantage of the label-correcting algorithm.

To simplify proofs as well as the algorithm, we define a residual network where the directions of all matched arcs are reversed and the costs of these arcs are equal to the negative of their original values. We will also remove artificial arcs connecting to node s where the left node adjacent to the arc is no longer a free left node. The directions and costs of unmatched arcs remain unchanged. In the residual network, an augmenting path (augmenting cycle) in the original network corresponds to a directed path (directed cycle) in the residual network.

For a given dual feasible flow, shortest paths from s to all other nodes in the network are found using the shortest path algorithm. A free right node is chosen with the smallest shortest path label among all free right nodes. This shortest path (not counting the artificial arc) is called an *augmentation path*.

The augmentation path contains either one arc connecting a free left node to the free right node, or contains an alternating sequence of matched and unmatched arcs starting from a free left node and ending with a free right node. An augmentation is performed along the augmentation path by pushing one unit of flow from the free left node to the free right node along this augmentation path. Equivalently, all the unmatched arcs are changed into matched arcs and all the matched arcs are changed into unmatched arcs along the augmentation path.

We must now show that the new flow is still dual feasible after the shortest path augmentation.

Lemma 1: A shortest path augmentation on a dual feasible flow produces another dual feasible flow.

Proof. A flow is dual feasible if and only if it contains no negative augmenting cycle with respect to it. Hence showing that the new flow is dual feasible is equivalent to showing that there is no negative augmenting cycle after the shortest path augmentation.

Since we reverse the direction of an arc if it is matched, an augmenting cycle is a directed cycle. Then to show there is no negative augmenting cycle, we can show that there are no negative directed cycles.

Let π_u be the shortest path label of node u before the augmentation and $\bar{c}_{u,v} = \pi_u + c_{u,v} - \pi_v \geq 0$ (the reduced cost of arc (u,v) before the augmentation) for all arcs $(u,v) \in A$.

We know $\bar{c}_{u,v} = 0$ for all arcs (u,v) on the augmentation path. After the augmentation, the directions of arcs on the augmentation path are reversed while the directions of arcs not on the augmentation path remain unchanged. Since there were no directed negative cycles before augmentation, and since $\bar{c}_{u,v} \geq 0$ for all arcs $(u,v) \in A$, there is no negative directed cycle with respect to $\bar{c}$ after the augmentation. ∎

Before continuing to the shortest path augmentation procedure that produces an optimal assignment, we prove another lemma:

Lemma 2: Let π^i and π^{i+1} be the shortest path labels of stage i and stage $i+1$, then $\pi_t^{i+1} \geq \pi_t^i$ for all nodes t in the network.

Proof. There are two cases:

1. The shortest path from s to t in stage $i+1$ contains no arcs which are in the augmentation path of stage i.

2. The shortest path from s to t in stage $i+1$ contains some arcs which are in the augmentation path of stage i.

In case 1, $\pi_t^{i+1} \geq \pi_t^i$ since the shortest augmenting path in stage $i+1$ is also a valid augmenting path in stage i.

In case 2, let $\ell^i(u,v)$ and $\ell^{i+1}(u,v)$ denote the shortest path cost from node u to node v in stages i and $i+1$, respectively. Clearly $\ell^i(s,t) = \pi_t^i$. Let the shortest augmenting path from s to t in stage $i+1$ be $s, ..., v_1, u_1, ..., v_2, u_2, ..., v_p, u_p, ..., t$, where (u_k, v_k), $k = 1, 2, ...p$ is an arc on the augmentation path in stage i. We first show that $\ell^i(s, u_p) \leq \ell^{i+1}(s, u_p)$ by induction.

Since arc (u_1, v_1) is on the shortest augmentation path in stage i, we have $\ell^i(s, u_1) \leq \ell^i(s, v_1) - c_{u_1, v_1} = \ell^{i+1}(s, u_1)$. Suppose $\ell^i(s, u_{p-1}) \leq \ell^{i+1}(s, u_{p-1})$. Then:
$$\begin{aligned}\ell^i(s, u_p) &\leq \ell^i(s, v_p) - c_{u_p, v_p} \\ &\leq \ell^i(s, u_{p-1}) + \ell^i(u_{p-1}, v_p) - c_{u_p, v_p} \\ &\leq \ell^{i+1}(s, u_{p-1}) + \ell^{i+1}(u_{p-1}, v_p) - c_{u_p, v_p} \\ &= \ell^{i+1}(s, u_p)\end{aligned}$$
The first inequality follows from the fact that arc (u_p, v_p) is on the shortest augmentation path at stage i, the second inequality follows from the triangle inequality for shortest path labels, and the last inequality follows from our induction assumption. Hence $\pi_t^i = \ell^i(s, t) \leq \ell^i(s, u_p) + \ell^i(u_p, t) \leq \ell^{i+1}(s, u_p) + \ell^{i+1}(u_p, t) = \ell^{i+1}(s, t) = \pi_t^{i+1}$. ∎

Procedure Short-Path-Augmentation (SPA)

Input: A partial matching such that all matched arcs' costs are equal to zero and unmatched arcs' costs are greater than or equal to zero.

Output: An optimal assignment.

1. *Define threshold costs thresh[1], thresh[2], thresh[3] and thresh[4], where thresh[4] is set to infinity. Construct an initial subnetwork $\bar{G} = (N_1, N_2, \bar{A})$ of good arcs with $\bar{A} = \{(u, v) \in A | \pi_u^o + c_{u,v} \leq thresh[1]\}$, where π^o is the set of initial shortest path labels. Add an artificial node s and arcs from s to the free left nodes to $\bar{G}$ and set the costs of the artificial arcs to zero. Reverse the directions of matched arcs. Set thresh_label[v] $\leftarrow$ thresh[1] for all nodes $v \in N_2$. Set stage $i = 0$;*

2. *Find shortest paths from s to all nodes in the network $(N_1, N_2, \bar{A})$ and let π^i be the shortest path labels. If $\pi_v^i \leq$ thresh_label[v] for all $v \in N_2$ go to (4). Otherwise go to (3).*

3. *For each right node v such that $\pi_v^i >$ thresh_label[v], set thresh_label[v] $\leftarrow$ thresh[k] such that thresh[k] $> \pi_v^i$ and $\bar{A} \leftarrow \bar{A} \cup \{(u, v)\}$ for all $(u, v) \in A$ and $\pi_u^i + c_{u,v} \leq$ thresh_label[v]. Go to (2).*

4. *Let $t \in N_2$ be the right node with the minimum shortest path label among all the free right nodes and let P be the shortest path from s to t. For all arcs a in P, if a is marked matched, mark it unmatched; if a is marked unmatched, mark it matched. Reverse the direction of arc a and set the cost $c_{u,v} = -c_{u,v}$ for arc a. If all nodes are matched, stop. Otherwise set $i \leftarrow i + 1$ and go to (2).*

Theorem 1: Procedure SPA produces an optimal assignment.

Proof. Lemma 1 shows that the flow is always dual feasible if the augmentation is made along a shortest augmenting path. Then when all nodes are matched, we have a solution which is both primal and dual feasible. Therefore, we only need to show that the shortest augmentation path produced using the good arcs is indeed the shortest augmentation path in the original network.

For this purpose we need to show that those arcs which are not in the set of good arcs will not be on the shortest augmenting paths immediately before the execution of step (4). In other words, $\pi_u^h + c_{u,v} \geq \pi_v^h$ at stage h for all arcs (u, v) not in the set of good arcs.

If the last time that arc (u, v) is examined (in step (1) or step (3)) was in stage j with $j \leq h$ and $\pi_u^j + c_{u,v} > thresh\_label[v]$, then this arc (u, v) is not in the set of good arcs at stage h. We have $\pi_u^h + c_{u,v} \geq \pi_u^j + c_{u,v} > thresh\_label[v] \geq \pi_v^h$, where the first inequality follows from Lemma 2 and last inequality follows from step (2) of the procedure SPA. Hence, all arcs not in the set of good arcs at stage h will not be part of the shortest augmentation path at stage h. ∎

The threshold costs are chosen in the following way. Let the cost vector c be the final cost vector from the initialization phase. For each free left node the minimum cost of its out arcs is calculated. Let c_{max} be the maximum of these minimum cost values and c_{min} be the minimum. Then $thresh[1]$ is set to $c_{min} + c_{max}/h$, $thresh[2]$ to $thresh[1] \times 2$, $thresh[3]$ to $thresh[2] \times 2$, ..., and and $thresh[k]$ is set to infinity. In this implementation, we chose h as 6 and k as 4. $thresh[1]$ is the most important parameter. If $thresh[1]$ is too big, unnecessary arcs are put in the set of good arcs. If $thresh[1]$ is too small, many corrections in the set of good arcs will be needed later.

This set of threshold costs may not be the best in all cases, but it seems to be effective according to our computational results. For random dense assignment problem instances, shortest path augmentations are performed only on a very small subset of arcs. The following table shows that only about $n \log n$ out of a total of n^2 arcs are used on the average.

Problem	End-of-Step1	End-of-Step2	End-of-SPA
128x128	175	231	1,022
256x256	350	472	2,292
512x512	703	954	5,256
1024x1024	1,404	1,905	10,445
2048x2048	2,828	3,839	22,875

Table 7. Good Arcs in DHK for Assign-Xhi-Degree Network

We also note that the threshold labels used here are different from what are used by Glover et al. [12]. In their algorithm, they use threshold labels to choose the eligible nodes to be scanned in their shortest path algorithm. In our algorithm, all the nodes are eligible to be scanned but only a portion of arcs out of each node are eligible to be scanned, which are determined by their reduced costs and the threshold labels.

Since the label of a left node can be obtained from the label of its matched right node, we need only keep track of the labels of right nodes. Furthermore, if the label of a node is not affected by the augmentation path, then its label need not be updated in the next shortest path calculation. For this purpose, we need to keep track of the nodes whose labels are not going to change after an augmentation.

We do not calculate π^o in step (1) since this is expensive. Instead we set $\pi_u^o = 0$ for all $u \in N_1$ for the purpose of selecting the initial set of good arcs. Since the costs of all unmatched arcs are greater than or equal to zero, and the costs of all matched arcs are equal to zero with respect to the final costs of the initialization phase, $\pi_u^o \geq 0$ for all $u \in N_1$. Hence, setting $\pi_u^o = 0$ for all $u \in N_1$ does not affect the correctness of procedure SPA nor of Theorem 1.

We note that the subnetwork of good arcs generated for the random geometric matching problem instances are not so sparse (Table 8). We also note that the idea of good arcs do not apply well for the problems which are already very sparse(Table 9).

Problem	End-of-Step1	End-of-Step2	End-of-SPA
128x128	182	267	5286
256x256	364	547	17431
512x512	728	1142	72968
1024x1024	1454	2375	226337
1024x1024	1461	2368	247293

Table 8. Good Arcs in DHK for Assign-Geometric Network

Nodes(Arcs)	End-of-Step1	End-of-Step2	End-of-SPA
512(4608)	349	384	4,608
1024(10240)	700	767	10,240
2048(22528)	1,396	1,533	22,528
4096(49152)	2,800	3,069	49,152
8192(106496)	5,599	6,135	106,496
16384(229376)	11,205	12,280	229,376

Table 9. Good Arcs in SHK for Assign-High-Cost Network

4 Computational Results

Our computational results are based on averages of over 30 runs for each problem instance. The random problem instances are generated by DIMACS *assign.c* and *dcube.c*. We have run our algorithm on an HP-9000/720 work-station and the CPU time do not include problem setup or input and output time.

All algorithms were coded in C and were compiled with the default optimization. Arrays inside the program are dynamically allocated using malloc.

4.1 Algorithms:

DHK: Algorithm presented in this paper using matrix format to store network data. This algorithm is for dense assignment problems.

SHK: Algorithm presented in this paper using linked-list format to store sparse network data. This algorithm is for sparse assignment problems.

DJV: Jonker and Volgenant's algorithm using matrix format to store network data. This algorithm is for dense assignment problems.

SJV: Jonker and Volgenant's algorithm using linked-list format to store sparse network data. This algorithm is for sparse assignment problems.

4.2 Assignment Problems:

This describes the testbed instances for assignment problem instances tested in this experiment.

1. **Assign-Xhi-Degree.** Use assign.c with the following commands for problem size $N = 256, 512, 1024, 2048, 4096$.

    nodes (N)

    sources (N/2)

    complete

    maxcost 100000

    seed (X)

2. **Assign-Xhi-Degree-Lo-Cost.** Use assign.c with the following commands for problem size $N = 256, 512, 1024, 2048, 4096$.

    nodes (N)

    sources (N/2)

    complete

    maxcost 100

    seed (X)

3. **Assign-Lo-Cost.** Use assign.c with the following commands for problem size $N = 512, 1024, 2048, 4096, 8192, 16384$.

    nodes (N)

    sources (N/2)

    degree ($2 \log_2 N$)

    maxcost 100

    seed (X)

4. **Assign-Hi-Cost.** Use assign.c with the following commands for problem size $N = 512, 1024, 2048, 4096, 8192, 16384$.

    nodes (N)

    sources (N/2)

    degree ($2 \log_2 N$)

    maxcost 100000000

5. **Assign-Two-Cost.** Use the instances generated for Assign-Lo-Cost for problem size $N = 512, 1024, 2048, 4096, 8192, 16384$. A graph g.asn is transformed using the "awk" command awk -f twocost.a <g.asn >gt.asn. The resulting graphs have all costs either 100 or 100000000. These instances are hard for auction algorithms.

6. **Assign-Geometric.** Use dcube.c with the following commands to generate random points in 2-space with coordinates in the range $1..10^6$.

nodes (N)

dimension 2

maxloc 1000000

seed (X)

The resulting graph g.geom can be converted to an assignment problem with awk -f geomasn.a <g.geom >g.asn. Half the points are red and half blue, arcs connect red points to blue, and arc costs correspond to Euclidean distances.

4.3 Results of the experiments

In all the tables, the column "Init" indicates the average CPU time (in milliseconds) in initialization stage. Column "SPA" indicates the average CPU time spent on shortest path augmentations. Column "DHK/DJV" ("SHK/SJV") indicates the time ratio of algorithm DHK(SHK) to DJV(SJV).

Problem	DHK			DJV			DHK/
	Init	SPA	Total	Init	SPA	Total	DJV
128x128	64	46	110	128	7	134	0.82
256x256	258	162	420	359	56	416	1.00
512x512	1,218	612	1,880	1,450	526	1,976	0.95
1024x1024	5,679	2,336	8,015	5,525	4,161	9,686	0.82
2048x2048	25,193	9,959	35,153	21,696	29,000	50,696	0.69

Table 10. CPU time (in ms) for Assign-Xhi-Degree network

Problem	DHK			DJV			DHK/
	Init	SPA	Total	Init	SPA	Total	DJV
128x128	46	48	95	31	24	55	1.72
256x256	143	146	289	125	134	259	1.11
512x512	493	729	1,222	660	578	1,238	0.98
1024x1024	1,915	0	1,915	3,203	1,972	5174	0.37
2048x2048	7,666	0	7,666	16,196	11,562	27,759	0.27

Table 11. CPU time (in ms) for Assign-Xhi-Degree-Lo-Cost network

| Problem | SHK | | | SJV | | | SHK/ |
Nodes(Arcs)	Init	SPA	Total	Init	SPA	Total	SJV
512(4608)	12	31	44	4	18	22	2.00
1024(11264)	43	188	232	22	92	114	2.03
2048(22528)	115	689	805	62	366	428	1.88
4096(49152)	285	2,439	2,725	158	1,461	1,620	1.68
8192(106496)	669	8,117	8,786	364	5,518	5,882	1.49
16384(229376)	1,515	28,284	29,799	823	23,309	24,132	1.23

Table 12. CPU time (in ms) for Assign-Lo-Cost network

| Problem | SHK | | | SJV | | | SHK/ |
Nodes(Arcs)	Init	SPA	Total	Init	SPA	Total	SJV
512(4608)	11	37	49	27	1	29	1.68
1024(10240)	42	186	228	133	1	134	1.70
2048(22528)	129	838	967	523	11	535	1.80
4096(49152)	318	2,988	3,306	1,437	56	1,493	2.21
8192(106496)	724	11,514	12,239	4,956	1,782	6,739	1.81
16384(229376)	1,712	41,676	43,399	13,535	13,502	27,038	1.60

Table 13. CPU time (in ms) for Assign-Hi-Cost network

| Problem | SHK | | | SJV | | | SHK/ |
Nodes(Arcs)	Init	SPA	Total	Init	SPA	Total	SJV
512(4608)	22	0	23	2	9	11	2.09
1024(10240)	57	0	58	16	39	56	1.03
2048(22528)	136	0	136	56	132	188	0.72
4096(49152)	320	0	320	146	500	647	0.49
8192(106496)	710	0	710	330	1,873	2,204	0.32
16384(229376)	1,671	0	1,671	744	7,192	7,937	0.21

Table 14. CPU time (in ms) for Assign-Two-Cost network

| Problem | DHK | | | DJV | | | DHK/ |
	Init	SPA	Total	Init	SPA	Total	DJV
128x128	377	402	779	1,469	322	1,790	0.43
256x256	1,571	2,025	3,597	4,663	2,103	6,766	0.53
512x512	6,876	14,531	21,407	15,990	15,957	31,947	0.67
1024x1024	29,704	73,935	103,639	51,609	89,885	141,449	0.73

Table 15. CPU time (in ms) for Assign-Geometric network

4.4 Discussion:

We have considered the successive shortest augmenting path algorithm for solving the assignment problem. We have proposed an algorithm, in which we developed a new initialization procedure and a technique to compute the shortest augmenting paths by using a small subset of the arcs. This approach is relatively efficient for dense assignment problems as we could see from the running times shown in Tables 10 and 11. This algorithm solves the dense assignment problem in an empirical running time proportional to something between n^2 and $n^2 \log n$ as shown in the following table.

Problem	CPU	CPU$\times 10^3/(n^2)$	CUP$\times 10^4/(n^2 \log n)$
128x128	110	6.47	9.58
256x256	420	6.41	8.01
512x512	1,880	7.17	7.96
1024x1024	8,015	7.64	7.64
2048x2048	35,153	8.38	7.62

Table 16. Algorithm DHK running time analysis on Assign-Xhi-Degree network

However, our algorithm for sparse random networks(SHK) is not as efficient as Jonker and Volgenant's(SJV) as shown in Tables 12 and 13. But the performance of our algorithm improves as the problem size increases.

Our initialization procedure matches about 90% of the nodes for both dense and sparse networks. We note that our procedure performs better than Jonker and Volgenant's on random problems with small cost range while theirs performs better on those with large cost range.

References

[1] Ahuja, R. K., T. L. Magnanti, and J. B. Orlin. 1988. Network Flows, in *Handbooks in Operations Research and Management Science, Vol 1. Optimization.*

[2] Akgul, M. 1985. A genuinely polynomial simplex algorithm for the assignment problem, Research Report, Department of Computer Science and Operations Research, North Carolina State University, Raleigh, NC.

[3] Akgul, M. 1988. A Sequential Dual Simplex Algorithm for the Linear Assignment Problem. *Operations Research Letters* 7, 155-158.

[4] Balinski, M. L. and R. E. Gomory. 1964. A Primal Method for the Assignment and Transportation Problem. *Management Programming* 13, 1-13.

[5] Balinski, M. L. 1986. A Competitive (Dual) Simplex Method for the Assignment and Transportation Problem. *Mathematical Programming* 34, 125-141.

[6] Barr, R., F. Glover, and D. Klingman. 1977. The Alternating path basis algorithm for the assignment problem. *Math. Prog.* 12, 1 -13.

[7] Bellman, R. 1958. On a Route Problem. *Quart. of Appl. Math.16,* 87-90.

[8] Bertsekas, D. 1981. A New Algorithm for the Assignment Problem. *Mathematical Programming* 21, 152-157.

[9] Chang, S.F. and S.T. McCormick 1989. A fast implementation of a bipartite matching algorithm. Technical Report, Columbia University, New York.

[10] Engquist, M. 1982. A Successive Shortest Path Algorithm for the Assignment Problem. *INFOR* 20, 370-384.

[11] Ford, L. R. 1956. *Network Flow Theory.* The Rand Corporation Report P-923, Santa Monica, CA.

[12] Glover, F., R. Glover, and D. Klingman. 1986. The Threshold assignment algorithm. *Math. Prog. Study* 26, 12-37.

[13] Goldfarb, D. 1985. Efficient Dual Simplex Algorithms for the Assignment Problem. *Math. Prog.* 33, 187-203.

[14] Hung, M. S. 1983. A Polynomial Simplex Method for The Assignment Problem. *Oper. Res.* 31,595-600.

[15] Hung, M. S. and W. O. Rom. 1980. Solving the assignment problem by relaxation. *Oper. Res.* 28, 969-892.

[16] Jonker, R. and A. Volgenant. 1987. A Shortest Augmenting Path Algorithm for Dense and Sparse Linear Assignment Problems. *Computing* 38, 325-340.

[17] Kuhn, H. W. 1955. The Hungarian Method for Assignment Problem. *Naval Research Logistics Quarterly* 2, 83-97.

[18] Orlin, J. B. 1985. On the Simplex Algorithm for Networks and Generalized Networks. *Mathematical Programming Study* 25, 166-178.

[19] Roohy-Laleh, E. 1980. Improvements to the theoretical efficiency of the network simplex method, Ph.D. Thesis, Carleton University, Ottawa, Canada.

[20] Weintraub, A., and F. Barahona. 1979. A dual algorithm for the assignment problem. Departmente de Industrias Report No. 2, Universidad de Chile-Sede Occidente.

GTE Laboratories Incorporated, 40 Sylvan Road, Waltham, MA 02254
E-mail: jh04@gte.com for Jianxiu Hao and *gak0@gte.com* for George Kocur.

The Assignment Problem on Parallel Architectures

M. BRADY, K. K. JUNG, H.T. NGUYEN,
R. RAGHAVAN AND R. SUBRAMONIAN

ABSTRACT. This paper presents a wide variety of implementation results for the Assignment Problem on parallel architectures. In the first part, we consider implementation on simple bit-serial massively parallel SIMD arrays of processing elements. We evaluate SIMD implementations of two different algorithms. The first is a parallelization of Munkres' algorithm. We have implemented the algorithm on the AP2S (a Lockheed invention) and on the commercial WaveTracer Zephyr 8K processor machine. Surprisingly, we find great discrepancies in the performance on these two seemingly similar machines, which we explain. The second algorithm takes advantage of recent theoretical work on the problem for the PRAM model to obtain the asymptotically best SIMD mesh algorithm that we know of. We compare this more complicated but asymptotically good algorithm to the Munkres implementation using the WaveTracer.

Finally, as a result of discussions at the DIMACS workshop, we have implemented the Auction algorithm on a variety of parallel architectures. These include SIMD machines, (an 8K-processor WaveTracer Zephyr and an 8K-processor MasPar MP-1), MIMD machines (a 128-processor CM-5), shared memory multiprocessors (the Silicon Graphics IRIS 4D/340 DVX) and distributed network of workstations (SUN SPARCstation-2's over Ethernet). We discuss the architectural features that aided and hindered the implementations and suggest what architectural features are crucial to a successful implementation. We found that the Auction algorithm, while easily parallelizable and very fast, is not as parallel as we had anticipated. We also found that given the poor cache utilization of Auction, the SIMD machines were much more efficiently utilized than the RISC-based machines. We provide the technological reasons for these findings.

---

[0]The work was performed with Lockheed Independent Research funds under project RDD506. Thanks are due to National Science Foundation Infrastructure Grant number CDA-8722788, which provided access to the CM-5.

1991 *Mathematics Subject Classification.* Primary 90B10 90C27; Secondary 68Q22.
Key words and phrases. Parallel Architectures, Assignment Problem, Bipartite Graph Matching.

1. Introduction

The Assignment Problem is a classic combinatorial optimization problem. Its practical importance has resulted in a large number of theoretical algorithms and practical implementations. It embodies a fundamental linear programming structure, and hence can be used to solve other problems such as linear network flow and general matching. In this paper we evaluate the practicality of parallelizing this problem, considering a wide variety of parallel algorithms and architectures.

First, let us more formally describe the problem. The input to the Assignment Problem is a set of n objects and n persons and n^2 values, $w(i,j)$ such that $w(i,j)$ is the cost of assigning person i to object j. The problem is to find a permutation, $\Pi = \{\pi_1, \pi_2, .., \pi_n\}$ such that $\sum_{i=1}^{i=n} w(i, \pi(i))$ is minimum. In other words, we seek a matching of minimum cost.

It is easy to see why the Assignment Problem is also called the minimum weight bipartite matching problem. Define $G = (V, E)$ as an undirected bipartite graph such that V can be partitioned into two disjoint sets S and T and all edges have one endpoint in S and the other in T. Let the number of edges $|E| = m$ and the number of vertices $|S| = |T| = n$. A matching, M, is then a vertex-disjoint subset of edges of E, and the goal is to find a maximum matching of minimum cost.

The paper is organized into two major parts. In the first part, Sections 2, 3, 4, and 5, we present our parallel implementations of two different algorithms on simple bit-serial massively parallel Single-Instruction Multiple-Data (SIMD) arrays of processing elements (PEs). First, we give a parallelization of the algorithm of Munkres [**Mun57**]. We have implemented the algorithm on the AP2S (a Lockheed invention) and on the commercial WaveTracer Zephyr 8K processor machine. Our second algorithm takes advantage of recent theoretical work on the problem for the PRAM model to obtain the asymptotically best SIMD mesh algorithm that we know of. We compare this more complicated but asymptotically good algorithm to the Munkres implementation using the WaveTracer.

A summary of our implementation results is given in Table 1. It shows the times for a problem size of 512×512. The "optimal" times, without loss of cycles due to architectural bottlenecks, are shown in parentheses.

As a result of discussions at the DIMACS workshop, we have subsequently implemented the Auction algorithm on a variety of parallel machines: a shared-memory multiprocessor, SIMD machines, and a distributed network of workstations. The Auction parallelizations are the subject of the second part of the paper. Our decision to implement Auction on a variety of parallel architectures was influenced by discussions we had at the DIMACS workshop and the suggestions of the referees. It provides an appropriate baseline against which to compare the algorithms and implementations of the previous sections. In this paper, we detail our our insights into the behavior of the algorithm, the features

Algorithm	Machine	Uniform Random	$i \times j$
Munkres	AP2S	(0.0946)	N/A
Munkres	WaveTracer	1310 (6.98)	$\gg$ 4 days
Brady	WaveTracer	1040 (3.25)	217 (61.7)
Auction	SUN SPARCstation-2	3.97	10.9

TABLE 1. Summary of Implementation Results for 512×512 data. (Figures in parentheses are ideal times (in seconds) based on cycle counts.)

of the various architectures that helped and hindered our implementations and, most importantly, how the behavior of the algorithm strongly influenced the method of parallelization on the different architectures.

Part 1. Assignment on SIMD Machines

We now describe the results of our implementations of the Assignment Problem on massively parallel bit-serial arrays of SIMD processors. First, we describe the SIMD model and give details of the hardware platforms on which we obtained performance results in Section 2. In Section 3, we describe our parallelization of Munkres's algorithm [**Mun57**] and present the results of our implementations. We then describe the architectural limitations of current SIMD architectures which prevent the algorithm from achieving the times expected by a simple count of parallel machine cycles. In Section 4, we describe an SIMD algorithm [**Bra88**] based on more recent theoretical work on Assignment for the PRAM model [**GPV88**]. Finally, in Section 5 we summarize the results of Part 1.

2. SIMD Computing Model

The SIMD model consists of an array of identical processors operating in lock-step synchrony. In each cycle, they perform the same instruction on their own data. In one step, a processor can perform one instruction or send/receive one bit/nibble from a neighboring processor. Typically, the PEs have a limited instruction set and a very small finite precision (typically 1 to 4 bits). Thus, the execution time of an operation is dependent upon the word length. An $n \times n$ mesh-connected array contains n^2 processors positioned on an n-row, n-column square lattice. Each processor can communicate with its four nearest neighbors, except those on the border, which have only two or three neighbors. (Our algorithms could also be implemented on a mesh with "wraparound" connections, yielding a speedup of a small constant factor.) The first commercially available chip embodying this architecture was the GAPP invented by W. Holsztynski [**Hol86**].[1]

---

[1]The DAP [**Red78**] was invented earlier, but was not commercially available at that time. Further, our remark pertains to the availability of *chips*. The importance of this is made clear in the Appendix.

Our early results on the Assignment Problem were implemented on an internally constructed machine AP2S (Adaptive Parallel Processing System, see [**RJN90**]) based on the GAPP chip. The AP2S contains a two-dimensional (108 x 96) array of approximately 10,000 processing elements running at 10 MHz. Each processing element contains only 128 bits of local memory. Our algorithms were constructed for an "enhanced" SIMD model, in which the existence of row and column highways is assumed. Specifically, a row-OR or column-OR is assumed to take only one cycle. (This type of hardware enhancement is present in the AMT DAP machine for all the sizes manufactured, and we expect that our implementations for this model would also be efficient on the DAP [**Red78**]. It is, however, clearly an assumption that cannot hold in the limit for very large n.) The AP2S has special features discussed in Appendix A.

The AP2S system has limitations which restricted its usefulness for further investigations for the DIMACS project. Primarily, the limited local memory severely restricted the size of problem which was feasible. Thus, the remainder of our implementations (i.e. for larger arrays) were done on the commercial WaveTracer Zephyr machine (see [**Jac90**]). In the WaveTracer Zephyr's largest configuration, there are 16K simple bit-serial processors with 2 Kbits of on-chip memory and 32 KBytes of off-chip memory per PE, each running at 9 MHz. While the hardware interconnection lattice is three-dimensional, our implementations configured the machine to a two-dimensional array. The Zephyr does not contain row and column highways. However, it does have a global bus that can perform the NAND of a bit from each PE in one cycle.

It is important to note that some of our results have a computer hardware/software flavor rather than a purely mathematical one. For example an early invention of ours called the Distributed Macro Controller (DMC) [**HR88**] allows, among other things, a run-time allocation of memory. This significantly enhances the speed of execution of the algorithm as discussed in the next section. We are also interested in the bit-level performance of these implementations, since massively parallel machines generally use bit/nibble serial processors. Further, the bit complexity is an inherently important quantity in the cost scaling techniques such as in [**GPV88**].

The architectures we have chosen represent the simplest and least expensive massively parallel computing models (around $10 per PE, total systems cost of a hundred thousand dollars for about 8,000 processors in the WaveTracer). They represent a different way to arrange about the same amount of hardware as a high-end workstation, as opposed to a multimillion dollar supercomputer. Thus, we view our results as an evaluation of the amenability of Assignment Problem to this alternate architecture, and believe a comparison to sequential implementations on high-end workstations would be valid and interesting. One such comparison is make on our Cost Scaling Algorithm, given in Section 4.

3. Munkres' Algorithm

3.1. Description of Munkres' Algorithm. As mentioned above, our first parallel implementation was on the Kuhn-Munkres algorithm. The method of Munkres has been described both in his original paper [**Mun57**] and in textbooks (e.g. [**BM76**] to which we refer for details). The algorithm operates on the $n \times n$ matrix of edge weights. The general strategy of the algorithm is to adjust the dual variables in order to obtain a matching on a set of edges of weight zero, while ensuring that all other weights are at least zero. In matrix terminology, adjusting a dual variable amounts to adding or subtracting a constant from a row or column of the matrix.

Initially, we reduce the weight as much as possible by subtracting the minimum weight in each row (column) from every element in that row (column). We then begin by selecting a maximal set of independent edges with zero weight. The algorithm then proceeds by searching for augmenting paths consisting only of zero-weight edges. If one can be found, the augmentation can be performed. Otherwise, if the search fails, the information gathered is used to reduce the total weight of the matrix.

The advantage of Munkres' formulation is that the algorithm is described entirely in terms of row and column operations on the matrix. Our parallelization of the Munkres steps involves what we might call both "local" as well as "global" parallelizations [**Hol87**]. By "local" we here mean faithfully following the sequential algorithm, and parallelizing the steps as far as possible. However, we also make some changes that allow a single parallel step to correspond to several simultaneous Munkres steps. Each one of the latter, in turn, allows a "local" parallelization. We do not intend to attach any geometric meaning to the terms "local" and "global" here; these refer only to the structure of the algorithm. The pseudo-code of the parallel algorithm is shown in Figures 1, 2 and 3. We note that the "Maximal Matching" step, which takes a step along several potential augmenting paths at once is an example of a "global" parallelization.

It can be seen from the above that the algorithm consists largely of the following kinds of basic row/column computations, each of which admits a parallel implementation:

(i) Broadcast a value from one selected element in each row (column) to all elements in the respective rows (columns).

(ii) Given a n-vector V and a $n \times n$ matrix Q, subtract vector V from matrix Q, i.e. compute $\forall i \forall j : 1 \leq i, j \leq n, Q(i,j) - V(j)$. A simpler problem is that of subtracting a constant from a matrix.

(iii) Find the minimum of every row (or column) of the real matrix Q; find the minimum of the entire matrix Q.

(iv) Among possibly many elements (of a matrix) which possess a given property, extract one.

We now discuss the ease of parallelization of each of the above four steps.

Munkres()
{
    1. *Reduce the edge weights and find a maximal matching of zero-weight edges.*
        Find the minimum cost C_j in each column j.
        Subtract C_j from all $c(i,j)$ in each column j.
        Find the minimum cost C_i in each row i.
        Subtract C_i from all $c(i,j)$ in each row i.
        Mark the eligible edges $E \leftarrow 1$ (unmatched edges with $c(i,j) = 0$).
        call Maximal_Matching().
    2. *Search for augmenting zero-weight paths; If none, reduce matrix total weight.*
        while number of matched edges $\neq n$
            $tagged\_row \leftarrow 0$
            2.1. *Mark the matched columns:*
            $matched\_columns \leftarrow$ column broadcast of matched edges M.
            2.2. *Tag eligible edges in unmatched columns:*
            if $E = 1$, matched_column $= 0$, tagged_row $= 0$
                $tag \leftarrow 1$.
            endif
            2.3. **if** any tagged edge is in a matched row
                call Find_Matched_Edge().
            else if any tagged edge is in an unmatched row **or**
                all eligible edges are tagged
                call Augment_Path().
            else (no new tagged edges found)
                call Reduce_Matrix().
            endif
        endwhile
}

FIGURE 1. Pseudo-code for Parallel Munkres Algorithm.

Maximal_Matching()
{
    $E \leftarrow$ eligible edges.
    $A, R, C \leftarrow 0$.
    $A(0,0) \leftarrow 1$ (Activate the top left element).
    for $k = 1$ **to** $2n - 1$
        if $A = 1$, $E = 1$, $R = 0$, and $C = 0$ (Select eligible edges on the
            active diagonal whose rows and columns are so far unmatched.)
            $M, R, C \leftarrow 1$ (Match them, and mark their row and column
            as matched).
    endif
        $A \leftarrow NORTH(A) \vee WEST(A)$ (Move active diagonal down and
            right one step).
        $R \leftarrow WEST(R)$ (Propagate row-matched flag right).
        $C \leftarrow NORTH(C)$ (Propagate column-matched flag down).
    endfor
}

FIGURE 2. Procedure Maximal_Matching(). E, M, A, R, C are one bit array variables. E marks eligible edges and M marks matched edges. A marks the active diagonal (e.g., if $A(i,j) = 1$, then element (i,j) is on the active diagonal). Flags R and C are used to mark matched rows and columns, respectively.

Find_Matched_Edge()
Find matched edge(s) whose row(s) contains a tagged edge. Unmark the column as matched (but leave the edge in M for now).
{
    1.   *tagged_row* ← row broadcast of tagged edges.
    2.   **if** (i, j) is matched **and** its row is tagged
              Unset *matched_column* of all elements in column j.
        endif
}

Augment_Path()
Follow a newly-discovered augmenting path by successively matching tagged edges and unmatching matched edges.
{
    1.   Select one tagged edge on an unmatched row (call it *new*).
    2.   **while** *new* exists
              Untag *new*.
              Column broadcast *new*.
              Unmatch the matched edge (set $M = 0$) in *new's* column.
              Row broadcast the newly unmatched edge.
              Set a tagged edge in the newly unmatched row as *new*
                    (if one exists).
        endwhile
    3.   Add all *new* visited to M.
}

Reduce_Matrix()
Use the matched_column and tagged_row information to reduce the total weight of the matrix, keeping all weights ≥ 0.
{
    1.   *min* ← global minimum of edge costs in unmatched cols. and
              untagged rows.
    2.   Subtract *min* from all untagged rows.
    3.   Add *min* to all unmatched columns.
}

FIGURE 3. Major Subroutines for the Parallel Munkres Algorithm.

(i) In general, column broadcast can be viewed as a generalization of column-OR. The broadcasting elements must be marked with a one bit flag, F. To broadcast a single bit, A, we perform the column-OR of $A \wedge F$. As a result, is every element in a column j receives $A(i,j)$ from element for which $F(i,j) = 1$. To broadcast longer words, this operation is simply repeated for each bit in the word.

(ii) The second part of problem 1 is easy on SIMD processors. Let the terms of matrix Q and the constant be rational b-bit numbers of the form $\frac{x}{2^E}$, where all x's are integers and integer E is fixed. Then it takes $O(b)$ instructions for this task. A sequential processor, on the other hand, requires $O(n^2)$ instructions. The subtraction of a vector from a matrix can be computed in $O(b)$ steps on a bit-serial mesh, although row or column highways may be needed to first broadcast the vector elements $V(j)$ to all (i,j).

(iii) A sequential processor needs to execute $O(n^2)$ instructions in order to find the minimal value among n^2 items. On the other hand the bit-serial parallel computer needs only $O(b)$ instructions if a global operation is available (all commercial machines have this facility). The same ideas apply to the enhanced geometric SIMD matrix processors, which provide row and column operations (row-OR/column-OR). Such arrays are able to perform the computations of row-minima in $O(b)$ instructions.

(iv) This problem is reduced to (3) if together with the elements of he relevant matrices we also store unique addresses (e.g., the matrix coordinates of the elements). Now we can pick one place of interest (the event that the matrix entry is 1, for example) by performing the global minimum algorithm on the array of bits which have the data in the most significant bit positions, and the addresses in the least significant bits. This can be done in $O(\log n)$ instructions. The number of instructions needed by a sequential processor for the same task is variable and random, it increases on average when the probability of the event that the entry is 1 decreases.

The pseudo-code clearly shows that *the majority of operations such as tagging, uncovering etc. are single-bit operations, so that a bit-serial machine should be very advantageous for this.* Further, since the weights are decreasing with the repetition of the main steps of the algorithm, a machine that offers dynamic memory allocation, or deals with only as many bits as necessary at each step without overhead, offers an advantage. Such is the case with the machine augmented with the DMC that we constructed. In this case the average running time is less than that indicated by (1) above.

The sequential Munkres algorithm may require $O(n^3)$ row/column operations, and therefore $O(n^4)$ total time. The performance of our parallelization can be shown in the worst case to be $O(n^3 \log nC)$ with n^2 bit-serial processors and weights with a magnitude $< C$, (so that $\log C$ is the number of bits needed to specify the weights, and $\log n$ bits are required to represent a processor's row or column number). Thus, the parallel work (processor-time product) is $n^5 \log nC$, which appears poor. However, we have given some reasons for thinking that the constants hidden within the asymptotic bounds are quite small, for the model we are considering. (This model allows the row-OR/column-OR to occur in one cycle). Further, the results for random inputs are dictated by average-case bounds which are quite a bit better than the above. Of course this would be true for the sequential algorithm as well.

3.2. Performance Results of Parallel Munkres Algorithm. We now justify the above assertions by displaying the results of the runs in Table 2. In all cases, we display the running time in terms of machine cycles. The AP2S runs at 10 MHz, thus 10,000 cycles = 1 msec. The cycle count does not include any input-output time; further the cycle count assumes the "enhanced" GAPP model, in that a row-OR or column-OR is counted to take only one cycle. Table 2 shows the averaged cycle-count (constructed as specified by the DIMACS consortium) for

(i) random 11-bit samples (obtained using "assign.c" in the DIMACS library),
(ii) 21-bit numbers (obtained using "dcube.c") and
(iii) $i \times j \times 100$.

We feel that the times are surprisingly fast. Even for the 256×256 matrix with all entries having 21-bits, the algorithm completes (with the caveats mentioned above) in less than 37 msec on the average, with a 10 MHz clock. (It should be pointed out that more modern chips or machines run at higher clock speeds). For the 11-bit case, the times have dropped to about 15 msec. We should note that our actual array size is approximately 100×100, so that most of the time would be spent in input and output for the larger sizes of problem, due to the very limited amount of local memory per processor. As stated, this has not been included in the cycle count. The small size of our array is also why we are not reporting larger sized problems.

In order to investigate larger problems, and allow comparison with other algorithm, we have also implemented the Munkres algorithm on the WaveTracer Zephyr. The results of this work are given in Table 3. It is obvious that these results are several orders of magnitude slower than one would expect, given the previous table. The reasons for this are tied up in the architectures, and are explained in detail in the next subsection. In order to analyze the problem, we have measured both the actual elapsed time (*response time*) and counted the number of parallel array machine cycles used, multiplied by the clock rate (*cycle time*). It indicates that part of the problem is due to a large percentage of wasted parallel cycles, due to bottlenecks elsewhere in the machine. The cycle times represent the idealized best case in which the parallel array wastes no idle cycles, and the array is as large as the input size.

Size	Cost	Ideal Time (in msec)
16	10^3	0.41
32	10^3	1.03
64	10^3	2.65
128	10^3	6.56
256	10^3	14.5
512	10^3	34.1
1024	10^3	75.1
16	10^6	0.68
32	10^6	1.90
64	10^6	5.27
128	10^6	14.0
256	10^6	36.5
512	10^6	94.6
1024	10^6	239
16	$i \times j \times 100$	3.20
32	$i \times j \times 100$	18.4
64	$i \times j \times 100$	110

TABLE 2. Performance of Munkres' Algorithm on AP2S. The ideal time is calculated based on number of machine cycles and a clock rate of 10 MHz.

Size	Resp. Time (sec.)	Cycle Time (sec.)	Cycles (millions)
	Uni. Rand. Data, $0 - 100K$		
128×128	30.2	0.34	3.1
256×256	193	0.98	8.9
512×512	1300	2.66	23.9
	Geom. Rand. Data, $0 - 1M$		
128×128	30.2	0.34	3.1
256×256	200	1.02	9.2
512×512	1430	2.93	26.3
	$i \times j$ Data		
128×128	3480	39.3	334
256×256	67400	344	3100
512×512	$\gg$ 4 days	N/A	N/A

TABLE 3. Performance of the Munkres Algorithm on the WaveTracer.

Remark. We did not implement Munkres' algorithm on the MasPar because the MPL compiler [**Chr90, Nic90**] provided did not have virtualization, which greatly expedites debugging.

3.3. Limitation of Current SIMD Architectures.

In commercially available bit-serial machines, the cycle-count for operations is a very poor indicator of actual elapsed time. We describe here the problems, the reason for these, and also how we had solved these in the AP2S which was operational in early 1987. Consider 4-Byte (32-bit) addition, $a \leftarrow b + c$, performed n times in a loop or simply repeated n times, and also the ideal time computed from cycle-count and clock speed. One would not expect a difference in the time taken by these two dummy programs. However, Table 3 indicates a significant difference. This slowness is due to the controller bottleneck and command transfer bottleneck for small but significant parallel slackness. To understand this, let us explain how a particular machine (the WaveTracer) works. On the WaveTracer, a program compiled by multiC [**Jac90**] runs on the host which produces appropriate commands (DTC commands) to the 18 MHz control processor (CP) via a SCSI bus each time a parallel operation is invoked. To repeat somewhat, the part of the multiC program which has parallel instructions (the program can also have serial instructions to be run on the host) is interpreted by the controller into a set of "nano-instructions" and broadcasts these nano-instructions to each processor element (PE) of the processor array (PA) only once and then repeatedly executes these $n = \frac{v}{p}$ times, where $v =$ number of virtual processors, and $p =$ number of physical processors.

Sometimes, as in our implementation of our Assignment algorithms, v is the size of the problem as well. The mapping of virtual to physical processors is all done automatically by the multiC compiler, and is implemented in WaveTracer hardware and in its software system to ensure efficiency and ease of use. The host, CP, and PA are running in a pipelined fashion.

The multiC compiler has a primitive *optimizer* which looks ahead for repeated instructions. As shown in Table 4, the fact that loop time is greater than repeated time indicates that the compiler catches repeated commands and therefore saves some SCSI transfers between the host and the CP. When $n = 1$, for addition and memory transfer operations, the significant difference between the timing for instructions that are unfolded (i.e. repeated, instead of occurring in a loop) and the ideal timing is due entirely to the CP bottleneck. The PA executes faster than the CP can interpret DTC commands to nano-instructions. As n gets higher (4 and 16), the difference becomes insignificant since the PA now repeatedly executes the same set of nano-instructions n times, therefore the CP has plenty of cycles to keep the PA busy while the CP interprets the next command.

A third slowdown arises as follows. Whenever a pipelined flow of information from the host to the PA is broken such as in a global maximum operation, a

1-bit operation	Array Size							
	128 × 64			256 × 128			512 × 256		
	Loop	Seq.	Ideal	Loop	Seq.	Ideal	Loop	Seq.	Ideal
$A \leftarrow B + C$	44.7	22.9	0.44	44.7	32.2	1.78	44.8	23.1	7.11
$A \leftarrow B \times C$	44.8	19.5	0.44	44.8	19.9	1.78	44.8	19.8	7.11
$A \leftarrow B$ (Regis.)	42.0	16.4	0.33	42.0	16.7	1.33	42.2	16.6	5.33
$A \leftarrow B$ (Memory)	129	76.8	1.00	349	257	4.00	1130	1180	16.0
Global Max.	3300	3280	2.11	3340	3410	8.44	3350	3370	33.8
32-bit operation	128 × 64			256 × 128			512 × 256		
	Loop	Seq.	Ideal	Loop	Seq.	Ideal	Loop	Seq.	Ideal
$A \leftarrow B + C$	43.6	20.2	14.2	61.5	56.1	56.9	228	227	228
$A \leftarrow B \times C$	145	139	139	564	557	558	2240	2250	2231
$A \leftarrow B$ (register)	41.5	17.2	7.2	48.4	28.5	28.9	116	114	116
$A \leftarrow B$ (memory)	129	77.1	46.3	349	297	171	1230	1230	683
Global Max.	3370	3320	50.4	3350	3350	202	3350	3313	807

TABLE 4. Timing for some Basic Operations on WaveTracer (in μ-seconds)

delay is incurred. The global maximum operation finds the maximum number over the entire PA and then broadcasts this single piece of information all the way back to the host through the CP and SCSI bus. While the PA send the maximum value to the CP then to the host, both the CP and the host are idling while waiting for this value. Hence the pipelined flow is broken. As Table 4 indicates in this case, there is no significant difference between the loop time vs. repeated time and also among different problem sizes. The time it takes to complete this instruction depends only on the time to transfer the maximum value from the PA to the host, and dominates the computation time on the PA to find the maximum. In practice, getting information from the PA to host should be avoided where possible.

Consider the code segments in Figure 4, which select the processor with the highest v value. Variables $maxi$ and j are resident on the host while v, w, $processor\_id$ and p are resident on the array processors. The code on the left is about 1.25 times slower than the code to the right. This is because in assigning the maximum to $maxi$, a variable resident on the host, we force the host to wait for the value of $maxi$ and hence break the instruction pipeline.

These three reasons for the time discrepancy were completely absent in the AP2S, as discussed in Appendix A. However, the AP2S did not allow serial and parallel code to be interleaved in one program. Another feature of our controller which we emphasize lest it be misunderstood is that dynamic reallocation of memory did not cost any machine cycles. These features are available on an actual, not a hypothetical, machine.

4. Parallel Cost-scaling Algorithm

4.1. Description of Parallel Cost-scaling Algorithm.
Recently, we have designed Assignment algorithms for SIMD implementation [**Bra88**], building

code for finding j: object with maximum $v = w(i) - p$
  { =>? = is the max-finding operator }
  { =<? = is the min-finding operator }

$v \leftarrow w[i] - p$ $\qquad\qquad\qquad\qquad$ $v \leftarrow w[i] - p$
$maxi =>? = v$
if $maxi == v$ **then** $\qquad\qquad$ **if** $v == (>? = v)$ **then**
  $j =<? = processor\_id$ $\qquad\qquad$ $j =<? = processor\_id$
endif $\qquad\qquad\qquad\qquad\qquad$ **endif**

FIGURE 4. Effect of Breaking Pipeline in SIMD Machines

upon theoretical advances in algorithms based on the PRAM model. The technique is an adaptation of the sublinear time parallel PRAM algorithm of [**GPV88**], which uses a two-phase cost scaling approach. We obtain an $O(n^{5/3} \log^2 nC)$ time algorithm for an $n \times n$ mesh, where weights are integral and C represents the weight of maximum magnitude. One factor of $\log nC$ is due to use of $lognC$-bit operations on a bit-serial machine and the second is incurred by the cost scaling iterations. This algorithm is memory efficient, using only a small constant number of $\log n$-bit registers per processor. The subroutine which dominates the computation time (all pairs shortest paths) fully utilizes the n^2 processors throughout its execution, so the processor utilization is high for any number of processors less than or equal to n^2.

To adapt the PRAM algorithm to the SIMD model, we employ a novel technique in which the data is permuted in each pass to make optimal use of the local communication. A step-by-step description of the algorithm is given below. The outer loop of the algorithm performs cost scaling, in which a solution is obtained through iterative improvement. Steps 1 through 4 in "Match()" (Figure 5) begin with a solution which is 2ϵ-optimal and improve it to an ϵ-optimal solution (see [**Ber86**] for a discussion of the cost scaling technique). The algorithm has two major phases. The first is a very efficient "Push and Relabel" strategy which takes one step on many augmenting paths in parallel. The second is a more powerful but slower (in constant factor) routine which searches for complete disjoint augmenting paths. Each pass of the second phase finds all of the one-step paths of the first phase, some multiple-step paths, and at least one complete augmenting path. Thus, while each pass of the second phase takes longer, it may require fewer total passes. In general, the first phase is more efficient at the beginning when there are many free vertices, and later the second phase becomes more efficient because it is guaranteed to increase the matching size by at least one during every pass. There is, therefore, an optimal "cutoff" point at which one should switch from phase 1 to 2.

Let us describe the computation required in each phase. Phase 1 is described as "Push_Relabel_Matching()" (Figure 6). Phase 1 begins by marking eligible edges from free S-vertices, using simple row and column operations. Next, a

Match()
{
$\forall i \forall j : c(i,j) \leftarrow n \times w(i,j)$
$\epsilon \leftarrow C$
while ($\epsilon > 1$) **do**
    1. Find the minimum unmatched edge cost C_j in each column j.
    2. Subtract C_j from all $c(i,j)$ in each column j.
    3. Unmatch arcs (i,j) with $c(i,j) \geq \epsilon$.
    4. **while** there are unmatched vertices (number of matches $< n$) **do**
        if (number of matches $<$ Crossover)
            call Push_Relabel_Matching().
        else
            call Shortest_Path_Matching().
        endif
    endwhile
    5. $\epsilon \leftarrow \epsilon/2$;
endwhile
}

FIGURE 5. Procedure Match().

maximal matching of eligible edges is computed, by sweeping the matrix diagonally from the top left to the bottom right. Define a diagonal k as the set of elements (i,j) such that $i+j = k$. When diagonal k becomes active, each eligible edge (i,j) on diagonal k is matched if its row and column does not yet contain a matched edge. If edge (i,j) is matched, then that information will be passed to the row i and column j elements in the next diagonal, $k+1$, and so on. (This simple subroutine, described in Figure 2 in the Section 3, is not much more time consuming than the row/column operations.) Finally, after selecting new matched edges and unmatching the old matched edges with which they conflict, we update the dual variables. This consists of increasing the costs in each column until the column's matched edge has cost $= \epsilon$. We then decrease the costs in each row until the cost of some unmatched edge in the row is $= -\epsilon$. Again, these require only simple row and column operations. The time to execute each of these steps is $O(nb)$ with very small constants, where b is the bit length of the variables in the operation.

Push_Relabel_Matching() works best when there are many unmatched edges, and therefore many edges get selected by the maximal matching pass. Each unmatched edge accounts for a fixed amount of progress in each pass (whether it gets matched or not). In general, at the beginning many matches are found in each pass, while later it may take many passes of phase 1 just to generate a single new match.

Push_Relabel_Matching()
{
1. $matched\_cost(i,j) \leftarrow$ column broadcast of the cost of the matched edge. (If the column is unmatched, broadcast ϵ.)
2. For all unmatched edges, mark eligible if $c(i,j) < matched\_cost(i,j)$.
3. Find a maximal matching of the eligible edges.
4. Column broadcast the newly matched edges new_M.
5. Unmatch the old edges from newly matched columns.
   Match the new edges in these columns.
6. Modify the column costs (increase them as high as possible without making the matched edge in the column have $c(i,j) > \epsilon$).
   Column broadcast the cost of the newly matched edges.
7. Modify the row costs (decrease them as low as possible without making the unmatched edge in the row have $c(i,j) < \epsilon$).
}

FIGURE 6. Procedure Push_Relabel_Matching() implements Phase 1 of the Match() procedure.

The second phase, called "Shortest Path Matching" (Figure 7), works much harder at the dual variable update stage, finding all feasible augmenting paths. In place of the maximal matching step, this allows us to take as many steps as possible along augmenting paths from the free S-vertices. Furthermore, by reordering the rows and columns, we can find these augmenting paths in one diagonal sweep of the matrix, in time not much greater than a simple maximal matching.

The time-consuming portion of phase 2 is in finding all feasible edges for shortest augmenting paths. In order to compute shortest paths on the cost data, ϵ is temporarily added to all costs, making all values non-negative, as in [**GPV88**]. Furthermore, in the case of a zero-length path, we must be able sort the vertices in path order, so that we can search the path in one sweep of the matrix. We add an extra $1/2n^2$ to the costs, which ensures that all edges temporarily have strictly positive cost. The value $1/2n^2$ is small enough not to affect the resulting shortest paths, since the highest precision computation is $1/n$, and the longest possible path is of length $2n$.

The heart of the computation is the calculation of the shortest path from every S-vertex to some free T-vertex. By carefully pipelining the operations, the distances between all pairs of vertices in a general graph can be done in $O(n \log nC)$ steps on a bit-serial $n \times n$ array, using the algorithm of Van Scoy [**Sco80**] (operating on integers of $O(log nC)$ bits). The algorithm implements a pipelined version of the following, in which $d_k(i,j)$ represents the length of the shortest path between i and j using only intermediate vertices in the range 1 through k.

Shortest_Path_Matching()
{
   1.   Compute the distance labels, dS, dT (shortest paths to free T-vertices).
   2.   Permute rows and columns to order S-vertices and T-vertices by distance label, with the largest distance labels on top/left.
   3.   Mark eligible edges (those which appear in shortest augmenting paths).
   4.   Set reduced costs $cb(i,j)$ as:
$cb(i,j) \leftarrow c(i,j) - d_s(i) + d_t(j)$ where arcs (i,j) are unmatched.
$cb(i,j) \leftarrow c(i,j) - d_t(j) + d_s(i)$ where arcs (j,i) are matched.
   5.   Find and augment along maximal S-paths, starting at the largest distance labels.
   6.   Set $c(i,j) \leftarrow cb(i,j)$.
}

FIGURE 7. Procedure Shortest_Path_Matching() implements Phase 2 of the Match() procedure.

for $k = 1$ **to** n
   for $i = 1$ **to** n
      for $j = 1$ **to** n
         $d_k(i,j) = min(d_{k-1}(i,j), d_{k-1}(i,k) + d_{k-1}(k,j))$

Of course, in a general graph every vertex is represented by both a row and a column. A small modification is required in order apply the algorithm to the bipartite graph. Let $dS_k(i,j)$ represent the shortest path from an S-vertex i to a T-vertex j using only intermediate S- and T-vertices in the range 1 through k. $dT_k(i,j)$ is defined similarly for $T \to S$ paths. Note that if (k,k) is a matched edge, we can compute the length of an (i,j) path using $dS(i,k) + dT(k,k) + dS(k,j) = dS(i,k) + dS(k,j) + c(k,k)$. Thus, we begin by permuting rows and columns so that the r matched edges are on the first r diagonal elements. We can then implement the following algorithm, using a direct adaptation of Van Scoy's algorithm.

for $k = 1$ **to** r
   for $i = 1$ **to** n
      for $j = 1$ **to** n
         $dS_k(i,j) = \mathbf{min}\ (dS_{k-1}(i,j), dS_{k-1}(i,k) + dS_{k-1}(k,j) + c(k,k))$

A description of the procedure is given in Figure 8. The subroutine which finds the shortest paths distances between all pairs in the bipartite graph is given in Figure 9. See [**Sco80**] for a more detailed description of the all pairs shortest paths portion of the algorithm.

Next, we sort the rows and columns by distance label, so that the highest distances reside in the upper left corner and the lowest reside in the lower right.

Distance()
{
   Initialization:
   1. Permute rows and columns so that matched edges are on the first r elements of the diagonal.
   2. Set $dS(i,j) \leftarrow c(i,j) + \epsilon + 1/2n^2$. (If (i,j) matched, $dS \leftarrow nC$).
   3. $dT(i,i) \leftarrow c(i,j) + \epsilon + 1/2n^2$.
      (If (i,i) unmatched (i.e., $i > r$), $dT \leftarrow 0$).
   4. Row broadcast $dT(i,i)$ to all $dT(i,j)$.
   5. $A, N, W, S, E, NW, SW, SE, NE \leftarrow 0$.
   6. $A(0,0) \leftarrow 1$.

   Shortest path computation:
   7. **call** All_Shortest_Paths().

   Mark the edges on the shortest augmenting paths:
   8. $Y \leftarrow Column\text{-}OR(M)$.
   9. $lS \leftarrow Row\text{-}Minimum(dS + (Y \times nC))$.
   10. $lT \leftarrow Column\text{-}Maximum(M \times (lS + dT))$.
   11. $G \leftarrow 0$.
   12. **if** $((lS - lT = c + \epsilon + 1/2n^2)$ **or** $M = 1)$ $G \leftarrow 1$ **endif**
}

FIGURE 8. Computation of Edges on Shortest Augmenting Paths.

All_Shortest_Paths()
{
    for $i = 1$ **to** $3r+$ **min** $\{2r, 2(n-r)\}$
        Propagate $N, W, S, E, NW, SW, SE, NE$ to corresponding neighbors.
        if $(A = 1)$ $N, W, S, E \leftarrow 1$ **endif**
        if $(A = 2)$ $NW, SW, SE, NE \leftarrow 1$ **endif**
        if $(NORTHWEST(A) = 3)$ $A \leftarrow 1$ **endif**
        if $(A = 3)$ $A \leftarrow 0$ **endif**
        if $(NW = 1)$
            $N\_data \leftarrow SOUTH(N\_data)$
            $W\_data \leftarrow EAST(W\_data)$
            $new\_data \leftarrow N\_data + W\_data$
        endif
        if $(NE = 1)$
            $N\_data \leftarrow SOUTH(N\_data)$
            $E\_data \leftarrow WEST(E\_data)$
            $new\_data \leftarrow N\_data + E\_data$
        endif
        if $(SW = 1)$
            $S\_data \leftarrow NORTH(S\_data)$
            $W\_data \leftarrow EAST(W\_data)$
            $new\_data \leftarrow S\_data + W\_data$
        endif
        if $(SE = 1)$
            $S\_data \leftarrow NORTH(S\_data)$
            $E\_data \leftarrow WEST(E\_data)$
            $new\_data \leftarrow S\_data + E\_data$
        endif
        if $(NW \vee NE \vee SW \vee SE = 1)$ $dS \leftarrow$ **min** $\{dS, new\_data\}$ **endif**
        if $(W \vee E = 1)$ $N\_data, S\_data \leftarrow dS + dT$ **endif**
        if $(N \vee S = 1)$ $W\_data, E\_data \leftarrow dS$ **endif**
    endfor
}

FIGURE 9. All_Shortest_Paths() Algorithm. $N, W, S, E, NW, SW, SE, NE$ are one bit flag array array variables which propagate in the directions of their names. A is a 2-bit array variable representing one of three states. These nine variables control the activity at each of the elements. $N\_data, W\_data, S\_data, E\_data$ contain the current (i, j) distance data flowing in each of the four respective directions.

| | Push Relabel Matching ||| Shortest Path Matching |||
Size	RTime (sec.)	CTime (msec.)	Cycles 1000's	RTime (sec.)	CTime (msec.)	Cycles 1000's
128 × 128	0.16	4.15	37.4	3.33	246	2220
256 × 256	0.31	7.92	71.2	42.7	500	4500
512 × 512	6.67	15.5	140	316	1060	9540

TABLE 5. Comparison of a Single Pass of Push_Relabel_Matching() versus a Single Pass of Shortest_Path_Matching(). RTime = Response Time, CTime = Cycle Time.

This has the effect of ordering the edges into a level graph, such that on a diagonal sweep of the matrix, the unmatched edges of level i are visited, followed by matched edges at level $i - 1$, unmatched edges at level $i - 2$, and so forth. We find a maximal (but not necessarily *maximum*) set of disjoint augmenting paths by doing a maximal matching at the highest level, i, then unmatching the necessary matched edges at level $i - 1$, followed by maximal matching at level $i - 2$, etc. This procedure, called "Maximal_S-paths()", proceeds in one $O(n)$ diagonal sweep of the array, as shown in the pseudo-code description in Figure 10.

The asymptotic complexity of a pass of the Shortest_Path_Matching() algorithm is $O(n \log nC)$, the same as for the phase 1 Push_Relabel_Matching(). Furthermore, it performs strictly more work than a pass of phase 1. In each pass, a maximal set of edges take one step along an augmenting path (as in phase 1), while some multiple step augmentations are performed, and at least one complete augmenting path is found. It efficiently finds many augmenting paths per pass when unmatched arcs are numerous, and finds at least one augmenting path even when few unmatched arcs remain. Phase 2 by itself will achieve the $O(n^{5/3} \log^2 nC)$ bound. But while phase 2 has the same asymptotic complexity as phase 1, the constant for the distance computation is nearly two orders of magnitude greater than that of the row/column operations. A comparison of the time for a single pass of phase 1 vs. phase 2 is given in Table 5. The superior performance of phase 1 is due both to its simplicity and because it uses less memory per processor. (The importance of minimizing memory per processor is that it allows the machine to avoid the delay of accessing off-chip memory.) This means that it is worthwhile to run the much faster phase 1 up until the last possible moment, when hundreds of phase 1 passes begin to be required in order to obtain one additional match, and then switch to phase 2. The optimal crossover point is explored in the experiments described in Section 4.2, along with the times required to run phase 1 of phase 2 individually to to completion.

4.2. Performance Results of the Parallel Cost-Scaling Algorithm.
We have implemented the above algorithm on the WaveTracer, and report the results in this section. (As with Munkres', we did not implement this algorithm

Maximal_S-Paths()
{
    $E \leftarrow$ eligible edges
    $A, R, C \leftarrow 0$
    $A(0,0) \leftarrow 1$ (Activate the top left element.)
    for $k = 1$ **to** $2n - 1$
        if $A = 1$ (Consider the edges on the active diagonal.)
            if $E = 1$, $M = 0$, $R = 0$, and $C = 0$ (Select eligible unmatched
            edges whose rows and columns are so far unmatched.)
                $M, R, C \leftarrow 1$ (Match them, and mark their row and column
                as matched.)
            else if $M = 1$
                if $R = 1$ or $C = 1$ (If new match selected in a matched
                edge's row or col.)
                    $M \leftarrow 0$ (Unmatch the edge)
                else
                    $R, C \leftarrow 1$ (Otherwise, mark the matched edge's row
                    and column.)
                endif
            endif
        endif
        $A \leftarrow NORTH(A) \vee WEST(A)$ (Move active diagonal down and
        right one step.)
        $R \leftarrow WEST(R)$ (Propagate row-matched flag right.)
        $C \leftarrow NORTH(C)$ (Propagate column-matched flag down.)
    endfor
}

FIGURE 10. Procedure Maximal_S-Paths(). E, M, A, R, C are one bit array variables. E marks eligible edges and M marks matched edges. A marks the active diagonal (e.g., if $A(i,j) = 1$, then element (i,j) is on the active diagonal). Flags R and C are used to mark matched rows and columns, respectively.

on the MasPar because the MPL compiler did not provide virtualization.) As discussed in Section 4.1, the controller bottleneck in the machine caused a great discrepancy in the clock time required versus the time one would calculate by counting the number of WaveTracer machine cycles used and scaling by the machine's clock rate. We therefore report here both the actual times ("response time") and the expected times, giving the number of cycles and the cycles times clock rate ("cycle time").

In order to tune the algorithm, we must first decide where the crossover from phase 1 to phase 2 should be made. Table 6 shows the times versus crossover point for the case $n = 128$. For random data at $n = 128$, it is clearly advantageous to run Push_Relabel_Matching() alone, with no crossover to Shortest_Path_Matching() at all. However, for the difficult case represented by the $i \times j$ data, a crossover at $n - 1$ appears optimal in terms of response times. The cycles used by both crossover $n - 1$ and no crossover are similar, however. This input represents a case in which it is in general hard to find many disjoint paths in parallel, and augmenting paths tend to be long. In such an instance, many passes of phase 1 may be required to complete the work of one pass of phase 2. We show the performance of both the $n-1$ crossover algorithm and no for the $i \times j$ data and phase 1 alone (no crossover) for random data in Table 7. In addition, we have included a sequential benchmark on the SUN SPARCstation-2. While this machine is not the "high-end workstation" that would qualify for an even comparison with the WaveTracer, the $50 - 250$ factors of speedups in response time demonstrate the massively parallel machine's favorable price-performance. The best achievable "cycle times" show an even more dramatic advantage.

The speed of these "theoretically good" algorithms does not appear competitive with the Munkres implementation on the AP2S given the previous section. However, it must be noted that we no longer have the use of row and column busses, providing row and column OR in one cycle. The Push Relabel pass would be much improved by row/column broadcast, since row/column minimum and maximum are heavily used. For the 16K processor machine, this would mean a factor of 128 improvement in row and column operations. The Shortest_Path_Matching() algorithm could probably also be improved by using the busses. The majority of the time is spent in the distance computation, which, as written, uses only local communication. However, with broadcast we could write a much simpler non-pipelined algorithm with significant constant-factor improvement in the running time (perhaps a factor of ten).

Once the Munkres algorithm is implemented on the WaveTracer, the speeds measured here become more in line with those for Munkres. The Munkres algorithm is still running faster on random data, by a factor of around 3. With random data, it is often possible to complete a large portion of the matching with very little work, giving the edge to the simpler implementation. However, for the difficult $i \times j$ data, notice that the scaling algorithm, with its better worst-case asymptotic bound, is about 8 times faster in cycle count (over 20 times faster in actual response time).

Crossover	Resp. Time (sec.)	Cycle Time (sec.)	Cycles (millions)
	128×128 Uni. Random Data, $0 - 100K$		
123	202	13.6	123
124	193	12.8	115
125	178	11.5	103
126	169	10.5	94.2
127	155	8.95	81.0
128	68.0	2.31	20.8
	128×128 Geom. Random Data, $0 - 1M$		
123	215	14.5	130
124	201	13.3	120
125	196	12.7	114
126	180	11.2	101
127	168	9.70	87.3
128	70.2	2.38	21.4
	$128 \times 128, i \times j$ Data		
123	243	16.8	151
124	225	15.1	136
125	204	13.1	118
126	189	11.2	101
127	163	8.41	75.7
128	243	8.29	74.6

TABLE 6. Determination of the Optimal Crossover Point from Phase 1 to Phase 2.

Size	Resp. Time (sec.)	Cycle Time (sec.)	Cycles (millions)
	Push Relabel Matching alone, Uni. Rand. Data, $0-100K$		
128×128	64.1	2.05	18.4
256×256	197	6.14	55.3
512×512	6160	17.2	155
	Push Relabel Matching alone, Geom. Rand. Data, $0-100K$		
128×128	82.7	2.69	24.2
256×256	233	7.26	65.4
512×512	6960	19.4	175
	Two-Phase Alg., Crossover at $n-1$, $i \times j$ Data		
128×128	163	8.41	75.7
256×256	2380	23.8	214
512×512	28900	85.1	766
	Sequential Timing for Push Relabel Matching alone Uni. Rand. Data, $0-100K$, on SUN SPARCStation-2		
Size	Time (sec.)		
32×32	19.9		
64×64	277		
128×128	3410		
256×256	48800		

TABLE 7. Timing for Optimally Tuned Crossover (crossover at $n-1$ for $i \times j$ data, and for no crossover (phase 1 alone) for random data). For comparison, the best sequential times we obtained using the algorithm are also included.

A major source of slowdown is in the cost scaling. To simplify implementation, we used only a scale factor of 2. Therefore, a great many cost scaling passes were required, often only to bring a few new edges into the lower ϵ range. For instance, consider a 128×128 problem with data in the range 0 to 1,000,000. The number of passes with scale factor 2 is about 27. This means that if we run the crossover at $n - 1$, the slow Shortest Path Matching phase will be invoked 27 times in all. If we choose to scale ϵ by a larger factor, then we risk requiring more passes of the fast phase 1, but we are guaranteed to reduce the number passes of the slow phase 2 (since it runs exactly once per ϵ value). Taking this approach to its extreme, we can do away with cost scaling altogether. For random data, it is generally quite easy to get the first 90% or so matches. If, in the scenario above, we were to get all but 12 matches quickly, we could then apply phase 2 at most 12 times in order complete the matching without using any scaling at all.

In Tables 8 and 9 we explore the best choice of factor for ϵ scaling, for the two phase algorithm and for each of the phases alone. Notice that for random data the Push Relabel Matching phase works best in a single pass, without any ϵ-scaling. This is in sharp contrast to the theoretical bounds for this algorithm. On the other hand, the $i \times j$ data behaves much as would be predicted by the theory, preferring a smaller scale of perhaps 16 (line 4 in the logarithmic scale shown in Tables 8 and 9). This is not surprising, since the theoretical bounds target the worst case, which this input tends to approximate.

The very best performance was obtained using the $n - 1$ crossover for all three data types. However, the proper choice for the scale divisor is again highly dependent upon the type of data. Table 10 shows the performance of the algorithms for various data types when the ϵ scaling has been optimized for the data type. The performance of the optimally tuned algorithm is about a factor of two better than the parallel Munkres algorithm on the WaveTracer. That the factor is only two is probably explained by the fact that the optimal crossover doesn't occur until around $n - 1$. This means that any *simple* method is good enough until the very last pass, at which point the fast Shortest Path computation has value. A weaker observation is that the performance advantage of the optimized algorithm appears to grow with the problem size (perhaps demonstrating its asymptotic superiority).

5. Summary and Discussion of Results of Part 1

Our SIMD implementations squeezed the most out of the machine but came up short against the following inherent limitations: low memory-to-processor bandwidth, slow controller, breaks in the host-to-array-processor pipeline.

The algorithm on the GAPP was written in assembler code, while the Wave-Tracer implementation was written in MultiC in a few days. For reasonable comparisons we have included a count of the machine cycles used by each of

| | Push Relabel Matching alone, 128 × 128 ||||||
| | Uniform 0-100K || Geometric 0-1M || $i \times j$ ||
$\log_2(scale)$	Response	Cycle	Response	Cycle	Response	Cycle
none	15.3	0.51	11.6	0.39	2600	86.7
1	68.0	2.31	70.2	2.38	243	8.29
2	41.9	1.40	41.9	1.40	122	4.05
3	32.6	1.09	31.4	1.05	150	5.00
4	27.7	0.92	28.8	0.96	128	4.29
5	27.9	0.93	26.9	0.89	128	4.26
7	25.9	0.86	24.4	0.81	204	6.79
10	29.0	0.97	26.7	0.89	182	6.07
15	59.5	1.98	48.1	1.60	470	15.7
20	72.2	2.40	52.1	1.74	470	15.7
30	71.2	2.37	49.9	1.66	470	15.7

| | Shortest Path Matching alone, 128 × 128 ||||||
| | Uniform 0-100K || Geometric 0-1M || $i \times j$ ||
$\log_2(scale)$	Response	Cycle	Response	Cycle	Response	Cycle
none	22.0	1.62	21.2	1.57	339	25.0
1	363	26.8	382	28.2	1370	101
2	197	14.6	195	14.4	754	55.6
3	136	10.0	144	10.6	529	39.0
4	102	7.50	113	8.33	558	41.2
5	89.4	6.59	93.4	6.89	454	33.5
7	68.8	5.07	67.2	4.95	376	27.7
10	47.4	3.49	47.6	3.51	281	20.7
15	33.4	2.46	33.9	2.50	177	13.1
20	32.7	2.41	32.3	2.38	345	25.4
30	22.0	1.62	21.3	1.57	339	25.0

TABLE 8. Performance Improvements obtained by increasing the ϵ scaling factor for Push Relabel Matching and Shortest Path Matching individually. All times are given in seconds. The table lists the base 2 logarithm of the scale divisor used. At a (log) scale of 30 ($scale = 2^{30}$), only two passes are made, with $\epsilon = C$ and $\epsilon = 1/n$. "None" means that the algorithms were run in a single pass (by setting $\epsilon = 1/n$ at the start).

	Two-Phase Algorithm, Crossover at 127, 128×128					
	Uniform 0-100K		Geometric 0-1M		$i \times j$	
$\log_2(scale)$	Response	Cycle	Response	Cycle	Response	Cycle
none	8.50	0.43	8.97	0.46	2510	85.6
1	155	8.95	168	9.70	163	8.41
2	87.2	4.95	90.8	5.20	120	5.51
3	62.6	3.50	68.2	3.83	108	4.65
4	50.9	2.80	55.8	3.14	93.1	3.96
5	45.3	2.43	49.1	2.71	188	6.89
7	39.7	2.09	39.6	2.12	196	7.15
10	37.8	1.89	38.8	1.94	161	5.78
15	40.0	1.82	59.1	2.47	448	15.4
20	41.0	1.84	69.1	2.82	437	15.0
30	34.2	1.46	62.1	2.42	437	15.0
	Two-Phase Algorithm, Crossover at 126, 128×128					
	Uniform 0-100K		Geometric 0-1M		$i \times j$	
$\log_2(scale)$	Response	Cycle	Response	Cycle	Response	Cycle
none	9.72	0.58	10.1	0.60	2500	84.5
1	167	10.4	179	11.2	191	11.2
2	94.5	5.83	99.0	6.09	101	6.11
3	69.8	4.27	71.3	4.35	90.3	4.95
4	55.8	3.34	59.2	3.59	112	5.20
5	50.8	3.02	53.2	3.17	174	6.96
7	44.4	2.57	42.8	2.46	209	7.83
10	44.3	2.36	41.9	2.26	138	5.28
15	46.9	2.29	55.3	2.54	433	15.0
20	54.3	2.54	60.1	2.70	429	14.7
30	48.1	2.16	53.7	2.31	429	14.7

TABLE 9. Performance Improvements obtained by increasing the ϵ scaling Factor for the two-phase Algorithm. All times are given in seconds.

Size	Response Time (sec.)	Cycle Time (sec.)	Cycles (millions)
	Crossover at $n-1$, no scaling, Uni. Rand. Data, $0-100K$		
128×128	9.32	0.47	4.2
256×256	114	1.13	10.2
512×512	1040	3.25	29.3
1024×1024	11,428.97	8.96	80.6
	Crossover at $n-1$, no scaling, Geom. Rand. Data, $0-1M$		
128×128	12.0	0.57	5.0
256×256	1273	1.24	11.2
512×512	1082	3.23	29.1
1024×1024	10,251.88	8.06	72.5
	Crossover at $n-1$, $scale = 2^4$, $i \times j$ Data		
128×128	93.1	3.96	35.6
256×256	1690	15.4	139
512×512	21700	61.7	555
1024×1024	220000	170	1530

TABLE 10. Performance with Optimally Tuned Epsilon Scaling.

the machines. However, the controller at the front end of the array has also proven to be a major issue. In particular, the response times on the WaveTracer are much slower than the computation time. This is almost entirely due to the fact that the controller is a bottleneck, and is unable to decode instructions for the parallel machine to keep pace with the speed of the processor. This is in contrast to our controller, called the DMC, which could keep pace with the machine, and allowed subroutine calls, memory allocation, and reprogrammability without any penalty. This investigation demonstrates the necessity of such a controller to realize the performance promised by massively parallel processors.

Part 2. Auction Algorithm on Parallel Architectures

The second part of the paper is organized as follows. In Section 6, we describe the Auction algorithm. In Section 7, we describe the computationally important aspects of Auction. We then describe our experiences with the Auction algorithm on a variety of parallel architectures as follows: Section 9 for SIMD machines, Section 10 for shared memory multiprocessors, Section 11 for Multiple-Instruction Multiple-Data (MIMD) machines, and Section 12 for a distributed network of workstations.

6. Description of the Auction Algorithm

We refer the reader to [**Ber91**] for an excellent tutorial. We provide a brief description of the algorithm, summarized from [**Ber91**]. We start with a weight matrix w where $w(i,j)$ is the cost of assigning person i to object j.

Initially, all persons are unassigned. Each object, j, has a price, $price(j)$, initially 0. The Auction algorithm proceeds in "rounds" or "bids". In a bid, an unassigned person bids for the object, j_i with maximal value m_i where $m_i = max_j\{w(i,j) - price(j)\}$. It also computes the second best object value, $nm_i = max_{j, j \neq j_i}\{w(i,j) - price(j)\}$. It then raises the price of j_i by $m_i - nm_i + \epsilon$, unassigns any person assigned to that object and assigns itself to that object. This process is repeated until all objects have received at least one bid. If $\epsilon < \frac{1}{n}$ and $w(i,j)$ are integers, then the assignment obtained upon termination is optimal. In our implementations, we shall define two vectors $s[1..n]$ and $t[1..n]$ where $t[j]$ is the object to which the jth person is matched and $s[i]$ is the person to which the ith object is matched.

We use ϵ-scaling in our implementations, which reduces the number of bids. ϵ-scaling consists of repeating the algorithm several times, starting with a large value of ϵ and progressively decreasing it until it is less than $\frac{1}{n}$ for integer costs. Also, for a given ϵ, rather than wait for all objects to be matched before we reduce ϵ, we set a threshold. Once the number of unmatched objects falls below this threshold, we move to the next ϵ value. The threshold is gradually reduced until it becomes 1 when ϵ is 1.

Gauss-Seidel vs. Jacobi. The variation of the Auction algorithm when only one person bids at a time, is called the *Gauss-Seidel* implementation. If all unassigned persons bid at the same time, we get the *Jacobi* implementation. Both yield correct results.

Remark. Note that Auction produces a maximum weight bipartite matching, whereas the Assignment Problem is to find a minimum weight bipartite matching. This is easily fixed by $\forall i \forall j, w(i,j) \leftarrow maxcost - w(i,j)$ where $maxcost = max_{i,j}\{w(i,j)\}$.

Complexity. The complexity of the Auction algorithm is $O(n^3 \log(nC))$, where $C = maxcost$.

7. Computationally Important Aspects of Auction

In this section, we shall consider the aspects of the Auction algorithm which are important from the viewpoint of efficient implementations.

7.1. Jacobi vs. Gauss-Seidel. The decision to use the Gauss-Seidel variation of the Auction algorithm rather than the apparently more parallel Jacobi variation was based on the following:

(i) the Jacobi variation requires more bids than the Gauss-Seidel [**WZ91, BC89**], and

(ii) *A reasonably high proportion of the total number of bids required by the Auction algorithm is spent with very few unmatched objects.* Therefore, the Auction algorithm is simply not as parallel as one would expect at first sight. Parallelizing across the bids of the different unassigned persons is going to run up against Amdahl's Law because a significant fraction of the bids will have very few unassigned persons. We summarize our results in Table 11. n is the problem size. Consider the values of α_3 and U in a given row. This is to be interpreted as: the fraction of bids in which the number of unassigned objects is less than or equal to U_3 is α_3 for Data Set III. (see Section 8.1 for explanation of Data Sets). α_1 and α_2 represent the same information for Data Sets I and II respectively. We noticed that for a given x, the fraction of bids in which less than or equal to x objects are unassigned seems to increase as the problem size increases.

n	$\min(\alpha_1)$	$\max(\alpha_1)$	$\text{avg}(\alpha_1)$	$\min(\alpha_2)$	$\max(\alpha_2)$	$\text{avg}(\alpha_2)$	α_3	U
1024	30.8	74.1	58.1	45.7	74.2	53.6	67.3	≤ 32
1024	19.2	71.5	53.5	39.8	72.3	50.8	64.5	≤ 16
1024	11.7	69.6	46.5	19.9	70.5	46.7	61.0	≤ 8
1024	7.3	67.4	22.1	2.64	68.0	21.8	55.3	≤ 4
1024	1.114	16.1	8.35	0.861	32.6	10.7	48.5	≤ 2
1024	0.554	15.3	6.93	0.307	32.1	8.10	31.1	≤ 1
2048	52.2	74.2	61.4	53.7	78.2	61.1	40.6	≤ 32
2048	48.5	71.9	58.4	48.8	76.7	57.9	35.0	≤ 16
2048	26.0	70.0	49.9	34.7	74.3	48.2	28.4	≤ 8
2048	1.88	51.2	14.4	2.09	68.6	15.3	21.9	≤ 4
2048	0.911	4.02	2.27	1.24	39.8	7.88	14.7	≤ 2
2048	0.114	3.37	1.12	0.469	32.0	6.01	8.1	≤ 1
4096	47.0	67.1	57.8	44.3	66.4	61.4	51.3	≤ 32
4096	42.7	64.3	54.6	41.1	64.5	58.9	46.8	≤ 16
4096	6.40	61.9	37.6	9.35	62.7	49.7	40.9	≤ 8
4096	2.21	6.03	3.43	2.26	3.89	2.79	36.7	≤ 4
4096	1.02	4.75	2.27	.483	2.34	1.61	31.3	≤ 2
4096	0.36	2.51	0.991	.145	1.43	0.894	21.3	≤ 1
8192	50.6	81.2	68.9	59.0	78.8	69.3	52.7	≤ 32
8192	47.3	80.1	67.0	56.5	77.7	67.6	48.7	≤ 16
8192	18.6	79.0	57.9	41.0	74.4	62.2	43.6	≤ 8
8192	2.08	61.7	24.8	2.50	61.5	18.5	37.8	≤ 4
8192	0.64	3.58	1.73	0.47	2.06	1.40	28.7	≤ 2
8192	0.30	2.60	0.971	0.26	1.30	0.877	17.7	≤ 1

TABLE 11. Fraction of Bids with at most a Given Number of Unassigned Persons.

7.2. Other Computationally Important Aspects.
Once we had restricted ourselves to the Gauss-Seidel variation of Auction, the following features were important.

(i) Memory requirement and processor-to-memory bandwidth. The weight matrix grows as $O(n^2)$. Therefore, memory requirements and processor-to-memory bandwidth are of concern. We felt that that memory would be the constraining factor rather than CPU cycles. In other words, we suspected that a problem which fully utilized the memory of the machine would be solvable in a relatively short amount of time.

(ii) Efficient global maximum computation. In each bid, each processor performs a few computations involving $\frac{n}{p}$ numbers, creating a partial result. Then, a few associative functions (e.g., find global maximum, find next global maximum) are performed on these p partial results. Therefore, we must have an efficient means of computing an associative function of the partial results of each processor.

The implication of this for $p \ll n$, is that most of the work is local to a processor since the weight matrix need not be shared. However, since even for the highest n that fills the memory of the system, the local work is of the order of milliseconds, the ability to disseminate information to and coalesce information across the processors must be reasonably fast. For a fixed n, as p increases, the ability to perform these global operations must also increase since the local work decreases.

(iii) Locality of reference. The Auction algorithm does not exhibit any regularity of access patterns across bids. In other words, even when the number of unmatched persons is small, the identity of the persons in this small set fluctuates rapidly and without any regularity. *This is possible the most damning idiosyncracy of the algorithm since it makes for very poor cache utilization.*

Since there is no way to predict which row of the weight matrix will be used in the next iteration while the current iteration is in progress, pre-fetching cannot be exploited.

(iv) Regularity of access patterns within a bid. Block transfers are useful since the data necessary to calculate a bid are in contiguous memory locations. Since each processor will need to work on $\frac{n}{p}$ contiguous elements of a row of the weight matrix during a bid, block transfers of these elements can be gainfully exploited, provided $\frac{n}{p}$ is reasonably large.

(v) Independence of computations within a bid. During a bid, each processor computes a value for each of the $\frac{n}{p}$ objects in its possession for the unassigned person under consideration. This operation is amenable to efficient vectorization.

(vi) Computation to communication ratio. The amount of computation done per datum read is extremely small, consisting of a few integer arithmetic operations and comparisons. This makes a high processor-to-memory bandwidth essential.

7.3. Basic Approach to Parallelization of Auction. In light of the above discussion, we adopted the following basic strategy on each of the architectures. Processor k, is assigned $\frac{n}{p}$ distinct columns of the weight matrix, and $\frac{n}{p}$ elements of the price vector. The s and t vectors are either distributed across the processors or given solely to a master processor in different architectures. In a bid, the kth processor calculates m_i^k, nm_i^k, j_i^k and $price[j]^k$, which are the values m_i, nm_i, j_i and $price[j_i]$ would have had had the columns assigned to k been all the columns of the weight matrix. Given these partial results, we still need to compute j_i and $price[j_i]$ and u, where $t[u] = j_i$ at the end of the previous bid. (In other words, u is the person unassigned as a result of i's successful bid. Note that u may be undefined, if the object that i bid for was unassigned.) This requires performing an associative function on the partial results of all processors. The new price of the object which has been successfully bid for, the identify of the successful bidder and that of the person unassigned as a result of the successful bid must be sent to the processor(s) which store this information.

A consequence of this approach is that the speed-up of the parallel implementations will be $\frac{n}{\frac{n}{p}+A(p)}$ where $A(p)$ is the time to perform an associative function of p numbers, one to a processor.

8. Results

We shall adopt a slightly unusual style of presentation. In this section, we shall present our results as follows. We shall first present the number of bids required by Auction for different problem sizes and different types of inputs. We shall then show how the time for a bid varies with problem size for the different architectures. The total time can be calculated by multiplying the number of bids by the time per bid. While the number of bids varies significantly as a function of the type of data used, the time per bid is only marginally affected by the characteristics of the input. We show how the time to load the problem varies with problem size on the different architectures.

In the following sections, we analyze the computational considerations of the previous section in the context of a variety of architectures. For each architecture, we shall (i) describe the relevant architectural features, (ii) present implementation details relevant to that architecture, (iii) conclude with a discussion of the results and what lessons we learned.

We feel that presenting the results in the format we have chosen will give the reader a better appreciation for the design choices we made and the architectural obstacles and support we had to contend with. It will also throw into contrast the differences in the architectures. By divorcing the discussion of the algorithm from that of the architecture, it will throw into relief performance bottlenecks/improvements due to the algorithm and those induced by the architecture.

Among the important architectural features we shall consider are the memory, processor-to-memory bandwidth, architectural support for and speed of computing an associative function of p variables, and the communication-to-computation ratio.

8.1. Test Cases. In this section, we describe the different test cases on which the algorithm was tested.

(i) Data Set I. $w(i,j)$ was set to a random value, distributed uniformly in the range $\{1,..,131072\}$.
(ii) Data Set II. $w(i,j)$ was set to a random value, distributed uniformly in the range $\{1,..,1048576\}$.
(iii) Data Set III. $w(i,j)$ was set to $i \times j$.

Each data point we have presented was the average of 10 runs. For Data Sets I and II, the random numbers were generated with the UNIX random number generator, $random()$, using different random seeds for each run. If s was the seed for a given run, processor i, used $i \times s$ as its seed. The time per bid in each case was the average of the different times observed.

We did not test our algorithms for sparse problems for the following reason. As we have argued above, we felt that Auction was best suited for exposing the parallelism within a bid, rather than across bids. In making the problem sparse, we would further reduce what little parallelism existed. Auction is normally considered an appropriate algorithm for sparse problems. However, this is true under the uniform access model (eg. PRAMs [**KR90**]). When this assumption breaks down, as it does for large problems on massively parallel processors, Auction will utilize the resources better on dense problems than on sparse ones.

8.2. Number of Bids vs. Problem Size. In this section, Tables 12 summarize our results showing how the number of bids varies as a function of problem size for the different test cases. The number of bids for Data Set I is nb_1. The number of bids for Data Set II is nb_2. The number of bids for Data Set III is nb_3.

n	$\min(nb_1)$	$\max(nb_1)$	$\text{avg}(nb_1)$	$\min(nb_2)$	$\max(nb_2)$	$\text{avg}(nb_2)$	nb_3
128	2584	2773	2706.8	2562	2909	2741.1	3172
256	5322	5962	5564.0	6148	6816	6649.0	11122
512	12226	15221	13042.9	12889	18587	14033.9	35720
1024	36674	79158	56284.7	41419	88864	60280.2	97232
2048	101061	191728	133413.0	106416	237651	133652.3	124861
4096	202679	312266	258081.9	208180	352964	304628.4	364751
8192	484827	1319080	865613.9	623361	1229350	865107.9	895926

TABLE 12. Number of Bids versus Problem Size.

Remark. Contrary to Bertsekas's suggestion that "For non-sparse problems, sometimes $\Delta = 1$, which in effect bypasses ϵ-scaling, works quite well" [**Ber91**], we found that ϵ-scaling had a dramatic impact in reducing the running time for fully dense problems. Intriguingly, we did not observe significant changes in the number of bids as the range varied. Also, while the structured $i \times j$ problem performed very badly without ϵ-scaling, with scaling it seemed to perform similarly to random data.

8.3. Time per Bid on Different Architectures. In this section, Table 13 summarizes our results on the time taken per bid as a function of problem size. The first column, n, is the problem size. T_{seq} is the sequential time on a SUN

SPARCstation-2. T_{maspar} is the time taken on the MasPar MP-1. T_{wvt} is the time taken on the WaveTracer Zephyr. T_{cm5} is the time taken on the CM-5 [2]. T_{sgi_4} is the time taken on the Silicon Graphics, using all 4 processors. T_{sgi_2} is the time taken on the Silicon Graphics, using 2 of the 4 processors. T_{sgi_1} is the time taken on the Silicon Graphics, using 1 of the 4 processors. T_{dist} is the time taken on the distributed network of workstations.

For a given n, the time per bid is reasonably constant across bids. The exception is the Silicon Graphics when $n = 4096$ (See Section 10 for details).

n	T_{seq}	T_{maspar}	T_{wvt}	T_{cm5}	T_{sgi_4}	T_{sgi_2}	T_{sgi_1}	$T_{dist_{16}}$
128	80	298	10409	603	161	124	147	37349
256	157	288	10488	605	138	166	262	41597
512	304	291	10602	615	169	277	484	40900
1024	609	291	10885	607	313	533	1004	43405
2048	1227	295	10759	620	603	1058	2037	44351
4096	N/A	281	10223	613	N/A	N/A	N/A	42583
8192	N/A	N/A	N/A	625	N/A	N/A	N/A	N/A

TABLE 13. Time per Bid (in μ-sec) versus Problem Size on Different Architectures.

8.4. Time to Load Data. In this section, Table 14 summarizes our results showing how the time to load the problem description varies as a function of problem size on the different architectures.

Both MasPar and WaveTracer provide parallel functions to load data to the processor array. MasPar loads data directly from disk to processor array memory, meanwhile WaveTracer loads from the front end. Therefore the timing given in Table 11 is from disk for the MasPar and from the front end memory for WaveTracer (data is loaded sequentially from disk to host memory). The time to load the Silicon Graphics is roughly the same as the time to load the SUN SPARCstation-2. It remains the same no matter whether 1, 2 or 4 processors are used.

9. SIMD Implementation

We implemented the Auction algorithm on an 8K MasPar MP-1208 and an 8K WaveTracer Zephyr.

9.1. Relevant Architectural Features. Speed. On the WaveTracer, each processor is a 9 MHz processor capable of 0.947 MIPS and 0.008544 MFLOPS. On the MasPar, each processor is a 12 MHz processor capable of 1.83 MIPS and 0.092 MFLOPS.

---

[2] At the time the experiments were performed, the CM-5 did not have the vector units

n	T_{seq}	T_{maspar}	T_{wvt}	T_{cm5}	$T_{dist_{16}}$
128	0.0283	0.690	0.480	0.0506	$T_{seq}/16$
256	0.0638	0.861	1.04	0.0995	$T_{seq}/16$
512	0.160	2.43	2.46	0.198	$T_{seq}/16$
1024	0.424	7.67	6.49	0.396	$T_{seq}/16$
2048	1.47	23.6	18.0	N/A	$T_{seq}/16$
4096	5.00	68.7	64.9	N/A	$T_{seq}/16$
8192	N/A	N/A	N/A	N/A	N/A

TABLE 14. Time to Load (in seconds) Data versus Problem Size on Different Architectures.

Bandwidth and Memory. The processor-to-memory bandwidth is 1 MByte/sec on the WaveTracer and 1.5 MByte/sec on the MasPar for reads from cache. The processor-to-memory bandwidth is 0.25 MByte/sec on the WaveTracer and 1 MByte/sec on the MasPar for reads from memory (32 KBytes for WaveTracer and 16 KBytes for MasPar), for a total memory of $\frac{1}{2}$ GBytes for WaveTracer and 128 MBytes for MasPar.

There is no support for block transfers. For these machines $\frac{n}{p}$ (the ratio of problem size to number of processors) was less than 1 and so block transfers would not have helped. While the processors have registers, there is no hardware support for caching. This is not a drawback for Auction since there is no locality of reference across bids.

Associative Function. The time take to perform a 4-Byte global maximum on the MasPar is 0.51 msec and 0.34 msec from memory and register (benchmark using MPL compiler) respectively and WaveTracer is 2.9 msec (multiC compiler). Global-OR of a 4-Byte word on the MasPar takes 0.1 msec and 0.07 msec from memory and register respectively, and 2.9 msec on WaveTracer. The global maximum is used to compute the highest value among the values of the different objects and the global-OR is used to locate the object with the maximum value.

On the MasPar, global-OR hardware which uses an inclusive-OR reduction tree is provided. The logical OR-tree consolidates a status responses from all the processors back to the host. It extract a scalar value through the OR tree from a register of the PE array only takes a few cycles plus a few clocks of pipeline overhead. The WaveTracer has a global bus that can perform the NAND of a bit from each PE in one cycle.

Implementation Details. The WaveTracer can be configured either as a 1-D, 2-D or 3-D array. We configured it as a 1-D array and got terrible performance. It turns out that the straight-forward configuration we had adopted ends up giving us $\sqrt{p}$ processors of a $\sqrt{p} \times \sqrt{p}$ 2-D processor array. The problem was easily solved in the following fashion.

On WaveTracer, function *multi_perform* establishes a parallel storage space and initializes parallel objects; it then calls another function to perform parallel processing. This function is declared as follow:

int multi_perform($func$, $xdim$, $ydim$, $zdim$);

where $func$ is the parallel function, and $xdim$, $ydim$, and $zdim$ are the dimensions. Instead of setting $xdim$ to p and $ydim$ and $zdim$ to 1 as one would for a 1-D configuration, we set $xdim$ and $ydim$ to $\sqrt{p}$, and $zdim$ to 1 to configure the machine as a 2-D array. This is a speed-trap for the unwary and while the fix was not onerous, the need for such a fix and the solution should be made more apparent.

A small part of each bid - determining which object had to be assigned to which person, unassigning objects and updating the price - was performed on the front end.

9.2. Results and Discussion. We learned the following lessons:

(i) **Controller Bottleneck.** On the WaveTracer, for our particular application, the cycle-count for operations is a very poor indicator of actual computation time. This is because of the controller bottleneck. The problem is described in detail in Section 3.3.

Most of the controller (microsequencer) problems mentioned in Section 3.3 did not quite plague the MasPar implementations This is because the MasPar has a sequencer and although the time to load and unload the sequencer is considerable, if the program fits within the memory of the sequencer, then there is no delay once the program starts up. We concluded that for the WaveTracer the array processor is poorly utilized unless

(a) it is executing a long operation (eg., floating-point multiplication) such that the computation time subsumes the time for the decoding and broadcast process

(b) sufficient parallel slackness exists ($p \ll v$), where v is the number of virtual processors. In this case, the same decoded instruction is executed $\frac{v}{p}$ times on each processor.

Neither of the two conditions are met in the case of Auction because the amount of computation performed per datum read is small and the memory is exhausted well before v even reaches p.

(ii) **Breaking the Instruction Pipe.** The flow of information from the host to the array processors is pipelined. If this pipe is broken, performance degrades (see Section 3.3).

10. Shared Memory Implementation

We implemented the algorithm on a 4-processor workstation from Silicon Graphics, the IRIS 4D/340 DVX, chosen because it was representative of shared memory systems.

10.1. Relevant Architectural Features. Speed. Each processor is a 25 MHz R3000 RISC based processor capable of 20 MIPS and 3.3 double precision

MFLOPS.

Bandwidth and Memory. The processor-to-memory bandwidth is 200 MBytes/sec for the primary cache (64 KBytes/sec/processor), and 16 MBytes/sec/processor for main memory (48 MBytes) and 0.5 MBytes/sec for disk (2 GBytes). It has block transfer capability, transferring data from main memory to the secondary cache.

Associative Function. There is no hardware support for finding an associative function. The global maximum is computed by one of the processors, designated as the master, reading the partial results of the other processors. Broadcasts are performed by the master writing the new value of X into its copy. There is hardware support to keep the caches consistent.

Implementation Details. A single processor was designated as the master processor. It was responsible for deciding which person would bid and also for coalescing the partial results of the other processors. Had there been more processors, the work of coalescing bids would have been distributed among the processors.

Computing an Associative Function. Reading the partial results of the other processors is a simple read from shared memory. This is fast, provided that the partial results are available (i.e., no busy waiting). Since the processors are identical and the work is load balanced, busy waiting was not an overhead.

We had essentially a producer-consumer relationship between the master processor and the rest of the processors. We chose not to incur the overhead of using system calls for the synchronization. Instead we used the following tagged scheme and busy-waiting (justified by the fact that we were running in single user mode and that the workload was evenly balanced, thereby making the waiting time minimal).

Details of Tagged Scheme to avoid Synchronization. For simplicity of exposition, let each processor have a local variable called b, representing the bid number, initialized to 0. (The first bid is numbered 1.) We also have the shared variables, B, B_i, and P_i. For simplicity of exposition, assume that each of these variables is a tagged variable, consisting of a value field and a tag field. Assume that a write is atomic and succeeds only if the tag of the tuple being written is greater than the existing tag. Since this architectural feature does not exist, we simulate it as follows. A write to a variable consists of writing a value and then a tag. A read to a variable consists of reading the tag and then the value. This protocol has been suggested by Lamport [**Lam77**]. The master processor writes $B \leftarrow (j, b)$ indicating that the bth bid will consist of computing the bid for person i. The slave processors read from B to determine which "person" is bidding. The ith slave processors writes $P_i \leftarrow (p_i, b)$ indicating that the partial result for the bth bid is p_i. The master processor accumulates the results by reading partial results from P_i. Detailed protocols are in Figure 11. Essentially, we exploit the fact that the algorithm is "knowledgeable" [**Sub92a**].

$b \leftarrow 0$
repeat
    $B_i \leftarrow B.$ { find out which person to bid for }
    if $B_i.tag > b$ **then** {if processor i has not worked on this already }
        $b \leftarrow B_i.tag$
        compute, p_i, partial result for person $B_i.value$
        $P_i \leftarrow (p_i, b)$
    endif
until computation terminates.
Protocol followed by ith slave processor.

repeat
    $B \leftarrow (j, b)$ { decide which person to bid for in bth bid }
    repeat
        $alldone \leftarrow true$
        for $i \leftarrow 1$ **to** $p-1$ **do**
            if $P_i.tag < b$ **then**
                $alldone \leftarrow false$
            endif
        endfor
    until $alldone = true$
until computation terminates.
Protocol followed by master processor.

FIGURE 11. Protocols followed to avoid Explicit Synchronization.

10.2. Results and Discussion. The machine was used essentially as a distributed memory machine with no dynamic reallocation of work. The shared memory was used to provide a low-latency inter-processor communication mechanism. While a single slow processor could degrade the entire performance, this was not observed because of even load balancing.

Given a static allocation of columns to processors, a shared memory system is not the best architectural choice. This is because there really is very little information to share. Most of the information, the weight matrix, is in memory and accesses to the main memory are serialized, albeit through a very high speed bus. This must not be read as a criticism of shared memory machines in general or the Silicon Graphics in particular. The point is that for problems where most of the problem description can be easily distributed across the processors and the amount of shared information is small, true shared memory systems might be an overkill.

Dynamic Task Allocation. When the problem is dense, there is not much reason to do dynamic task allocation since the load is evenly balanced. When the problem is sparse, we would like to use the shared memory to dynamically assign processors to work. Our first observation in implementing a dynamic scheme was that computing a bid is fairly similar to computing an associative function. We had developed efficient algorithms for this purpose [**MPS89**] and had found the implementations thereof to be eminently practical [**Sub92b**]. The essential idea behind these algorithms is to use randomization to achieve load balancing and exploit the fact that the computation is idempotent i.e., if two physical processors compute the partial result of the same virtual processor, this may lead to some inefficiency but no loss in correctness. We used the same technique in this case as well and achieved speedup, comparable ot the static allocation. Note that since we confined ourselves to dense problems, the randomization was more pedagogic in nature than critical to the success of the implementations.

Processor-Memory Bandwidth Limitations. The MPLink bus did not seem to impose a limitation presumably because the processors and bus are well matched and the memory reference patterns permit block transfers.

We ran into severe bandwidth limitations when the problem became too large to fit in memory ($n = 4096$). This is because the SCSI, which connects main memory to disk has a bandwidth of 2-4 MBytes/sec and essentially serialized accesses to memory.

Figure 12 shows how the average time per bid fluctuates when $n = 4096$ and the weight matrix could not fit in main memory. We observed that the average time per bid increased at around the points that ϵ changed. This is because when ϵ changes, all objects become unmatched.

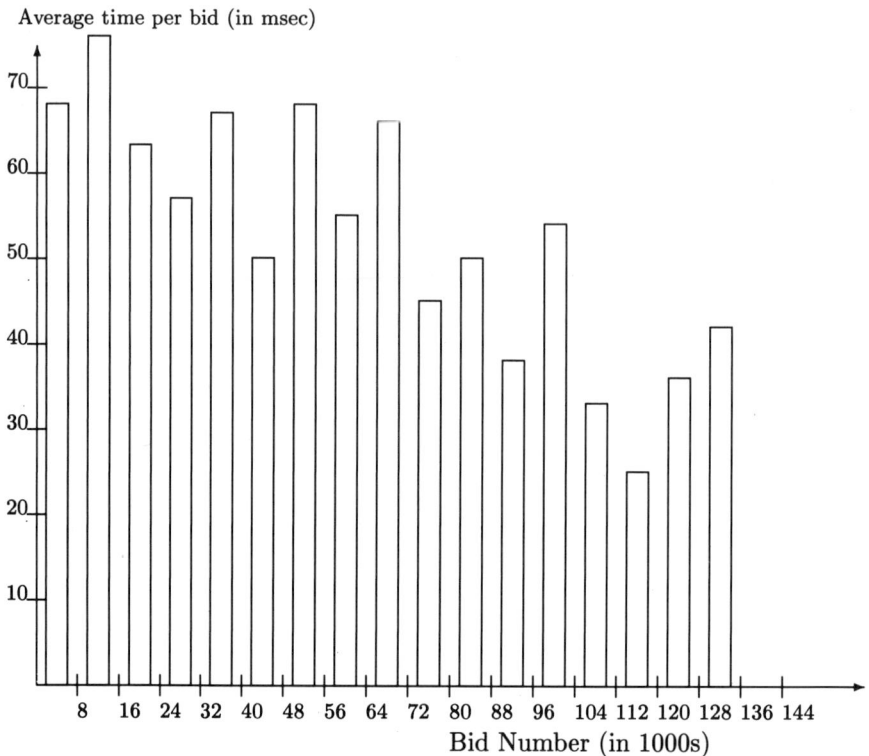

FIGURE 12. Variation of Average Bid Time for Large Problem Sizes.

11. MIMD Implementation

We implemented the Auction algorithm on an 128-processor CM-5 at UC, Berkeley. Each processor is a 33 MHz SPARC-2 processor capable of 25 MIPS and 4.2 MFLOPS. (At the time of writing, the vector units have not been installed.)

11.1. Relevant Architectural Features. Bandwidth and Memory. Each processor has 64 KBytes of cache and 8 MBytes of local DRAM memory. The host processor is a Sun 4. The host processor decides which person is bidding and broadcasts this information. The nodes, 128 in all, compute their partial results, based on the columns of the weight matrix that they possess.

Associative function. The partial results are accumulated (the computation of an associative function) and the broadcasts are performed using the control network [**Lei92**]. The control network is organized as a binary tree with the processors as leaves. Table 15 summarizes the timing information for 64 processors.

Operation	Time (in msec)
Broadcast (4 Bytes)	43.70
Broadcast (8 Bytes)	68.7
Broadcast (16 Bytes)	119
Broadcast (32 Bytes)	225
Broadcast (64 Bytes)	560
Broadcast (128 Bytes)	1020
Broadcast (256 Bytes)	1640
Reduce (4 Byte integers)	6.82
Reduce (8 Byte doubles)	41.3

TABLE 15. Time to Broadcast Data to 16 Processors versus Data Size.

Implementation Details. The machine does not support 64 bit integers. Since we start by multiplying all costs by n, we needed more than 32 bits to represent the numbers for Data Sets II and III ($n = 2048, 4096, 8192$). For these problems, we had to use double precision floating point numbers. The surprising result (see Figure 13) that there was no difference in performance whether we used 32-bit integers or 64-bit double precision floating point numbers shows that the time per bid is dominated by the time to perform the associative function and the broadcasts. Even for the biggest problem that we could attempt ($n = 8192$), each processor had only 64 columns. This is insufficient local work to compensate for the overhead of the broadcasts and maximum finding.

12. Distributed Implementation

We implemented the Auction algorithm on 16 SUN SPARCstation-2 processors connected on a 10 Mbits/sec Ethernet using Parallel Virtual Machine (PVM), a public-domain software package developed at Oak Ridge National Laboratories was used to implement a distributed solution [**Sun91**].

12.1. Relevant Architectural Features. Speed. Each processor is a 40 MHz SPARC-2 chip capable of 28.5 MIPS and 4.2 MFLOPS.

Bandwidth and Memory. Each SUN SPARCstation-2 has 64 KBytes of primary cache and 40 MBytes of main memory. The processor to memory bandwidth is 80 MBytes/sec peak (30 MBytes/sec average) for main memory. The bandwidth falls to 4 MBytes/sec for disk accesses. It has block transfer capability, transferring data in chunks of 512 Bytes from main memory to the cache.

Associative Function. Broadcasting is implemented in PVM using point-to-point messages with recursive doubling. Broadcast proceeds in "rounds", with the number of processors contributing to the broadcast effort doubling in each round. In round r, processor k transmits to processor $(k+2^r) \bmod p$ and receives no acknowledgement. A processor k joins the broadcast effort at round r_k, where r_k = number of 1's in the binary representation of $k \mod p$.

In the implementations of PVM, the UDP protocol was used [**Pos80**]. Therefore, the size (in Bytes) of the data message, or datagram, exchanged is computed as follows: size = Ethernet header + IP header + UDP header + data + Ethernet trailer = 14 + 20 + 8 + data + 4 = 46 + data. Hence, to broadcast 8 Bytes of data, 54 Bytes of data are sent. With a 10 Mbits/sec Ethernet, this should take 0.041 msec ideally. Table 16 gives the measured times.

PVM does not provide any support for computing an associative function. Two approaches were implemented. (i) Serial: All processors send their partial information to the processor designated as master, and (ii) Recursive doubling: The processors are laid out as the nodes of a binary tree and a node sends its partial information is sent to the parent. The performance results are presented in Table 17.

Size of message (in Bytes)	Time (in msec)
4	2.73
8	2.70
16	2.65
32	2.61
64	2.76
128	2.76
256	2.76
512	3.16
1024	3.43
2048	4.70

TABLE 16. Time to Broadcast Data to 16 Processors versus Data Size.

p	Serial Accumulate	Recursive doubling Accumulate
2	6.85	7.05
4	12.4	14.9
8	23.7	23.5
16	47.2	34.5

TABLE 17. Time (in msec) to Accumulate Partial Results versus Number of Processors.

Implementation Details. One processor was designated as the master. The problem descriptor, the identity of the unassigned person, is small. The information to be updated in each bid, the price of the object for which a bid was submitted (if at all), is small. A cute optimization was to combine this with the broadcast of the identity of the bidder. This was done by fixing the identity of the first bidder and combining the price of bid i with the identity of the bidder of bid $i+1$. Since the cost of the broadcast is virtually unaffected by this additional information, this update introduced no additional overhead (see Table 16).

12.2. Results and Discussion. We believed the Auction algorithm to be amenable to a distributed solution for the following reasons: One of the characteristics that makes a problem amenable to a distributed solution is that the problem descriptor should be small, the computing to be done should be large and the description of the solution should be small. This is in order that the communication costs can be subsumed by the computing time. We found that while there was no obvious flaw in this rationale, the performance was little short of disastrous.

Why was this so? The bottleneck was the "accumulate" or "gather" routine, where one had to perform some associative function of the partial results of each processor. Since the partial results of each processor could be coalesced using a binary tree (as is done in the broadcast), we felt that this should not take significantly longer than the broadcast time. Note that it would be overly optimistic to expect the same time as the broadcast since the broadcast consists of non-blocking sends whereas the accumulate involves both sends, receives and some small local computation. However, we did not expect the substantial difference between scatters and gathers.

Given current hardware and software technology, the overhead associated with transmissions makes it impractical to run applications which require small (of the order of a few msecs) latency. This raises the following possibility to decrease latency. Any information broadcast on an Ethernet is received by all processors. Each message has the id of the receiving processor and a processor simply ignores a message that does not belong to it. This suggests designating an id for the PVM and have a processor read all messages that had either its id or the PVM id.

Given that the memory is distributed, slow response from even one processor would degrade the entire system performance. There is simply no way to dynamically reassign work. This makes this approach increasingly treacherous as the number of processors is increased, unless one is in a position to guarantee the workload on *all* the participating processors. The same criticism does not hold for an MIMD implementation (as on a CM-5 or Intel Delta) where the machine is time-shared among users on an all-or-nothing basis. On these machines, imbalances in processing speeds are solely due to workload imbalance.

13. Summary and Discussion of Results of Part 2

We reached the following conclusions regarding the Auction algorithm. Its extreme simplicity makes it amenable to easy coding as well as keeps the constant factors, hidden in the big-O notation, small. The algorithm performs very little computation for each datum read. Since a significant fraction of the bids have very few objects unmatched, a situation which seems to worsen as the problem size increases, we do not believe that it is amenable to "massively" parallel implementations as has hitherto been claimed. However, for reasonably small numbers of processors, it performs very well. The Auction algorithm also has very poor locality of reference, thereby making for very poor cache utilization. While Auction has been considered good primarily for sparse algorithms, we believe that for a given p, the efficiency of utilization of resources decreases as the problem becomes sparser.

We have implemented the Auction algorithm on a variety of architectures and for each of these, we have outlined the architectural features that aided or hindered the implementation. We now outline what we believe would be the "ideal" architecture for problems like Auction. We would have relatively few numbers of powerful processors, support for block transfers, lots of memory and a high processor-memory bandwidth to this memory. It would have hardware support for 64-bit integers. Two hardware features which are not of critical importance are the size of the data cache and the bandwidth of the bus. The data cache need only be of size $= O(\frac{n}{p})$, and the bus needs to be able to accept a word every G cycles where $\frac{n}{p}C > pG \Rightarrow p < \sqrt{n} \times \sqrt{(C/G)}$. (where a processor requires $\frac{n}{p}C$ cycles to compute its portion of the bid.)

14. Acknowledgements

The work was performed with Lockheed Independent Research funds under project RDD506. We are also grateful to the people with WaveTracer, in particular Mary Cacciatone, Mike Lee, Tom McLaughlin and Robert Shuffleton. They made their machine available to us and gave freely of their time in advice and in assisting our debugging efforts. We would like to thank the people at MasPar, Don Rector and Jonathan Becker in particular, for the unlimited use of their machine and their technical assistance. Thanks to Mark Noga at Lockheed PARL for his technical assistance. We thank the referees of DIMACS for their comments, and for bringing the Auction method to our attention. Thanks are due to National Science Foundation Infrastructure Grant number CDA-8722788, which provided access to the CM-5.

References

[BC89] D. P. Bertsekas and D. A. Castanon. Parallel Synchronous and Asynchronous Implementations of the Auction Algorithm. Technical report, M. I. T., 1989. To appear in Parallel Computing.

[Ber86] D. P. Bertsekas. A Distributed Algorithm for the Assignment Problem. In *Proceedings of the 25th IEEE Conference on Decision and Control*, 1986. MIT LIDS Tech Report P-1606.

[Ber91] D. P. Bertsekas. The Auction Algorithm for Assignment and Other Network Flow Problems: A Tutorial. *Interfaces*, 20(4):133–149, July-August 1991.

[Bla90] T. Blank. The Maspar MP-1 Architecture. In *Proceedings of the 35th IEEE International Computer Conference*, pages 20–24, March 1990. San Francisco.

[BM76] J. Bondy and U. Murty. *Graph Theory with Applications*. North Holland, 1976.

[Bra88] M. Brady. Parallel Bipartite Matching on SIMD Arrays. Technical report, Lockheed Palo Alto Research Laboratory, 1988.

[Chr90] P. Christy. Software to support Massively Parallel Computing on the Maspar MP-1. In *Proceedings of the 35th IEEE International Computer Conference*, pages 29–33, March 1990. San Francisco.

[Lei92] C. Lesiserson et al. The network architecture of the connection machine cm-5. In *Proceedings of the 4th Symposium on Parallel Algorithms and Architectures*, pages 272–285, 1992.

[GPV88] A. Goldberg, S. Plotkin, and P. Vaidya. Sublinear-Time Parallel Algorithms for Matching and Related Problems. In *Proceedings of the 29th Symposium on Foundations of Computer Science*, pages 174–185, 1988.

[Hol86] W. Holszytnski. Geometric-Arithmetic Parallel Processor, GAPP, February 1986. Canadian Patent Number 1202208.

[Hol87] W. Holszytnski. The Assignment Problem on GSIMD Arrays. Technical report, Lockheed Palo Alto Research Laboratory, 1987.

[HR88] W. Holszytnski and R. Raghavan. The Distributed Macro Controller for GSIMD Arrays. In Stuart K. Tewksbury, editor, *Concurrent Computations: Algorithms, Architecture and Technology*, chapter 35, pages 689–696. Plenum Press, 1988.

[Jac90] J. H. Jackson. The Data Transport Computer: A 3-D Massively Parallel SIMD Computer. In *Proceedings of IEEE COMPCON*, pages 264–269, Spring 1990.

[KR90] R. M. Karp and V. Ramachandran. A survey of parallel algorithms for shared memory machines. In *Theoretical Computer Science*. North Holland, 1990.

[Lam77] L. Lamport. Concurrent reading and writing. *Communications of the Association for Computing Machinery*, 20(2):806–811, November 1977.

[MPS89] C. U. Martel, A. Park, and R. Subramonian. Work-optimal asynchronous algorithms for shared memory parallel computers. Technical Report CSE 89-8, University of California, Davis, Division of Computer Science, 1989. To appear in SIAM Journal of Computing.

[Mun57] J. Munkres. Algorithms for the Assignment and Transportation Problems. *SIAM Journal on Computing*, 5(1):32–38, 1957.

[Nic90] J. R. Nickolls. The Deisgn of the MasPar MP-1. In *Proceedings of the 35th IEEE International Computer Conference*, pages 25–28, March 1990. San Francisco.

[Pos80] J. B. Postel. User Datagram Protocol. In *Internet Request for Comments RFC-768*, 1980.

[Red78] S. F. Reddaway. DAP- A Flexible Number Cruncher. In *Proceedings of the LASL Workshop on Vector and Parallel Processors*, 1978. Los Alamos.

[RJN90] R. Raghavan, K. K. Jung, and H.T. Nguyen. Fine Grain Parallel Processors and Real-Time Applications: MIMD Controller/SIMD Array. In *Proceedings of 10th International Conference on Pattern Recognition*, pages 324–331, June 1990. Atlantic City, New Jersey.

[Sco80] F. Van Scoy. The Parallel Recognition of Classes of Graphs. *IEEE Transactions on Computers*, 29(7):563–570, July 1980.

[Sub92a] R. Subramonian. Designing synchronous algorithms for asynchronous processors. In *Proceedings of the 4th Symposium on Parallel Algorithms and Architectures*, 1992.

[Sub92b] R. Subramonian. Writing sequential programs for parallel processors: Implementation experience. In *Proceedings of the International Conference on Computing and Information*, 1992. To appear.

[Sun91] V. S. Sunderam. PVM: A Framework for Parallel Distributed Computing. Technical report, Emory University and Oak Ridge National Laboratory, 1991.

[WZ91] J. M. Wein and S. A. Zenios. On the Massively Parallel Solution of the Assignment Problem. *Journal of Parallel and Distributed Computing*, 13:228–236, 1991.

Appendix A. The Adaptive Parallel Processing System: Real-Time Control

The system including the controller has been described in [**HR88**] and [**RJN90**]. We shall re-emphasize here, very briefly, some of the salient points of the controller, called the Distributed Macro Controller (DMC). The DMC can broadcast commands and addresses to the PE array during program execution. It has the ability to reconfigure the instructions in response to conditionals generated either internally by the GSIMD computer or externally. It also dynamically allocates or releases memory. All this is designed to be accomplished without any interruption of machine execution. The DMC accomplished this by a hierarchical MIMD architecture comprising of three main units: the Flow Control Unit (FCU), and two identical Macro Generator Units (MGU). (In current technology, this may be unnecessary: a 200 MHz chip like the Alpha may well be able to control even a 40 MHz parallel computer: however, if both controller and parallel array are made with the same technology, our DMC design becomes necessary).

In addition to the FCU and the MGU, there is a Dynamic Re-interpretation Unit (DRU). The controller employs Very Long Instruction Words (VLIW) to include many subfields with each subfield controlling a particular function within the controller. These functions can be performed simultaneously thus enabling the controller to fully realize the processing speed of the array processor and support the development of high level programming. The functions of the separate units are as follows.

Flow Control Unit. The FCU provides the required signals to control execution of programs stored in the microprogram memory. To support structure programming, instruction like IF_THEN_ELSE, LOOP, GO-TO, or SUBROUTINE CALL are included as part of the flow control functions. Such conditionals are *not* part of the GSIMD programming environment.

Macro Generator Unit. The MGU has several important functions [**HR88**], and we shall describe only one: the Memory Management Unit (MMU). The MMU carries the tasks of keeping track of the available memory, allowing the programmer to work with logical addresses as opposed to physical ones, as well as to allocate and release memory banks at run-time. The length of a memory bank can be defined by users. This feature further increases the flexibility of the MMU. These features are achieved through a linked-list architecture and a rich set of stack operations.

We shall give some further detail on the MMU to clarify any misconceptions that our assumption of penalty-free memory allocation referred just to a model rather than, as it indeed does, to an actual machine. A memory bank can be dynamically allocated and released and its length pre-defined as part of the initialization phase. Symbolic programming and stack operations are provided to allow programs to be developed in a modular fashion. The MMU consists of three modules (see Figure 13): (i) Symbolic RAM (S-RAM), (ii) Circular Linked-List Logic (CLL), and (iii) Stacks.

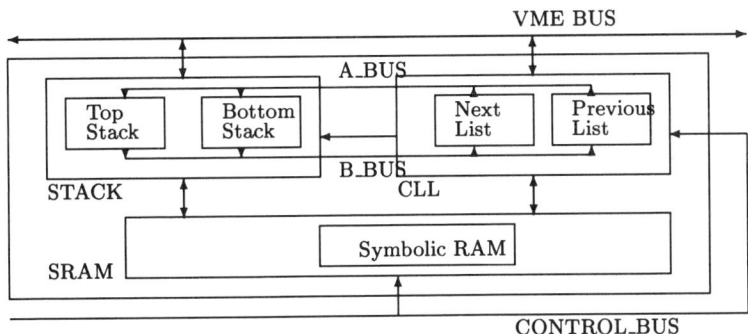

FIGURE 13. MMU

The S-RAM serves as a temporary storage area for the symbolic addresses. The starting address of a memory bank can be stored in the S-RAM and retrieved at a later time by referring to its symbolic name. The heart of the MMU is the circular linked-list architecture employed to perform the house keeping task of the array processor memory. The Circular Linked-List logic performs the following specific tasks: maintain the order of the linked lists and provide the appropriate address to the array processor upon request. The hardware architecture contains the following:

(i) *Next* and *Previous* Linked-List Tables

(ii) *Avai*(lable) and the *Occu*(pied) Pointers

The *Next* and *Previous* lists contain the current status of the circular linked list. The *Avai* pointer always points to the next available bank while the *Occu* pointer points to the first occupied bank. Upon allocation of the three banks (Banks 1-3), the *Occu* pointer is pointed to Bank 1 whereas the *Avai* pointer is pointed to Bank 4 (see Figure 14).

Subsequently, if memory another bank is allocated, the *Avai* pointer then will point to the next available bank which in this example is Bank 5. During the

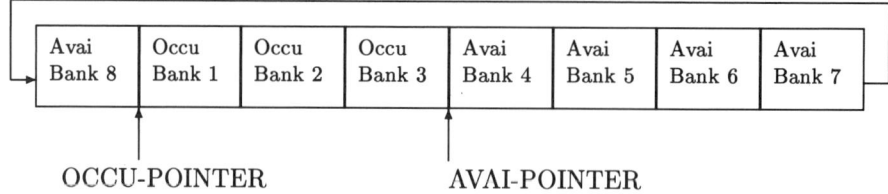

FIGURE 14. Example of Memory Bank Allocation

memory bank de-allocation, the released bank, say 2, is always pushed into the end of the list (see Figure 15).

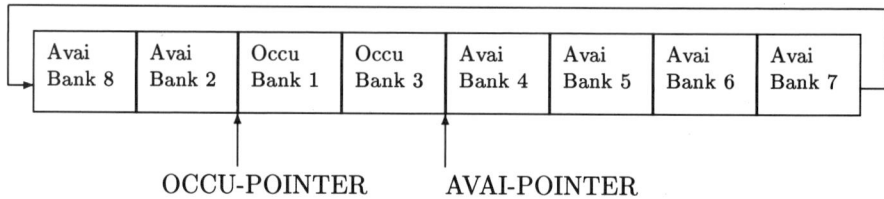

FIGURE 15. Example of Memory Bank De-allocation

An error signal is provided to indicate the occurrence of an overflow or an underflow situation. The *Next/Previous* address can be retrieved from these lists through a look-up table. Five different sources of address can be used to perform a memory de-allocation. Dual port memories, in conjunction with the registers, allow multiple memory accesses within clock cycle thus improving performance of this unit.

The DMC also provides the hardware necessary to support conditional and unconditional looping, branching, subroutine calling, and interrupt as directed by a AP2S program. In brief, it transfers many constructs that are implemented in software on sequential machines, in which case the sacrifice of speed is not an issue, into hardware so that a massively parallel machine can realize the algorithm speedups that their designers envisioned. None of the bottlenecks described above were present in the DMC.

Dynamic Re-interpretation Unit. The purpose of the DRU was to provide an adaptive programming environment by reinterpreting commands based on conditional bits, allowing functions such as adaptive template matching, and so on. It accomplishes this by a fast data communication path between the array processor and the host. Extracted information can be exchanged between these units through a dedicated input/output port without interrupting machine code execution. Therefore the problem with breaking the pipeline discussed in Section 3.3 does not occur in AP2S.

MARTIN BRADY, DEPT. OF ELECTRICAL AND COMPUTER ENGINEERING, PENNSYLVANIA STATE UNIVERSITY, UNIVERSITY PARK, PA 16802
E-mail address: brady@corvette.ece.psu.edu

KENNETH K. JUNG, ARTIFICIAL INTELLIGENCE LABORATORY, LOCKHEED RESEARCH AND DEVELOPMENT, 3251 HANOVER STREET, BLDG. 254G, PALO ALTO, CA 94304
E-mail address: kkj@aic.lockheed.com

H. T. NGUYEN AND RAGHU RAGHAVAN, INSTITUTE OF SYSTEM SCIENCE, NATIONAL UNIVERSITY OF SINGAPORE, SINGAPORE 0511
E-mail address: htn,raghu@iss.nus.sg

RAMESH SUBRAMONIAN, ROOM 571, DIVISION OF COMPUTER SCIENCE, UNIVERSITY OF CALIFORNIA, BERKELEY, CA 94720
E-mail address: subramon@cs.berkeley.edu

An Experimental Comparison of Two Maximum Cardinality Matching Programs

STEVEN T. CROCKER

ABSTRACT. We report results of experiments comparing the run time behavior of Micali and Vazirani's $O(m\sqrt{n})$ algorithm (MV) and Gabow's $O(n^3)$ algorithm (G) for maximum cardinality matching. These experiments used randomly generated graphs with vertex set sizes between 1024 and 8196 as input. Within the sequence of graphs of an experiment, some parameter was kept constant, such as the expected degree of a vertex or the number of vertices in the graph. The results show the dependence of run time on expected degree of a vertex, the correlation of blossom formation and run time, and the significant cost of beginning a new phase in algorithm MV. Also, they show thresholds and characteristics of blossom formation in both algorithms and of the number of phases in algorithm MV.

1. Introduction

In this paper, we report the results of experiments investigating the timings and combinatorial features found by two algorithms that solve the maximum cardinality matching problem on nonbipartite graphs. For a given graph $G = (V, E)$, let n be the size of the vertex set V, and m be the size of the edge set E. The experiments used two algorithms, one discovered by Micali and Vazirani[4] (referred to as MV), which has a bounded running time of $O(m\sqrt{n})$ and one discovered by Gabow[2] (referred to as G), which has a bounded running time of $O(n^3)$.

The emphasis of the project consisted of two experiments, *constant number of vertices (CED)* and *varying expected degree (VED)*. They were designed to investigate the dependence of algorithm performance on parameters of random graphs. The purpose of experiment *CED* was to look at algorithm performance as it depended on the size of a graph. This experiment gathered results of the algorithms running on graphs with different sized vertex sets but with constant expected degrees for their vertices. It was used to determine how significantly

---

1991 *Mathematics Subject Classification.* Primary 05C85; Secondary 05C70, 68Q20.

the initialization phases of the algorithms contributed to the run times, how rapidly the algorithms moved towards their asymptotic performance, and how their asymptotic performance compared to their theoretical bounds.

The purpose of experiment VED was to investigate the combinatorial and timing characteristics of the algorithms on random graphs with a constant number of vertices but over a range of edge densities. The multiple runs of this experiment using different vertex set sizes gave us a basis for claiming the observed performance characteristics were due to the density of the graph and not to the size of the graph. We also performed a bipartite version of experiment VED to investigate more thoroughly the effects of blossom formation on run time.

In the next section we give a brief description of the two algorithms used in this experiment and some details of the implementations. Section 3 discusses the design of the experiments and sections 4 and 5 contain the discussion of their results. Section 6 contains the results of running the two algorithms on another of the DIMACS Implementation Challenge's core experiments for maximum cardinality matching.

2. Algorithms and Their Implementations

In this section we give high level descriptions of the Micali-Vazirani and the Gabow matching algorithms. We follow each description with some implementation details.

2.1. The Micali-Vazirani Algorithm (MV).

Algorithm MV developed by Micali and Vazirani[4] proceeds in phases, where each phase increases the current partial matching by finding a maximum set of vertex disjoint shortest length augmenting paths. The algorithm halts when either there are no more free vertices or a phase finds no augmenting paths. A phase operates in two alternating modes, incrementing the depth of the breadth first search (*bfs*) trees and processing bridges. First, the algorithm synchronously builds breadth first search trees along alternating paths from each free vertex. All trees are built to the same level before the synchronous bfs continues onto the next level. At each level, the synchronous bfs discovers a (possibly empty) set of bridges. Bridges are edges, (u, v), which are used in both directions, as (u, v) and (v, u), by the same level of the synchronous bfs. Thus, a bridge is the edge at which two bfs tree's meet or at which a bfs tree rejoins itself. Before the phase extends the bfs to the next higher level, the phase switches to its other mode and examines each bridge discovered at the current level. For a given bridge, a synchronous double depth first search seeks disjoint alternating paths from the two vertices of the bridge, down their respective bfs trees. If the two depth first searches (dfs) find two vertex disjoint alternating paths to free vertices, then an augmenting path has been found. This path is augmented, increasing the matching by one, and then topologically erased from the bfs trees for the remainder of the phase. This erasing forces all augmenting paths found from bridges of this level to be vertex disjoint. If instead, in the double depth first search, the two dfs trees become bottlenecked at some vertex because neither dfs can find another path around that vertex, then a blossom has been discovered and the bottleneck is the blossom's base vertex. Defining a blossom requires traversing the ddfs tree to the bottleneck, and adding to the new blossom each vertex that is not already in a blossom. After all bridges have been examined, if no augmenting paths were found then the phase returns to building the synchronous bfs trees to the next higher level. If, however, one or more augmentations were done, then the phase is over. A new phase is begun by restarting the bfs trees at level 0 from the unmatched vertices in the original graph with the new matching.

Each phase of the algorithm examines a vertex or edge $O(1)$ times. By a result of Hopcroft and Karp[3], there are at most $\sqrt{n}$ phases. This gives the $O(m\sqrt{n})$ run time of the algorithm.

2.2. Implementation of Micali and Vazirani's Algorithm.
Algorithm MV was implemented in slightly more than 1000 lines of Pascal code by the author. The graph to be matched is represented as a modified adjacency list. Each vertex is represented by a record which contains a pointer to a list of incident edges. Each edge is represented by a unique record that contains pointers to both of its vertices and pointers to the next edge in each of their respective edge lists. The vertices' and edges' records also contain fields for the data needed to build and retain the bfs and the dfs trees, the blossoms, and the augmenting paths. Thus, the algorithm allocates almost all the storage space it needs dynamically, but it does so during initialization when reading the problem graph.

2.3. Gabow's Algorithm.
Gabow's algorithm, which is a variant of Edmonds' matching algorithm, is simpler. [1, 2] Algorithm G builds a search tree rooted at each free vertex, but it only builds one tree at a time. The search tree for this implementation is built in a breadth first manner in the sense that unexamined edges which would extend the current alternating search tree are examined in a first in, first out order; it is not a bfs where the alternating paths all have equal length, as in algorithm MV (however, in a bipartite graph where no blossoms occur, the two would be the same). Each time the current search tree is grown, it is extended by two edges: first by an unmatched edge that leaves the current search tree and then by the matched edge outside of the search tree that shares an endpoint with the unmatched edge. The vertex through which the matched edge was reached is called an *inner* vertex. The other vertex of the matched edge is *outer*. The algorithm has found an augmenting path when it reaches a free vertex; it has found a blossom when it reaches an outer vertex via an unmatched edge. A scheme using labels kept at each vertex allows the algorithm to reconstruct an augmenting path through blossoms after it reaches a free vertex.

2.4. Implementation of Gabow's Algorithm.
Algorithm G was implemented in slightly under 300 lines of C code by Ed Rothberg and modified slightly by the author. It was made available by the DIMACS Implementation Challenge organizing committee. The implementation uses array structures to store the graph and the data structures of the algorithm. The problem graph is represented as an adjacency list with two records allocated for each edge. The space requirements of this algorithm are considerably smaller than those of MV because of the simpler data structures involved. Besides the storage needed to store the original graph, algorithm G requires only five additional arrays of integers of length n. These arrays are allocated dynamically during program initialization.

3. Experiment Designs

The main project consisted of a number of separate experiments. Each experiment was made up of a number of individual runs. A run involved executing both algorithms on a sequence of related graphs and plotting the data gathered. A data point was generally the average of the statistics gathered by executing an algorithm on twenty random graphs chosen from $G_{n,m}$ where $G_{n,m}$ is the class of all graphs with n vertices and m edges. An algorithm's run time did not include the time to read the graph, to build its adjacency list, nor to print the maximum matching.

The design of an experiment was fixed by choosing the way in which n and m varied throughout a run. In the first experiment, called CED for Constant Expected Degree, the ratio $2m/n$ was fixed and n increased by factors of 2. The seven runs in this experiment used graphs with expected degrees of 2^i for $0 \leq i \leq 6$. In the second experiment, called VED for Varying Expected Degree, n was fixed and m increased by factors of $2^{1/16}$. The four runs in this experiment had n fixed at values of 2^{10+i} for $0 \leq i \leq 3$. Every data point shown is the mean of the statistics gathered from executing the algorithm on twenty random graphs generated with the same parameters n and m. Data points from each run are connected by lines in the figures of the experimental data that follow.

The random graphs of these experiments were generated by a random graph generator supplied by DIMACS. This generator worked by fixing the number of vertices and edges a graph would have and then, for each potential edge in the graph, randomly deciding whether or not that edge is in the graph with probability $\frac{\text{number of edges left to choose}}{\text{number of potential edges remaining}}$. The two experiments and the values of their key parameters are summarized in figure 1.

Lines connect data points of a particular run. In a figure with plots of multiple runs, the shape of the symbol used to denote data points for a particular run is the key to the value of the invariant for that run. Symbols are only printed at the data points which have a value on the abscissa that is an integral power of two. The experiments were conducted on a Sun 4/40 ipc workstation with 24 meg of

Experiment Name	Run Size	Symbol	Data Points Minimum Size	Maximum Size	Increment Factor
CED	deg = 1	□	$\|V\|$=256	$\|V\|$=32768	2
	2	○	256	32768	
	4	△	256	32768	
	8	+	256	32768	
	16	×	128	16384	
	32	◇	128	16384	
	64	▽	128	4096	
VED	$\|V\|$ =1024	□	$\|E\|$=512	$\|E\|$=92681	$2^{1/16}$
	2048	○	512	220435	
	4096	△	2048	193570	
	8192	◇	2048	65536	

FIGURE 1. Parameters of the two experiments. Each experiment kept the invariant parameter constant while increasing the variant parameter by the increment factor.

memory. The cpu was rated at 15 MIPS. The programs used for the experiments described in this paper are available via anonymous ftp from DIMACS.

4. Experiment Constant Expected Degree (CED)

The main purpose of this experiment was to study the effect of graph size on run time. We attempted to eliminate systematic structural changes within the graphs of a run and to isolate the effects of graph size by fixing the degree of all the graphs within a run. For example, one effect graph size could have on run time is the overhead of initialization of an algorithm after it has read the graph. A log-log plot of runtime versus number of vertices for polynomial time algorithms such as these two, produce linear graphs. The slope of these graphs give a means to compare actual performance with asymptotic bounds. Figure 2 shows a log-log plot of runtime vs. number of vertices for algorithm MV.

Figure 3 shows a log-log plot of runtime vs. number of vertices for algorithm G. Each line in the graph connects the data points for one run of the experiment. Vertices within all graphs of a run have the same expected degree. the symbol marking each data point denotes the expected degree of the graphs of that run.

The near linearity of all lines in figure 2 is also evident from the 0.99 or larger correlation coefficients of the data from these lines. The slopes of these lines, given in figure 4, which are all between 1.05 and 1.25 indicate the algorithm's runtime is growing superlinearly, but not as fast as the asymptotic bound of $O(m\sqrt{n})$ which would have yielded slopes for these sparse graphs of 1.5. The bunching of the lines of the runs with expected degrees of 4, 8, and 16 indicates that in this range, the runtime of the algorithm is not very affected by the

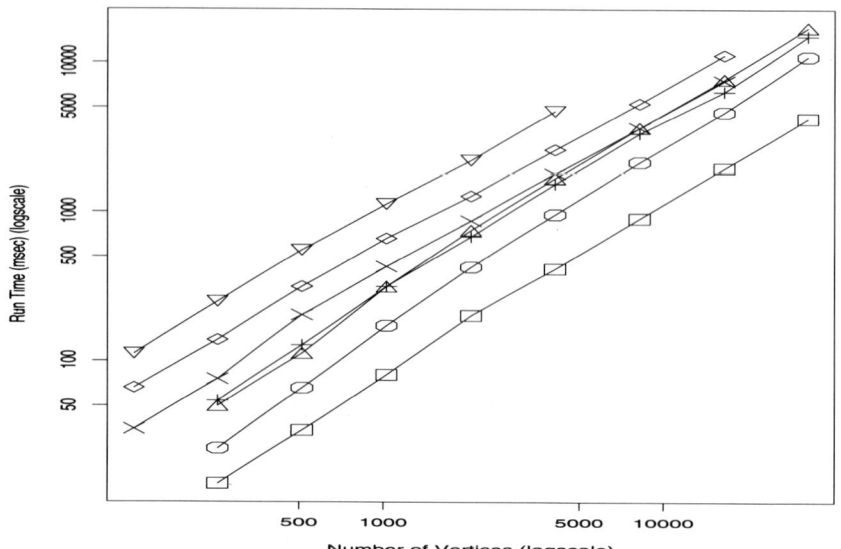

FIGURE 2. Experiment CED for algorithm MV showing run times versus number of vertices on a log-log scale. Symbols mark data points at vertex set sizes which are powers of 2. The symbol denotes the expected degree of a random vertex from any graph in the run. They are: expected degree $1 : \square$, $2 : \bigcirc$, $4 : \triangle$, $8 : +$, $16 : \times$, $32 : \Diamond$, and $64 : \triangledown$.

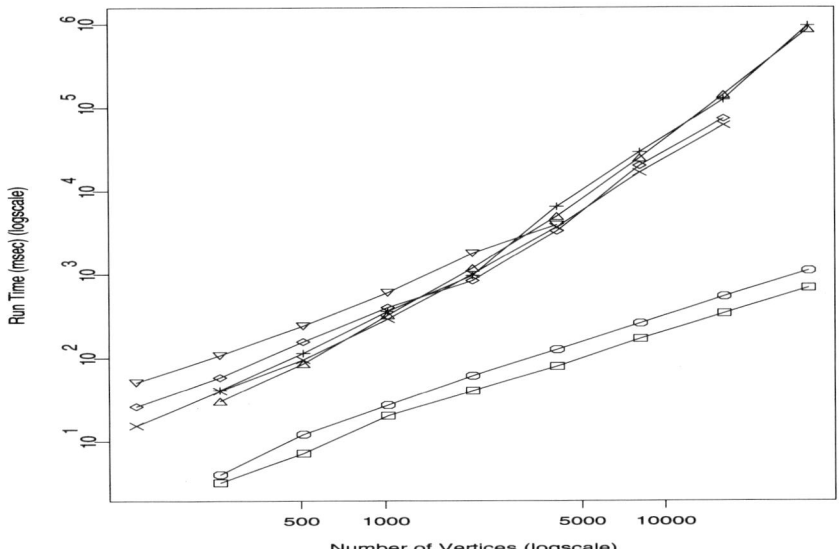

FIGURE 3. Experiment CED for algorithm G showing run times versus number of vertices on a log-log scale. Each run has an expected degree and symbol from $1: \square; 2: \bigcirc; 4: \triangle; 8: +; 16: \times; 32: \Diamond; 64: \triangledown$.

FIGURE 4. Slopes of the line fit by least squares through the points in Figures 2 and 3.

Expected Degree	Algorithm	
	MV	G
1	1.16	1.10
2	1.23	1.13
3	1.20	2.12
4	1.15	2.07
5	1.10	1.72
6	1.05	1.62
7	1.07	1.28

expected degree. This phenomenon and others will be clearly seen in the results of the next section.

The corresponding graph for algorithm G shows more complex shape. The data for the degree 1 and 2 runs show linear behavior. These runs have slopes similar to those of the MV algorithm (figure 4). The data from the runs of higher expected degree show a slightly convex shape. The slope of the next line, for graphs of expected degree 3, and the gap between that line and the line for graphs of expected degree 2 indicates a threshold exists for the algorithm on random graphs with expected degree between 2 and 3. The slopes of the lines from the runs with expected degree 3 and 4 show the algorithm runtimes are more than quadratic in the number of vertices over that range but significantly less than cubic. The gradual reduction of the slopes as the expected degree increases further indicates something else is affecting the average length of an augmenting path. Again, these phenomena will be more clearly seen in the next section.

5. Experiment Varying Expected Degree (VED)

In this experiment the densities of graphs in a particular run varied by keeping the number of vertices constant and increasing the number of edges. We looked at the relations of run time, blossom formation, and number of phases to the expected degree of the random graphs for algorithm MV and at the relations of run time and blossom formation to expected degree for algorithm G. Also, we compared the run times of the two algorithms and the number of blossoms they found.

Figure 5 shows the runtimes versus expected degree plotted on a log-log scale for the four runs of algorithm MV with 2^i, $10 \leq i \leq 13$, vertices. The relatively even spacing between all four lines is predictable from the linearity results of experiment CED in the previous section. The spike in the runtime of all four runs occurs at a degree in the range between 2.7 and 2.8. The period of flat runtimes between degree's 4 and 10 corresponds to the bunching of lines for degrees 4 and 8 (and at larger graph sizes, 16) in figure 2. By degree 20, the log-log plots have near linear growth with a slope slightly less than 1.

Figure 6 shows the run times versus the expected degree of algorithm G on a log-log plot for runs with graphs having vertex sets of size 2^i, $10 \leq i \leq 13$. As in the previous plot, the nearly even spacing of the runs corresponds to the near linearity found in experiment CED. The jump in run time in that experiment that occurred from expected degree 2 to 3 occurs here as a very sharp rise in run time as the expected degree increases through that range. This jump in run time is centered between 2.7 and 2.8, roughly the same location as the peak in run time which occurs in algorithm MV. From expected degree 3 up to much larger expected degrees, the run times are almost flat, which again corresponds with the bunching of the lines from the different runs of experiment CED. For the

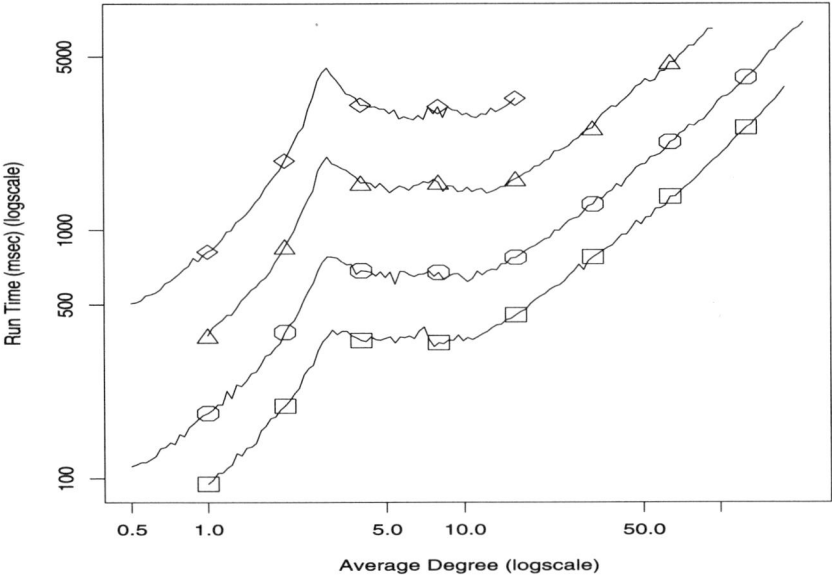

FIGURE 5. Experiment VED for algorithm MV showing run time versus expected degree on a log-log scale for graphs with $2^{10}, 2^{11}, 2^{12}$ and 2^{13} vertices.

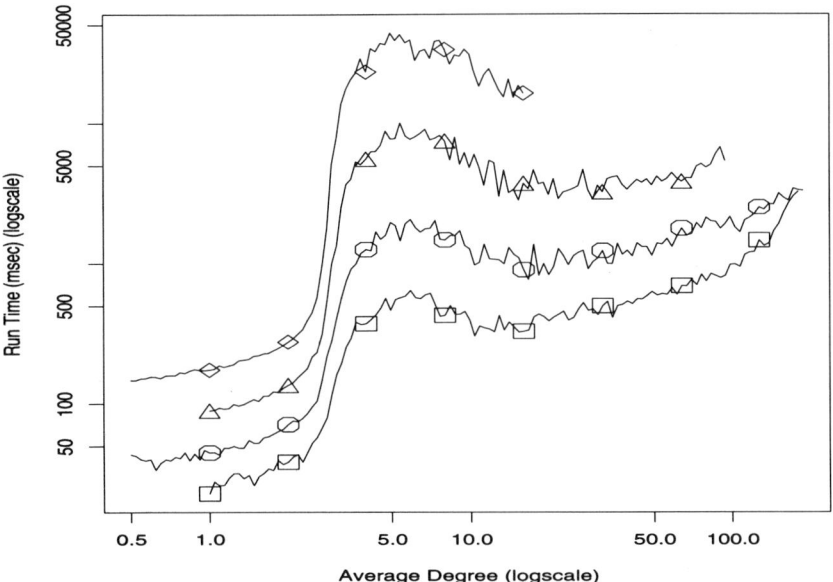

FIGURE 6. Experiment VED for algorithm G showing run time versus expected degree on a log-log scale for graphs with $2^{10}, 2^{11}, 2^{12}$ and 2^{13} vertices.

FIGURE 7. Mean runtimes and variances of the data from graphs which had 4096 vertices and average degrees that were a power of two.

Average Degree	Algorithm			
	MV		G	
	Mean	Var	Mean	Var
1	376	28	89	11
2	858	52	135	12
4	1547	171	5574	2144
8	1559	271	7428	4104
16	1608	75	3702	3580
32	2569	135	3203	1676
64	4749	483	3818	2136

smallest run, with vertex sets of size 1024, run times start rising more sharply as the graphs approach the size limit of the execution environment.

Figure 7 lists the mean runtimes and their variances from graphs which had expected vertex degrees which were a power of two. For very small degree graphs there is small variance in the runtimes because almost all the work of the algorithm is accomplished in the initial phase (algorithm MV) or greedy matching (algorithm G). With larger degrees, the variance in MV in comparison to the variance in G is quite small.

Figure 8 shows the relationships between the runtime, number of blossoms found, and number of phases for algorithm MV. The data in this figure is from the run with 1024 vertices; however the other three runs had similar plots. Unless otherwise noted, the discussion of this plot will refer generally to all of the four runs. The peak in the number of phases consistently occurs between the expected degrees of 2.7 and 2.8 for all runs. The simultaneous peaks in the number of phases and in the run time indicate there is a high overhead with initializing a phase. The initialization of a phase requires that the whole breadth first search structure and the depth first search structure from the previous phase must be cleared. The large number of data items stored with these structures leads to the large overhead. Blossoms are first discovered with graphs of expected degree near 2.5. Eventually, at higher expected degrees, blossoms are no longer found. For the graphs with vertex sets of 1024 vertices, this upper limit to blossom formation is around expected degree 10, while for the 8192 vertex run, the upper limit is near 20.

Figure 9 plots run times and blossoms found versus expected degree for algorithm G on graphs with 1024 vertices. Here again, blossoms discoveries begin with a sharp threshold at expected degrees near 2.7. They have a broad peak in the range of expected degrees 5 to 7 then decline rapidly until an expected degree near 10 from which point the number of blossoms discovered declines very gradually. From the sharp rise in run time between degrees 2.7 and 4 and

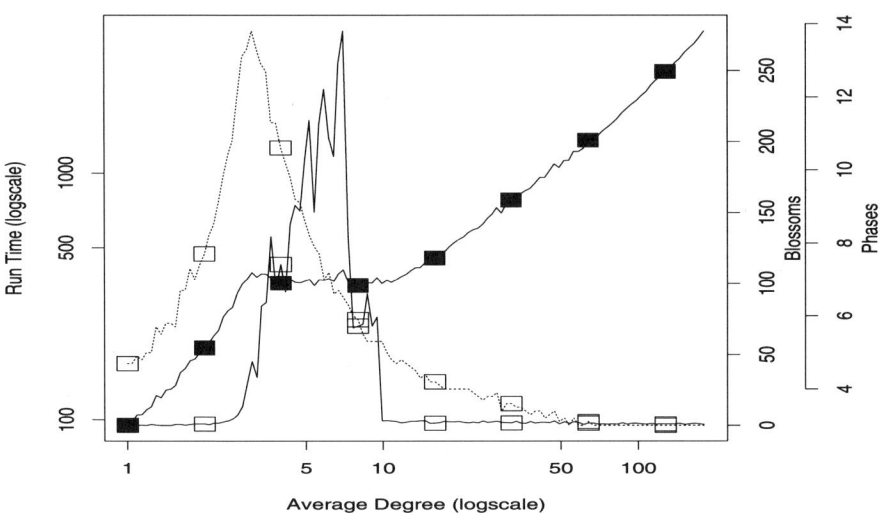

FIGURE 8. Experiment VED for algorithm G with 1024 vertices showing run times and number of blossoms discovered versus expected degree on a log-log scale. Run time: solid □, blossoms: hollow □ and solid line, phases: hollow □ and dashed line

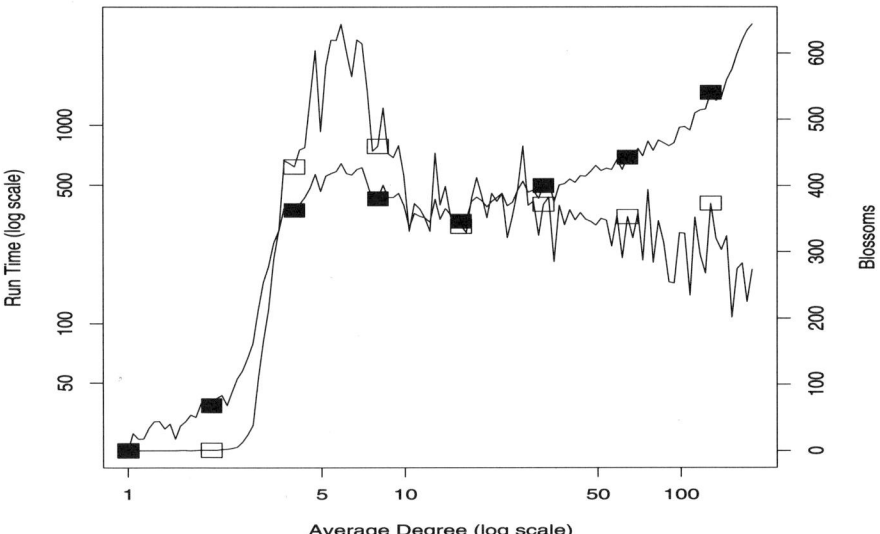

FIGURE 9. Experiment VED for algorithm G with 1024 vertices showing run times versus expected degree on a log-log scale. Run time: solid □, blossoms: hollow □

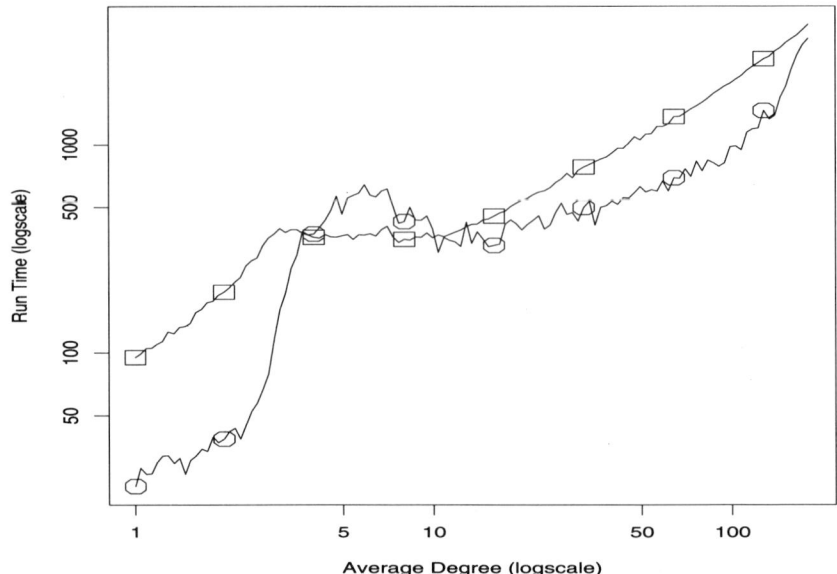

FIGURE 10. Experiment VED for vertex set sizes of 1024. Run time versus expected degree. Algorithm MV: □; algorithm G: ○.

the run time's local maxima near degree 6, it is apparent that their is a strong positive correlation between run time and blossom formation. For degrees larger than 100, the run time seems to rise rapidly with expected degree but the high density graphs for which these timings were obtained were very close to a size limit of what either the machine or compiler could handle. Beyond that limit, the programs either slowed way down due to disk operations (thrashing) or they crashed. Therefore this final rise in runtime is likely an artifact of the execution environment on which the experiment was run.

Figures 10 and 11 show the run times for both algorithms plotted against expected degree for the runs with vertex set sizes of 1024 and 8192. The two intermediate runs with vertex set sizes of 2048 and 4096 had similar plots with curves that fell, as expected, between the curves of these plots. In all four runs, algorithm G significantly outperformed algorithm MV for expected degrees less than 3. This is approximately where the local peak in algorithm MV's run time occurs. Eventually, with higher expected degree graphs, algorithm G's run times fall below those of algorithm MV. The interval of degrees where algorithm MV's run times are faster than algorithm G's grows with the size of the vertex set of the random graphs. Most of the increase in this interval occurs on the higher end. At the lower end, the crossover point seems to be bounded at graphs of expected degree 2.7. This is where algorithm MV has a peak in its run time and algorithm G has a sharp rise in its run time.

Experiment VED also led to some surprising discoveries of when blossoms are

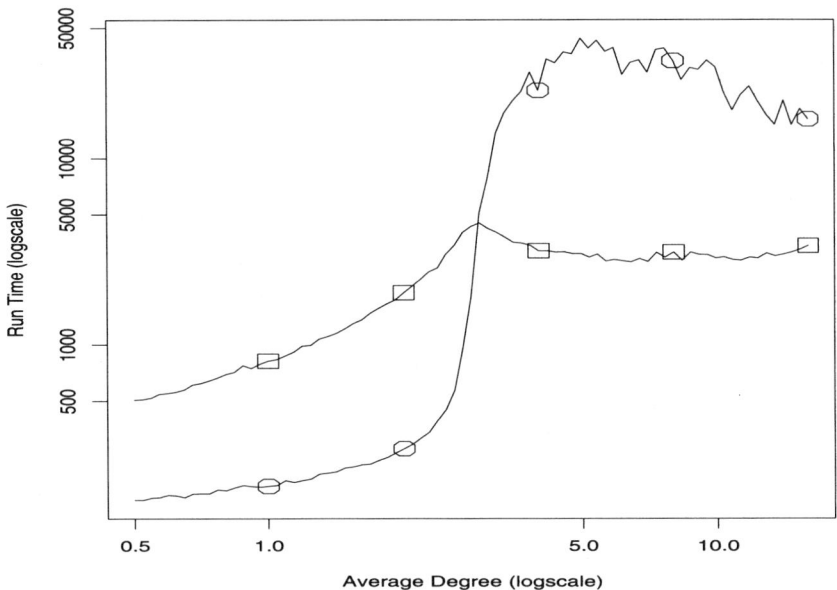

FIGURE 11. Experiment VED for vertex set sizes of 8192 showing run time verses expected degree. Algorithm MV: □, algorithm G: ○.

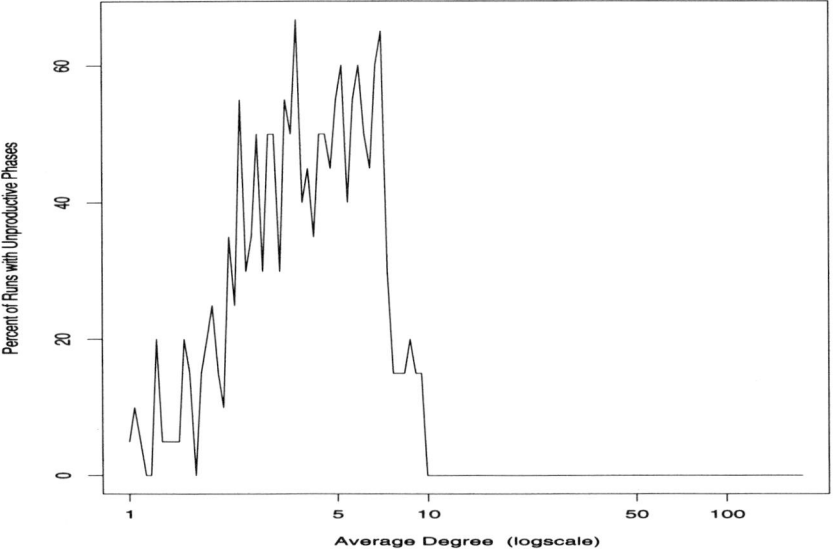

FIGURE 12. Experiment VED on MV for vertex set sizes of 1024, Fraction of blossoms found in unproductive phases versus Expected Degree.

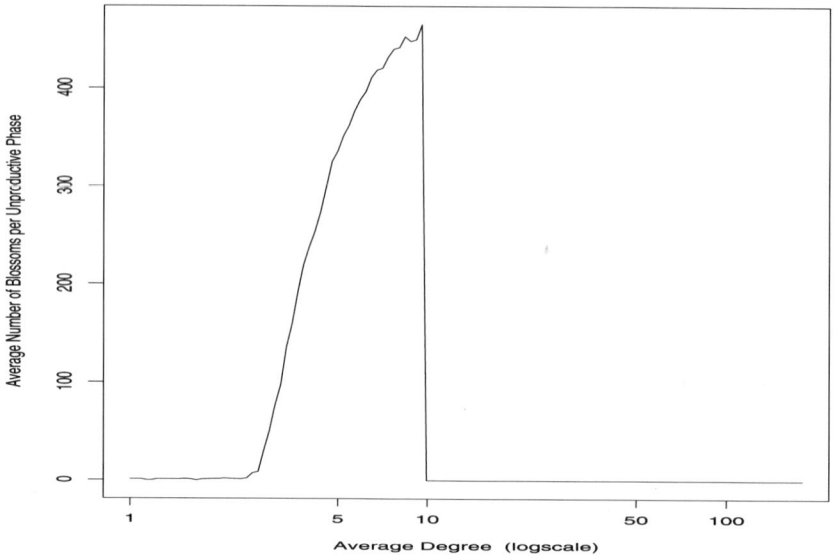

FIGURE 13. Experiment VED on MV for vertex set sizes of 1024, Average number of blossoms found in an unproductive phase versus Expected Degree.

discovered. Define an *unproductive phase* as a phase of the MV algorithm for which no augmenting paths were found. A *productive phase* is one that is not unproductive. If an unproductive phase occurs, it must be the final phase of the algorithm on a graph without a perfect matching. The way blossoms were found by algorithm MV on a random graph can be characterized in two ways. Either there were very few blossoms discovered or many blossoms were discovered but almost all of them came in an unproductive phase. Figure 12 shows the percent of all blossoms found by the algorithm which were found in an unproductive phase. Figure 13 shows the average number of blossoms discovered in a final phase for final phases in graphs of varying expected degree. The lack of blossoms discovered at degrees less than 2.5 is an indication that most of the components in the graph at these degrees are small trees. Between degrees 2.7 and 10 the average rises along a smooth curve. The smoothness of this curve is a result of a small variance in the raw data. After expected degree 10, unproductive phases cease to occur. Figure 14 shows the average number of blossoms discovered in a productive phase. The low values in this curve compared to the values in the curve in figure 13 show that when the algorithm starts finding many blossoms, it is almost surely in the final, unproductive phase.

To see the effects of blossom formation on run time, experiment VED was rerun using bipartite graphs. These graphs were generated in a similar manner as the general graphs except the set of possible edges to choose from was restricted to the edges in a complete bipartite graph with the partitions of the vertex set

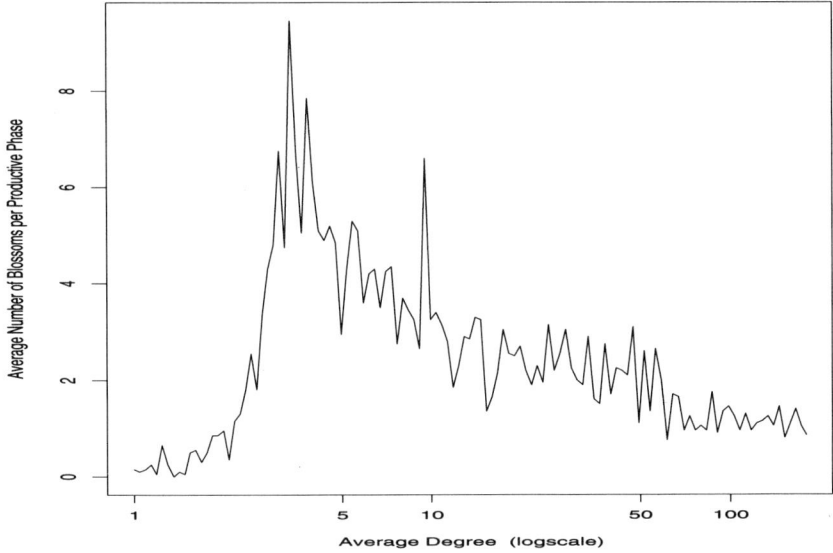

FIGURE 14. Experiment VED on algorithm MV for vertex set sizes of 1024 showing average number of blossoms found in productive phases versus expected degree.

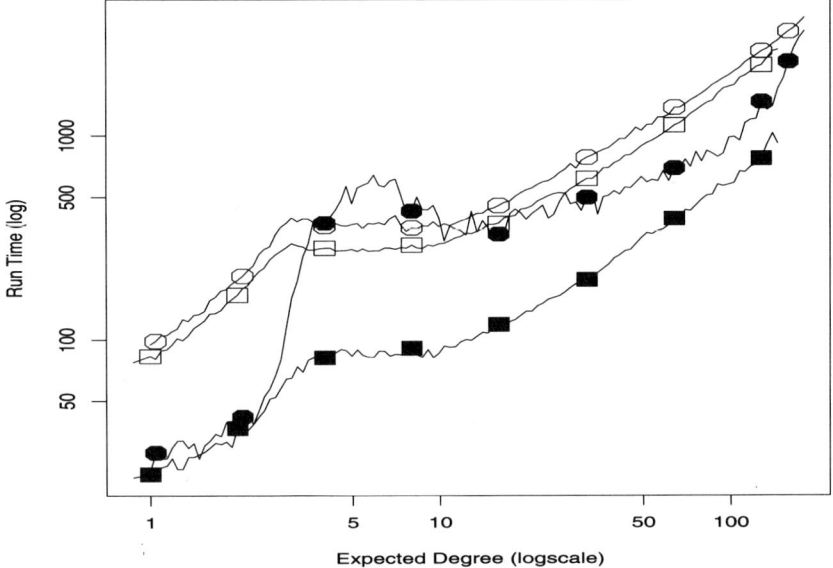

FIGURE 15. Experiment VED for both general and bipartite graphs with vertex set sizes of 1024. Run time verses expected degree. Algorithm MV: □; Algorithm G: ○; bipartite graphs: solid; general graphs: hollow.

having equal size. Figure 15 shows the results of these runs in comparison with the previous runs on general graphs. Algorithm MV is shown with hollow symbols; algorithm G has solid symbols. Algorithm MV shows no significant change in its run time on bipartite graphs from its run time on general graphs. This is a consequence of the main contribution of the MV algorithm to the development of matching algorithms, which is its efficient handling of blossoms. In contrast, algorithm G's run time is significantly faster for the bipartite graphs above average degrees of 2.7, where blossoms are first discovered in general graphs. The discovery of blossoms by algorithm G considerably lengthens its execution time. It is interesting to note that without the formation of blossoms, the shape of algorithm G's run time plot takes on a similar shape to that of the plots for algorithm MV on general and bipartite graphs.

6. Triangle Graphs

This section contains the results of another of the "core" experiments for the DIMACS Implementation Challenge. The experiment consisted of running both algorithms on two types of triangle graphs: singly and triply connected. The structure of both types on a vertex set of size $3k$ begins with k vertex disjoint triangles labeled $0 \ldots k - 1$. To complete the triply connected triangles, we add edges from each vertex to the corresponding vertex in the next larger triangle for triangles 0 to $k - 2$. The edge set of a triply connected triangle consists of these edges on the vertex set $\{0 \ldots 3k - 1\}$:

$(3i, 3i + 1), (3i + 1, 3i + 2), (3i + 2, 3i)$ $\quad 0 \leq i \leq k - 1$
$(3i, 3i + 3), (3i + 1, 3i + 4), (3i + 2, 3i + 5)$ $\quad 0 \leq i \leq k - 2$

The singly connected triangle graphs have only one edge from a triangle to the next larger triangle. Such a graph on the same vertex set would have an edge set of:

$(3i, 3i + 1), (3i + 1, 3i + 2), (3i + 2, 3i)$ $\quad 0 \leq i \leq k - 1$
$(3i + (i \bmod 3), 3i + 3 + (i \bmod 3))$ $\quad 0 \leq i \leq k - 2$

Figure 16 shows the run times of the triangle graphs and triple triangle graphs for both the MV and G algorithms. These graphs were proposed as an example for which the MV algorithm would find many blossoms.

The run times and blossoms found by both the MV and G algorithms on these graphs are given in figure 17. For singly connected triangles, the general direction an alternating path may take from an unmatched vertex is either left or right. Once the direction is chosen, there is very little variation in the way the search may progress. Thus, algorithm G, with its fewer data structures and the simpler process by which it takes a step in the alternating search away from the unmatched vertex, outperforms algorithm MV. The performance of G is consistently better than the performance of MV by a factor of at least three for these graphs. With the triply connected triangles, where there is more freedom to choose an alternating search path, the performance of algorithm MV improves to slightly less than a factor of three worse than algorithm G. The structure in

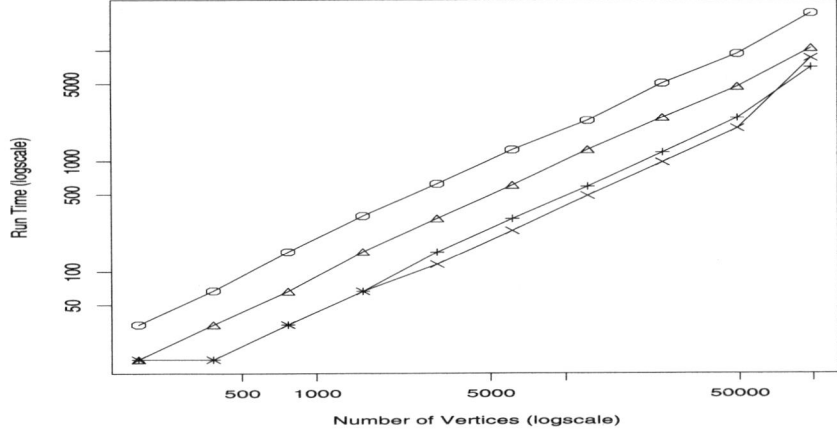

FIGURE 16. Experiment Tri, Triangle graphs' run times versus number of vertices on a log-log scale. MV on singly connected triangle graphs (*stg's*): ○; MV on triply connected triangle graphs (*ttg's*): △; G on STG's: +; G on TTG's: ×.

these graphs give quite a different outcome in the relative performance of the two algorithms than in the experiment using random graphs. There, for graphs of average degree four, algorithm MV had better performance than algorithm G (figures 10 and 11).

7. Conclusions

The experiments we ran have shown how the run times of the Micali-Vazirani and the Gabow algorithms for maximum matching in general graphs behave when the algorithms are executed on random graphs. We have also seen some of the combinatorial execution characteristics of these algorithms. For example, the number of phases which algorithm MV undergoes exhibits a sharp peak for random graphs which have an expected vertex degree of between 2.7 and 2.8. The number of blossoms discovered by algorithm MV takes on one of two classes. Either there are very few blossoms found, or there are a large number found, almost all of which are found during the last, unproductive phase.

The number of blossoms discovered by both algorithms running on random graphs show a sharp peak at expected degrees near 2.7. By expected degree 8, MV essentially discovered no more blossoms, while the blossom discoveries by G were reduced but almost level. The two algorithms exhibited opposite effects on their performance during the period for which large numbers of blossoms were being discovered. The peak in MV's run time near an expected degree of 3 is a result of the simultaneous peak in the number of phases. There is considerable overhead in initiating a phase, and the growth in the number of phases before

FIGURE 17. Experiment Triangles. Run times and blossom discoveries for algorithms MV and G on singly and triply connected triangle graphs.

Graph		Run Time (msec)		Blossoms	
Type	Triangles	MV	G	MV	G
Singly	64	33	16	41	31
Connected	128	67	16	82	64
Triangle	256	150	33	169	127
	512	317	66	338	256
	1024	617	149	681	511
	2048	1250	299	1362	1024
	4096	2300	583	2729	2047
	8192	4983	1183	5458	4096
	16384	9184	2416	10921	8191
	32768	21350	6966	21842	16384
Triply	64	16	16	0	0
Connected	128	33	16	0	0
Triangle	256	66	33	0	0
	512	150	66	0	0
	1024	300	116	0	0
	2048	600	233	0	0
	4096	1250	483	0	0
	8192	2433	966	0	0
	16384	4650	1949	0	0
	32768	10300	8432	0	0

average degree 3 explains the rise in run times while the decrease in the number of phases after degree 3 explains the corresponding plateau in run times.

The run times for algorithm G were highly affected by the formation of blossoms. At average degree 2.7 the run times rose sharply with the number of blossoms formed. By average degree 8 blossom discoveries were at half their peak level and had entered a long period of a slow decline. The run times over this same period declined at first and then entered an extended period of gradually increasing values.

References

1. J. Edmonds, *Paths, trees and flowers*, Canad. J. Math. **17** (1965), 449–467.
2. H. Gabow, *Implementation of algorithms for maximum matching on nonbipartite graphs.*, J. Assoc. Comput. Mach. **23** (1976), 221–234.
3. J. E. Hopcroft and R. M. Karp, *An $n^{5/2}$ algorithm for maximum matching in bipartite graphs*, SIAM J. Comput. **2** (1973), 225–231.
4. S. Micali and V. Vazirani, *An $o(|e|\sqrt{|V|})$ algorithm for finding maximum matching in general graphs*, Proc. 21st Annual IEEE Symposium on Foundations of Computer Science, 1980, pp. 17–27.

DEPARTMENT OF COMPUTER SCIENCE, UNIVERSITY OF CHICAGO, CHICAGO, ILLINOIS 60637
E-mail address: crocker@cs.uchicago.edu

Implementing an $O(\sqrt{N}M)$ Cardinality Matching Algorithm

R. BRUCE MATTINGLY AND NATHAN P. RITCHEY

ABSTRACT. This paper discusses an implementation of the $O(\sqrt{N}M)$ cardinality matching algorithm due to Micali and Vazirani. Differences in the original algorithm and this implementation are highlighted. Results from a series of computational experiments are presented, using two other cardinality matching programs for comparison.

1. Introduction

Let $G = (V, E)$ be a graph with a vertex set V and edge set E, with $|V| = N$ and $|E| = M$. A *matching* is a subset S of the edge set such that no two edges are adjacent to the same vertex. An edge is said to be *matched* if it is a member of S; otherwise it is unmatched. A vertex is *free* if it is not adjacent to a matched edge. The cardinality matching problem (CMP) is to find a matching that includes the largest number of edges.

This problem was first introduced by Berge [2], who developed the concept of an *alternating path*, a path in which only every other edge is matched. An alternating path between two free vertices is called an *augmenting path*, since a new matching with larger cardinality can be found simply by changing all unmatched edges in the path to matched and vice versa. Berge showed that a given matching has maximum cardinality if and only if it does not admit an augmenting path. The first effective algorithm for solving instances of the CMP was develped by Edmonds [5] and had complexity $O(N^4)$. Gabow [7] and Lawler [9] each developed $O(N^3)$ modifications of Edmonds' algorithm. Derigs and Kazakidis [3] developed a special data structure for handling blossoms and implemented Lawler's algorithm. This implementation has provided faster running times than other implementations [4]. Hopcroft and Karp [8] developed an

1991 *Mathematics Subject Classification.* Primary 68R10, 90-04; Secondary 90C35, 90-08.

$O(N^{2.5})$ algorithm for bipartite graphs. Even and Kariv [6] extended this algorithm to the non-bipartite case. Micali and Vazirani [10] developed an $O(\sqrt{N}M)$ algorithm using ideas from [6], but did not implement it.

The paper is organized as follows. In §2 a brief description of the Micali-Vazirani algorithm is given. Some differences between their algorithm and our implementation are also discussed. In §3 we discuss several test problems and compare the performance of our implementaion to the program of Derigs and Kazakidis and to an implementation of Gabow's algorithm. Concluding remarks appear in §4.

2. The Micali-Vazirani algorithm

The Micali-Vazirani algorithm consists of a main routine called SEARCH, two subroutines, BLOSS-AUG and FINDPATH, and a function called OPEN. We will attempt only a brief description of the major features of this algorithm. For more details, the reader is referred to [10].

The algorithm consists of a number of phases. During each phase, a set of disjoint augmenting paths of minimum length is found. The existing matching is increased along each path, and then a new phase begins. The algorithm terminates when no augmenting path can be found. Augmenting paths are found by SEARCH, which grows breadth-first search (BFS) trees from each free vertex, where each path in the tree is alternating. A *bridge* is an edge which joins two paths. Subroutine BLOSS-AUG is called to process bridges by conducting a double depth-first search (DDFS) from each vertex of the bridge to find where the two paths begin. If the paths originate from distinct vertices, an augmenting path has been detected, and subroutine FINDPATH is called to construct it. Otherwise, a cycle has been detected, and a blossom consisting of the set of vertices contained in the cycle is formed. The vertex at which the cycle begins and ends is not included in the blossom, but is called the base of the blossom. Most CMP algorithms handle blossoms by "shrinking" them; that is, the entire blossom is treated as a single node. The Micali-Vazirani algorithm delays the shrinking process so that the first augmenting path that is found has minimum length. FINDPATH may initially find an augmenting path containing "shrunken" nodes. In this instance, the function OPEN is called, which makes a recursive call to FINDPATH to find the actual path through the blossom. The algorithm includes a labeling procedure that enables an augmenting path through the blossom to be found quickly. After an augmentation occurs, a topological erase is performed to ensure that augmenting paths do not overlap. We will now discuss the major differences in the original algorithm and in our implementation.

2.1. Starting procedure. In theory, the algorithm could begin with an empty matching. In practice, it is more efficient to find an initial matching using a simpler procedure which requires less overhead. Then the Micali-Vazirani

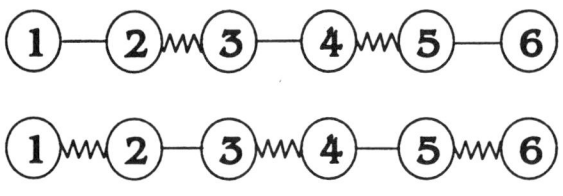

FIGURE 1

algorithm can be used until an optimal matching is found. For our implementation, we used the same starting procedure included in the code of Derigs and Kazakidis [3]. This procedure is based on the heuristic that, given an empty matching, vertices with the lowest degree should be matched first. In addition to reducing the overhead, the use of this starting procedure may change the order in which bridges are processed. This may affect the number of vertices matched initially, as shown in the following example.

EXAMPLE 2.1. Consider a graph with 6 vertices and 5 edges such that all edges are of the form $(i, i+1)$. In the first phase of the Micali-Vazirani algorithm, all edges are bridges and they may be processed in any order, perhaps as they are encountered. It is possible that edges $(2, 3)$ and $(4, 5)$ will be processed first, leaving vertices 1 and 6 free. This is shown in the top graph in Figure 1. Thus, another phase will be required. On the other hand, the other starting procedure attempts to process the adjacency lists of 1 and 6 first, since they have lowest degree. Thus, bridges $(1, 2)$ and $(5, 6)$ will be matched first. It is then possible to match $(3, 4)$ leaving no free vertices so that no additional phases are required. This is shown in the bottom graph in Figure 1.

2.2. Terminating the algorithm. The main routine in the algorithm, SEARCH, attempts to grow BFS trees from each free vertex. At each level i, SEARCH examines all vertices which are joined to free vertices by alternating paths of length i and attempts to extend these paths. (Free vertices are considered to be at level 0.) The algorithm terminates only when no vertices at level i exist. Our implementation checks the number of matched edges after the starting procedure and after each augmenting path is found, and terminates if the cardinality ever reaches $\lfloor N/2 \rfloor$, the theoretical maximum.

2.3. Detecting empty blossoms. If a bridge is found, SEARCH calls subroutine BLOSS-AUG which either forms a new blossom or discovers an augmenting path. There are some instances in which neither can occur, as shown in Example 2.2.

EXAMPLE 2.2. In Figure 2, $(2, 4)$ and $(3, 5)$ are matched edges. SEARCH begins growing trees from vertex 1. At level 1, two blossoms are formed: $B_1 =$

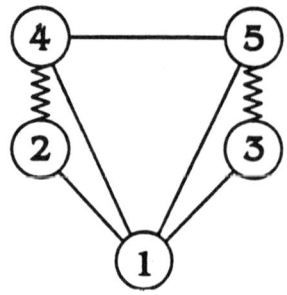

FIGURE 2

$\{2,4\}$ and $B_2 = \{3,5\}$. Vertex 1 is the base of each blossom. At level 2, edge $(4,5)$ is found to be a bridge. However, a given vertex can belong to only one blossom, so a new blossom cannot be formed. Our implementation of BLOSS-AUG detects cases such as this and returns immediately to the main routine.

2.4. The double depth-first search. When BLOSS-AUG conducts the DDFS from the vertices of the bridge, vertices in each path are marked L or R. If a path encounters a vertex contained in a blossom, the search immediately drops down to the base of the blossom. If this base is a member of another blossom, the process is repeated until a base which is not a member of any blossom is found. Only this base is marked L or R. (The algorithm specifies a function called *base** to handle this.) If the left and right searches meet at a common vertex, first the right search and then the left search backs up to try to find alternate paths. The following example shows that in the original algorithm, an error can occur when the searches attempt to back up.

EXAMPLE 2.3. In Figure 3, SEARCH begins at vertex 1. At level 3, blossom $\{6,7\}$ is formed with base 4. At level 5, three bridges are found: $(18,19)$, $(20,21)$ and $(22,23)$. If they are processed in that order, the following blossoms are formed: $\{18,19\}$ with base 13, $\{8,9,13,14,20,21\}$ with base 4, and $\{2,3,4,5,10,11,15,16,22,23\}$ with base 1. At level 6, bridge $(17,18)$ is found. The right search starts at 18, but since 18 is contained in a blossom, the search drops immediately to 13. Since 13 is in another blossom, the search then moves to 4, and finally to 1. The left search starts at 17, moves to 12, then to 6 and then drops through 4 to 1. When searches are conducted, each vertex maintains a pointer to the previous node in the search. When the right search occurred, 4 pointed back to 13. However, the left search caused 4 to point to 6. When the right search backs up, it will retrace the path found by the left search. This error occurs only when searches pass through nested blossoms. Our implementation maintains a pointer to the last node in a path which was not contained in a blossom. If backing up is necessary, the search backs up immediately to this node. In this example, the right search would immediately back up to 18.

2.5. Changes in FINDPATH and OPEN. When an augmenting path is found, FINDPATH is called twice to find paths from each vertex of the bridge

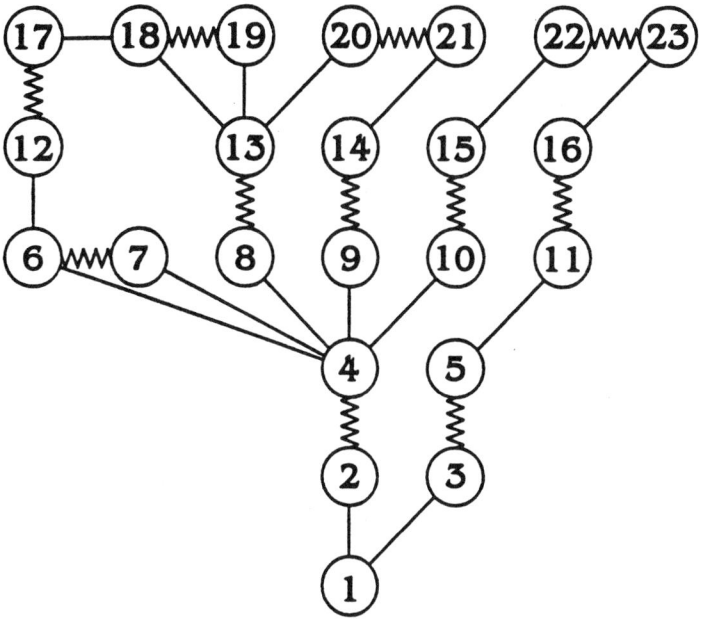

FIGURE 3

to a free vertex. If a path passes through a blossom, function OPEN is called. OPEN determines which direction the path should take through the blossom and then makes either one or two calls to FINDPATH. Since we implemented the algorithm in FORTRAN 77, a language which does not support recursion, we wrote a non-recursive version of FINDPATH and replaced calls to OPEN with explicit code.

Ordinarily, FINDPATH must check the left/right marks on vertices to ensure that two paths to the same vertex are not found. The following example describes one instance we discovered in which FINDPATH must allow a path to cross from left to right.

EXAMPLE 2.4. In Figure 4, bridge $(23, 24)$ is discovered at level 7. In the DDFS, the left and right searches initially reach 12. The right search can find no alternatives when it backs up, so it returns to 12. When the left search backs up to 23, another path is found starting at 18. Eventually both searches terminate at 6, which becomes the base of a blossom consisting of $\{8, 11, 14, 15, 18, 19, 23\} \cup \{9, 12, 16, 20, 24\}$ where the vertices in the first set are marked L and the vertices in the second set are marked R. At level 8, bridge $(25, 26)$ is found. FINDPATH

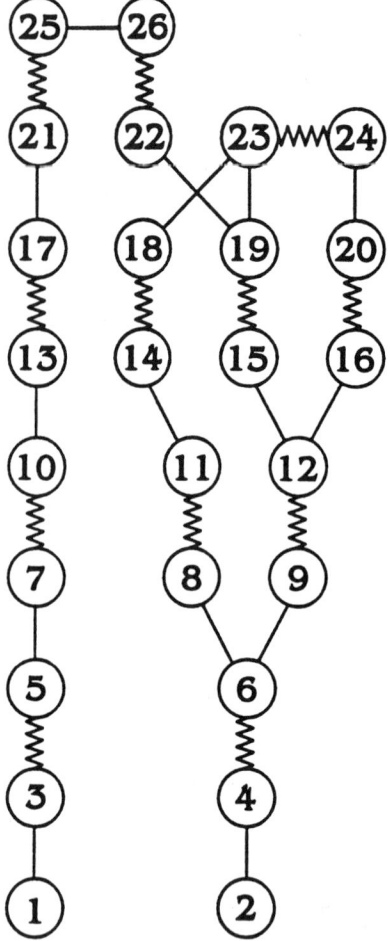

FIGURE 4

is called to find a path from 25 to free vertex 1, and again to find a path from 26 to 2. During this second call, the initial path is given by the sequence 26, 22, 19, 6, 4, 2. Since 19 is a member of a blossom, OPEN is called to find the true path from 19 to 6. From Figure 4, the only possible path would be 19, 15, 12, 9, 6. The original algorithm would not allow this since 15 is marked L while 12 is marked R. Our implementation allows this whenever OPEN determines that a path through a blossom can be found with only one call to FINDPATH instead of two.

2.6. Description of data structures. Parallel arrays are used to store much of the information about each vertex. For instance, MATCH$(i) = j$ and MATCH$(j) = i$ if (i, j) is a matched edge. If i is free, MATCH$(i) = 0$. EVLEV(i) and ODLEV(i) store the lengths of the shortest alternating paths of even and odd length from i to a free vertex. Every time a bridge is processed, a DDFS is conducted. During the jth DDFS, LR(i) is set to j if vertex i is found during

TABLE 1. Benchmark running times (seconds)

	min	max
Program 1	34.5	38.0
Program 2	59.1	67.4

the left search and $-j$ if it is found during the right search. The adjacency lists for each vertex are stored in a single array of length $2M$ called NGHBR. The adjacency list for vertex i is stored in locations NINDEX(i) through NINDEX($i+$ 1) $-$ 1. Bridges are stored using an array implementation of a linked list. The vertices of a bridge are stored in BRIDGE($1, i$) and BRIDGE($2, i$). The index of the next bridge to be processed is stored in BRIDGE($3, i$). The array BINDEX contains pointers to the array BRIDGE. BINDEX(i) points to the first bridge to be processed at level i.

3. Test problems and results

Several experiments were conducted to compare the running times of three programs: our FORTRAN implementation of the Micali-Vazirani algorithm, hereafter referred to as MV, the FORTRAN code of Derigs and Kazakidis [3], hereafter referred to as DK, and an implementation of Gabow's algorithm, hereafter referred to as G, written in C by Ed Rothberg and obtained from DIMACS. MV and DK used the same starting procedure to find an initial matching. This procedure was not incorporated into G.

Most of the experiments were conducted on a single processor of an Encore Multimax computer at the Department of Mathematical and Computer Sciences at Youngstown State University. This machine has 10 processors, each with a performance rate of 0.75 MIPS (millions of instructions per second.) There are 20 Mbytes of shared memory, and the operating system is UNIX. A few parallel experiments were conducted using all 10 processors of the Multimax. Other experiments were run on an 8-processor Cray Y-MP at the Ohio Supercomputer Center.

The running times reported include only the solution time and the time required to print out a single integer representing the optimal cardinality. The time required to read in the graph and print out the optimal matching was not included.

To facilitate the comparison of our timings with others, we measured the user time required to run the two **DIMACS** benchmark programs on a single processor of the Encore Multimax. We found slight variations in time due to the load on the system and due to the selection of different compiler options. The minimum and maximum times are reported in Table 1.

3.1. Hardcard. This problem was originally described by Gabow in [7], and generates fixed instances which are worst case inputs for Edmonds' algorithm.

TABLE 2. Running times (seconds) for one-connected triangles

N	DK	MV	G
2^9	1.16	1.15	0.69
2^{10}	4.23	2.32	1.95
2^{11}	16.1	4.33	6.54
2^{12}	63.4	8.81	23.9
2^{13}	253	23.3	92.0
2^{14}	1044	47.1	357

Given a graph with $6K$ vertices, a complete subgraph is formed by $4K$ vertices, while the remaining vertices are adjacent to only one edge. However, the starting procedure used by DK and MV destroys the worst case behavior by sorting the edges in increasing order according to degree. In all cases, the starting procedure found an optimal matching, so the implementations could not be compared further using this type of problem.

3.2. One-connected triangles. This generator was developed to compare the algorithms when a large number of disjoint blossoms occur in a CMP. Given an input parameter K, it generates a graph consisting of K triangles which are interconnected by only one edge. Table 2 shows the running times for problems of various sizes. G was the fastest algorithm for $N \leq 2^{10}$. For larger problems, MV was the fastest, with the difference in running times becoming more pronounced as the problem size increased.

3.3. Three-connected triangles. This problem generates instances similar to the previous test problem. However, successive triangles have three edges joining them. Running times for the programs are shown in Table 3 and show that DK was the fastest for all problems tested. In every instance, we observed that the cardinality of the initial matching was one less than the optimal cardinality, so only one augmenting path was needed to reach an optimal solution. In the previous example, the initial matching had cardinality which was about 89% of optimal. We developed other fixed-instance generators which will not be discussed in detail. Instead, we will simply note that DK was faster than MV in all instances where the initial matching was one less than optimal. This is not surprising, since the DK code is only about 400 lines long, and is much simpler than the MV code, which is over 1000 lines long.

3.4. Random problems. Test problems of a given size were generated by randomly selecting M edges from a complete graph with N nodes. Problems were tested with $2^9 \leq N \leq 2^{14}$. For each value of N, problems were generated such that the average degree of each node varied between 2 and 16.

For each data point, four sets of five problems of a given size were initially generated and run. The mean solution time for each set was calculated. Then the mean of the means and the variance of these means were computed. If the

TABLE 3. Running times (seconds) for Tri3

N	DK	MV	G
2^9	0.37	2.20	1.82
2^{10}	0.64	3.89	6.35
2^{11}	1.45	11.4	24.2
2^{12}	3.55	40.7	97.8
2^{13}	9.47	89.5	379
2^{14}	34.2	220	1563

variance was less than ten percent, the running time for that data point was taken as the mean of all running times. Otherwise, twenty more random problems were generated and the process was repeated. It was not feasible to run more than 60 problems for any given data point. In some of the larger problems, the variance of the means remained higher than ten percent. However, our results appeared to be consistent with the observed behavior on smaller problems.

Before examining the average running times of the algorithms as functions of the number of vertices and the average degree, it will be illuminating to examine the variation in running times on different random graphs of the same size. The following discussion focuses only on MV. For each data point, a series of graphs was generated. In addition to the solution time, the number of blossoms formed and the number of search levels required were recorded. (The further apart free vertices are in a graph, the more search levels are required to find an augmenting path.) This data was sorted according to solution time. Then, the data was divided into 5 groups. The first group contained the problems with the lowest solution times. The last group contained the highest solution times. For each group, the average number of blossoms formed and search levels required was found. Tables 4, 5 and 6 are representative of our findings. In each table, the graphs had $N = 2^{10}$. Table 4 shows the results for graphs with average degree 2. Virtually no blossoms were formed. However, the average number of search levels increased with average solution time. Table 5 shows the results for graphs with average degree 6. In the first three groups, the solution time varied only from 1.9 to 2.4 seconds. In these groups, fewer than 10 blossoms and 5 search levels were found on the average. In the last two groups however, the average number of blossoms was greater than 300, while the number of search levels was greater than 12. The solution times increased to an average of 6.3 to 7.5 seconds. In Table 6, results for average degree 12 are given. The number of blossoms and search levels in all groups was very low. The running times varied only from 1.4 to 2.1 seconds.

We considered whether the variation in solution times among problems of a given size could be attributed to differences in the cardinalities of the initial matchings found by the starting procedure. However, we found no data that would support this hypothesis, since little variation in the initial cardinalities

TABLE 4. Statistics for random graphs
with $N = 2^{10}$, average degree 2

Group	average time	average blossoms	average levels
1	1.6	0.3	3.3
2	2.0	0.0	4.0
3	2.4	1.3	5.5
4	2.7	0.0	6.0
5	3.0	1.8	8.5

TABLE 5. Statistics for random graphs
with $N = 2^{10}$, average degree 6

Group	average time	average blossoms	average levels
1	1.9	1.6	3.8
2	2.3	4.8	4.4
3	2.4	8.3	4.5
4	6.3	307	12.1
5	7.5	347	14.3

TABLE 6. Statistics for random graphs
with $N = 2^{10}$, average degree 12

Group	average time	average blossoms	average levels
1	1.4	1.9	2.0
2	1.6	1.5	2.1
3	1.7	2.4	2.1
4	1.9	2.6	2.0
5	2.1	3.3	2.4

was observed.

Tables 7 through 9 contain the mean running times for each algorithm on all problem sizes. (Algorithm G was not tested on a few of the largest problems.) Figures 5 through 7 compare the running times of all three algorithms for $N = 2^{10}, 2^{12}$ and 2^{14}.

Algorithm G was the fastest on all problems of average degree 2, regardless of the number of vertices. However, for average degree larger than 3, G was the slowest. For all problems with $N = 2^{12}$, the running times for MV and DK were very close. DK proved to be faster for $N < 2^{12}$, while MV was faster for

TABLE 7. Running times (seconds) for DK on random problems

degree	N:	2^9	2^{10}	2^{11}	2^{12}	2^{13}	2^{14}
2		0.46	1.37	4.58	13.8	60.2	231
3		0.73	2.45	8.93	33.8	133	504
4		0.68	2.33	7.44	25.5	104	409
5		0.66	1.99	6.38	22.2	87.8	330
6		0.63	1.73	5.66	19.6	71.7	277
7		0.58	1.61	5.23	17.3	62.0	237
8		0.53	1.51	4.64	15.5	54.9	205
12		0.52	1.28	3.81	12.5	40.5	149
16		0.52	1.31	3.39	10.5	33.1	126

TABLE 8. Running times (seconds) for MV on random problems

degree	N:	2^9	2^{10}	2^{11}	2^{12}	2^{13}	2^{14}
2		1.10	2.57	6.09	13.8	35.2	78.4
3		2.36	5.93	17.3	44.1	115	277
4		2.08	5.73	11.2	25.5	53.4	120
5		2.63	5.32	10.3	22.4	49.5	107
6		2.63	4.56	11.0	23.0	45.2	101
7		2.11	4.00	12.0	23.2	41.2	94.5
8		1.37	3.95	9.35	24.4	38.5	103
12		1.38	2.38	5.39	11.1	24.6	52.9
16		1.34	2.55	5.39	12.6	25.4	80.3

TABLE 9. Running times (seconds) for G on random problems

degree	N:	2^9	2^{10}	2^{11}	2^{12}	2^{13}	2^{14}
2		0.17	0.34	0.78	1.49	3.04	6.15
3		0.48	1.66	5.55	20.3	74.0	464
4		1.51	5.68	25.4	97.5	473	2480
5		1.94	6.54	29.3	120	474	—
6		1.83	7.11	30.7	125	554	—
7		2.10	6.58	33.6	117	543	—
8		1.85	5.82	30.7	119	436	—
12		1.17	3.30	15.0	56.7	260	—
16		1.18	4.77	17.6	66.4	—	—

$N > 2^{12}$. These results are not surprising. The lower theoretical complexity of MV is important on large problems, while the decreased overhead of DK was more important on the small problems. DK and G had the same theoretical

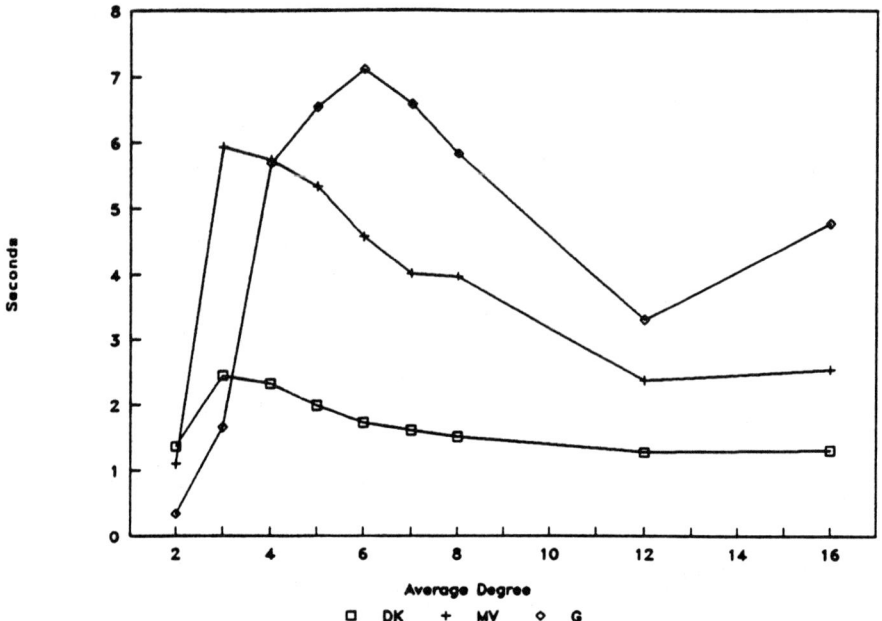

FIGURE 5. Running times for random problems with $N = 2^{10}$

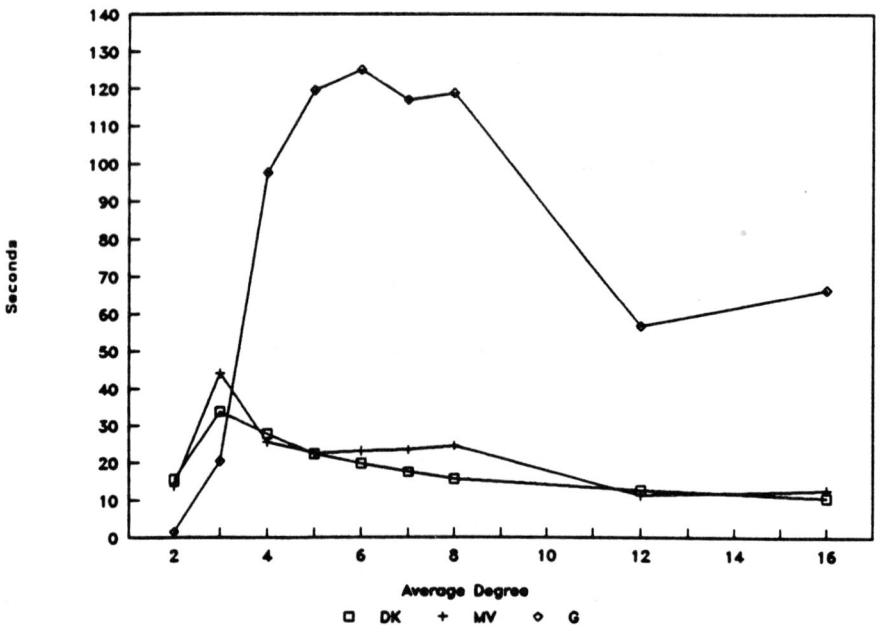

FIGURE 6. Running times for random problems with $N = 2^{12}$

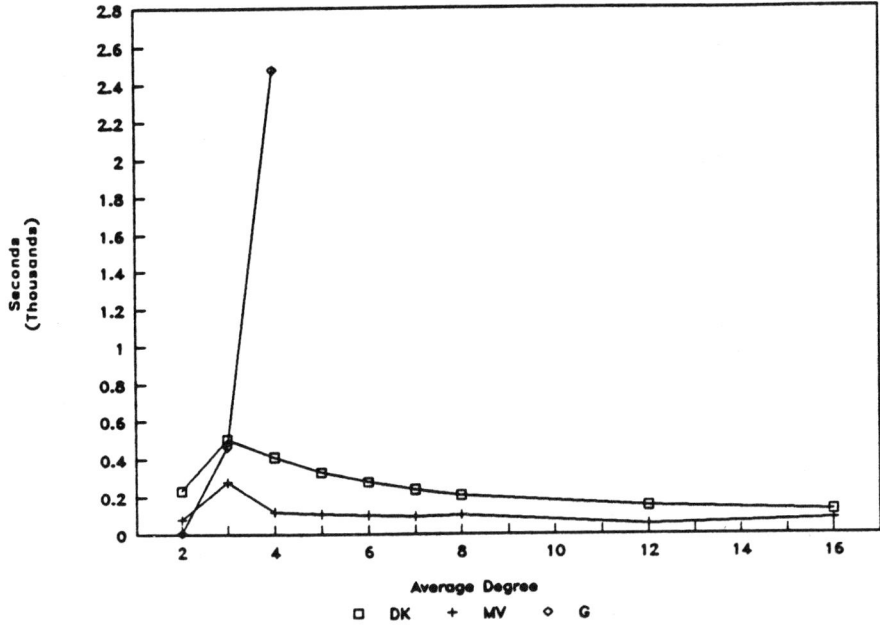

FIGURE 7. Running times for random problems with $N = 2^{14}$

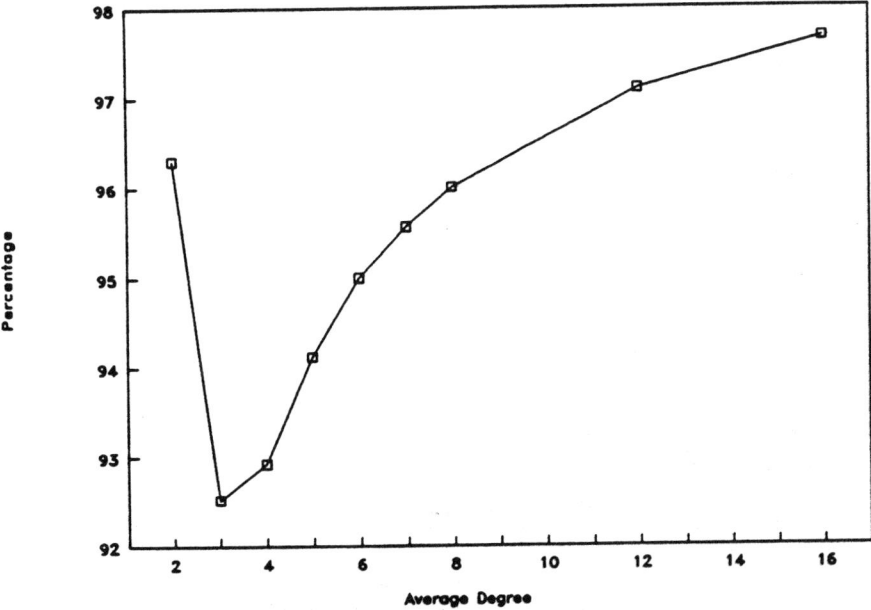

FIGURE 8. Percent of optimal solution found by starting procedure

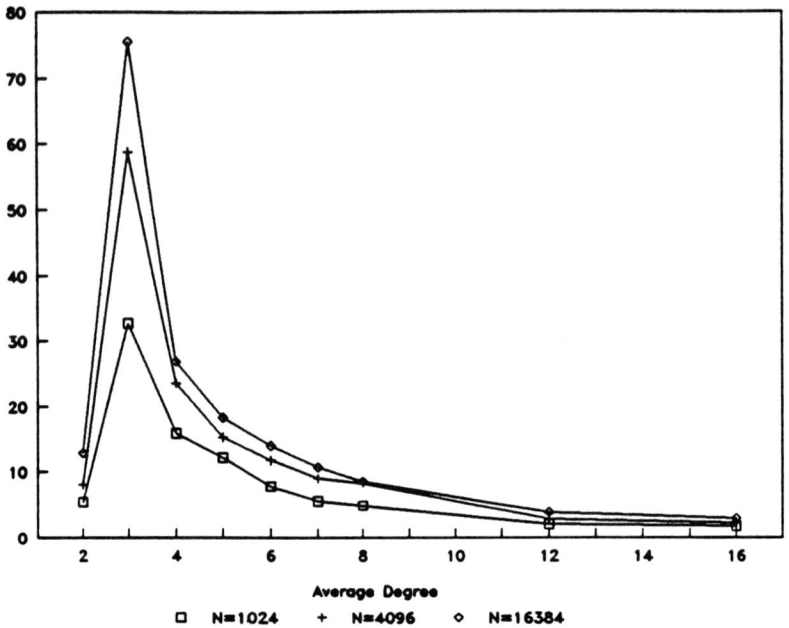

FIGURE 9. Average number of search levels in MV

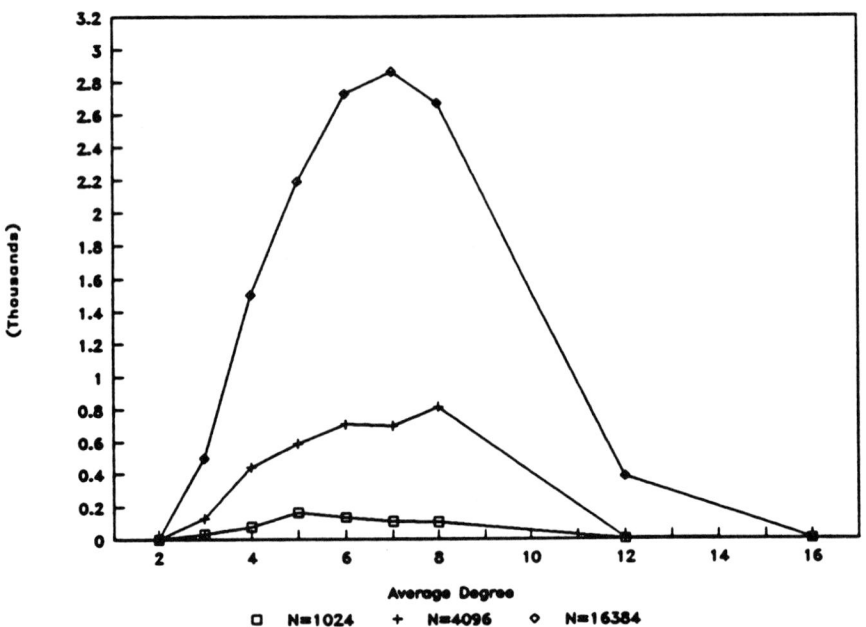

FIGURE 10. Average number of blossoms formed by MV

complexity, but DK incorporated an efficient starting procedure. We found that MV without the starting procedure was much slower in practice.

For problems with fixed N, the peak running time for both DK and MV occurred at average degree 3. We examined several other quantities as a function of average degree. First, we measured how close the initial matching found by the starting procedure was to being optimal. Figure 8 shows the average ratio of the initial cardinailty to the optimal cardinality, expressed as a percentage. The results turned out to be virtually independent of N. The minimum percentage occurred at average degree 3. This may explain the peak in solution times, since the less efficient main procedures were left with more work to do. Figure 9 shows the average number of search levels required by MV. The peak again occurs at average degree 3 for all values of N. Consider what happens as edges are added to a graph. The number of paths between two vertices cannot decrease, and shorter paths may be introduced. Therefore, one would expect the number of search levels to eventually decrease as average degree increases. Why the peak occurs at 3 has not been fully explored. Figure 10 shows the average number of blossoms found. Few blossoms occur in graphs with low or high average degree. At low degree, cycles are less likely to occur. At high degree, there is less opportunity to form blossoms for two reasons. First of all, the starting procedure matches a higher percentage of the vertices. Secondly, since the average number of search levels is low, the lengths of augmenting paths between free vertices is also low. Note that for MV, the peak running time and the peak number of blossoms do not occur at the same average degree, which leads us to believe that the number of search levels may be the most important factor influencing running time.

A functional relationship between problem size and solution time was found for each program. Figure 11 shows a log-log graph of the solution time for each method as a function of the number of vertices, with the average degree held constant at 4. Similar behavior was observed when the average degree was held constant at other values. Regression analysis techniques were used to find that the solution time for DK was proportional to $N^{1.28}$ with $R^2 = 0.999$. The solution time for MV was proportional to $N^{0.79}$ with $R^2 = 0.997$. The solution time for G was proportional to $N^{1.47}$ with $R^2 = 0.998$. Since the number of edges was proportional to N, the theoretical running times would be $O(N^{2.5})$ for both DK and G, and $O(N^{1.5})$ for MV. However, the theoretical complexity does not take into account the efficient starting procedure used by MV and DK, or the fact that all inputs were not worst-case.

3.5. Parallel experiments. Parallel implementations of the Micali-Vazirani algorithm were attempted using the parallel FORTRAN compilers available on the Encore Multimax and the Cray Y-MP. These compilers automatically modify FORTRAN programs for parallel execution. (The Cray FORTRAN compiler also attempts to vectorize code using vector pipelines which minimize delays between the end of one computation and the beginning of the next.) On both machines, additional optimization is usually required to obtain the best performance.

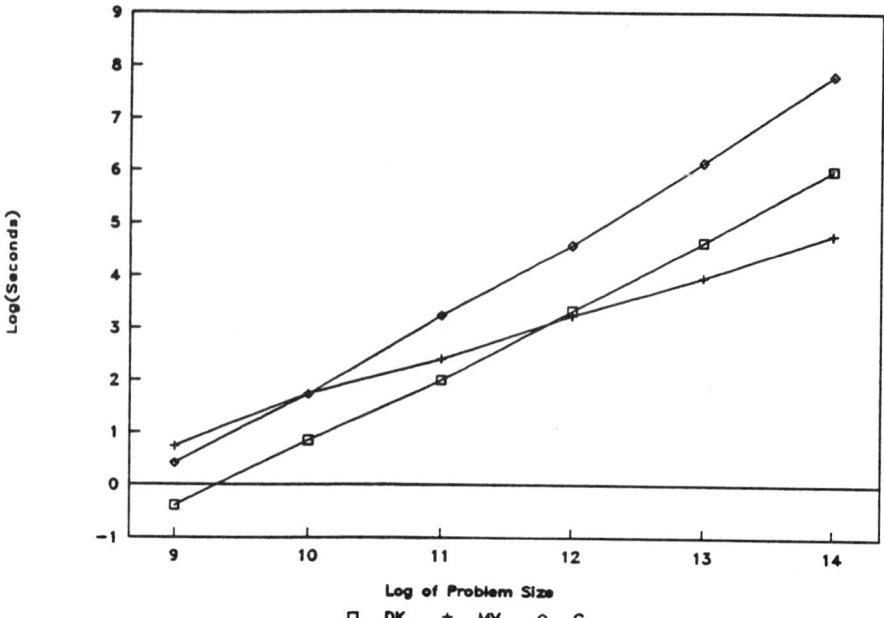

FIGURE 11. Running times for random problems with degree 4

Unfortunately, there are many factors which inhibit parallelism. Much of the Micali-Vazirani algorithm involves depth-first searches, which are inherently serial [1]. Furthermore, in this research, we were primarily interested in studying sparse graphs, since we found that it was fairly simple to find an optimal matching in dense graphs such as **hardcard**. Parallel breadth first search techniques [1] are better suited to dense graphs than sparse ones. From the viewpoint of the parallel compilers, parallelism (and vectorization) is easiest to achieve on programs which process arrays. Since we were only studying sparse graphs, representing the graph with an adjacency matrix was not feasible, and adjacency lists were used instead.

For each parallel experiment, the speedup, or ratio of serial and parallel exectuion times, was computed. The highest speedup obtained in any instance was only 1.25. In some small problems, the extra overhead incurred by starting parallel processes caused a slowdown; i.e., the speedup was less than 1.

4. Summary

An implementation of the Micali and Vazirani CMP algorithm has been presented. We discovered and corrected several logical errors in the original algorithm. We also compared the solution times of our implementation and two other programs on random and fixed instances. We observed that MV had the lowest solution times for all problems with $N > 2^{12}$ except for graphs having average degree 2. For algorithm MV, the maximum running time, the maximum num-

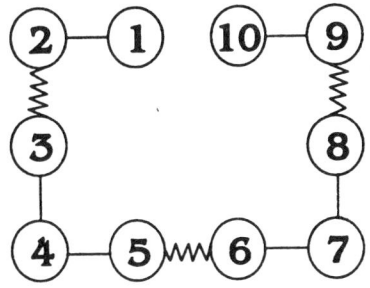

FIGURE 12

ber of search levels and the minimum initial cardinality all occurred at average degree 3.

We found that the MV running time depends on the number of search levels required. An open question at this point is how to minimize the number of search levels. The following example discusses this problem.

EXAMPLE 4.1. In Figure 12, vertices 1, 4, 7 and 10 are free, and edges $(2,3)$, $(5,6)$ and $(8,9)$ are bridges discovered at search level 2. If $(2,3)$ is processed first, then an augmenting path from 1 to 4 is found. Because of the topological erase, $(5,6)$ cannot be processed. At the same search level, $(8,9)$ can be processed and an augmenting path from 7 to 10 is found, producing an optimal matching. But consider what happens if $(5,6)$ is processed first. An augmenting path form 4 to 7 is found, and due to the topological erase, neither of the other bridges can be processed. Since 1 and 10 are still free, another phase must begin. The search will detect a bridge at level 5, and an augmenting path from 1 to 10 will be found. The MV algorithm does not consider the order in which bridges should be processed, and it is not clear how to construct a method for determining the best order.

4.1. Software. A listing of the DK code we used can be found in [3]. Our implementation of the Micali-Vazirani algorithm and the problem generators we developed are all available from **DIMACS** using the anonymous ftp facility. The programs (all written in FORTRAN 77) are as follows:

(a) **cardmp.f**: Our implementation of the Micali-Vazirani algorithm.
(b) **hardcard.f**: Generates fixed instances which are worst-case for Edmond's algorithm.
(c) **t.f**: Generates a series of one-connected triangles.
(d) **tt.f**: Generates a series of three-connected triangles.

In addition, we used two C programs provided by **DIMACS**: **match.c**, which is Rothberg's implementation of Gabow's algorithm, and **random.c**, which was used to generate the random graphs. We had to make modifications to generate random numbers on the Encore Multimax. These modifications, as well as the random seeds we used, are available upon request.

4.2. Acknowledgements. We would like to thank Doug Faires, Kriss Schueller and Matt Smith for their help with installing the $\mathcal{A}\mathcal{M}\mathcal{S}$-TEX Version 2.1 software, Bill Tomory who produced Figures 2, 3 and 4, and Reneé Vivacqua for Figures 1 and 12. We owe Steve Crocker special thanks for his insightful comments about the Micali-Vazirani algorithm and for a modification to **match.c** which improved its running time.

References

1. E. Arjomandi and D. G. Corneil, *Parallel computations in graph theory*, Proc. 16th Annual Symposium on Foundations of Computer Science, IEEE, 1975, pp. 13–18.
2. C. Berge, *Two theorems in graph theory*, Proc. Natl. Acad. Sci. U.S. **43** (1957), 842–844.
3. R. E. Burkard and U. Derigs, *Assignment and matching problems: Solution methods and FORTRAN-programs*, Lecture Notes in Economics and Mathematical Systems, vol. 184, Springer-Verlag, Berlin, 1980.
4. U. Derigs, *Programming in networks and graphs*, Lecture Notes in Economics and Mathematical Systems, vol. 300, Springer-Verlag, Berlin, 1988.
5. J. Edmonds, *Paths, trees, and flowers*, Can. J. Math. **17** (1965), 449–467.
6. S. Even and O. Kariv, *An $O(n^{2.5})$ algorithm for maximum matching in general graphs*, Proc. 16th Annual Symposium on Foundations of Computer Science, IEEE, 1975, pp. 100–112.
7. H. N. Gabow, *An efficient implementation of Edmonds' algorithm for maximum matching on graphs*, J. ACM **23** (1976), 221–234.
8. J. E. Hopcroft and R. M. Karp, *An $n^{2.5}$ algorithm for maximum matching in bipartite graphs*, SIAM J. on Comp. **2** (1973), 225–231.
9. E. L. Lawler, *Combinatorial Optimization. Networks and Matroids*, Holt, Rinehart and Winston Inc., New York, 1976.
10. S. Micali and V. V. Vazirani, *An $O(\sqrt{|v|}|E|)$ algorithm for finding maximum matching in general graphs*, Proc. 21st Annual Symposium on Foundation of Computer Science, IEEE, 1980, pp. 17–27.

DEPARTMENT OF MATHEMATICAL AND COMPUTER SCIENCES, YOUNGSTOWN STATE UNIVERSITY, YOUNGSTOWN, OHIO 44555

E-mail address: FR132601@YSUB.YSU.EDU

DEPARTMENT OF MATHEMATICAL AND COMPUTER SCIENCES, YOUNGSTOWN STATE UNIVERSITY, YOUNGSTOWN, OHIO 44555

E-mail address: nate@macs.ysu.edu

Solving Large-Scale Matching Problems

DAVID APPLEGATE AND WILLIAM COOK

ABSTRACT. We describe a new implementation of Edmonds' blossom algorithm for computing minimum weight perfect matchings. Combining this with specialized pricing techniques, we obtain a solution method for large-scale graphs. We report on the solution of a set of geometric test problems (complete graphs, described as points in the plane), the largest having 101,230 nodes.

1. Introduction

Let $G = (V, E)$ be a graph with node set V and edge set E. A *matching* $M \subseteq E$ is a subset of the edges such that each node in V is met by at most one edge in M. It is a *perfect matching* if each node is met by exactly one edge. Given a real weight w_e for each $e \in E$, the *minimum weight perfect matching problem* is to find a perfect matching M with minimum weight $w(M) = \sum(w_e : e \in M)$. Applications of minimum weight perfect matchings are discussed in Ball, Bodin, and Dial [5], Frederickson, Hocht, and Kim [15], Iri and Taguchi [23], and Reingold and Tarjan [31].

One of the cornerstones of combinatorial optimization is Edmonds' [14] polynomial-time *blossom* algorithm for min-weight perfect matching. This algorithm has been studied both from the viewpoint of obtaining good bounds on its asymptotic running time and from the viewpoint of developing fast computer implementations. Increasingly better time bounds have been obtained by Ball and Derigs [6], Gabow [16], Gabow [17], Gabow, Galil, and Spencer [18], Gabow and Tarjan [19], Galil, Micali, and Gabow [20], and Lawler [25]. The current best is $O(|V|(|E|+|V|log|E|))$ by Gabow [17]. (A nice overview of some of these results is given in Ball and Derigs [6].) Sophisticated computer implementations are described in Burkard and Derigs [7], Cunningham and Marsh [8], Derigs [9],

---

1991 *Mathematics Subject Classification.* Primary 90C27, 05C70; Secondary 90C06, 05C04.

The first author was supported in part by the National Science Foundation under contract CCR-9007602.

© 1993 American Mathematical Society
1052-1798/93 $1.00 + $.25 per page

[10], [11], Derigs and Metz [12], Derigs and Metz [13], and Lessard, Rousseau, and Minoux [26].

An important class of matching problems are those described by specifying n points in the plane to match so as to minimize the distance between the pairs. In terms of graphs, this means we have the complete graph on n nodes with the weight of each edge being the distance between its end points (under some metric). Applications of this class of problems include the routing of mechanical plotters as described in [23] and [31]. Fast heuristics have been proposed for finding good solutions to these geometric problems (see Assano, Edahiro, Imai, Iri, and Murota [3] and Avis [4]) and fast exact methods have been developed for special configurations of points (see Marcotte and Suri [27]). Moreover, Vaidya [34] has shown that the blossom algorithm can be implemented with an $O(|V|^{2.5}(\log|V|)^4)$ time bound for this class.

In this paper we describe a new computer implementation of the blossom algorithm for general graphs, together with specialized methods for the solution of geometric matching problems. Like the code of Derigs and Metz [12], we use a fractional matching "jump start" and work with a sparse subgraph of the original graph G, using linear programming pricing to handle the remaining edges. Our pricing technique makes use of the least-common ancestor algorithm of Aho, Hopcroft, and Ullman [2], and we employ a sweep method to reduce the number of edges that need to be priced explicitly when solving geometric problems.

The paper is organized as follows. In Section 2 we give a brief outline of the blossom algorithm, and in Section 3 we describe our implementation. The overall strategy for solving large problems is described in Sections 4, 5, and 6. Our computational tests are reported in Section 7, including the solution of a geometric problem containing 101,230 nodes.

2. Blossom algorithm

The blossom algorithm is a primal-dual method based on Edmonds' linear programming formulation of the min-weight perfect matching problem. To describe this, let $\mathcal{O}$ denote the set of all odd subsets of V containing at least 3 nodes (we refer to these sets as *blossoms*), and for each $S \subseteq V$, let $\delta(S)$ denote the set of edges that meet exactly one node in S. For a vector $(x_e : e \in E)$ and a set $H \subseteq E$, let $x(H)$ denote the sum $\sum(x_e : e \in H)$. Edmonds [14] used the blossom algorithm to show that the convex hull of the incidence vectors of the perfect matchings in G is precisely the solution set of the linear system

$$
\begin{align}
x(\delta(\{v\})) &= 1 \text{ for all } v \in V \tag{1}\\
x_e &\geq 0 \text{ for all } e \in E \tag{2}\\
x(\delta(S)) &\geq 1 \text{ for all } S \in \mathcal{O}. \tag{3}
\end{align}
$$

So the minimum weight of a perfect matching is equal to min $\{wx : x$ satisfies (1), (2), and (3) $\}$. The dual to this linear programming problem is

(4) $$\max \sum(y_v : v \in V) + \sum(Y_S : S \in \mathcal{O})$$

subject to

(5) $\quad y_u + y_v + \sum(Y_S : S \in \mathcal{O}, (u,v) \in \delta(S)) \leq w_{(u,v)}$ for all $(u,v) \in E$

(6) $\quad\quad\quad\quad\quad\quad\quad\quad\quad\quad\quad\quad\quad\quad\quad\quad Y_S \geq 0$ for all $S \in \mathcal{O}$.

Given a solution $(\bar{y}, \bar{Y})$ to this dual problem, the *reduced cost* of an edge $e = (u,v)$ is

$$w_{(u,v)} - \bar{y}_u - \bar{y}_v - \sum(\bar{Y}_S : S \in \mathcal{O}, (u,v) \in \delta(S)),$$

that is, the slack in the corresponding constraint (5). An edge is called *tight*, with respect to $(\bar{y}, \bar{Y})$, if its reduced cost is 0. Similarly, a blossom $S \in \mathcal{O}$ is called *full*, with respect to a (partial) matching $\bar{x}$, if $\bar{x}(\delta(S)) = 1$. With these definitions, the complementary slackness conditions for a primal-dual pair of solutions can be stated as: for all edges $e \in E$, if $\bar{x}_e > 0$, then e is tight, and for all blossoms $S \in \mathcal{O}$, if $\bar{Y}_S > 0$ then S is full. So we can prove that a given perfect matching is optimal by providing a dual solution such that these conditions are satisfied. The blossom algorithm produces just such a proof. At each step, the algorithm has a partial matching and a dual solution that satisfy the complementary slackness conditions. The matching is grown via augmenting paths until we reach a perfect matching (which we know is optimal). A good description of the blossom algorithm can be found in Pulleyblank [29]. We will not go into all of the details of the algorithm, but we do need to describe some of its structure to present our implementation below.

We start with a simplified version of the blossom algorithm, solving the *fractional matching problem* $\min\{wx : x$ satisfies (1) and (2)$\}$. This will serve two purposes. First, it builds a framework that will be useful in our discussion of the blossom algorithm below. Secondly, fractional matchings are used at several points in the scheme for solving large matching problems described in Section 4. So let us suppose there exists a solution to (1) and (2). The fractional matching algorithm finds a $0, 1/2, 1$ valued solution $\bar{x}$ and a dual solution $\bar{y}$ such that $\bar{x}_{(u,v)} > 0$ only if (u,v) is tight, that is, $w_{(u,v)} = \bar{y}_u + \bar{y}_v$ (there are no variables Y_S). So, by complementary slackness, $\bar{x}$ will be an optimal solution to the fractional matching problem. Moreover, $\bar{x}$ will have the property that the edges e with $\bar{x}_e = 1/2$ form node disjoint odd circuits. (Notice that the fractional matching problem can also be solved by reducing it to a bipartite matching problem.)

Initially, let $\bar{x}_e = 0$ for all edges e and for each node v let $\bar{y}_v = (1/2)\min\{w_e : e \in \delta(v)\}$. At each iteration, choose an *unmatched* node r (that is, $\bar{x}(\delta(r)) = 0$) and grow a tree T rooted at r having the following properties: each edge in T is tight and for each node v in T, the unique path in T from v to r alternates between matched edges ($\bar{x}_e = 1$) and unmatched edges ($\bar{x}_e = 0$). Such a tree T is called an *alternating tree*. The nodes of T are labeled "+" and "−" according to the parity of the number of edges in the path back to the root r, that is, node r and all nodes of even distance from r receive the label "+" and all nodes of odd distance receive the label "−". We *grow* T by appending matched edges that meet "−" nodes or tight unmatched edges that join "+" nodes to nodes not yet in T.

If we reach an unmatched node v in T (other than r), then $\bar{x}$ can be *augmented* along the path from v to r, replacing $\bar{x}_e$ by $1 - \bar{x}_e$ for each edge e in the path. Such an augmentation increases the size of $\bar{x}$ by 1 (that is, $\sum(\bar{x}_e : e \in E)$ increases by 1). It is also possible to increase $\bar{x}$ if we reach a node v that meets an edge e with $\bar{x}_e = 1/2$. In this case, by the way we build $\bar{x}$, we know that v lies on an odd circuit C all of whose edges have value $1/2$. Thus, the size of $\bar{x}$ can be increased by $1/2$ by augmenting along the path in T from v to r and, starting with node v, alternately flipping the $1/2$-valued edges in C to 0 and 1 as we traverse the circuit back to v (see Figure 1).

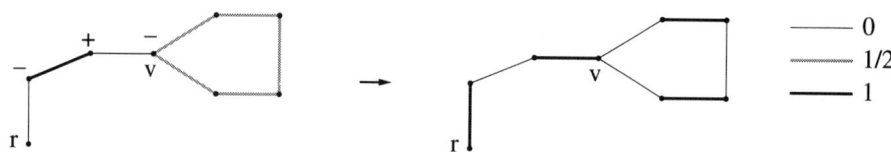

FIGURE 1. A 1/2 - augmentation

A third, and final, way to increase the size of $\bar{x}$ involves the formation of $1/2$-valued odd circuits. Suppose there exists no node in T meeting a $1/2$-valued edge and suppose there does exist a tight edge (u, v) joining two "+" nodes. Let P_1 denote the path in T from u to r and let P_2 denote the path from v to r. The union of P_1, P_2 and (u, v) forms an odd circuit C and a path, P, from w (the first node in P_2 we meet as we traverse P_1) to r. Again, $\bar{x}$ can be increased by $1/2$ by augmenting along P (so w is now unmatched) and setting $\bar{x} = 1/2$ for all edges in the circuit C (see Figure 2). Since we have assumed that no node of T meets a $1/2$-valued edge, we know that C is node disjoint from any other $1/2$-valued circuits that may exist. Now suppose we cannot grow T any further and we have reached neither an unmatched node, nor a node meeting a $1/2$-valued edge, nor a tight edge joining two "+" nodes. At this point we must alter the dual solution to create a new tight edge that allows us either to grow the tree or augment the fractional matching. The constraints on the dual change are that all edges in T, as well as all other edges having $\bar{x}_e > 0$, remain tight and that $\bar{y}$ remains a dual solution (that is, $\bar{y}_u + \bar{y}_v \le w_{(u,v)}$ for all edges (u, v)). These

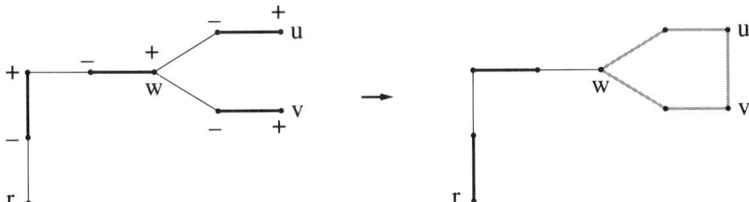

FIGURE 2. Creating a 1/2 - valued circuit

conditions can be met by adding some $\epsilon > 0$ to the $\bar{y}$-value of all "+" nodes and subtracting ϵ from all "−" nodes, making ϵ as large as possible subject to $\bar{y}$ remaining a dual solution. What stops us from making ϵ arbitrarily large is that some edge, joining a "+" node and a node not yet in T or joining two "+" nodes, becomes tight. (If no edge becomes tight, then there is no solution to (1) and (2).) In the first case we can grow T and in the second case we can create a 1/2-valued circuit. It is useful to remark that if all edge costs are integral, then the $\bar{y}$ values remain 1/2 integral throughout the procedure. This follows easily from the form of the dual changes. Indeed, suppose that at a general stage all $\bar{y}$ values (and, hence, all reduced costs) are 1/2 integral and that we are about to make an ϵ dual change. If ϵ is bounded by the reduced cost on an edge joining a "+" node and a node not yet in the tree, then it is certainly 1/2 integral. The only problem seems to be when ϵ is bounded by 1/2 of the reduced cost of an edge joining two "+" nodes. But notice that since all edges in the alternating tree are tight, we must have $\bar{y}_u \equiv \bar{y}_v$ (modulo 1) for any two nodes in the tree. It follows that the reduced cost of the edge joining the two "+" nodes is integral, and hence ϵ is again 1/2 integral.

So that is the algorithm: we repeatedly grow T, alter the dual solution, and grow T some more, until we can perform an augmentation. The full blossom algorithm works in the same way, but with two extra operations to handle the blossom variables Y_S. The key idea is that when we set $\bar{Y}_S > 0$ for some set S, we then treat S as a single node. The intuition is that the complementary slackness condition $\bar{x}(\delta(S)) = 1$ is the same as the constraint $\bar{x}(\delta(v)) = 1$ for individual nodes v. Thus, one of the new operations is to *shrink* a subset of nodes S into a *pseudonode* by contracting all edges having both ends in S. This operation will arise when we have a tight edge joining two "+" nodes (or pseudonodes) in the current alternating tree. Where in the fractional algorithm we performed a 1/2-augmentation around the odd circuit C that is formed, we instead shrink the nodes of C into a pseudonode. The converse operation is to delete a pseudonode S, bringing a subset of the nodes (and pseudonodes) we previously shrunk back into the current alternating tree T. This occurs when the $\bar{Y}$-value corresponding to S goes to 0 after a dual change. In this situation, S is a "−" node and thus is met by exactly two edges in T: a matched edge e and an unmatched edge f. We need to match the nodes of S, other than the node $w \in S$ that is already matched by e, with edges that were previously contracted. This step is straightforward, since S is spanned by an odd circuit C of tight edges (and any dual changes made after S is shrunk will not effect the reduced costs of the edges that were contracted in the shrink operation). To maintain the alternating tree T, we replace S by the even length path in C from the node $u \in S$ that meets the unmatched edge f to the node w that meets the matched edge e, labeling the nodes "$-$" and "$+$" as we move from u to w. (So the internal nodes and edges of the odd path from u to w will not be part of the tree.) This entire operation will be referred to as *expanding* the pseudonode S (see Figure 3).

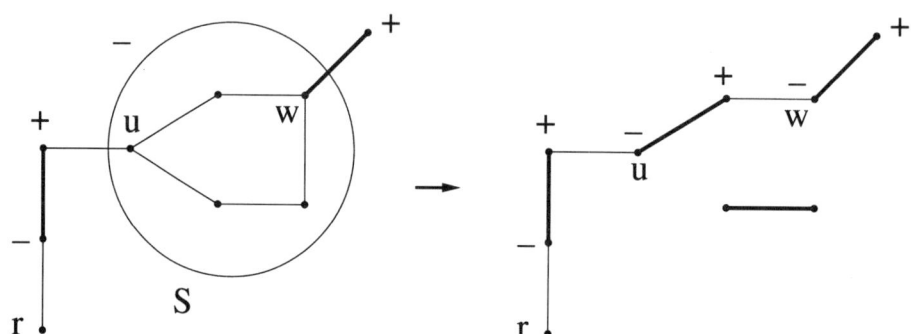

FIGURE 3. Expanding a pseudonode

At a general stage of the blossom algorithm we have carried out some number of shrinkings and expansions, resulting in a current *surface graph*. The procedure is such that a perfect matching amongst the tight edges in this graph (tight is still defined in terms of the original graph G) translates directly to an optimum matching in G. Initially, the surface graph is G itself, $\bar{x}$ and $\bar{Y}$ are 0, and $\bar{y}$ is set as in the start of the fractional algorithm. At each step we choose an unmatched node r in the surface graph and grow an alternating tree T rooted at r. If we reach an unmatched node in T, then we augment the matching in the surface graph and continue from another unmatched node r. If there exists a tight edge joining two "$+$" nodes, then we shrink the nodes of the odd circuit that is formed and continue growing the tree T. If neither of these operations are possible and we can no longer grow T, we perform a dual change, again adding as large as possible $\epsilon > 0$ to the $\bar{y}$-value of the "$+$" nodes (or the $\bar{Y}$-value of the "$+$" pseudonodes) and subtracting ϵ from the "$-$" nodes. If a new tight edge appears we can continue the algorithm as in the fractional case. But it may happen that ϵ is bounded by the $\bar{Y}$-value of a "$-$" pseudonode S (since these variables must remain nonnegative). In this case, we expand S and continue the algorithm. (If ϵ can be made arbitrarily large, then G has no perfect matching.)

When the algorithm terminates, we have a perfect matching of the final surface graph. To find the implied optimum matching of G, we just need to clean up the pseudonodes, matching their spanning circuits appropriately. Since all edges in this matching are tight and all pseudonodes correspond to full blossoms, we know that the matching is optimal. As in the fractional matching case, it again follows from the form of the dual changes that if all edge costs are integral, then $\bar{y}$ and $\bar{Y}$ remain 1/2 integral throughout the course of the algorithm.

3. Implementation

Our implementation of the blossom algorithm is straightforward. In particular, we do not make use of priority queues for handling the dual changes and determining what action should be taken while growing the alternating trees. It would be interesting to see a direct comparison of a sophisticated implementation of a priority queue approach with the implementation presented here. In our description below, we will assume that the reader is familiar with a the C (see Kernighan and Ritchie [24]) and with basic data structures (see Aho, Hopcroft, and Ullman [1] and Tarjan [33]).

The input to our implementation is a list of edges, giving their endpoints and costs. We assume that the costs are integral. Thus, by multiplying each of them by 2 at the start of the algorithm we may assume that the dual variables take on only integer values (using the argument that shows they are 1/2 integral on general integer input).

The graph, G, is stored internally as a list of edges and a list nodes, each node having an adjacency list consisting of a linked list of pointers to the edges it meets. (Linked lists are used to allow us to add edges to G in the procedures for large problems and to rebuild adjacency lists after a shrink or expand operation.) The adjacency lists can be implemented with an *edgeptr* structure having two fields: a pointer to an edge and a pointer to the "next" edgeptr. Since we need two such edgeptrs for each edge (the edge appears in two lists), we may as well include them directly in our edge structure. With this setup, however, the following trick can be made. We include the edgeptrs as the first two fields in our edge structure. Because the edgeptrs are now part of the edge structure, the pointer to the edge can be directly determined from a pointer to the edgeptr. Thus, we save the space for storing the pointer to the edge, and the time to look up that pointer, at the expense of some computation to convert edgeptr pointers to edge pointers. We refer to this convention as using *ugly edgeptrs* and incorporate it in our implementations.

As before, we first give the details of the fractional matching algorithm, and afterwards the changes needed to handle blossoms. The implementation of the fractional matching algorithm is simple. The alternating tree, T, is grown using depth-first search from an unmatched node r. For each node v in T, we store a pointer to the first edge in the path from v to r. Thus, it is an easy matter to trace an augmenting path from v back to r. To make 1/2-valued circuits easy to traverse, we orient the circuits and use a field in the edge structure to allow each 1/2-valued edge to point to the next edge in its circuit. We also keep a pointer from each node to the matching edge that meets it, if one exists, to allow us to skip over "−" nodes in the grow steps. Dual changes are made by traversing T in depth-first order, computing the bound on ϵ obtained from the edges meeting each "+" node and keeping a linked list of the nodes meeting edges that give the minimum bound. After making the change (by again traversing the tree) we need to grow T from each node on the linked list. And that is the implementation, except for minor details that we do not want to discuss.

Naturally, the full blossom algorithm requires some additional data structures. Our implementation follows the basic framework of the Blossom II code of Pulleyblank [29]. To begin with, we maintain more information about the alternating tree T, since in practice we need to traverse T more often than in the fractional algorithm. Besides the "parent edge" pointers we used above, we also keep parent, child, and sibling node pointers, that is, for each node we keep a pointer to its parent p (in the rooted tree T), a pointer to its first child, under some ordering of its children, and a pointer to the next child of p, under some ordering of p's children. We use the same system to record the current set of blossoms that have been shrunk to pseudonodes, where "parent", "sibling", and "child" refer to the nesting of the blossoms. When a pseudonode is formed from an edge (u,v) joining two "+" nodes, we call (u,v) the *blossom forming* edge. Tracing the path from u back to r, we call the first node, w, that also lies on the path from v to r, the *blossom root* (see Figure 4). Given the blossom form-

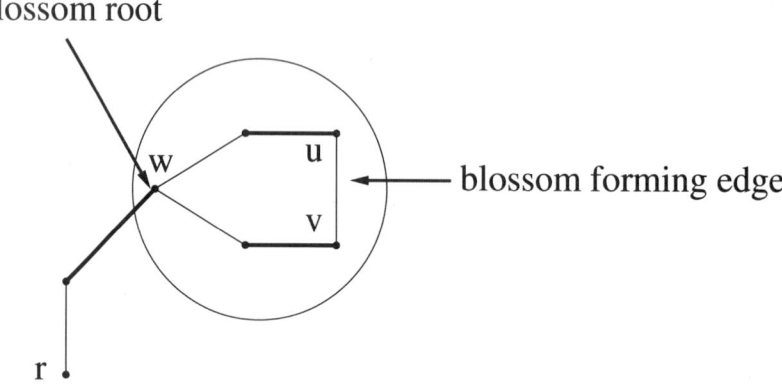

FIGURE 4. A blossom

ing edge and the blossom root, it is an easy matter to use the alternating tree information to traverse the spanning circuit of the blossom when needed during an expand operation. Thus, for each pseudonode v we let its "nest" child be the blossom root, w, of the blossom we shrunk to obtain v and we let w's "parent edge" pointer be the address of the corresponding blossom forming edge.

The surface graph is maintained by keeping a pointer from each original node to the surface node that contains it, and keeping pointers from each edge to its surface ends (if the edge appears in the surface graph). But this information is not sufficient to grow the alternating tree in a depth-first manner, since we must have access to adjacency lists for the surface nodes. This complication is avoided in some earlier codes by growing the tree (or multiple trees) by running through the edge list at each grow step, checking for tight edges joining a "+" node and an unlabeled node, or joining two "+" nodes. Instead we form explicit adjacency lists for the pseudonodes. Thus, when we shrink a set S we must gather the edgeptrs from the nodes in S, and when we expand S we must distribute them back to the individual nodes. Note that we do not make new edgeptrs, but rather reuse the existing ones. (Indeed, with ugly edgeptrs it is not possible to create new edgeptrs as we go along.)

As we mentioned above, T is grown using depth-first search from an unmatched node r. Whenever we reach a tight edge that joins a "+" node and an unlabeled node, we immediately add the edge to T and continue to grow from the unlabeled end of the edge (checking if it is matched, etc.). But when we find a tight edge e joining two "+" nodes, we do not shrink the blossom that is formed. Instead, we add e to a "shrink list" of potential pseudonodes. Only when we can no longer grow the tree do we go back and attempt the shrink operations, one at a time. After each shrink, we grow the tree from the newly created pseudonode, perhaps adding new edges to the shrink list in the process. (Notice that some of the edges on the list may have been contracted in earlier shrinks.) When the list is empty, we proceed with a dual change. The delayed shrinking is an attempt to cut down on the total number of blossoms used in the final dual solution, which is an important consideration in the procedure for large problems discussed in the next section.

A dual change is again carried out by first traversing T to compute the bound on ϵ, keeping a linked list of those nodes that meet edges giving the minimum bound, and then traversing T a second time to make the actual change. If ϵ is bounded by the $\bar{Y}$-value on some "−" pseudonodes, then we expand all such nodes and iterate the dual change procedure. Otherwise, we attempt to grow the tree from each node on the linked list we created during the first traversal of T, again delaying any shrinking until the tree can be grown no further. This grow-shrink-dual change loop continues until we find an augmenting path to match the node r. This completes the description of our blossom implementation.

Given the relative simplicity of the fractional algorithm, we should expect it to run much faster in practice than the full blossom algorithm. This is indeed the case, as is indicated in Table 1. The test problems reported in the table arise as sparse subgraphs (either the union of a set of Hamilton circuits, or one Hamilton circuit plus the 5 nearest neighbors to each node) of a set of geometric problems. (The source of the problems is given in Section 7.) The running times are reported in CPU seconds on a Dec 5000/200 workstation (which is based on a 25 MHz MIPS 3000 micropressor). The Dec 5000/200 performs the DIMACS benchmarks [28] wmatch test 1 in 1.9 CPU seconds and wmatch test 2 in 18.0 CPU seconds. The codes were written in the C programming language and compiled using cc with the -O optimizer provided with the Ultrix 4.2 operating system. To compare the performance of the blossom code, we have also included the running times of the SAP code of Derigs [11] in Table 1. The SAP code was shown in Derigs [9], [10], [11] to be superior to other codes in the literature for sparse problems. (For dense problems, Derigs [10] reports improved running times with a two-stage approach such as we describe in the next section. Also, it should be noted that Derigs and Metz [12], [13] have improved the SAP code by adding a "jump start", as we describe below.) This code is written in Fortran, and was compiled using f77 with the -O3 full optimizer, provided with Ultrix 4.2. The reported times do not include the time for reading the input file.

Nodes	Edges	Fractional (CPU Seconds)	Blossom (CPU Seconds)	SAP (CPU Seconds)
1002	3241	0.07	0.27	10.74
2392	7236	0.15	0.80	147.28
3038	5915	0.23	4.53	524.04
4040	9532	0.28	9.48	417.17
5934	10459	0.40	26.68	2072.07
7396	22148	1.50	9.90	3133.66
8064	26323	0.60	15.88	2934.30
11848	35453	0.82	78.15	10595.46
18092	56023	1.22	87.53	72246.99
20726	66872	1.27	878.20	106039.00
80864	255173	5.85	3733.87	****
101230	315677	7.37	4182.68	****

TABLE 1. Fractional, Blossom, and SAP (Derigs) Times (Dec 5000)

The times reported in Table 1 indicate that it is feasible to consider using fractional matchings to "jump start" the blossom algorithm, as in Derigs and Metz [13]. The idea is to solve the fractional matching problem to get good initial primal and dual solutions to begin the blossom algorithm. The dual solution is straightforward, we just let $\bar{y}$ be the optimal fractional dual solution and set $\bar{Y}_S = 0$ for all odd sets S. Thus, tight edges in the fractional problem are also tight at the start of the blossom algorithm. So any 1-valued edge in the optimal fractional matching can be set to 1 in the initial primal solution. Moreover, from each odd circuit of 1/2's in the fractional solution we can also include edges that match all but one of the nodes in the circuit in the initial primal solution. The running times using the jump start are reported in Table 2 (the jump start code of Derigs and Metz [13] was not available for a direct comparison). The last column in the table reports the total number of alternating trees grown in the blossom phase (so all but twice this number of nodes were matched in the initial primal solution). Again, the CPU times do not include the time for reading the input file. In 8 of the 12 cases, we see an improvement in the running time over the straight blossom implementation reported in Table 1.

Nodes	Edges	CPU Seconds	Number of Augmentations
1002	3241	0.20	33
2392	7236	0.70	84
3038	5915	4.08	56
4040	9532	4.97	64
5934	10459	22.05	86
7396	22148	20.65	70
8064	26323	17.00	215
11848	35453	66.37	387
18092	56023	69.20	646
20726	66872	1023.08	916
80864	255173	3491.32	3209
101230	315677	3475.58	8348

TABLE 2. Matching with fractional jump start (Dec 5000)

4. Strategy

When solving large matching problems, it quickly becomes impractical to work with the edge set of any graph, G, that is not extremely sparse. Simple storage requirements are one obvious reason, but equally important is the fact that the running time of the blossom algorithm is greatly influenced by the number of edges in G. For this reason, we follow a standard procedure in combinatorial optimization and use a two-stage approach to solve large problems. We first choose a sparse subgraph, G', of G and find the optimal matching contained in it. Next, we compute the reduced costs of the edges in G that have not been included in G' (we call this the *pricing* step). If any edge e has negative reduced cost, then we must *repair* the matching by adding e to G' and recomputing the primal and dual solutions. After repairing all negative edges, we repeat the pricing step, and again make repairs, continuing these price-repair rounds until we have no edges with negative reduced cost. At this point, the complementary slackness conditions hold for the entire graph G (setting $\bar{x}_e = 0$ for all edges not in G') and thus we have the overall optimal matching. This price-repair approach was used to solve matching problems by Grötschel and Holland [21] in a cutting-plane algorithm, and by Derigs and Metz [13] in a procedure based on the blossom algorithm. In the next two sections we discuss the price and repair phases in some detail. The remainder of this section is devoted to the choice of the sparse graph G'.

The criteria for selecting a sparse subgraph are that it should be easy to generate and it should have the property that, given an optimal dual solution, not too many of the non-included edges have negative reduced cost. An obvious choice is the *k-nearest neighbor* graph of G (for some positive integer k), where for each node v we include in the edge set of G' the k edges of minimum weight meeting v. This initial graph is used by Grötschel and Holland [21] (with $k = 5, 10, 15$) and Derigs and Metz [13] (with $k = 6, 8$), and works well, especially on problems where the distances are evenly distributed (that is, there is no clustering of the nodes) such as (uniform) randomly generated problems. A potentially better subgraph, however, can be constructed by making use of the fact that fractional matching problems can be solved very quickly in practice.

The idea is to first solve the fractional matching problem over the k'-nearest neighbor graph (for some k') and then build a new edge set consisting of the k edges of minimum reduced cost meeting each node. In other words, after finding a fractional dual solution $\bar{y}$, we select the k-nearest neighbors in G relative to the edge weights $w_{(u,v)} - \bar{y}_u - \bar{y}_v$. We refer to this sparse graph as the *fractional k-nearest neighbor* graph. In both cases, we take a greedy matching from G and add it to the sparse edge set, to avoid the complications that arise when G' does not contain a perfect matching.

The choices of sparse graphs are compared in Table 3, with $k = 5$, 10, and 15, on a set of five geometric test problems (described in Section 7). In the

Problem	Subgraph	Price-Repair Rounds	Edges Added	Price-Repair Time (CPU Seconds)	Total Time (CPU Seconds)
7396	Nearest 5	13	651	646.2	707.0
	Nearest 10	5	53	250.2	358.3
	Nearest 15	4	15	244.8	401.5
	Fractional 5	11	613	592.1	725.2
	Fractional 10	5	35	250.1	439.5
	Fractional 15	2	9	107.2	360.9
8064	Nearest 5	13	687	495.2	524.2
	Nearest 10	6	84	170.8	200.8
	Nearest 15	5	23	92.2	158.3
	Fractional 5	9	332	239.1	281.0
	Fractional 10	5	35	102.7	128.0
	Fractional 15	2	4	66.7	112.6
11848	Nearest 5	10	1120	1237.5	1349.7
	Nearest 10	5	77	244.8	537.5
	Nearest 15	3	14	198.3	562.2
	Fractional 5	6	654	580.0	674.6
	Fractional 10	3	32	116.4	444.3
	Fractional 15	2	13	53.4	521.2
18092	Nearest 5	15	1408	1173.0	1263.8
	Nearest 10	7	73	851.2	1145.5
	Nearest 15	3	6	318.0	774.2
	Fractional 5	11	628	864.2	1062.0
	Fractional 10	2	6	260.2	585.0
	Fractional 15	1	0	56.2	472.2
20726	Nearest 5	10	3391	22431.9	23419.2
	Nearest 10	5	289	3727.55	6021.6
	Nearest 15	4	83	2427.7	7830.6
	Fractional 5	9	2029	14382.7	15396.5
	Fractional 10	4	182	3908.1	6697.4
	Fractional 15	3	49	2031.6	7504.9

TABLE 3. Choice of sparse subgraph (CPU Seconds, Dec 5000)

fractional matching graphs, the initial fractional matching was found over the 5-nearest neighbor graph (so $k' = 5$).

The tests indicate that the fractional k-nearest graph has an advantage over the corresponding k-nearest graph. Although there is not a clear winner between the fractional 10-nearest and the fractional 15-nearest, we will use the fractional 10-nearest in our procedure for geometric problems since this will allows us to keep our memory usage on very large problems at an acceptable level.

5. Making repairs

We begin our discussion with the easy case of fractional matching problems. Suppose we have optimal fractional primal and dual solutions, $\bar{x}$ and $\bar{y}$, for a graph G, and we need to introduce an edge (u,v) that is not yet contained in E. Let $t = w_{(u,v)} - \bar{y}_u - \bar{y}_v$ be the reduced cost of the new edge. If $t \geq 0$, we have no work to do. Otherwise, we alter the dual solution by replacing $\bar{y}_u$ by $\bar{y}_u - t$. So, to maintain complementary slackness, we must also alter the primal solution. If, u meets an edge e having $\bar{x}_e = 1$, then we simply set $\bar{x}_e = 0$. Otherwise, u lies in a 1/2-valued circuit C, and we replace the 1/2 values on the edges of C by a matching that leaves only u unmatched. In either case, all edges (including (u,v)) will have nonnegative reduced cost and all edges e with $\bar{x}_e > 0$ will be tight, so we can use the fractional algorithm (growing an alternating tree from u) to compute new optimal primal and dual solutions.

Repairing (integral) matchings is more difficult. Again, suppose we have a graph G and optimal primal and dual solutions, $\bar{x}$ and $(\bar{y}, \bar{Y})$ (with surface graph $\bar{G}$) and we need to add the new edge (u,v). If either u or v is itself a node in $\bar{G}$ (say node u), then we can follow the procedure used above: decrease the value of $\bar{y}_u$, set $\bar{x}_e = 0$ for the matching edge meeting u, and apply the blossom algorithm to compute a new optimal pair of solutions. The difficulty arises when both u and v are contained in pseudonodes of $\bar{G}$. In this case, we cannot simply alter the value of $\bar{y}_u$ (or $\bar{y}_v$), since this will disturb the structure of $\bar{G}$, for example, the minimal pseudonode containing u will no longer be spanned by a tight odd circuit. To handle this, we can make dual changes to bring u (or v) to the surface, expanding all pseudonodes containing u. Ball and Derigs [6] gave an elegant procedure for this task, simplifying earlier work of Weber [35]. (An alternative "primal" approach to this problem is described in Cunningham and Marsh [8].) Their idea is to add a new node p and an edge e joining p and u (giving e reduced cost 0), and grow an alternating tree from node p. Of course, we will find no augmenting path, but during the search we will perform dual changes that allow us to expand pseudonodes. We stop the process as soon as u reaches the surface (that is, u is no longer contained in any pseudonodes). Then, deleting p and e, we proceed as above.

This repair strategy is very simple to describe and implement, but it does cause some problems in the overall price-repair strategy. The trouble is that it introduces many changes in the dual solution. Besides being time consuming, these changes affect the number of price-repair iterations we need to make. In this regard, we would like to keep the dual solution as stable as possible, to avoid introducing new negative reduced cost edges. For this reason, we employ a slightly more complicated repair procedure. The main idea is to allow matched edges to have negative reduced costs. This can be justified by either adding the redundant constraints

(7) $$x_e \leq 1 \text{ for all } e \in E$$

to the linear system (1), (2), (3) and considering the new primal-dual pair of linear programming problems, or simply noting that the reduced cost of a matched edge e can always be raised to 0 by increasing the $\bar{y}$-value of one of its ends. (Note that this modification cannot be done during the course of making repairs, since this would disturb the structure of the surface graph, as we mentioned above.)

The added freedom of having negative reduced costs can be an advantage in making repairs. For example, if the maximal pseudonodes containing u and v are joined by an alternating path of tight edges in $\bar{G}$ (starting and ending with matched edges) then we can set $\bar{x}_e = 1 - \bar{x}_e$ for each edge in the path and insert the edge (u, v) in the graph as a matched edge with negative reduced cost (see Figure 5). The price of this freedom is the additional overhead needed

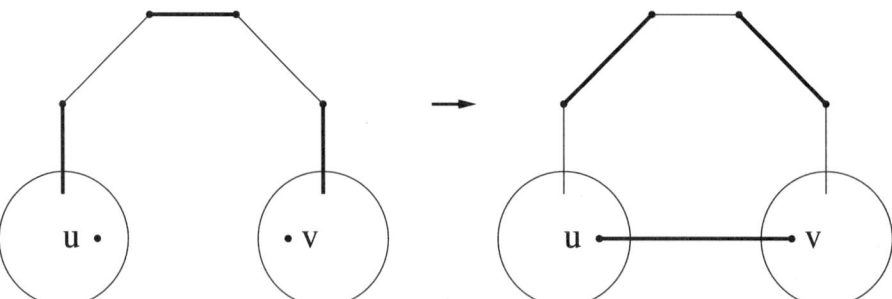

FIGURE 5. Inserting an edge

to maintain the negative edges. This manifests itself in two ways. First, we must modify the grow step of the blossom algorithm to only bring matched edges into the alternating tree when they have 0 reduced cost, since augmenting paths through the tree will flip some matched edges to unmatched edges, and these cannot be negative. Secondly, when computing the dual change bound ϵ we must take into consideration the reduced costs of the matched edges joining either two "−" nodes or a "−" node and an unlabeled node, since we do not want the reduced cost of such edges to become positive after the dual change. These complications slow down the algorithm somewhat, so in our implementation we compute the initial optimal matching and dual solution without allowing negative reduced costs, and switch to the negative version only when we begin the repair phase.

To make use of the negative reduced costs, we need a different mechanism for carrying out the repairs. To start off, we add a new node p and new edges (p, u) and (p, v), giving (p, u) reduced cost 0 and setting the reduced cost of (p, v) to

0 if $\bar{y}_u \equiv \bar{y}_v$ (modulo 2) and otherwise setting the reduced cost of (p, v) to 1. This choice of reduced costs ensures that the dual variables will continue to take on only integer values in the modified graph (recall the 1/2-integral argument and the fact that we have doubled the integral edge costs at the start of the algorithm). We grow an alternating tree from p, with the following two stopping rules. First, we stop if either u or v becomes a surface node, as in the Ball and Derigs procedure. Second, we stop whenever we find a blossom forming edge where p is the blossom root. This latter case corresponds to the situation depicted in Figure 5 (namely, we have a tight alternating path from u to v, starting and ending in matched edges) and we proceed as described above.

Nodes	Negative (Pricing Time)	Negative (Rounds)	BD (Pricing Time)	BD (Rounds)
7396	249.75 (seconds)	5	510.87 (seconds)	11
8064	102.55	5	354.40	9
11848	116.58	3	1437.33	7
18092	260.17	2	577.92	4
20726	3910.67	4	26541.10	8

TABLE 4. Price-repair Mechanism (Dec 5000)

We tested both the Ball and Derigs method and the negative reduced cost method on a series of geometric problems. The results are reported in Table 4, and show the advantage of negative reduced costs, both in total running time and in the number of price-repair rounds required.

6. Pricing

At the start of the pricing step, we have optimal primal and dual solutions, $\bar{x}$ and $(\bar{y}, \bar{Y})$, for a sparse subgraph of our original graph G. Let $\mathcal{B}$ denote the set of blossoms $S \in \mathcal{O}$ with $\bar{Y}_S > 0$, and let B be the rooted tree representing the nested structure of $\mathcal{B}$. The nodes of B consist of a root node r_V (representing the entire node set V), the original nodes V, and the pseudonodes $\mathcal{B}$. The children of each node S are the maximal members of $V \cup \mathcal{B}$ that are properly contained in S. (So each original node is a leaf of the tree, the children of each pseudonode are the nodes (and pseudonodes) that were shrunk to form it, and the children of r_V are the nodes (and pseudonodes) of the surface graph.) For each original node $v \in V$ we compute

(8) $$sum(v) \equiv \bar{y}_v + \sum \{\bar{Y}_S : S \in \mathcal{B}, v \in S\}$$

by traversing the path in B from v to the root. For any pair of nodes u, v let $lca(u, v)$ denote the least common ancestor of u and v in the tree B. Then, the reduced cost of an edge (u, v) is

(9) $$w_{(u,v)} - sum(u) - sum(v) + 2 \cdot sum(lca(u, v)).$$

So the work in computing the reduced costs can be reduced to least common ancestor calculations. We carry these out with an off-line version of Aho, Hopcroft, and Ullman's [2] algorithm, using path compression to implement the set union operations in near-linear time (Tarjan [**32**]). (We chose this method, over the linear-time least common ancestor algorithm of Harel and Tarjan [**22**], for its low overhead and programming ease.)

Although the least common ancestor algorithm is quite efficient, for large problems we need to take care not to overwhelm the method by simply pricing every edge on each round. A first step in this direction is to take advantage of the fact that we know exactly what occurs between the pricing rounds, namely a series of repair operations. Thus, if the reduced cost of an edge goes down between rounds, then we know it must go down during the dual changes in one of the operations. For this to happen, one of the ends of the edge must be contained in a "+" node in the operation's final alternating tree. (Newly formed pseudonodes are always labeled "+", so a "−" pseudonode does not hide any nodes labeled "+" duing the operation, and only "−" pseudonodes are expanded, so any nodes that leave the tree were not labeled "+" earlier in the operation.) So if we start with all nodes unmarked, and, after each repair operation, mark every node contained in a "+", then during the next pricing phase we need only price those edges that meet at least one marked node.

In the special case of geometric problems, the idea of pricing only when necessary can be carried much further. Suppose we have coordinates (v_x, v_y) for each node $v \in V$ and that the weight of an edge (u, v) is the distance from (u_x, u_y) to (v_x, v_y) under some norm $\|\cdot\|$. (Higher dimensional problems can be handled in the same manner.) Suppose further that $\|\cdot\|$ is such that the weight of edge (u, v) is at least $|u_x - v_x|$. (This holds for any L_p norm, such as the Euclidean and max norms.) Then the reduced cost of (u, v) is bounded below by

$$(10) \qquad |u_x - v_x| - sum(u) - sum(v).$$

To make use of this fact we maintain two sorted lists, the first, LIST1, sorted by increasing values of $v_x - sum(v)$, and the second, LIST2, sorted by decreasing values of $v_x + sum(v)$, where each node in V appears in exactly one of the lists. With this setup, we can readily determine which edges meeting a given node u make the expression (10) nonnegative (and so need not be explicitly priced). The process is to scan from the top of LIST1, letting v be the next node in the list, and consider pricing each edge (u, v) until we reach the point when $v_x - sum(v) \geq u_x + sum(u)$ (so (10) is nonnegative). We then do the corresponding scan of LIST2, stopping when $v_x + sum(v) \leq u_x - sum(u)$. When we consider pricing an edge (u, v), we compute $w_{(u,v)}$ and only add (u, v) to a "checkout list" if $w_{(u,v)} - sum(u) - sum(v) < 0$. (This is a stronger condition than (10) being negative.) Once the checkout list gets too long (in our implementation, 100,000 edges), we price all edges on the list with the least common ancestor algorithm, making repairs on those having negative reduced cost, and then come back and continue the pricing phase. To maintain the sorted lists, we alternate the pricing rounds with forward passes and backward passes. A forward pass begins with all nodes on LIST1, and moves each node from LIST1 to LIST2 after it is processed. A backward pass begins with all nodes on LIST2, and moves nodes from LIST2 back to LIST1. At the end of a pass, we price the edges in the checkout list, repairing those that are negative. The entire process stops when we go through a pass without finding any negative edges. Node marking can be included in the scheme by simply not processing a node if it is unmarked (just move it from one list to the other). During the procedure we need only recompute the value of $sum(v)$ when we start the processing of v. This will mean that some of the $sum()$ values will be out of date in the middle of the procedure, but whenever $sum(v)$ is less than it should be we know that v has been marked by one of the repair operations, and thus will be processed at a later point.

7. Computational tests

Our test bed of geometric problems is listed in Table 5. The majority of the problems are described in Reinelt [30], and are available over the Internet (see [30]). In two cases, the original data sets contained an odd number of points. For these instances, we sorted the x, y coordinates and deleted the last point.

Nodes	Norm	Source
1002	Euclidean	TSPLIB [30]
2392	Euclidean	TSPLIB [30]
3038	Euclidean	TSPLIB [30]
4042	Max	Circuit board (University of Bonn)
5934	Euclidean	TSPLIB [30]
7396	Euclidean	TSPLIB [30]
8064	Max	Circuit board (University of Bonn)
11848	Euclidean	TSPLIB [30]
18092	Euclidean	Locations in USA
20726	Max	Map of Tokyo [3]
80864	Max	VLSI (Bellcore)
101230	Max	VLSI (University of Bonn)

TABLE 5. Geometric Test Problems

The solution times for the geometric problems are given in Table 6. The times are reported in seconds on a Dec 5000/200 workstation (as are the times in the remaining tables). The second column reports the time for computing the fractional 10-nearest graph (including the input time); the third column gives the time needed to compute the fractional matching "jump start" on the fractional 10-nearest graph; the fourth column gives the time needed to compute the integral matching; the fifth column reports the time in the price-repair phase; and the last column gives the total running time of the process (including the input time).

The extremely long pricing times for the two largest examples suggests that for problems of this size we probably should spend more time selecting our initial sparse graph, perhaps using an "integral nearest k-neighbor graph" (where we first compute the optimum matching on a sparse graph, then compute the k-nearest graph with respect to the reduced costs derived from the matching's dual solution).

Nodes	Fractional Nearest	Jump start	Matching	Price-Repair	Total
1002	4.08	0.12	0.17	0.47	6.2
2392	15.10	0.32	2.65	2.87	24.2
3038	22.34	0.57	14.23	3.20	45.5
4040	8.21	0.47	13.47	5.25	33.1
5934	39.82	0.72	81.88	118.73	261.2
7396	123.62	2.98	47.65	249.75	438.9
8064	25.25	1.17	20.10	102.25	164.0
11848	120.08	1.68	168.02	116.9	445.1
18092	58.83	2.43	230.85	260.23	584.7
20726	77.60	2.63	2678.13	3919.33	6710.2
80864	991.30	12.52	10207.27	136166.37	147535.1
101230	1069.90	15.42	7977.35	457839.98	467544.9

TABLE 6. Solution Times of Geometric Problems (Complete Graphs)

In our final two tables, we report on the solution times for two common classes of randomly generated data. The first type are (Euclidean) geometric problems with integral coordinates drawn uniformly from the 100,000 by 100,000 square. (The generator we use is described in McGeoch [28] as dcube.c.) The second class are graphs with $2|V|\log|V|$ edges drawn uniformly from the set of possible edges, with each edge having a random integral edge cost drawn uniformly from (0, 100000). (This generator is described in McGeoch [28] as random.c.) For most of the problem sizes in each class, we made a series of 10 independent runs, and report the average, the minimum, and maximum CPU times. (The times include the time spent to generate and read the input files.) The last column in each of the tables gives the average number of blossoms that had positive value in the final dual solution. These numbers are a good indicator of the complexity of a problem instance. From this viewpoint, one can see that the sparse random graphs do not really provide a good test of a matching code.

Nodes	Trials	Average Time	Min Time	Max Time	Average Blossoms
2^{10}	10	11.9	7.2	15.6	299.5
$2^{10.5}$	10	21.5	14.5	28.5	392.0
2^{11}	10	43.9	33.3	63.0	575.0
$2^{11.5}$	10	83.9	39.8	128.6	747.2
2^{12}	10	156.6	92.4	212.7	1079.7
$2^{12.5}$	10	324.7	150.2	522.2	1526.1
2^{13}	10	586.6	366.8	741.7	2132.8
$2^{13.5}$	10	1867.8	1185.4	3578.7	2999.3
2^{14}	10	2883.3	2200.4	3664.9	4170.5
2^{15}	10	9407.5	5555.7	17117.9	8279.1
2^{16}	10	49060.5	26969.1	70670.4	16183.8
2^{17}	10	254955.3	160094.4	398992.2	32001.0

TABLE 7. Random Geometric Problems with Euclidean Norm

Nodes	Trials	Average Time	Min Time	Max Time	Average Blossoms
2^{10}	10	9.8	7.9	23.8	23.9
$2^{10.5}$	10	12.8	10.8	23.2	12.5
2^{11}	10	23.6	20.9	29.8	7.4
$2^{11.5}$	10	50.7	36.5	91.5	41.7
2^{12}	10	81.5	61.8	129.1	43.3
$2^{12.5}$	10	155.3	103.7	424.9	72.6
2^{13}	10	194.8	169.5	300.2	25.8
$2^{13.5}$	10	667.9	298.2	2631.7	276.3
2^{14}	10	884.1	500.0	2795.8	170.6
2^{15}	10	1919.2	1459.7	4004.8	164.4
2^{16}	5	6101.8	5573.2	6970.3	105.6

TABLE 8. Random Costs (0,100000) with $2|V|\log|V|$ Edges

Acknowledgements

We would like to thank Joel Gannet, Olaf Holland, Hiroshi Imai, David Johnson, Bernhard Korte, Allen McIntosh, and Gerd Reinelt for kindly providing us with test data. We would also like to thank the two referees for a number of helpful comments. A portion of this work was carried out while the authors visited the Institute for Operations Research at the University of Bonn, West Germany, with support provided by SFB (303).

References

1. A. V. Aho, J. E. Hopcroft, and J. D. Ullman, *The design and analysis of computer algorithms*, Addison-Wesley, Reading, MA, 1974.
2. _____, *On finding lowest common ancestors in trees*, SIAM Journal of Computing **5** (1976), 115–132.
3. T. Asano, M. Edahiro, H. Imai, M. Iri, and K. Murota, *Practical use of bucketing techniques in computational geometry*, Computational Geometry (G. T. Toussaint, ed.), North Holland, 1985, pp. 153–195.
4. D. Avis, *A survey of heuristics for the weighted matching problem*, Networks **13** (1983), 475–493.
5. M. O. Ball, L. D. Bodin, and R. Dial, *A matching based heuristic for scheduling mass transit crews and vehicles*, Transportation Science **17** (1983), 4–31.
6. M. O. Ball and U. Derigs, *An analysis of alternative strategies for implementing matching algorithms*, Networks **13** (1983), 517–549.
7. R. E. Burkard and U. Derigs, *Assignment and matching problems: Solution methods with FORTRAN-programs*, Springer Lecture Notes in Mathematical Systems 184, 1980.
8. W. H. Cunningham and A. B. Marsh, III, *A primal algorithm for optimum matching*, Mathematical Programming Study **8** (1978), 50–72.
9. U. Derigs, *A shortest augmenting path method for solving minimal perfect matching problems*, Networks **11** (1981), 379–390.
10. _____, *Solving large-scale matching problems efficiently: A new primal matching approach*, Networks **16** (1986), 1–16.
11. _____, *Solving non-bipartite matching problems via shortest path techniques*, Annals of Operations Research **13** (1988), 225–261.
12. U. Derigs and A. Metz, *On the use of optimal fractional matchings for solving the (integer) matching problem*, Computing **36** (1986), 263–270.
13. _____, *Solving (large scale) matching problems combinatorially*, Mathematical Programming **50** (1991), 113–122.
14. J. Edmonds, *Maximum matching and a polyhedron with 0,1 - vertices*, Journal of Research of the National Bureau of Standards **69B** (1965), 125–130.
15. G. N. Frederickson, M. S. Hocht, and C. E. Kim, *Approximation algorithms for some*

routing problems, SIAM Journal of Computing **7** (1978), 178–193.
16. H. N. Gabow, *A scaling algorithm for weighted matching on general graphs*, Proceedings 26th Annual Symposium of the Foundations of Computer Science, 1985, pp. 90–100.
17. _____, *Data structures for weighted matching and nearest common ancestors with linking*, preprint, 1990.
18. H. N. Gabow, Z. Galil, and T. H. Spencer, *Efficient implementation of graph algorithms using contraction*, Journal of the ACM **36** (1989), 540–572.
19. H. N. Gabow and R. E. Tarjan, *Faster scaling algorithms for general graph matching problems*, Journal of the ACM, to appear.
20. Z. Galil, S. Micali, and H. N. Gabow, *An $O(EV \log V)$ algorithm for finding a maximal weighted matching in general graphs*, SIAM Journal of Computing **15** (1986), 120–130.
21. M. Grötschel and O. Holland, *Solving matching problems with linear programming*, Mathematical Programming **33** (1985), 243–259.
22. D. Harel and R. E. Tarjan, *Fast algorithms for finding nearest common ancestors*, SIAM Journal of Computing **13** (1984), 338–355.
23. M. Iri and A. Taguchi, *The determination of an XY-plotter and its computational complexity*, Proceedings of the 1980 Spring Conference of the Operations Research Society of Japan, (in Japanese), pp. 204–205.
24. B. W. Kernighan and D. M. Ritchie, *The C programming language*, Prentice-Hall, Englewood Cliffs, New Jersey, 1978.
25. E. L. Lawler, *Combinatorial optimization: Networks and matroids*, Holt, Rinehart, and Winston, New York, 1976.
26. R. Lessard, J.-M. Rousseau, and M. Minoux, *A new algorithm for general matching problems using network flow subproblems*, Networks **19** (1989), 459–479.
27. O. Marcotte and S. Suri, *Fast matching algorithms for points on a polygon*, to appear in SIAM Journal of Computing, 1991.
28. C. McGeoch, *The first dimacs international algorithm implementations challenge: the core experiments*, draft.
29. W. R. Pulleyblank, *Faces of matching polyhedra*, Ph.D. thesis, University of Waterloo, Waterloo, Ontario, 1973.
30. G. Reinelt, *TSPLIB - A traveling salesman problem library*, to appear in ORSA Journal on Computing.
31. E. M. Reingold and R. E. Tarjan, *On a greedy heuristic for complete matching*, SIAM Journal of Computing **10** (1981), 676–681.
32. R. E. Tarjan, *Efficiency of a good but not linear set union algorithm*, Journal of the ACM **22** (1975), 215–225.
33. _____, *Data structures and network algorithms*, SIAM, Philadelphia, PA, 1983.
34. P. M. Vaidya, *Geometry helps in matching*, SIAM Journal of Computing (1989), 1202–1255.
35. G. M. Weber, *Sensitivity analysis of optimal matchings*, Networks **11** (1981), 41–56.

SCHOOL OF COMPUTER SCIENCE, CARNEGIE MELLON UNIVERSITY
Current address: Math Sciences Research, AT&T Bell Laboratories
E-mail address: david@research.att.com

COMBINATORICS AND OPTIMIZATION RESEARCH GROUP, BELL COMMUNICATIONS RESEARCH
E-mail address: bico@bellcore.com

Appendix A

Electronically Available Materials

Several participants in the DIMACS Challenge have contributed bibliographies, problem instances, instance generators, problem-solving codes, and programming tools for the network and matching problems. These files should remain available for some years in the public ftp directory `pub/netflow` on the Internet site `dimacs.rutgers.edu`. Please send inquiries regarding files in this directory to `netflow@dimacs.rutgers.edu`.

With few exceptions, the instances, instance generators, and problem solvers have input/output formats that conform to the DIMACS Challenge standard. In some cases translation programs are provided for converting to and from other formats. The Latex document `pub/netflow/general-info/specs.tex` contains specifications of instance formats for each problem.

Anonymous ftp

To obtain the files by anonymous ftp, type the following command sequence (your actions are displayed in `typewriter font`). The password you provide when logging in should be your email address.

```
$ ftp dimacs.rutgers.edu
Connected to dimacs.rutgers.edu
Name (. . . loginid): anonymous
Password: your email address
Guest login ok, access restrictions apply.
ftp  cd pub/netflow
```

Typing the command ? at the prompt will produce a list of commands. Typing `help commandname` will produce a short command description. Some useful commands appear in the list below.

cd directoryname Change to the named directory.

cdup Change to the parent of the current directory.

get Copy a file from the remote site to your local directory. The system prompts for remote and local filenames.

ls Get a listing of directory contents.

quit Terminate the ftp connection.

Directory Structure

Below is a summary of the structure of **pub/netflow**. Some file and subdirectory names are omitted here; check the **readme** file in each subdirectory for further information.

benchmarks/ This subdirectory contains instances and codes for comparisons across architectures and compilers.

biblio/ Contains bibliographies and pointers to the literature.

general-info/ Contains DIMACS documents produced during the Challenge, including descriptions of standard I/O formats and specifications of standard benchmark experiments.

generators/ Contains codes for generating problem instances. Interesting subdirectories are **matching/** and **network/**.

instances/ Contains real-world instances in DIMACS Challenge format.

matching/ Contains programs to solve matching problems. Subdirectories are **cardinality/** and **weighted/**.

maxflow/ Contains programs to solve maximum network flow problems.

mincost/ Contains programs to solve minimum-cost network flow problems.

program_tools/ Contains miscellaneous tools and software packages to support program development.

readme A file containing an annotated listing of main parts of the directory structure.

recursive.ls A file containing an exhaustive recursive directory listing, with no annotation.

submit/ A publicly-writable directory for submission of files through anonymous ftp. If you want to submit something: (1) Check that you don't overwrite an existing file; and (2) notify the administrator by sending a note to **netflow@dimacs.rutgers.edu**. A file placed in **submit** eventually gets moved to an appropriate place in the **pub/netflow** structure.

tar.notes Contains instructions for obtaining the entire directory structure as a Unix compressed **tar** file. The compressed file contained about 4.5 million bytes in March 1993.

Network Instance Generators

Instance generators and related tools for network flow problems are in the directory **pub/netflow/generators/network**. The codes listed below produce instances in the DIMACS Challenge formats .asn, .min and .max. In addition, several programs not listed here are available to perform conversions among the instance formats: for example, the awk program asnmin.a converts .asn files to .min files, so that assignment instances can be used as input to minimum-cost flow solvers.

In the list below a slash / indicates a subdirectory containing a collection of files.

ac.c This generator, by J. Setubal, is a C version of the ac-max.pas generator by Waissi (described below). It generates a complete acyclic network for maximum flows.

assign.c This generator by C. McGeoch produces random bipartite graphs for the assignment problem.

capt/ The CAPT generator produces instances for the capacitated transportation problem. The authors, R. G. Bland and D. L. Jensen, give a detailed description in "A report on the computational behavior of a polynomial-time network flow algorithm,' *Mathematical Programming*, 54 (1992), pp 1-39.

genrmf/ This generator, written by T. Badics to produce DIMACS-compatible networks, constructs the RMFGEN networks described by D. Goldfarb and M. D. Grigoriadis in "A computational comparison of the Dinic and Network Simplex methods for maximum flow," *Annals of Operations Research*, 13 (1988), pp 83-123.

geomasn.a This awk program by C. McGeoch converts geometric point sets into bipartite graphs for the assignment problem. The points are colored red and blue and all red-to-blue edges are generated with edge costs corresponding to Euclidean distances.

grid-on-torus/ This generator by A. Goldberg produces capacitated transportation instances laid out on a grid-on-torus network.

gridgen/ This generator of minimum-cost network flow instances was written by Y. Lee and J. Orlin. It constructs a grid-like network with sources and sinks selected at random in the grid.

gridgraph/ This generator by M. Resende produces grid-like networks for the minimum-cost flow problem. The generator uses subroutines described by D. Goldfarb and M. D. Grigoriadis (see above citation), and by L. Schrage in "A more portable FORTRAN random number generator, " *ACM Transactions on Mathematical Software*, June 1979.

mesh/ This generator by A. Goldberg produces instances of the minimum-cost circulation problem. The network is laid out as a grid on a torus. A version that fixes a small formatting problem was contributed by L. Liu.

netgen/ The NETGEN generator is described by Klingman, Napier, and Stutz, in "NETGEN: A program for generating large scale capacitated assignment, transportation, and minimum-cost flow network problems," *Management Science*, 20 (1974), pp. 814–820. A version of NETGEN written in C and producing instances in the Challenge format was contributed by N. Schlenker.

tr.c Contributed by J. Setubal, this is a C version of Waissi's one-way transit grid generator (see below).

twocost.a This awk program performs a transformation on assignment instances: arc costs greater than 50 are converted to 10^8 and arc costs less than or equal to 50 are converted to 10^2.

waissi/ G. R. Waissi contributed five network generators for maximum flow instances. The generators produce: complete acyclic networks; bipartite networks; one-way transit grid networks; two-way transit grid networks; and random networks. Standard Pascal versions as well as codes that compile under Turbo Pascal 5.5 and Sun Pascal are available. Files with the suffix .p were modified by C. McGeoch to compile under Sun Pascal.

washington R. Anderson and several seminar students at the University of Washington contributed this generator of instances for the maximum flow problem. The generator produces ten distinct types of networks. Most of the networks have vertices arranged in a rectangular grid such that the source has arcs to the first column and the sink node has arcs from the last column. Let (i, j) denote a vertex in row i and column j. The network types are listed below.

1. Mesh Graph. Each vertex (i, j) has three arcs, to vertices $(i + 1, j - 1)$, $(i + 1, j)$ and $(i + 1, j + 1)$, wrapping around column ends when necessary.

2. Random Level Graph. Each vertex in column i has three arcs to randomly-chosen vertices in column $i + 1$.

3. Random 2-Leveled Graph. Each vertex in column i has three arcs to vertices chosen randomly from columns $i + 1$ and $i + 2$.

4. Square Mesh. Vertices form a square grid. The vertex degree d is specified by the user. Each vertex (i, j) has arcs to vertices $(i + 1, j)$ through $(i + 1, j + d - 1)$, wrapping onto higher-numbered columns when necessary.

5. Basic Line. Vertices are arranged in a line. Vertex i has arcs to random vertices numbered i through n. Arc capacities are generated uniformly.

6. Exponential Line. Same as Basic Line except arc capacities depend upon the distance from the source vertex.

7. Double Exponential Line. Same as Exponential Line with a different distribution on arc capacities.

8. DinicBadCase. This network causes n augmentation phases to occur in Dinic's algorithm.

9. GoldBadCase. This network appears to be a hard instance for Goldberg's algorithm.

10. Cheriyan. This network, suggested by J. Cheriyan, is a hard instance for Goldberg's algorithm.

Matching Instance Generators

Several generators of instances for weighted and nonweighted matching are available. Instances generated in .edge format appear as a list of edges. Instances in .geom format appear as a list of vertex coordinates in two or more dimensions. Unless otherwise stated the codes listed below were written by C. McGeoch. A slash / denotes a directory containing several files.

clusters.c This generator produces instances in .geom format. The program takes a random walk: with high probability the next point is near the current point and with low probability it is distant. Adjusting program parameters produces various kinds of clustered point sets.

dcube.c This generator produces points uniform within a d-dimensional cube.

fractals/ This generator produces random point sets that converge to fractal images. Input sets to produce Serpenski triangles, ferns, broccoli, and nested boxes are provided.

hardcard.f This generator by R. B. Mattingly produces graphs in .edge format that Gabow has shown will be hard for Edmond's cardinality matching algorithm. See H. N. Gabow, "An efficient implementation of Edmond's algorithm for maximum matching on graphs," *Journal of the ACM*, 23 (1976), pp. 221–234.

neighbors.c This generator reads a file of d-dimensional points and produces a subgraph of the complete graph, where point i is connected to its next k neighbors (in input order). Distances are calculated according to the Manhattan metric. The input is in .geom format and the output is in .edge format.

random.c This generator produces random undirected graphs with uniform edge costs in .edge format.

t.f and tt.f These generators by N. Ritchey and R. B. Mattingly produce a sequence of connected triangles. The instances tend to generate a lot of blossoms.

CATHERINE C. MCGEOCH

Appendix B

Panel Discussion Highlights

Three panel discussions were conducted during the DIMACS Implementation Challenge Workshop, October 14-16, 1991. The first occurred after the technical talks on Monday, October 14, and covered the maximum flow problem, which had been the subject of the talks that day. The second occurred after Tuesday's technical sessions, and covered that day's topic, minimum cost flow. The third took place Tuesday evening, and was a general discussion of the Challenge, how it could be improved, and what topics might be appropriate for subsequent Challenges. All the panel discussions were videotaped, and these summaries are based on those tapes, together with partial transcripts prepared by Tamas Badics. The discussions were free-wheeling, and we shall only cover the highlights here, using paraphrases rather than quotes. We have also reorganized the presentations of each discussion somewhat so that comments on related issues could be put together. Comments placed within square brackets [like this] provide updates to issues under discussion, as of April, 1993.

B.1 The Maximum Flow Problem

[Panelists: Farid Alizadeh, Richard Anderson, Ulrich Derigs, Greg Shannon, Venkat Venkatesan; Chair: David Johnson].

The first question, addressed to all members of the panel, asked what the panelists had learned during the first day of the workshop. Greg Shannon observed that although overall timing analysis is important, much more can be learned by studying how the various operations and subroutines contributed to the whole. Especially with random instances, one can use regression techniques to estimate the growth rates of operation counts and thus determine the key bottlenecks in algorithmic performance and the key differences between competing algorithmic approaches. He thought this was especially important in the case of parallel algorithms.

Venkat Venkatesan said he was pleased with the format of the Challenge and the way participants were able to obtain instances and generators over the Internet from DIMACS. His one regret was that the DIMACS libraries did not contain more instances from real applications, and he thought that developing such a library would be a worthwhile goal.

Farid Alizadeh echoed the call for a collection of practically occurring instances, and then said that one thing in particular had struck him in his own work with Goldberg, and in the talks of the others. This was that for best results from the push-relabel approach of Goldberg-Tarjan, it seems to be critical to call a global relabeling procedure at certain intervals. He did not yet, however, see a clear theoretical or experimental resolution of the question of when to make the call or how to implement it. It still seems to be an art, with the most successful approach seeming to depend on the structure of the instance.

Richard Anderson said that the thing that struck him most in doing implementations was how much better the average case was than the worst case. Whereas the theoretical worst-case running time might grow proportionally to n^3, he and his co-author Setubal were getting runtimes that looked like $n^{1.5}$. They were originally interested in doing parallel implementations, but started playing around with sequential implementations, and by using global relabeling heuristics saw orders of magnitude improvements that were more than they could have hoped for by implementing the original algorithm on a 20-processor parallel computer.

Ulrich Derigs had two points to make. The first was that hardware, especially parallel hardware as it becomes available, would have a great impact on software and algorithms. Faster speeds in general mean larger instances can be attacked than have previously been contemplated, and new techniques will be needed to exploit parallelism fully. His second point was that our science like other sciences was developing in a revolutionary rather than evolutionary mode. This seemed to be especially true in the areas of maximum flow and matching, where he had been working. Someone has a new insight into the problem, and then people try to implement it, adding further heuristic ideas on top of the big idea, and they get very good codes. Then for a long time the idea undergoes a deep theoretical investigation. Better and better theoretical worst-case bounds are obtained, but there is no improvement on practical instances, because the bottlenecks with which the worst-case analysis has to cope do not typically arise. And then another big insight comes along. As old examples of such insights, he mentioned Edmonds in matching, and Dinic/Edmonds-Karp in max-flows, with the push-relabel approach being the new "big insight" in the latter area. He thought that the next big step would probably be something completely new that we haven't thought about yet.

The question was raised as to whether the Goldberg-Tarjan push-relabel approach was truly the best way to handle max-flow problems across the board, as most of the papers had seemed to indicate. Anderson pointed out that in his experiments the Dinic approach was competitive only for dense graphs, and even there he expected that all that push-relabel might well win with a better tuning of the global relabelings. This seemed to be the consensus.

There followed a lengthy discussion of the problem of getting meaningful test instances. David Johnson asked if anyone in the audience had actually had occasion to use their max-flow code in a real application. The only example appeared to be the one from David Applegate and Bill Cook, who used it in their Traveling Salesman Problem (TSP) experiments for finding violated subtour elimination constraints in the TSP polytope. They were using a fairly unsophisticated preflow-push code, and with faster codes might be able to get beyond the maximum size of $|V| \approx 11,000$ in their experiments.

One speaker expressed the desire to obtain "hard" test instances, practical or artificial. Derigs expressed the reservation that often supposedly hard instances are easily defeated by adding a few simple heuristics to your code. You could forbid people to modify their codes in this way, but that would seem to defeat one of the purposes of testing, which is to improve your algorithms. Johnson said that what he thought was needed was a wide variety of challenging instances, so that any algorithm that did well on the whole battery would have a greater chance of performing robustly in the real world.

The discussion then turned to the issue of "tuning" the codes for performance. Alizadeh mentioned the large amount of work left to do (and still ongoing at that time) in tuning the frequency and thoroughness of global relabeling. This was made difficult by the fact that clock time on his machine depended on what other jobs were currently active, and so was not always reliable. Even if you try to simply count operations, you find different results for different instance classes. (Moreover, everything they learned was very architecture dependent, since the CM-2 Connection Machine they were using was very different from a sequential machine or even from other parallel machines.) As far as other approaches, Mauricio Resende said that his interior point code with Geraldo Veiga had 3-4 critical parameters where tuning could be profitable. One conclusion of this part of the discussion was that the goal should be to learn enough about your algorithm's detailed performance so that, given the parameters of a machine, you would have a head start at tuning it for that machine.

One participant raised an important question of experimental methodology. He observed that in this day's talks everyone seemed to have performed some number of runs and then simply presented the mean as a data point. This may have been too simplistic. For instance, for one particular instance class he was measuring average running times and computing standard deviations also. The guidelines in the DIMACS document sent to participants said that if your variance was high, then do more experiments so as to narrow your 95% confidence interval. For one particular instance class he did this, but the confidence interval remained wide. When he looked at the actual running times, he noted that they fell into two different classes. They were either very small or very large for instances in the given class. In the case of $|V| = 8,000$ one was five times larger than the other. When he checked further, he found that for this generator the instances that came out either had the minimum cut very close to the source or very close to the sink. In the first case the algorithm was fast, in the second it was slow. Thus for this example the average didn't mean nearly as much as a more detailed analysis.

The panel concluded with a discussion of where research in max-flow algorithms was headed. Derigs did not think people should spend their time just working on getting the fastest code. If ten different groups tried this, you'd probably get five "best" codes, depending on the set of instances they used for tests and the collection of heuristics they came up with. And overall they would not differ by enough to be worth the effort. For now (until the next "new insight") all these "best" codes would probably be be based on the push-relabel approach. Anderson said that the biggest surprise for him at the workshop was the results of Badics and Boros that showed that the dynamic tree data structure *can* give a speedup in practice, and he would like to see follow-up work on this. He'd also like to see a theoretical analysis of the average case for the grid-based generators he used in his studies, in hopes of a rigorous explanation of the speedups he observed over the worst-case. Shannon hoped for a much better understanding of the effects and significance of global relabeling.

B.2 Minimum Cost and Multi-Commodity Flows

[Panelists: Tom McCormick, Cliff Stein, Jim Orlin, Anil Joshi, Andrew Goldberg, Bob Bland; Chair: David Johnson].

Here it was agreed that no dominant algorithm or algorithm-type had emerged from the experiments. This was true both for the case of minimum cost flows, on which the majority of the work had been done, and for the multi-commodity flow problem. Andrew Goldberg suggested, however, that with more polishing of algorithms and testing of still-larger instances, an asymptotic winner might still be identified. Anil Joshi noted that it would certainly take larger instances for his interior point code (with Arthur Goldstein and Pravin Vaidya) to prove competitive, although Mauricio Resende said that his interior point code (with Geraldo Veiga) appeared to cross-over with the NETFLOW network simplex code at about $|V| = 1500$ for some classes of instances. Joshi pointed out, however, that the real advantage of interior point approaches (an advantage they share with general simplex approaches) is their ability to handle generalizations and variations of the basic min-cost or multi-commodity flow problems, such as the addition of various types of side-constraints.

Given that there was no winner, the question of relative robustness arose. Bertsekas's RELAX code, which many of the participants used as a benchmark, seemed to have the most wide-ranging behavior, being by far the fastest on some tightly capacitated instances, but also being by far the worst on other classes. David Johnson asked whether any code was at least always within a factor of 10 of the best. Although it wasn't clear that even this criterion was met by any of the codes studied to date, Bob Bland suggested that some future implementation of the successive approximation approach or of network simplex might do the trick. Jim Orlin pointed out that Johnson's goal, if taken literally, might be impossible, since there would always be very special cases, e.g., the assignment problem, where special purpose algorithms could greatly outperform general approaches. He too, however, thought that if robustness meant consistent running times for different instances of the same size, then simplex (and perhaps interior point) codes might be the most consistent. (Farid Alizadeh pointed out from the audience that in this sense the ellipsoid method was exceedingly robust — it always yielded the same, horrible, running time!)

The question then arose whether we could learn to identify the best type of algorithm for each type of instance. This was deemed to be a difficult question, given an arbitrary instance, although it was suggested that we might already have a start on determining the *worst* types of instances for each type of algorithm.

Bob Bland made two interesting observations vis-a-vis the question of relative running time. First, he noted that the faster speeds of our computers themselves have made a major difference in what we can do. The amount of data collected in say the Lee and Orlin paper [not submitted to this proceedings] would have been impossible to collect 10 years earlier. We are now able to look in much more detail at the internal operations of our algorithms, based on large bodies of observations, and he noted, in a comment echoing the previous day's Panel, that such a study can be much more important than simply comparing running times.

Bland's second point questioned the importance of running times from a second angle. He noted that most of the codes studied were sufficiently fast that for small instances their running times were negligible, and hence which algorithm one chose was not necessarily important. For large instances, however, it might also not make much difference. Even here, the time to optimize might typically be a small part of the whole problem-solving process, which includes data collection, modelling, and solution interpretation. Where it does make a difference would be at the high end, where it was not a question of how fast an instance can be solved, but of whether the instance could be solved at all. The choice of algorithms also might make a difference in those cases where the min-cost flow code would be used as a subroutine call in some other algorithm, for example, in the multi-commodity flow code of Leong, Stein, and Shor. (Johnson pointed out that it also makes a difference when you are doing experiments: the amount of data you can obtain is inversely proportional to running time.)

The remainder of the panel discussion was devoted largely to methodological questions. Once again, dissatisfaction was expressed with the paucity of "real world" instances available to the participants, and hopes were expressed that DIMACS might do something about it. Johnson remarked that DIMACS was thinking of making a grant proposal to the NSF to set itself up as a clearing house for instances and codes for network as well as other important combinatorial problems. Preliminary feedback from the NSF, however, had been that there was no need; this was already being done elsewhere. No one in the audience knew any alternative sources that were even as good as DIMACS already was. The NETLIB instances and NETGEN generators were available from various sites (including DIMACS) but that was all, and they were very limited. Bland mentioned that NETLIB started out with 45-50 instances, and when after 10 years people realized these were too small, merely added 50 more. People kept tuning their codes to these instances, but he contended that after you've solved a problem, the question of how fast your algorithm is the 2nd, 3rd, 7th, or 12th time you solved it might not be all that relevant. Andrew Goldberg hoped we could find some really big instances, as opposed to the small ones in NETLIB. Jim Orlin suggested that one thing that was needed, assuming one could get an initial set of real-world instances, was a collection of operators that could take a given instance and create from it a bigger instance with similar properties, such as perhaps graph product constructions. Bland thought it was important, however, that we get a large and various collection of those real-world instances to start with. Johnson observed that even a small collection could at least serve as a sanity-check for the random instance generators we already have, and as a possible inspiration for new such generators.

The discussion then turned to the applications where such instances might be found. Bland pointed out that in the early 1970's IBM did a survey of the users of their linear programming package, and at that time it seemed that a majority of the users were using it for problems related to minimum cost flow. Jim Orlin said that he was almost finished with a book he was writing on network flows, and that book would have 500 references listing applications, with 150-200 of them covered in the text or exercises. Unfortunately, it appeared that most of the papers did not include the full descriptions of any real-world instance of significant size, possibly because these were proprietary. One interesting property that many applications have is that the graph structure of the instances encountered is fixed, or at least highly restricted, and what differs from instance to instance is more the numerical parameters (capacities, demands, costs, etc.).

There followed a lengthy discussion of the various input formats. Many of the benchmark codes against which people compared their implementations were written long before the Challenge began, and hence used input formats other than the common one specified in the DIMACS Challenge documentation. Some were more sparse than others, with some instances requiring 8 Megabytes of storage in one format but only 3 in another. The idea of describing instances simply by giving a generating program and a seed was suggested, but as Farid Alizadeh pointed out, even this might fail for generators such as NETGEN, which requires n^2 space even though the eventual output might be only $O(n)$ in size. Thus the ability to generate an instance might depend on how much memory your computer has. There was also the question of the portability of code involving random number generators, and Jon Bentley's recent work on portable random number generators (reviving ideas of Knuth) was mentioned.

Another key problem facing researchers, as Cliff Stein pointed out, was the cultural difference between the two groups of researchers working in this field. Most of the people at the Workshop were from the computer science community, programming in C on UNIX workstations, whereas traditionally the field has been dominated by operations researchers working on IBM mainframes or PC's and programming in FORTRAN. Running codes in two different languages poses special problems, especially when you want to call one as a subroutine of the other. Even *reading* the FORTRAN code proved hard for some of those with modern CS training in structured programming. Stein ended up having to use a FORTRAN to C converter to obtain a subroutine he could understand and get to run in his environment. Tom McCormick had related problems: When his department converted to a SPARCstation environment, they didn't even include a FORTRAN compiler, so he couldn't include any of the FORTRAN-based codes in his comparisons.

B.3 Present and Future Implementation Challenges

[Panelists: Bob Tarjan, Cathy McGeoch, Mike Grigoriadis; Chair: David Johnson].

This discussion took place after dinner on the second day of the workshop. The panel was made up of those members of the Organizing Committee who were present. Cathy McGeoch began with a brief history of the current Challenge. The idea for an implementation contest or challenge was first proposed informally by David Johnson in 1989, but didn't really get moving until Cathy obtained an NSF grant to direct the project while on sabbatical leave from Amherst, with DIMACS providing the additional needed support. Detailed planning began in late summer 1990, with the formation of the Organizing Committee, and the choice of the topics for the Challenge. The original plan had been that the Challenge would cover only one topic, say network flow, but the Committee decided that to encourage broader participation, a wider variety of topics should be allowed, which let to the current mix of max-flow, min-cost flow, bipartite and non-bipartite matching. (Papers on multi-commodity flow were accepted since the algorithms they examined used min-cost flow algorithms as subroutines.)

In October 1990, Cathy began publicizing the Challenge by a variety of means, including electronic mailing lists like TheoryNet and the DIMACS email list, publications such *SIGACT News*, and talks and handouts at various conferences, including the fall ORSA-TIMS meeting and the IEEE FOCS conference. Eventually some 150 different people requested more information. By February, there were 26 projects in the works, involving 57 people. The fact that there were 26 papers at the workshop involving 58 people is something of a coincidence, however, since in the interim 10 groups had dropped out and 10 new ones had joined.

David Johnson noted that although a decision had been made early on by the Organizing Committee that the Challenge was to be a challenge, not a contest, there had been, in response to popular demand, an unofficial "run-off" the previous night among codes for the maximum flow problem. The test instance had been one derived from the Traveling Salesman application of Applegate and Cook, and so was not of quite the same type as the randomly-generated instances most people had been studying. It seemed inappropriate to award a prize for the fastest code under such informal circumstances, and at any rate, the true "prize" that participants should be aspiring to would be a paper in the conference proceedings. However, in response to the perceived demand for some sort of awards, the Organizing Committee had come up with several, including the following:

The Parallel Slowdown Award to Farid Alizadeh and Andrew Goldberg, for their results of the previous night, where using 8000 processors on a Connection machine, they needed only 160 seconds to solve a problem that Anderson and Setubal solved in 10 seconds using a single processor (albeit a much more powerful one).

The Don Quixote Award to Tom McCormick for having implemented what proved to be the most impractical algorithm of the workshop.

The Conventional Wisdom Award to Endre Boros and Tamas Badics for their results indicating that dynamic trees might be practically useful after all.

The Rip Van Winkle Award was shared between Joshi/Goldstein/Vaidya and Resende/Veiga for having spent the longest time waiting for their codes to generate an individual data point.

The Statistician in the Closet Award to Bland et al. for the most impressive use of statistical analysis techniques.

The Altruism Award to Anderson and to Goldberg for making their instance generators available to the other participants.

The Stamina Award to everyone in the audience, for having that day endured 12 talks, one tutorial, and two panel discussions.

The discussion then turned to the current Challenge. The requests for more instance generators and for specific instances obtained from real-world applications were repeated [many new generators have been added to the DIMACS directories since the workshop, but we are still short on real-world instances]. It was also urged that the DIMACS directories be maintained beyond the life of the Challenge [which they have been; a list of what is currently available appears in Appendix A].

Support was again expressed for the idea, raised in the previous panel discussion, that DIMACS should become an official, long-term clearing house for codes and test instances for various problems. Cathy McGeoch asked whether people thought the codes maintained by DIMACS should be "certified," i.e., should DIMACS hire someone to do this. She noted that during the past year the only level of certification she had attempted was to make sure that codes actually compiled and ran on a Sun workstation, which sometimes required considerable work when there were missing libraries, etc. (She did have some student help, fortunately.) Andrew Goldberg suggested that if DIMACS became an official repository, then perhaps an organizing committee would have more luck at soliciting "real" instances from industry. Perhaps the process of encoding instances into DIMACS format would sufficiently disguise the instances so that concerns about their proprietary nature could be allayed.

As to the running of this Challenge, some participants thought the schedule had been too tight, and would have liked more time to finish their experiments, although most admitted that they hadn't really even started by January, even though the Challenge was announced in October. David Johnson stated the suspicion that no matter what the deadline was, people wouldn't start work until six or fewer months earlier. There was also some concern that the due date for proceedings submissions (January 1992) was too soon [although given the length of time the editorial process eventually took, people in the end had plenty of time to upgrade their papers].

There was also a discussion of the extended abstracts that the participants had to submit in August to the Organizing Committee. The purpose of this had been both to screen out projects that had failed to obtain significant results (only two submissions were eliminated because of this), and also to provide an

APPENDIX B: PANEL DISCUSSION HIGHLIGHTS 591

opportunity for the Organizing Committee to provide feedback, so as to improve the eventual papers and workshop presentations. In addition, participants were sent copies of all the accepted abstracts related to their particular problem area, to encourage communication between participants and help them put their own contributions in a more complete context. There was some feeling that participants should have received copies of *all* the extended abstracts ahead of time, or at least at the workshop. This had not been done partly because of expense issues, but David announced that a preliminary proceedings, including the abstracts (or revised versions) from those authors who wished to contribute, would be published soon by DIMACS. [It appeared as DIMACS Technical Report 92-4, a limited edition 387-page volume distributed mainly to participants and organizers.]

Cathy McGeoch asked how well the decision to rely primarily on the Internet for communication had worked. Some participants had only learned about the Challenge from handouts, or from direct conversations with Cathy. However, once participation began, all communication had been electronic, with material transferred either by email or anonymous FTP. (The only exception was that she had had to use Federal Express to send the introductory Challenge documents to a couple of participants who didn't have Latex, even though they did have a network connection). Participants seemed quite satisfied with the arrangements, although it was noted that the restriction to the Internet may well have prevented participation by whole classes of people who consequently weren't here to complain. Examples were "O.R. people who aren't yet used to email" and Eastern Europeans who did not yet have access to email, or at least did not have inexpensive access.

As to the quality of the material that was distributed, Cliff Stein said that the comments on testing were quite useful to him. Cathy said that some people had requested more exact rules for what to test and report, and Andrew Goldberg noted that this was especially important for those who were interested in comparisons of running time. Anil Joshi would have liked a statistical package or UNIX shell script for automatically creating tables etc., although Cathy pointed out that not all participants were UNIX-based.

On the topic of future Challenges, there were many ideas. Jim Orlin suggested that the next Challenge be on exactly the same topic, given that there was so much left to do. He thought that it was better to advance the state of the art further than to start over in a new domain. Andrew Goldberg suggested a combination of a follow-up plus a more-restricted list of new problems to study. Others thought that this Challenge profited from covering a wide range of problems, as it allowed participants to see how the issues they confronted were reflected in other domains. David Johnson said that the original idea had been for the Challenge to jump from topic to topic, and he proposed that the next Challenge cover NP-hard problems, so as to fit in with DIMACS's upcoming Special Year on Combinatorial Optimization. Both optimization and approximation algorithms could be studied. One possible problem domain was the Traveling Salesman Problem, although he thought that this had already been very well studied experimentally, and he proposed Graph Coloring, Maximum

Independent Set, and Satisfiability as possible topics. Additional suggestions from the audience included "Cryptography versus Cryptanalysis," the ground state problem in statistical mechanics, and special cases of linear programming analogous to (but different from) the network problems covered in this first Challenge. Cathy said that people had stopped her on the street to propose computational geometry as a domain. David said that computational geometry had been the initial idea for the first Challenge (back when the DIMACS Special Year had been devoted to that topic), but they hadn't been able to settle on specific problems within that domain and so had eventually moved over to network problems. He also noted that the choice of topic for the next Challenge would be strongly influenced by whoever might want to take a sabbatical to DIMACS and be coordinator.

[As it turned out, Mike Trick of CMU was that lucky person, and is currently directing the 2nd DIMACS Implementation Challenge (1992-93) on the topics of Graph Coloring, Maximum Independent Set, and Satisfiability. A third Challenge will take place during the 1993-94 DIMACS Special Year on Parallel Computing. It will cover an as-yet-unspecified application of massively parallel computation and will be coordinated by Sandeep Bhatt of Bellcore.]

DAVID S. JOHNSON